METEOROLOGY

JONES & BARTLETT LEARNING TITLES IN PHYSICAL SCIENCE

Astronomy Activity and Laboratory Manual
Alan W. Hirshfeld

Environmental Oceanography: Topics and Analysis
Daniel C. Abel & Robert L. McConnell

Environmental Science, Eighth Edition
Daniel D. Chiras

Environmental Science: Systems and Solutions, Fourth Edition
Michael L. McKinney, Robert M. Schoch, & Logan Yonavjak

Essentials of Geochemistry, Second Edition
John V. Walther

Igneous Petrology, Third Edition
Alexander McBirney

In Quest of the Solar System
Theo Koupelis

In Quest of the Stars and Galaxies
Theo Koupelis

In Quest of the Universe, Sixth Edition
Theo Koupelis

Invitation to Oceanography, Fifth Edition
Paul R. Pinet

Invitation to Organic Chemistry
A. William Johnson

Meteorology: Understanding the Atmosphere, Third Edition
Steven A. Ackerman & John A. Knox

Organic Chemistry, Third Edition
Marye Anne Fox & James K. Whitesell

Outlooks: Readings for Environmental Literacy, Second Edition
Michael L. McKinney

Principles of Atmospheric Science
John E. Frederick

Principles of Environmental Chemistry, Second Edition
James E. Girard

Restoration Ecology
Sigurdur Greipsson

THIRD EDITION

METEOROLOGY

Understanding the Atmosphere

Steven A. Ackerman
University of Wisconsin-Madison

John A. Knox
University of Georgia

JONES & BARTLETT
LEARNING

World Headquarters

Jones & Bartlett Learning
40 Tall Pine Drive
Sudbury, MA 01776
978-443-5000
info@jblearning.com
www.jblearning.com

Jones & Bartlett Learning Canada
6339 Ormindale Way
Mississauga, Ontario L5V 1J2
Canada

Jones & Bartlett Learning International
Barb House, Barb Mews
London W6 7PA
United Kingdom

Jones & Bartlett Learning books and

products are available through most bookstores and online booksellers. To contact Jones & Bartlett Learning directly, call 800-832-0034, fax 978-443-8000, or visit our Web site, www.jblearning.com.

Production Credits

Chief Executive Officer: Ty Field
President: James Homer
SVP, Chief Operating Officer: Don Jones, Jr.
SVP, Chief Technology Officer: Dean Fossella
SVP, Chief Marketing Officer: Alison M. Pendergast
SVP, Chief Financial Officer: Ruth Siporin
Publisher, Higher Education: Cathleen Sether
Acquisitions Editor: Molly Steinbach
Senior Associate Editor: Megan R. Turner
Editorial Assistant: Rachel Isaacs
Production Manager: Louis C. Bruno, Jr.
Senior Marketing Manager: Andrea DeFronzo
V.P., Manufacturing and Inventory Control: Therese Connell
Cover Design: Kristin E. Parker
Photo Research and Permission Supervisor: Christine Myaskovsky
Illustrations: Carolyn Arcabascio; Imagineering Media Services, Inc.
Composition: Circle Graphics, Inc.
Cover Image: Courtesy of Dr. Hank Revercomb, Space Science and Engineering Center—University of Wisconsin–Madison
Printing and Binding: Courier Kendallville
Cover Printing: Courier Kendallville

To order this book, use ISBN: 978-1-4496-3175-8

Library of Congress Cataloging-in-Publication Data

Ackerman, Steven A.
  Meteorology : understanding the atmosphere / Steven A. Ackerman, John A. Knox. — 3rd ed.
    p. cm.
  Includes index.
  ISBN 978-0-7637-8927-5 (alk. paper)
  1. Meteorology—Textbooks. 2. Atmosphere—Textbooks. I. Knox, John, 1965– II. Title.
  QC861.3.A34 2011
  551.5—dc22   2010035976

6048

Printed in the United States of America
15 14 13 12 11   10 9 8 7 6 5 4 3 2 1

Dedication

To Anne, Erin, and Alana, who are always lovingly patient with my meteorological distractions. I thank my parents and siblings for their good humor.

S. A. A.

To the memory of my mom and my late Aunt Dardy, who gave me my first book about weather when I was five years old; to the rest of my family; to my college mentors and my students; and to the two people who most inspired me to study and teach meteorology: the late Dr. Lyle Horn of the University of Wisconsin–Madison and J. B. Elliott of the National Weather Service (Birmingham, retired).

J. A. K.

Brief Contents

Contents

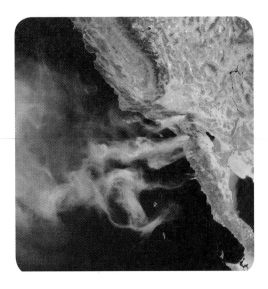

CHAPTER 2   The Energy Cycle   30

CHAPTER 3   Temperature   61

CHAPTER 4   Water in the Atmosphere   96

CHAPTER 5 Observing the Atmosphere 141

CHAPTER 6 Atmospheric Forces and Wind 179

CHAPTER 7   Global-Scale Winds   210

CHAPTER 8   Atmosphere-Ocean Interactions: El Niño and Tropical Cyclones   233

CHAPTER 9   Air Masses and Fronts   275

CHAPTER 11   Thunderstorms and Tornadoes   339

CHAPTER 16   Climate Forecasting 499

Preface

Change is the story of the atmosphere. The weather changes and the climate changes, and they, in turn, change lives. A century ago, storms struck without warning and killed thousands. Today, thanks to computer models and satellites, the public is warned of impending storms up to a week in advance. Two centuries ago, natural climate change caused famines. Today, the governments of the world study future climate change scenarios caused by fossil fuel emissions and discuss the best ways to avoid or adapt to the changes in advance. The pace of change in meteorology is remarkable. Today's 3-day weather forecasts are as accurate as the 36-hour forecasts just 15 years ago.

Mark Twain is supposed to have said, "Everyone talks about the weather, but nobody does anything about it." This, too, has changed in the 100-plus years since Twain made these remarks. In addition to thousands of degreed meteorologists and climatologists, the public plays an increasingly important role in the study of weather and climate today. Volunteer observers record and upload temperature and precipitation data every day. At the same time, thousands of people donate their free computer time to worldwide collaborative simulations of future climate.

We wrote this textbook for you, the student. You may not be able to change the weather, and most of you will not become professional meteorologists. But you can discover the processes that determine weather and climate and learn how they are relevant to your everyday life. Knowledge gained from reading this book can help you better understand severe weather, impending climate change, greenhouse warming, the depletion and recovery of the ozone layer, and the causes and effects of El Niño. You will experience and be influenced by these events throughout your lifetime. *Meteorology: Understanding the Atmosphere* will help you to grasp the fundamentals and gain an appreciation of the complexities involved with these issues.

Our goal in writing this book has been to provide you with a perspective on meteorology as a science in which observations play a key role. Thus, we provide many observations throughout this book, both personal and from scientific instruments, along with analysis of what these observations mean. This approach of observing and then analyzing the atmosphere to gain an understanding is a scientific way of thinking. In fact, it is how we, the authors, explore and understand the atmosphere. We hope that you find this approach exciting and that it inspires you to a lifetime of watching and understanding the weather.

AUDIENCE

The first edition of *Meteorology: Understanding the Atmosphere* earned recognition from the Society of Academic Authors, who awarded it the 2003 William Henry Fox Talbot Prize for visual excellence in a college textbook. This edition, like the previous editions, is designed for college-level introductory courses in weather and climate and meteorology. The study of weather and climate is an interdisciplinary science incorporating the atmosphere, land, oceans, cryosphere, and biosphere. Our book emphasizes the atmosphere, focusing on it as an important component of the

Earth as a total system. However, we also present oceanography in a unique, unified atmosphere-ocean interactions chapter, reflecting the latest understanding of the influence of oceanic processes on the atmosphere.

Our textbook is tailored to general education students who are not meteorology/atmospheric science majors, but it does include sufficient scientific rigor to be used in majors-only courses. Our emphasis is on clear, simple, and up-to-date explanations of weather phenomena motivated by observations of the atmosphere that appeal to business and education majors as well as meteorology majors. Many students better understand the world around them through observations and experiences rather than through mathematics. For this reason, our book does not avoid math, but approaches topics from the point of view of a non-science student. Therefore, the book is accessible to non-majors while still providing detailed mathematics when appropriate. The extensive library of online Java applets that accompany the textbook appeal to today's visually oriented students who are at home in an interactive, game-like virtual environment.

NEW AND KEY FEATURES IN THE THIRD EDITION

The observations we provide in our textbook range from those made by students themselves—a tornado or a spectacular cloud photograph—to scientific observations made using the latest satellite and radar technology. Difficult topics such as the extratropical cyclone and weather forecasting are enlivened with compelling narratives that combine the very latest science with page-turning mysteries. Water in the atmosphere, tropical cyclones, and weather observations are each examined in unique, unifying chapters that set these topics in broader interdisciplinary contexts. Throughout the text, weather phenomena come alive for the reader via a focus on conceptual models, visualization of life cycles, and the critical weather safety information needed to keep weather from becoming a killer. An instructor-friendly design places the tropical cyclone chapter relatively early in the text so that fall-semester courses can cover this subject during hurricane season. Similarly, the small-scale winds chapter is modularly designed around a geographical theme to facilitate selective coverage based on winds common to the region of a particular college or university. The accompanying Java applets have been acclaimed nationally by the Internet and meteorology education communities and extend the textbook treatment of dozens of key topics.

Areas of New and Improved Coverage

Because the atmosphere and the study of it are always changing, we have thoroughly updated each and every chapter in this book to reflect the current state of scientific understanding. The third edition of *Meteorology: Understanding the Atmosphere* includes one new chapter, Chapter 16, Climate Forecasting. This new chapter is in response to the explosion of work in climate prediction during the past decade and is intended as a companion to our unrivaled treatment of weather forecasting in Chapter 13.

The following list highlights the areas of key content updates:

Chapter 3, Temperature, debunks the widespread myth that the Gulf Stream is the primary source of Europe's relative wintertime warmth, as demonstrated recently by climate researchers.

Chapter 5, Observing the Atmosphere, includes an explanation of the new "dual-pol" radars that are being deployed throughout the United States.

Chapter 7, Global-Scale Winds, includes a one-of-a-kind "Beyond Conceptual Models" section that updates the conventional descriptive textbook treatment of this subject with the latest advances in our understanding of the Hadley Cell, subtropical highs, and the jet streams.

Chapter 10, Extratropical Cyclones and Anticyclones, contains multiple changes and enhancements, including Gordon Lightfoot's lyrics for the song "The Wreck of the Edmund Fitzgerald" and a discussion of their meteorological accuracy; the very latest understanding of the occluded front, replacing the nearly century-old speculations that still dominate textbook treatments of the subject today; recent research on Great Lakes windstorms, including the Fitzgerald storm; and the 2003 European heat wave, as a classic example of anticyclones and their sometimes devastating societal impact.

Chapter 11, Thunderstorms and Tornadoes, provides analyses and first-hand accounts of the 2007 Greensburg, Kansas, tornado and the 2008 Atlanta, Georgia, tornado, the latter by the co-author who observed the tornado's path of destruction from just two miles away.

Chapter 13, Weather Forecasting, now includes a description of the fascinating job of a "forensic meteorologist" and the very latest details on numerical weather prediction models and ensemble forecasting.

Chapter 15, Human Influences on Recent Climate, has been reorganized to emphasize the climate change that has already occurred, including an up-to-the-minute explanation of the impact of the urban heat island on precipitation.

Chapter 16, Climate Forecasting, covers climate prediction from monthly/seasonal scales to the multidecadal global climate model predictions associated with the Intergovernmental Panel on Climate Change (IPCC). We also examine a range of possible societal responses to climate change.

PEDAGOGICAL FEATURES

Our book contains numerous features to enhance student learning:

- An outline and a list of chapter goals begin each chapter
- Introductions focus on observations of the atmosphere
- Learning applets designed to support student learning are integrated into the text and address specific student learning difficulties
- Extended boxes delve into advanced and unusual topics in each chapter
- End-of-chapter summaries review the main ideas presented in each chapter
- A list of key terms is provided at the end of each chapter
- An annotated chapter-by-chapter bibliography at the end of the book gives inspired students a roadmap to more advanced resources

ANCILLARIES

For the Student

A student companion Web site has been developed to accompany *Meteorology: Understanding the Weather, Third Edition*. The site hosts study quiz questions, an interactive glossary, crossword puzzles, interactive flashcards, and Java applets tied to the text to further explain this text's dynamic concepts. These features are available by visiting http://physicalscience.jbpub.com/ackerman/ meteorology/ and using the free access code contained in each new copy of the textbook.

For Instructors

An Instructor's Media CD-ROM containing PowerPoint™ Lecture Outlines and a PowerPoint Image Bank is available to all qualifying instructors who adopt *Meteorology: Understanding the Atmosphere, Third Edition*. An instructor's manual and learning management system-compatible test bank are also available for download through the Jones & Bartlett Learning catalog page, http://www.jblearning.com/catalog/9781449631758/.

Steven Ackerman
John Knox

Acknowledgments

This book would not have been possible without the efforts of many. Tom Whittaker has created some of the world's best meteorological Java applets to support our student technology and instructor lecture presentation CD. Aneela Qureshi did a great job overseeing the editing and updating of the glossary and appendix, as well as creating the annotated bibliography. Pam Knox provided valuable editing and proofing during every stage of this process, which considerably improved the content of the entire book. Evan Knox provided input on graphics and figures. Anne Pryor and Erin Pryor-Ackerman provided editorial comments on several chapters. We owe Dick Morel a particular debt of gratitude for first connecting us with Jones & Bartlett Learning. At Jones & Bartlett, Carolyn Arcabascio, Lou Bruno, Christine Myaskovsky, Caroline Perry, and Molly Steinbach deserve special mention for helping make the new edition a reality. Finally, we thank our students for their comments and our many colleagues who took the time out of their busy schedules to review or comment on all or parts of text, including the following:

Stephen Jascourt, *UCAR/COMET*
Peter Lynch, *University College, Dublin, Ireland*
Anders Persson, *European Centre for Medium-Range Weather Forecasts*
Jean Phillips, *University of Wisconsin–Madison*
David M. Schultz, *University of Helsinki, Finland*
J. Marshall Shepherd, *University of Georgia*
Shawn Trueman, *St. Cloud Technical and Community College*
Steven Vavrus, *University of Wisconsin–Madison*

About the Authors

Steven A. Ackerman is Professor of Atmospheric and Oceanic Sciences at the University of Wisconsin–Madison and is Director of the Cooperative Institute for Meteorological Satellite Studies. He received his B.S. degree in physics from the State University of New York–Oneonta, and he was honored by Oneonta with its 2009 Distinguished Alumnus Award. He earned his Ph.D. in atmospheric sciences at Colorado State University. Renowned for his ability to inspire active student participation in his classes, Dr. Ackerman has won numerous teaching and academic awards, including the American Meteorological Society's prestigious national Teaching Excellence Award for "his abundant energy and steadfastness in the promotion and practice of excellence in teaching and mentoring and for the development and wide dissemination of highly regarded learning materials for undergraduate and graduate students in the atmospheric sciences." He has also won the Chancellor's Award for Distinguished Teaching from the University of Wisconsin. Dr. Ackerman's research interests center on interpreting satellite observations of clouds, aerosols, water vapor, and land surfaces. He was a recipient of the NASA Exceptional Public Service Medal in 2010.

John A. Knox is an Assistant Professor of Geography at the University of Georgia (UGA), where he has taught a wide range of weather-related courses at the undergraduate and graduate levels. He has also taught meteorology at Valparaiso University and Barnard College of Columbia University. A National Science Foundation Graduate Research Fellow and Rhodes Scholar finalist, Knox received his B.S. in mathematics from the University of Alabama at Birmingham and his Ph.D. in atmospheric sciences from the University of Wisconsin–Madison. He was a postdoctoral fellow in climate systems at Columbia University in conjunction with the NASA/Goddard Institute for Space Studies in New York City. He has published over 30 journal articles in atmospheric science research and geoscience education. Dr. Knox received the 2010 T. Theodore Fujita Research Achievement Award from the National Weather Association for his work to improve clear-air turbulence forecasting methods. He also won a 2011 Sandy Beaver Excellence in Teaching Award from UGA. Knox serves as the manager for the University of Georgia's "WxChallenge" weather forecasting contest team, which perennially places among the top college forecasting teams in North America.

Introduction to the Atmosphere

1

AFTER COMPLETING THIS CHAPTER, YOU SHOULD BE ABLE TO:
- Describe the components of the atmosphere
- List the different layers of the atmosphere in order of ascending altitude
- Explain the relationship between atmospheric pressure and altitude
- Read a weather map
- Distinguish between weather watches and warnings

INTRODUCTION

It's a hot, muggy summer night at the baseball stadium. The Atlanta Braves are on their way to another winning baseball season, and they are taking a night "off" to play an exhibition game against a minor league all-star team. The standing-room-only crowd, the largest in years, applauds as future Hall of Fame players take the field.

Midway through the game, however, the **weather** turns violent. High winds suddenly blow chairs off the stadium roof. Then the sky explodes with light and sound as lightning strikes an electric transformer on a pole out beyond center field. A fireball dances along the power lines, and the stadium lights go dark.

Frightened, the baseball players run off the field into the dugouts, and panicked fans shriek as the thunder crashes. One little boy dives under his stadium seat in terror, only to peek out and observe ominous purple and green **clouds** racing overhead. Reports of a funnel cloud—a tornado in the thunderstorm clouds above the stadium—spread among the crowd. Flooding rains descend, and sopping-wet spectators splash through puddles and duck lightning bolts as they flee to their cars.

Conversations on the way home focus on the rain, the wind, the lightning, and the possible tornado, not on the game everyone eagerly anticipated just a couple of hours before. The American pastime of baseball has been upstaged by the universal spectacle of the weather.

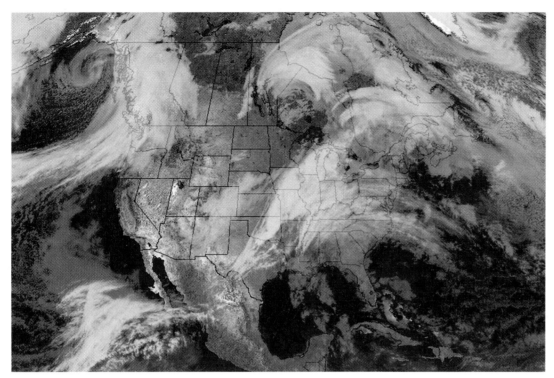

FIGURE 1-1 This weather satellite image, taken on November 29, 2006, presents a view of North America as you would see it if you were above Earth looking down. A storm is developing over the central United States, and a large weather system is moving into the Pacific Northwest.

As this true story illustrates, weather affects every facet of our lives, even when we least expect it. We all experience weather through our senses: the flash of lightning, the crack of thunder, the stickiness of a summer night, and the peculiar smell of rain. In some cases, an event is so memorable that we experience long-ago weather in stories told at family gatherings. In this book, we will pay close attention to how you sense the atmosphere, using sight, sound, touch, and even stories. We hope that years from now you'll remember how observations of the atmosphere help to explain the weather around you.

Meteorology, the study of weather and climate, is a science—a young and exciting science. Meteorologists sense the **atmosphere** like everyone else. As we'll see, young scientists using only their eyes and brains have made some of the greatest discoveries in meteorology. Today, meteorologists also use a variety of specialized techniques and tools, including weather satellites, to supplement their senses (FIGURE 1-1). In this book, you will learn about the atmosphere in the context of how it is sensed by meteorologists. Our goal is to help you understand the science behind weather forecasts for tomorrow and predictions of global warming in future years.

We begin our exploration of meteorology in this chapter with the basics: what the atmosphere is made of and how our observations of it are turned into weather maps.

WEATHER AND CLIMATE

Weather is the condition of the atmosphere at a particular location and moment. Each day current weather conditions are given in local weather reports. These reports usually include current temperature, relative humidity, dew point, pressure, wind speed and direction, cloud cover, and precipitation. Such weather information is important to us because it influences our everyday activities and plans. Before going out for the day, we want to know how cold or hot it will be and whether it will rain or snow. Meteorology is the study of these weather variables, the processes that cause weather, and the interaction of the atmosphere with the Earth's surface, ocean, and life.

The fundamental cause of weather is the effect of the Sun on the Earth. At any time, only half of the Earth is warmed by the Sun, while the Earth's other side is shadowed. This causes uneven heating of the Earth's surface by the Sun every day, with some regions warmer than others. For reasons we explore in later chapters, temperature differences cause weather: winds, clouds, and precipitation. Seasonal weather patterns result from variations in temperature caused by the Earth's tilt toward the Sun in summer and away from the Sun in winter. The distribution of water and land and the topography of the land contribute to the shaping of Earth's weather patterns on smaller scales. In Chapter 2, we explore in more detail the uneven heating of the Earth by the Sun and its effects on weather and climate.

The **climate** of a region, in contrast to the weather, is the condition of the atmosphere over many years. On average, Florida will have a mild climate all year long. Minnesota, however, will have a climate with warm, even hot, summers and very cold winters. The climate of a region is described by long-term averages of atmospheric conditions such as temperature, moisture, winds, pressure, clouds, visibility, and precipitation type and amount. The description of a region's climate must include extremes as well as averages—for example, record high and low temperatures. We learn more about this in Chapter 3.

Climatology is the study of climate. Climatologists examine the long-term averages and extremes of the atmosphere. Increasingly, climatologists also investigate the changes of climate in the past and possible climate changes in the future. The study of climate also includes these kinds of variations and the frequency of the variations.

A close relationship exists between meteorology and climatology. Both fields study the atmosphere. However, climatology has been more concerned than meteorology with how oceans, landforms, and living organisms affect the atmosphere. The atmosphere, the thin ocean of air that we live in, is the main focus of meteorology. In this chapter, we examine the basics of the atmosphere that are essential for both meteorology and climatology.

THE EARTH'S MAJOR SURFACE FEATURES

Our atmosphere receives energy from the Sun. Much of the energy transfer occurs at the bottom of the atmosphere, where the surface of the Earth exchanges energy and water with the atmosphere. The distribution of land and water therefore plays a major role in determining climatic conditions and weather patterns. Approximately 70% of Earth's surface is water. The four major water bodies are, from largest to smallest in surface area, the Pacific, Atlantic, Indian, and Arctic Oceans.

Asia, Africa, North America, South America, Antarctica, Europe, and Australia are the seven continents in order from largest to smallest in surface area. More than two-thirds of these landmasses are located in the Northern Hemisphere (**FIGURE 1-2**). Differences in current climate and weather patterns between the northern and southern hemispheres can often be attributed to differences in the amount of land in each.

Surrounding Earth's surface is the atmosphere—a thin envelope of gases no taller than the distance of an hour's drive on the highway. The atmosphere protects us from the Sun's high-energy radiation and provides the air we breathe and the water we drink.

MAKING AN ATMOSPHERE: GASES AND GRAVITY

In our everyday lives we encounter matter in three forms: solid, liquid, and gas. For example, we are all familiar with water as solid ice, liquid water, or a vapor. The atmosphere is made primarily of a mixture of gases that includes liquid and solid particles suspended in air, such as water droplets, ice crystals, and dust particles.

The **molecules** of gases and liquids are in constant motion. They naturally spread or **diffuse** from areas of high to low concentration. If someone peels an orange, its aroma will soon permeate the room. The aromatic molecules diffuse from an area of high concentration, right around

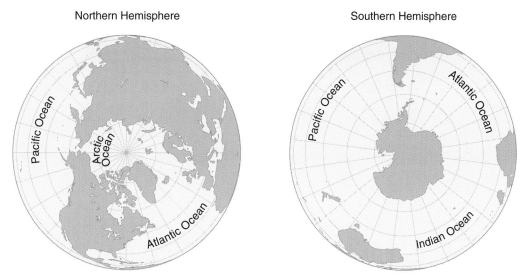

Northern Hemisphere

Southern Hemisphere

FIGURE 1-2 The global distribution of land and water strongly influences weather and climate patterns.

the orange, to areas where there are few, if any, "orange aroma" molecules. If our atmosphere is an area of concentrated molecules, then why don't they eventually diffuse into empty outer space?

Gravity, the mutual attraction between objects, is the force that holds the atmosphere in place. Gravity keeps the Moon orbiting the Earth, the planets orbiting the Sun, and you from floating into space. The Earth's gravity exerts a pull on gas molecules in the atmosphere and keeps them from diffusing out to space.

Gravitational attraction between objects depends on the masses of the objects. If an object has a large mass, it will have a strong gravitational attraction and can attract objects that do not have much mass. For example, the Sun is large and is composed mostly of light gases such as hydrogen and helium. The Earth is much smaller than the Sun, and its atmosphere contains little hydrogen and helium. The Moon is much smaller than the Earth and does not even have an atmosphere! Gravitational attraction also depends on the distance between the objects and weakens rapidly with distance. An object twice as far from Earth as another feels only one-fourth as much "pull" as a result of gravity.

By itself, gravity would turn the atmosphere into a layer cake with the heaviest molecules near the Earth's surface and the lightest molecules at the top of the atmosphere. However, weather processes "stir" the atmosphere, helping to keep it well-mixed almost to its outer edge. This is how heavier-than-air molecules, such as **chlorofluorocarbons** (**CFC**s), are able to reach the stratosphere.

ATMOSPHERIC EVOLUTION AND COMPOSITION

Gravitational attraction plays an important role in the evolution of the concentration of gases in our atmosphere. Since their formation approximately 4.5 billion years ago, the Earth and its atmosphere have undergone extraordinary changes. In the beginning, the Earth's atmosphere was hot and consisted mostly of hydrogen (H), helium (He), methane (CH_4), and ammonia (NH_3).

The **permanent gases** composing today's atmosphere are mostly **nitrogen** (**N$_2$**, 78%), **oxygen** (**O$_2$**, 21%) and **argon** (**Ar**, 1%), with much smaller "trace" amounts of some of the gases found in the early Earth's atmosphere.

Some gases in the atmosphere, however, experience changes in their concentrations in space and time. These **variable gases** include **water vapor** (**H$_2$O**, 0% to 4%) and the **trace gases**, which include **carbon dioxide** (**CO$_2$**), **methane** (**CH$_4$**), **nitrous oxide** (**N$_2$O**), **ozone** (**O$_3$**), and CFCs. **TABLE 1-1** lists the major permanent and variable gases composing our current atmosphere. In particular, the amount of water vapor in the atmosphere varies from day to day

TABLE 1-1 **Composition of the Atmosphere**

	Symbol	Percentage by Volume (%)
Major Permanent Gas		
Nitrogen	N_2	78.08
Oxygen	O_2	20.95
Argon	Ar	0.93
Variable Gas		
Water vapor	H_2O	0 to 4
Carbon dioxide	CO_2	0.039
Methane	CH_4	0.00018
Nitrous oxide	N_2O	0.00003
Ozone	O_3	0 to 7×10^{-6}
CFCs	CFCs	2×10^{-9} to 5×10^{-8}

and place to place. This variability and the movement of water in all three phases underlie many aspects of weather, including changes in the weight of air (BOX 1-1).

The gases in today's atmosphere are largely a result of emissions by volcanoes over billions of years. A volcanic eruption throws ash and rock and large amounts of gases into the atmosphere. The major gases in a volcanic plume are water vapor, carbon dioxide, and nitrogen. What happened to these gases after their release into the atmosphere over billions of years?

After its formation, the Earth began to cool. During the cooling process, the water vapor from volcanic eruptions condensed and formed clouds. Precipitation from the clouds eventually formed the oceans, glaciers, lakes, and rivers. The development of the oceans affected atmospheric concentrations of carbon dioxide. Some carbon dioxide from the atmosphere dissolved and accumulated in the oceans as they formed.

Box 1-1 Moist Air Is Lighter Than Dry Air

For now we define moist air as a volume of air with many water vapor molecules and dry air as a volume of air that contains only a few water vapor molecules. To understand why moist air is lighter than dry air at the same temperature, we have to define a few concepts.

- A molecule of water has the properties of water and is composed of two hydrogen atoms and one oxygen atom.
- The weight of an individual atom is represented by its atomic weight. The (rounded) atomic weight of hydrogen (H) is 1, oxygen (O) is 16, nitrogen (N) is 14, and carbon (C) is 12. The weight of a molecule is determined by summing the atomic weights of its atoms. A water molecule (H_2O) has a molecular weight of 18 ($1 + 1 + 16$). Free nitrogen (N_2) has a molecular weight of 28 ($14 + 14$), and an oxygen molecule (O_2) has an atomic weight of 32. Therefore, a water molecule is lighter than either a nitrogen or oxygen molecule.
- A fixed volume of a gas at constant pressure and temperature has the same number of molecules. It does not matter what the gas is—the same number of molecules will exist in that volume. This is known as Avogadro's Law.

To make a given volume of air moister, we need to add water vapor molecules to the volume. To add water molecules to the volume, we must remove other molecules to conserve the total number of molecules in the volume (Avogadro's Law). Dry air consists mostly of nitrogen and oxygen molecules, which weigh more than water molecules. And so this means that when a given volume of air is made moister, heavier molecules are replaced with lighter molecules. Therefore, moist air is lighter than dry air (if both are at the same temperature and pressure). As we shall see later, severe thunderstorms can form when heavier dry air overlies lighter moist air, a condition that can lead to an unstable atmosphere.

What happened to the nitrogen outgassed by volcanoes? Nitrogen is a chemically stable gas, which means that it does not interact with other gases or the Earth's surface. After nitrogen enters the atmosphere, it tends to stay there. This accounts for the high concentration in today's atmosphere; nitrogen has been accumulating over billions of years.

Volcanoes emit very little oxygen. Then how did oxygen come to comprise such a large amount of today's atmosphere? Approximately 3 billion years ago, tiny one-celled green-blue algae evolved in the ocean. Water protected the one-celled organisms from the Sun's lethal ultraviolet light. The algae produced oxygen as a by-product of photosynthesis, the process plants use to convert solar energy, water, and carbon dioxide into food. Today's oxygen levels are the result of billions of years of accumulation.

As the oxygen from plants slowly accumulated in the atmosphere, ozone began to form. Ozone is both caused by and provides protection from damaging ultraviolet energy emitted by the Sun. The development of an atmospheric "ozone layer" allowed life to move out of the oceans and onto land. We cover ozone in more detail in Chapter 2.

VARIABLE GASES AND AEROSOLS

Despite their small concentrations, variable gases are vitally important to meteorology and climatology. The three major variable gases in the atmosphere are carbon dioxide, water vapor, *most variable* and ozone. These gases play important roles in the energy cycles of the atmosphere. Methane and CFCs, as well as other variable gases, also matter despite their very small concentrations. These gases are important because they interact with other gases and modify the energy balance of the atmosphere (discussed in the next chapter). In addition to gases, small particles suspended in the atmosphere are also important in determining the quality of the air we breathe and the transfer of energy in the atmosphere. For climatic predictions, it is important to know how and why the concentrations of these gases and particles change over time.

To know how the concentration of a gas changes, we have to know how it enters and departs the atmosphere. A **source** is a mechanism that supplies a gas to the atmosphere, and a **sink** removes a gas from the atmosphere. The routes by which a gas enters and leaves the atmosphere are known collectively as a cycle. In this section, we first consider the carbon dioxide and hydrologic (water) cycles before moving on to discuss methane, CFCs, and aerosols. (The formation and destruction of ozone are discussed in the next chapter.)

◼ Carbon Dioxide Cycle

The atmospheric carbon dioxide cycle (FIGURE 1-3) describes how carbon dioxide moves between the atmosphere, the ocean, and the land. Nearly half of the carbon dioxide that enters the atmosphere moves between the ocean and plants. Here we review how carbon dioxide enters and leaves the atmosphere.

As mentioned earlier, volcanoes inject carbon dioxide into the atmosphere and are therefore an atmospheric source of carbon dioxide. Plants, through the process of photosynthesis, use sunlight, water, and carbon dioxide to manufacture food. Plants remove carbon dioxide from the atmosphere during photosynthesis, where it becomes incorporated into their tissues in the form of other chemicals such as sugars. Photosynthesis is a process that allows plants to become a temporary sink of atmospheric carbon dioxide. When the plants die and decay, they release the stored carbon dioxide into the atmosphere. Dead plant tissue and other dead organisms are therefore a source of atmospheric carbon dioxide. Plant decomposition and geological forces over millions of years have generated coal and oil fields underground. The burning of these fuels returns carbon dioxide into the atmosphere. Through respiration, animals inhale atmospheric oxygen and exhale carbon dioxide and are, therefore, another source of atmospheric carbon dioxide.

The atmospheric concentration of carbon dioxide is monitored throughout the world. The concentrations that have been carefully measured at Mauna Loa, Hawaii, since 1958, are shown in FIGURE 1-4. The steady increase in carbon dioxide concentration is attributed to the

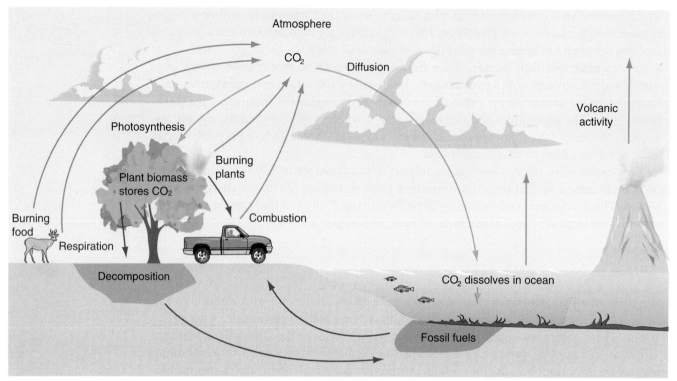

FIGURE 1-3 The carbon dioxide cycle. The blue lines represent processes by which carbon dioxide enters the atmosphere. Red lines represent the primary processes by which carbon dioxide is removed from the atmosphere. Black arrows represent processes that store carbon in the Earth.

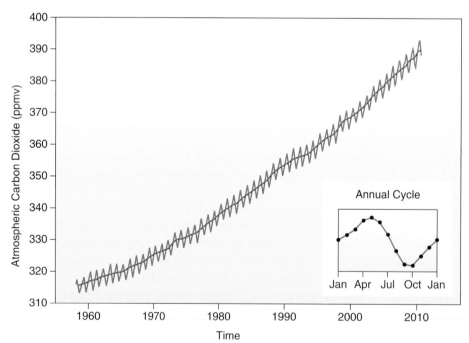

FIGURE 1-4 Carbon dioxide measurements made monthly since 1958 at Mauna Loa, Hawaii. Currently, human activity is causing a significant year-to-year change in the amount of carbon dioxide in the atmosphere. This accounts for the rise of the curve from left to right. The shorter term seasonal oscillations in the curve are caused mostly by the worldwide effect of plants on the carbon dioxide cycle. The blue curve is the annual mean atmospheric concentration of carbon dioxide in parts per million by volume (ppmv).

burning of fossil fuels and, to a lesser extent, deforestation. Imposed on this increasing trend is a repetitive cycle of peaks and valleys.

Because most of the land surface of the world is in the Northern Hemisphere (Figure 1-2), the life cycle of plants in the Northern Hemisphere drives this seasonal cycle of peaks and valleys of carbon dioxide. During winter, dormant plants stop removing carbon dioxide from the atmosphere. However, at the same time decaying plants continue to release the carbon dioxide stored in their tissues into the atmosphere. Because the source is greater than the sink, atmospheric concentrations of carbon dioxide increase throughout the winter until late spring. In summer, decomposition also occurs, but photosynthesis is at a maximum, and carbon dioxide is removed from the atmosphere in large quantities. The sink is now larger than the source. This causes a decrease in carbon dioxide concentrations throughout the summer, leading to minimum yearly values in early autumn.

The amount of carbon dioxide in the atmosphere is an important factor that influences atmospheric temperature. Warm periods in the Earth's long-term history are associated with high levels of atmospheric carbon dioxide. As we discuss in the next chapter and Chapter 15, the increase of atmospheric carbon dioxide caused by burning of fossil fuels plays a vital role in the planet's warming. When discussing predictions of global climate warming, you should keep in mind that large quantities of carbon dioxide are dissolved or stored in the oceans. The ocean contains 50 times more carbon dioxide than the atmosphere. Marine organisms use some of the carbon dioxide in the oceans to build shells. When they die, their shells accumulate on the bottom of the ocean and form carbonate rocks, removing carbon from the cycle. Scientists do not completely understand how much carbon dioxide the oceans will be able to absorb as the amount of carbon dioxide in the atmosphere changes.

Hydrologic Cycle

In the atmospheric sciences, water is very important because it couples, or connects, the atmosphere with the surface of the Earth. Water is also the only substance that exists naturally in the atmosphere in all three phases: gas, solid, and liquid. Changing from one phase of water to another, such as from liquid to gas, is an important means of transferring energy in the atmosphere.

The **hydrologic cycle** (FIGURE 1-5) describes the circulation of water from the ocean and other watery surfaces to the atmosphere and the land. A major source of atmospheric water vapor (i.e., water in the gas phase) is evaporation from the oceans. **Evaporation** is the change of phase of liquid water to water vapor. Evaporation from lakes and glaciers supplies a relatively small amount of water to the atmosphere. **Transpiration**, the process by which plants release water vapor into the atmosphere, is also a source of atmospheric water vapor. The surface of the Earth is the major source of atmospheric water vapor, so the amount of water vapor in the atmosphere is generally largest near the surface and rapidly decreases with distance from the surface.

Often atmospheric water vapor changes phase to form solid and liquid particles. The change of phase from water vapor to liquid is called **condensation**. You see condensation occurring whenever a cold drink glass becomes wet on a hot humid day. Similarly, clouds form in the atmosphere via condensation. Usually more than 50% of the globe is covered with clouds. The occurrence of clouds is more frequent in some areas of the world and less frequent in others. FIGURE 1-6 depicts the cloud cover on a single day, as measured from several different weather satellites. Global patterns in cloud cover are evident. A lack of clouds is observed at about 30° north and south latitude, particularly over the deserts of Africa. A band of clouds is observed in the vicinity of the equator. The reasons for the global cloud patterns shown in Figure 1-6 are discussed in Chapter 7.

Precipitation, such as rain, snow, sleet, freezing rain, and hail, falls from clouds. Precipitation is a sink of atmospheric water because it removes water from the atmosphere. Precipitation returns water to the Earth's surface after it has evaporated and completes the hydrologic cycle. Precipitation on land may collect in lakes, run in rivers directly back to the sea, or percolate into the soil.

Precipitation may fall on glaciers, which then store the water. The occurrence and extent of glaciers are functions of the temperature of the atmosphere and the amount of precipitation. The presence of glaciers will also affect the atmospheric temperature. They reflect most solar energy

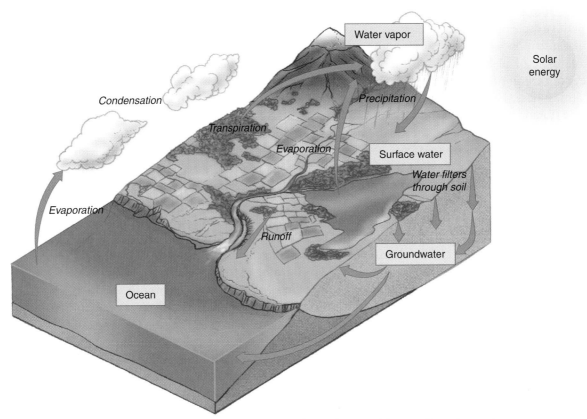

FIGURE 1-5 The hydrologic cycle describes how water enters and leaves the atmosphere.

back to space, reducing the amount that can be absorbed to heat the Earth. Like rivers, glaciers flow, but usually extremely slowly. Evaporation from lakes returns water to the atmosphere. Lake water and precipitation seep below the surface of the Earth and are stored there. Water stored in the soil ultimately flows back to the oceans, as does precipitation and river and glacier flows, completing the cycle.

Because water plays a major role in weather and climate, it is important to understand the hydrologic cycle. A change in one component of the hydrologic cycle can affect weather. For

FIGURE 1-6 Satellites are used to determine global distributions of cloud amounts. This is a view of Earth from a combination of weather satellites on the same day as shown in Figure 1-1 (November 29, 2006). Can you find the growing storm over the central United States and the large weather system moving into the Pacific Northwest? Also, notice the very large storm over the north Atlantic. The global cloud patterns provide insight regarding atmospheric wind patterns.

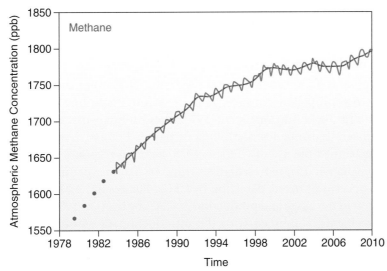

FIGURE 1-7 Atmospheric methane concentrations since 1978. Rapid increases in methane are presumably caused by agricultural, industrial, and animal sources tapered off in the mid-1990s, but the most recent observations suggest another rise in methane. (*Source*: NOAA/ESRL Global Monitoring Division—THE NOAA ANNUAL GREENHOUSE GAS INDEX [AGGI]. [n.d.]. Retrieved from http://www.esrl.noaa.gov/gmd/aggi/.)

example, a decrease in the amount of cloud cover over land during the day will allow more solar energy to reach the surface and warm the ground and the atmosphere above.

Methane

In addition to adding carbon dioxide to the atmosphere, human activities are changing the atmospheric concentration of other trace gases, such as methane. For example, the concentration of methane has doubled since the beginning of the Industrial Revolution. We do not fully understand either those earlier increases in atmospheric methane or the leveling off of methane concentrations since about the year 2000 (**FIGURE 1-7**). Human activities that contribute to the increased methane concentration include the decay of organic substances in landfills and rice paddies (the cultivation of rice has doubled since 1940), natural gas production, the burning of forests, coal mining, and even cattle raising. Methane is a by-product of cows' digestive process and accounts for 28% of human-related methane emissions globally!

Methane and other trace gases play key roles in the global warming debate. Not only do they affect the Earth's energy balance, their concentrations can be affected by warming temperatures. For example, some recent scientific observations have found that warmer land and ocean temperatures are causing bubbles of methane to escape from underground into the ocean (**FIGURE 1-8**) and atmosphere. An understanding of how trace gases increase the global

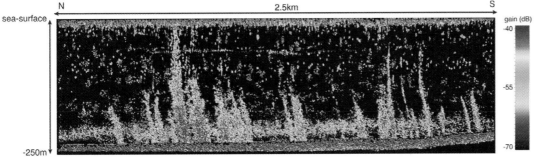

FIGURE 1-8 Methane bubbles observed by sonar rising from the bottom of the Arctic Ocean off Norway, as reported in a 2009 research paper. This region of the Arctic Ocean has warmed 1° C (1.8° F) in the past 30 years, helping to release the bubbles from the sea floor.

average temperature, and how that warming can in turn affect the concentration of trace gases, requires us to revisit so-called greenhouse warming throughout this book.

Chlorofluorocarbons

Chlorofluorocarbons (CFCs) do not occur naturally. They were invented by chemists in 1928 and were used as propellants in spray cans, in Styrofoam puffing agents, and as coolants for refrigerators and air conditioners. In 1974, these human-made, or **anthropogenic**, gases were first linked to ozone destruction (see Chapter 2). In response to public concerns, the United States banned the use of CFCs in 1978. Later, in response to the discovery of the "ozone hole" (see Chapter 15), representatives from 23 nations met in Montreal, Canada, in 1987 to address concerns of ozone depletion by CFCs. The resulting Montreal Protocol called for a 50% reduction in the usage and production of CFCs by the year 1999. This and subsequent international agreements, combined with the introduction of substitute chemicals, largely eliminated the use of CFCs worldwide. Although the global use of these chemicals has declined for over two decades, their concentrations in the atmosphere have decreased only very slowly in recent years (**FIGURE 1-9**). This is because CFCs are very stable molecules and will stay in the atmosphere for nearly 100 years after their release before they decompose.

Aerosols

Clouds are not the only liquid and solid particles present in the atmosphere. Smoke, salt, ash, smog, and dust are examples of particles suspended in the atmosphere. These particles are collectively known as **aerosols**.

The size of an aerosol particle is measured in **microns** (one millionth of a meter) and varies with the type of aerosol (**FIGURE 1-10**). The amount and type of aerosol can influence the climate of a region by modifying the amount of solar energy that reaches the surface. The largest airborne particles are those thrown into the atmosphere during volcanic eruptions, forest fires, and tornadoes. Most atmospheric aerosols are too small to be visible to the naked eye, but they serve an important purpose (discussed in Chapter 4)—they are needed to form most clouds.

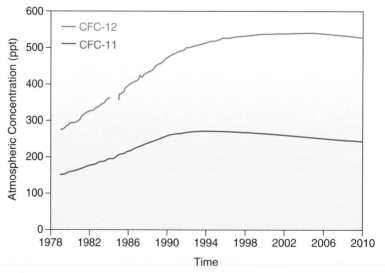

FIGURE 1-9 Concentrations of CFC-11 and CFC-12 in parts per trillion per volume since 1978. Although the usage of CFCs has declined in recent years, the concentration in the atmosphere has not declined as quickly. This is because CFCs are very chemically stable, and thus, after these molecules enter the atmosphere, they remain there for approximately a century. (*Source*: NOAA/ESRL Global Monitoring Division—THE NOAA ANNUAL GREENHOUSE GAS INDEX [AGGI]. [n.d.]. Retrieved from http://www.esrl.noaa.gov/gmd/aggi/.)

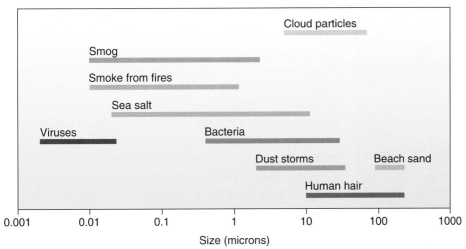

FIGURE 1-10 Aerosols are liquid and solid particles that are suspended in the atmosphere. Typical sizes of different aerosols are shown. For reference, the diameter of a human hair, beach sand, viruses, and bacteria is also shown.

Aerosols are more prevalent in certain regions of the world than others. For example, dust derived from soil is very common over deserts. Atmospheric winds can transport this dust to other regions of the globe where it may be deposited. Dust deposits are found on the ice sheets of Greenland. Dust is not generated in large quantities over Greenland, so these deposits must have originated in another location and then been transported toward the Arctic. An analysis of the Greenland ice sheet indicates an increased amount of dust present in the atmosphere at the end of the last "ice age," suggesting persistent dry, windy conditions. Dust deposits over glaciers are therefore a window into the weather of centuries past.

Sea salt is an aerosol that commonly occurs over the oceans and along shorelines. Sea salt enters the atmosphere from the oceans when spray from waves evaporates, leaving behind tiny salt particles in the atmosphere. Sea salt particles are important in the formation of clouds.

Other primary sources of aerosols are wind erosion, fires, volcanoes, and human activity. Wind erosion is a natural way of getting aerosols into the atmosphere. Winds lift soil particles off the bare ground and transport them into the atmosphere. Fires produce large amounts of aerosols. Air heated by fire can lift particles several thousand feet above the ground, where winds can carry the aerosols far away from their original source (**FIGURE 1-11**).

Volcanic eruptions spew huge amounts of particles into the atmosphere. The eruption of Mount St. Helens in Washington State on May 18, 1980, injected approximately 472 million metric tons (520 million tons) of ash into the atmosphere! Chapter 3 presents a discussion of how aerosols generated in the largest volcanic eruptions can lower Earth's temperature.

Airborne particles generated by human activity are referred to as anthropogenic aerosols. Sources of these aerosols include fuel combustion, construction, crop spraying, and industrial processes. Large anthropogenic aerosols are generated in mechanical processes such as grinding and spraying. The smaller particles are generated in processes where combustion is incomplete, such as in automobiles.

Click on "Mexican Smoke over the Midwestern States" to view a 10-day cycle of northward smoke movement from Central America.

ATMOSPHERIC PRESSURE AND DENSITY

Basic Concepts

Pressure is the force exerted on a given area. The atmosphere is made up of gas molecules that are constantly in motion. These molecules exert a pressure when they strike an object. The pressure exerted by the molecules hitting you is a function of their speed, number, and mass. Because the molecules that compose the air are moving in all directions, the pressure is exerted in all directions.

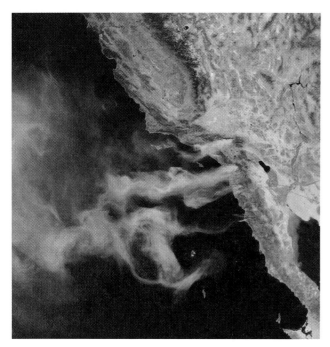

FIGURE 1-11 Smoke from fires in southern California and northern Baja California on October 23, 2007. The locations of fires are denoted in red. In this picture, taken by a weather satellite, the smoke plumes extend for hundreds of miles over the Pacific Ocean. Hot, dry "Santa Ana winds" that blow toward the Pacific coast from the deserts of the western United States helped to fan the flames, which forced 1,000,000 people from their homes, the largest evacuation in California's history.

The concentration of molecules is measured in terms of **density**, or mass per unit volume. The atmospheric density at sea level for a standard temperature of 15° C (59° F) is 1.225 kilograms per cubic meter. In comparison, the density of liquid water is much greater, approximately 1025 kilograms per cubic meter. The pressure, temperature, and density of a gas are related to one another through a mathematical formula known as the ideal gas law (**BOX 1-2**). Changing one of these variables will cause a change in one or both of the others.

The ideal gas law states that the ratio of pressure to the product of the density times the temperature is always the same:

$$\frac{\text{Pressure}}{\text{Density} \times \text{Temperature}} = \text{Constant}$$

If one of the three variables changes, then the other two also have to change to keep this ratio constant. For example, air that is warmed at constant pressure has to have a lower density. In other words, warmer air is less dense than colder air. Or, if the air pressure is reduced, then either the density, temperature, or both need to decrease too. As we'll see shortly, this agrees with the vertical distribution of density and temperature in the troposphere.

Air has mass and, because of gravity, has weight. You may find it useful to think of atmospheric pressure as the weight of a column of air above you. **FIGURE 1-12** depicts how pressure and density vary with altitude. As you move upward from the surface, you are decreasing the amount of air above you and therefore its weight or pressure. Because air is compressible, the air near the surface is more compressed by the weight of the air above it than is air at higher altitudes. As a result, *atmospheric pressure and density decrease rapidly as you go up from the Earth's surface.*

Barometric Pressure and Sea-Level Pressure

Television weather reports show atmospheric pressure in terms of inches or millimeters of mercury. In meteorology, it is more common to report atmospheric pressure in millibars or Pascals (named after the 17th-century scientist Blaise Pascal, who in 1648 discovered the variation of pressure with altitude) rather than inches of mercury. One **millibar** (mb) equals 100 Pascals and 0.76 millimeters (0.03 inches) of mercury. The average surface pressure at sea level is 1013.25 mb, or 29.92 inches, of mercury. Pressure is also sometimes expressed in terms of pounds per square inch (psi); air pressure in tires is frequently measured in psi. One millibar is equivalent to 0.0145 psi. A common pressure in automobile tires is 32 psi, which converts to 2206 mb, or 65.3 inches, of mercury—more than twice the atmospheric pressure at sea level!

First we need to explain the science behind the terminology of atmospheric pressure. What does atmospheric pressure in "inches of mercury" really mean? The TV meteorologist is reporting atmospheric pressure in terms of the height of a column of mercury in an atmospheric pressure measuring device known as a **barometer**. To understand how a barometer works, consider **FIGURE 1-13**, in which a tube with no air in it is placed upside down in a liquid. Because gravity pulls objects toward the center of the Earth, the weight of an object exerts a downward pressure on a surface it rests on. Air has mass and because of gravity exerts a downward pressure on objects. The molecules in the air collide with and exert a pressure on

Box 1-2 The Ideal Gas Law

Mathematics is a convenient and powerful tool for expressing physical ideas. Most atmospheric processes can be explained in descriptive terms; however, mathematical equations provide further insight. The ideal gas law describes the relationship between pressure, temperature, and volume of a gas. It is written as follows:

$$\text{Presssure} \times \text{Volume} = k \times \text{Temperature}$$

In this equation, k is a constant of proportionality that depends on the size of the gas sample. Although this law is an approximation of how a gas behaves, it is an excellent one for atmospheric studies. The ideal gas law can also be written as follows:

$$\text{Presssure} = R \times \text{Density} \times \text{Temperature}$$

where R is a constant that depends on the gases that compose the atmosphere. For dry air, the value of R is 287.05 Joules per kilogram per degree Kelvin.

We will use the ideal gas law to understand the relationship between pressure, volume, and temperature of a gas. Changing one of these properties changes the others. Let us start by considering a fixed volume of a gas. You can envision this as a gas stored in a thick metal container. If we double the temperature of the gas, then to make the left-hand side of the equation to equal the right-hand side, the pressure inside the container must double. If we fix pressure at a constant value, then doubling the temperature of the gas would also double the volume of the gas.

The study of how one variable changes with respect to another variable is a fundamental concept of the mathematics subject known as *calculus*. Because weather variables change with respect to each other and across time and space, students planning to be meteorologists must study calculus.

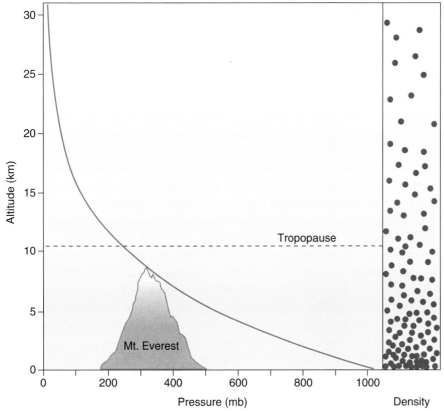

FIGURE 1-12 Atmospheric pressure and the density of air both decrease rapidly with increasing distance from the Earth's surface. (Adapted from Rauber, R. M., Walsh, J. E. and Charlevoix, D. J. *Severe & Hazardous Weather: An Introduction to High Impact Meteorology, Third edition.* Kendall/Hunt, 2008.)

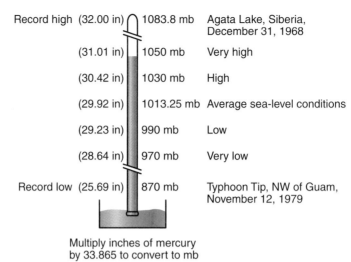

Record high (32.00 in)	1083.8 mb	Agata Lake, Siberia, December 31, 1968
(31.01 in)	1050 mb	Very high
(30.42 in)	1030 mb	High
(29.92 in)	1013.25 mb	Average sea-level conditions
(29.23 in)	990 mb	Low
(28.64 in)	970 mb	Very low
Record low (25.69 in)	870 mb	Typhoon Tip, NW of Guam, November 12, 1979

Multiply inches of mercury
by 33.865 to convert to mb

FIGURE 1-13 A summary of observed sea level atmospheric pressure. The average sea level pressure of the planet is 1,013.25 mb (29.92 inches of mercury). The lowest sea level pressure ever measured is 870 mb (25.69 inches), and the highest is 1,083.8 mb (32.00 inches). If these pressure units are unfamiliar, you can use this figure as a reference to determine whether observed pressures are considered to be low or high values.

the surface of the liquid. This pressure pushes the fluid up the tube. The greater the pressure, the higher the column of fluid is. The height of this column is a measure of the **barometric pressure.**

As we go higher in the atmosphere, the number of molecules colliding with the fluid surface decreases, and so the pressure exerted on the fluid is less. Thus, the column of fluid in the barometer is lower. This property of the atmosphere helps explain how the human body reacts to airplane travel and mountain climbing (BOX 1-3).

The decrease of atmospheric pressure with altitude creates a problem for interpreting pressure readings at locations with different altitudes. For example, Denver is approximately 1.6 kilometers (1 mile) above sea level, whereas San Francisco is near sea level. Because pressure always decreases with altitude, the barometric pressure measured at Denver will always be lower than that measured at San Francisco regardless of the weather situation. To avoid this apples-and-oranges situation when comparing the pressure of different locations, meteorologists always adjust all surface barometric pressure measurements to a single altitude: sea level. Standardizing to **sea-level pressure** removes the effect of altitude on pressure and allows the meteorologist to focus on the smaller, but important, surface pressure differences resulting from weather systems.

Let's learn how to do sea-level pressure conversions. Assume that on a given day that the barometric pressure in Denver is 850 mb and the barometric pressure at sea level in San Francisco reads 1013 mb, which is the average pressure at sea level. Denver's pressure sounds very low compared with San Francisco's. How can we adjust Denver's pressure to sea level? A city at sea level, such as San Francisco, has 1 more mile of air above it than Denver has. To adjust Denver's barometric pressure for sea level we have to add the pressure caused by the difference resulting from that extra mile of air. The lowest mile of air has a pressure of approximately 170 mb. The pressure in Denver adjusted to sea level is then 850 mb + 170 mb = 1020 mb.

This calculation shows that in our example Denver's weather is being affected by higher-than-average pressure, despite our initial impressions to the contrary. In a similar way, atmospheric pressure readings across the world are standardized by calculating what the pressure would be if the weather station were located at sea level.

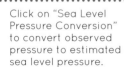

Click on "Sea Level Pressure Conversion" to convert observed pressure to estimated sea level pressure.

Box 1-3 Why Do Your Ears Pop?

Have you ever felt pressure in your ears when you were on an airplane flight or climbing a mountain? Have you noticed that the discomfort is worse if you have a cold and that chewing gum or yawning helps to alleviate it? Have your ever had trouble hearing an airline crew member make an announcement shortly after takeoff or shortly before landing because of loudly crying babies? A combination of meteorology and biology explains all of these observations.

As discussed in this chapter, atmospheric pressure decreases rapidly as you go up. In most of the western and eastern United States, it is possible to drive from sea level to 1 kilometer (3300 feet) elevation or higher in a few hours. This is equivalent to a pressure decrease of roughly 100 mb.

Pressure changes in an airplane are even quicker. Commercial airplanes cruise at an altitude of about 10 kilometers (33,000 feet) where atmospheric pressure is so low the interior of the plane must be pressurized to keep the crew members and the passengers conscious. Although the plane's cabin is pressurized, during flight, it is not kept at the same pressure as at the ground. To do so would place too much stress on the plane's structure when the far less dense air at cruising altitude surrounds it. Instead, cabin pressure during the middle of a long flight is usually about 750–800 mb. Because of this, cabin pressure decreases some during takeoff and increases some during landing.

If you chart the changes in surface atmospheric pressure throughout the day, you will observe that pressure does not usually change more than a few millibars a day. Because pressure varies slowly with time, our bodies have had no need to adapt to rapid changes in pressure. Yet, this is precisely what happens in an airplane or when climbing a mountain. If there is a rapid change in pressure, our bodies are slow to react to it. To understand how a rapid pressure change hurts our ears, we have to review some biology.

The pinna (the part of the ear that is visible) collects sounds and directs the sound waves into the auditory canal to the eardrum (tympanic membrane). Sound waves traveling through air are variations in pressure. When these pressure variations strike the eardrum, they cause the eardrum to vibrate. Tiny bones (the hammer, anvil, and stirrup) sense these vibrations and pass them to the inner ear. To hear sounds clearly requires the tympanic membrane to vibrate freely. Air on either side of the membrane exerts a pressure that maintains the proper tension for the eardrum to vibrate. The ear is connected to the nasal passage and throat by the Eustachian tube. This tube is a passageway for air to enter and leave the middle ear to balance the pressure exerted by air on the auditory canal side of the eardrum.

A rapid change in altitude results in a pressure imbalance on the eardrum. When the air pressure rapidly drops, the eardrum bulges outward, causing discomfort. To balance this pressure difference at the eardrum, air molecules must escape through the Eustachian tube. When the pressure inside your middle ear adjusts to the outside pressure, the eardrum returns to its normal tension. This movement of the eardrum makes a popping sound, and we say that your ear "pops."

Now we can explain the usefulness of yawning and chewing gum on an airplane. When you yawn or swallow, you are allowing the Eustachian tube to open wider and equalize this pressure difference more rapidly. A cold may block the Eustachian tube, making it more difficult to adjust to changes in air pressure. This increases the distortion of the eardrum and causes pain.

Babies on airplanes are too young to understand that they need to yawn or swallow. Furthermore, their Eustachian tubes are tiny and are easily blocked during colds or ear infections, which babies frequently endure. Therefore, during airplane takeoffs and landings, babies feel intense pressure and pain in their ears, and they respond by crying. Remember this with some sympathy the next time you fly on an airplane accompanied by the serenade of screaming infants!

DIVIDING UP THE ATMOSPHERE

Meteorologists find it useful to divide the atmosphere into layers. We can divide the atmosphere vertically according to pressure. For example, in this book we will often discuss the winds at an altitude where the atmospheric pressure equals 500 mb. Why are we interested in 500 mb? The average atmospheric pressure at the Earth's surface is a little over 1000 mb, so if we were at an altitude of 500 mb, approximately half the atmosphere would lie above us and half below us. We learn in later chapters that the winds at about 500 mb help "steer" atmospheric storms. This makes the 500-mb level especially important for weather forecasting.

We can also divide the atmosphere according to temperature. Unlike pressure, temperature does not always decrease with increasing distance from the surface. If we were to average temperature over many years at many different locations across the globe for many different altitudes, however, a distinct pattern would emerge between temperature and distance from the surface. FIGURE 1-14 presents an example of this pattern. Based on this temperature profile, the atmosphere can be divided into four main layers: the troposphere, stratosphere, mesosphere, and thermosphere.

▦ The Troposphere

From the surface up to approximately 10 to 16 kilometers (6 to 10 miles), temperature generally decreases with altitude. This is because the atmosphere is nearly transparent to the Sun's energy and is instead heated from below by the Earth's surface, like a pot of water on a stove. The region of the atmosphere closest to Earth, where the temperature decreases as you go up, is called the **troposphere**. The word *troposphere* is derived from the Greek word *tropein*, meaning "to change." Almost all of what we normally call "weather" occurs in the troposphere, and approximately 80% of the atmosphere's mass is located in the troposphere. For these reasons, the troposphere is the focus of this book.

The top of the troposphere is referred to as the **tropopause**. It acts as an upper lid on most weather patterns, just as a lid on a pot of water on a stove keeps the water from escaping. The height

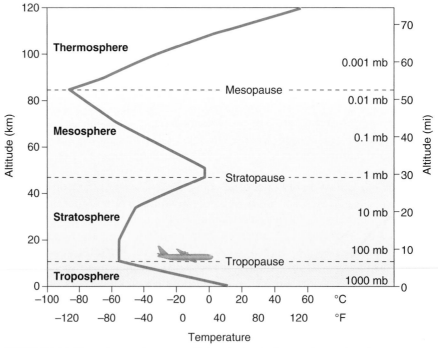

FIGURE 1-14 The atmosphere is subdivided into layers based on how air temperature changes with altitude. (Adapted from Rauber, R. M., Walsh, J. E. and Charlevoix, D. J. *Severe & Hazardous Weather: An Introduction to High Impact Meteorology, Third edition.* Kendall/Hunt, 2008.)

of the tropopause is a function of latitude. It is higher in the equatorial regions, where the tropospheric air is warmer and expands upward to about 16 km, than in the cold polar regions, where the tropopause height is closer to 10 km. Weather systems can also affect the height of the tropopause, as we'll see in Chapter 10.

The Stratosphere

Above the tropopause lies the **stratosphere**, where temperature increases with altitude. Temperature is increasing because ozone molecules in the stratospheric "ozone layer" (see Chapter 2) are absorbing solar energy within the stratosphere. Air flow in the stratosphere is much less turbulent than in the troposphere. For this reason, jet aircraft like to cruise at stratospheric altitudes, where the flight is less "bumpy." The increasing temperature with altitude of the stratosphere makes it difficult for tropospheric air to mix into the stratospheric air, as explained in later chapters. The lack of mixing and turbulence makes the stratosphere very layered or "stratified"—hence its name. In the vicinity of storms with strong up-and-down wind motions, such as thunderstorms and low-pressure systems, tropospheric air can penetrate into the stratosphere, and stratospheric air can descend into the troposphere.

The **stratopause** marks the top of the stratosphere. On average, the stratopause occurs at an altitude of approximately 50 kilometers (31 miles). Its average temperature is close to 0° C, and during the winter the stratopause can be warmer than the ground far beneath it. But don't consider the stratopause a likely holiday vacation destination. Nearly 20% of the Earth's atmosphere is in the stratosphere, meaning that at the stratopause, the pressure is only 0.1% of that at the Earth's surface. Human beings could not survive at such low pressures; the liquids in their bodies would literally boil away!

The Mesosphere and Thermosphere

Above the stratopause lies the **mesosphere**. Temperature decreases with altitude in the mesosphere, just as it does in the troposphere. The **mesopause** separates the mesosphere from the **thermosphere** at an average altitude of about 85 kilometers (53 miles). In the thermosphere, the temperature again increases with altitude. The density of the atmosphere in the thermosphere is so low that above about 120 kilometers (75 miles) the atmosphere gradually blends into interplanetary space. The entire Earth's atmosphere is extremely thin, less than 2% as thick as the Earth itself.

From the mesosphere on up, the atmosphere becomes more and more affected by high-energy particles from the Sun. These particles break apart atmospheric molecules, which then form ions. For this reason, the region of the mesosphere and thermosphere is sometimes called the *ionosphere*. Some interesting visual phenomena occur in the mesosphere and thermosphere as a result of the interaction of solar particles and the atmosphere, such as the aurora (**FIGURE 1-15**). In addition, the ionosphere reflects radio waves, permitting radio stations to be heard far beyond the horizon.

AN INTRODUCTION TO WEATHER MAPS

Basic Concepts

An important part of studying meteorology is simply paying attention to weather conditions and applying your knowledge to what you observe. However, many different variables including temperature, moisture, cloudiness, precipitation, and others are used to describe weather. All of these must be considered and analyzed numerically. Furthermore, the weather at one location is often caused by larger weather patterns. So it is not enough to consider the weather locally. We must also analyze the weather for many other locations. This means even more numbers.

To identify key weather-making patterns, we need to see all of the numbers describing weather at many locations, all at once. It is also vital to see how the observations relate to each other geographically. A list of numbers doesn't help much; we need a picture. So, we plot weather

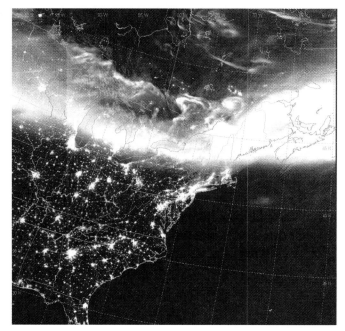

FIGURE 1-15 A visible satellite image of eastern North America at night on November 8, 2004. In addition to the lights of major cities, a bright band is seen extending from Minnesota to Michigan and across the entire state of Maine into Canada. This bright band is an aurora caused by a strong flow of solar particles near the end of the last "solar maximum." The next peak in solar activity, bringing with it more auroral activity, is expected by 2014.

variables on a weather map to understand the relationships among them. To paraphrase the saying "A picture is worth a thousand words," a weather map is worth a thousand numbers— or more. In this section, we learn how to interpret weather maps, in particular the **surface chart** that depicts weather at the Earth's surface.

Perhaps the most obvious features on surface weather maps are fronts. A **front** is a boundary between two regions of air that have different meteorological properties, such as temperature and humidity. A **cold front** denotes a region where cold air is replacing warmer air. A **warm front** indicates that warm air is replacing cooler air. We will discuss fronts in detail in Chapters 9 and 10 of this book.

All surface weather maps today depict frontal locations because they are regions of rapidly changing and sometimes dangerous weather conditions. The frontal lines drawn on a weather map represent the locations where the fronts meet the Earth's surface (**FIGURE 1-16**). On a weather map, a blue line with blue triangles indicates a cold front. The triangles point in the direction the front is moving. A warm front is shown as a red line with red semicircles pointing in the direction of frontal movement. A **stationary front** is a front that is not moving and is represented as shown in the lower-left corner of Figure 1-16. The **occluded front**, represented as a purple line with alternating triangles and semicircles, has characteristics of both cold and warm fronts and is discussed in more detail in Chapters 9 and 10.

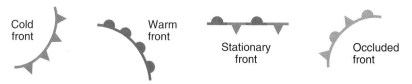

Cold front   Warm front   Stationary front   Occluded front

FIGURE 1-16 The colors and symbols for the four types of fronts: cold, warm, stationary, and occluded.

To locate the position and type of a front on any given day requires the analysis of the weather conditions from locations over a wide geographic area. To help in this analysis, we often draw lines on weather maps connecting locations that have the same temperature. Lines of constant temperature are called **isotherms** (from *iso*, "same," and *therm*, "temperature"). Similarly, **isobars** connect locations with the same sea level atmospheric pressure. Both isotherms and isobars are often shown on television and newspaper weather maps. **Isotachs** connect locations with the same wind speed. **Isopleth** is a more general term describing contours along which any particular variable is constant.

Click on "Learning to Contour Weather Maps" to practice how to contour weather maps.

The Station Model

Television weather reports sometimes represent local weather conditions with smiling suns, rainy clouds, or flashing bolts of lightning. However, one smiling sun can cover several states and gives you no information about temperature, wind, and pressure. Meteorologists need a way to condense all the numbers describing the current weather at a location into a compact diagram that takes up as little space as possible on a weather map. This compressed graphical weather report is called a **station model.**

A simplified example of a station model plot used to represent meteorological conditions near the surface for a specific location is shown in **FIGURE 1-17.** The station model depicts weather

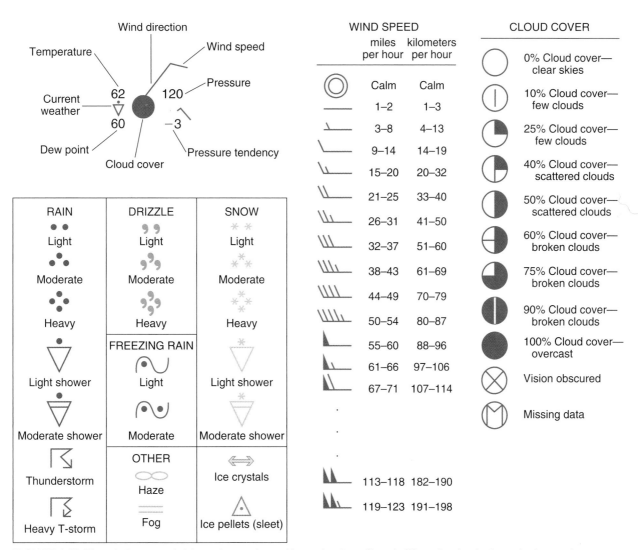

FIGURE 1-17 The station model for a typical weather situation (top left) and selected symbols used in the station model and their meaning. See the text and Web site for an explanation of how to decode a station model.

conditions at a particular time in a specific location, plus cloud cover, wind speed, wind direction, temperature, dew point temperature, atmospheric pressure adjusted to sea level, and the change in pressure over the last 3 hours. In Figure 1-17, nine weather variables commonly reported on the evening news are plotted—much more information than a smiling sun can convey, in much less space.

Let's decode the station model at the top of Figure 1-17. First, cloud cover is depicted graphically by the type of shading inside the circle, which represents the location of the weather station. The circle is completely shaded, so this means 100% cloud cover or "overcast" conditions.

Next, the temperature reading is located at the 10 o'clock position with respect to the circle (i.e., just above and to the left of it). In the United States, temperature is reported on weather maps in degrees Fahrenheit. Therefore, in our example the temperature is 62° F. The dew point temperature, which is a measure of the amount of moisture in the atmosphere, is located at the eight o'clock position. In our example it is 60° F.

Current weather conditions are symbolically represented in between the temperature and dew point temperature. Some symbols for common weather conditions are shown in the lower part of Figure 1-17. In our example, the station is currently reporting a light rain shower. If there is no precipitation, haze, or fog occurring at the time of observation, the current weather condition location is left blank.

Wind speed and direction are also depicted graphically on a station model by the position of the "flagpole" extending out from the circle at the center. The pole points to the direction from which the wind is coming. So, the wind blows from the "flagged" end toward the "pointed" end. The "flags" are long and short lines that indicate wind speed, as explained in Figure 1-17.

The measured atmospheric pressure is adjusted to sea level and printed in the upper-right corner of the station model. The units used are millibars. Sea level pressures outside of tornadoes and hurricanes are always between 900 and 1099 mb. To save space on the map, the leading 9 or 10 is dropped, as is the decimal point. To decode the value of pressure on the station model, add a 9 if the first number is a 7, 8, or 9; otherwise add a 10. (This rule does not work in situations with very low or high pressure, but in those rare cases, the pattern of isobars helps you to know whether the leading digit(s) should be a 9 or a 10.) In our example, the "120" means that the station's sea level atmospheric pressure is 1012.0 mb. If it had been "831" instead, the sea level pressure would be 983.1 mb.

The change in surface pressure during the last 3 hours is plotted numerically and graphically on the lower right of the station model. The change in pressure is represented by a value (in tenths of a millibar) and a line that describes how the pressure was changing over time from left to right. In our example, the line goes up and then goes down, indicating that the pressure rose and then fell over the past 3 hours, a total change of 0.3 mb. A more complete explanation of how to decode pressure tendency symbols is provided on this text's Web site.

Many other weather variables can be depicted on the station model: visibility (next to current weather conditions), weather conditions during the past hour (lower right), precipitation amounts (near pressure tendency), and peak wind gusts during the past hour (number in knots [1 knot = 1.15 mph] next to the "flags" on the "flagpole"). In this text, we will not look at all station model variables at once; instead, we will focus on different variables in different chapters. This text's Web site contains a complete guide to the station model and examples of current weather maps that use surface station models.

An example of an actual surface weather map is given in FIGURE 1-18. This map corresponds to the satellite image shown in Figure 1-1. The surface weather map shows that much of the central United States is cloudy, in agreement with Figure 1-1.

Focusing on the station model for Dallas, Texas, in north-central Texas, we see that the temperature is 77° F and the dew point temperature is 63° F. The sea level pressure at Dallas is 1006.0 mb; the winds are strong and from the south, and Dallas is experiencing broken cloud conditions. Interestingly, just to the north of Dallas in Oklahoma City, the temperature is

Click on "Decoding the Surface Station Model" for a complete guide to station models and for practice deciphering several examples.

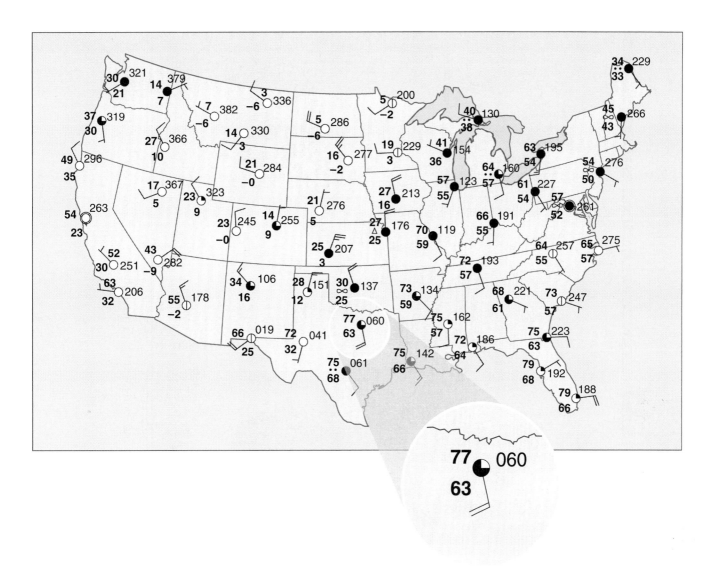

FIGURE 1-18 A surface weather map for 2200 UTC on November 29, 2006, using station models as explained in Figure 1-17. Compare this map and the weather described by it to the satellite image in Figure 1-1. (Adapted from Plymouth University Weather Center [http://vortex.plymouth.edu/make.html.]. Accessed June 10, 2010.)

only 30° F with strong winds from the *north.* This is a sign that a front lies between Dallas and Oklahoma City, as we learn in Chapter 9.

The weather observations in Figure 1-18 were all made at the same time. Unlike television programming, Eastern Time is not the reference for these maps. In the following, we learn how to decode meteorological time.

TIME ZONES

To depict current weather patterns using a weather map and to predict future weather, it is important to coordinate the time of global weather observations. To aid in this coordination, weather organizations throughout the world have adopted the Coordinated Universal Time (**UTC**, for **Universel Temps Coordonné**) as the reference clock. UTC is also denoted by the abbreviations GMT (Greenwich Meridian Time) or, often with the last two zeroes omitted, Z (Zulu).

The reference **time zone** for UTC is centered on Greenwich, England (**FIGURE 1-19**). The International Date Line (180° longitude, halfway around the world from Greenwich) separates

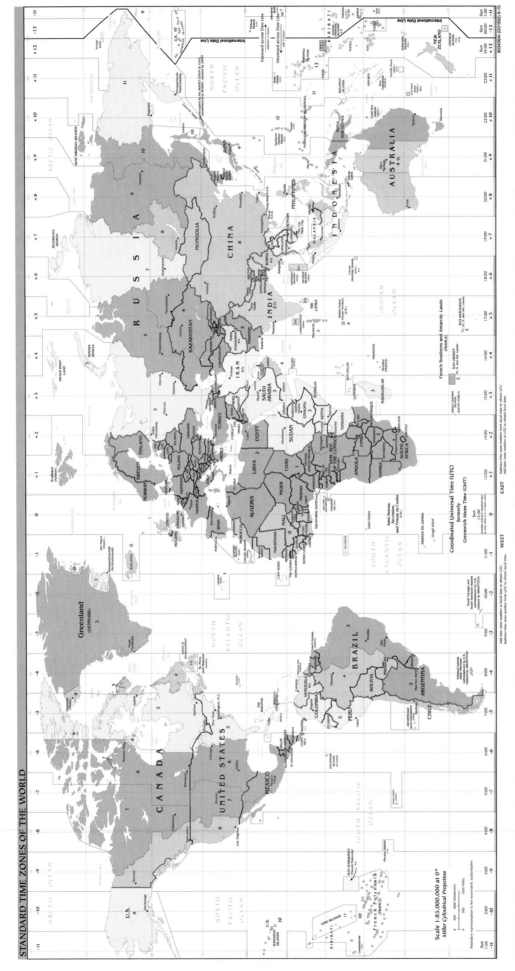

FIGURE 1-19 A map depicting the conversions between UTC (Universal Coordinated Time) and local times throughout the world. (Courtesy of *The World Factbook*, 2009.)

one day from the next. Just to the east of the dateline is 24 hours earlier than just to the west of the dateline.

Meteorology also uses the 24-hour military-style clock. For example, 1:30 PM UTC is 1330 (1200 + 130) UTC, and 0130 UTC is 1:30 AM. Observations of the upper atmosphere are coordinated internationally to be made at 0000 UTC (midnight at Greenwich) and 1200 UTC (noon at Greenwich). The Eastern Standard Time zone of the United States is 5 hours earlier than Greenwich time. So, an observation made at 1300 in New York City would be recorded as 1800 UTC time.

Why is Eastern Standard Time 5 hours earlier than Greenwich time? Since the Earth has 360° of longitude and rotates around its axis once every 24 hours, this means that local time should be 1 hour earlier for every (360°/24) = 15° of longitude west of Greenwich. (In practice, the exact lines dividing time zones are sometimes drawn along rivers and state and province boundaries, not along longitude lines.) New York City is at approximately 74° west longitude, or 74°W. This is almost 5 × 15° of longitude west of Greenwich. Using the 15° rule, this explains why Eastern Standard Time is 5 hours earlier than Greenwich or UTC time.

Using the same reasoning, New Orleans (90°W) and the rest of the Central Standard Time zone are 6 hours earlier. Denver (105°W) and the rest of Mountain Standard Time are 7 hours earlier. Los Angeles (118°W) and the rest of Pacific Standard Time are 8 hours earlier. Juneau (134°W) and the rest of Alaskan Standard Time are 9 hours earlier, and Honolulu (158°W), and the rest of Hawaiian Standard Time is 10 hours earlier than Greenwich.

In our example in Figure 1-19, the weather observations were taken at 22Z or 2200 UTC on November 29, 2006. Although the observations were all taken at about the same moment, the clock read 5:00 PM local time in hazy New York City, 4:00 PM in rainy Chicago and warm Dallas, 3:00 PM in frigid Denver, and 2:00 PM in clear Los Angeles. (In most of North America, Europe, and northern Asia, Daylight Saving Time moves local time forward 1 hour. In the U.S., Daylight Saving Time has begun on the second Sunday in March and has ended on the first Sunday in November since 2007.)

WEATHER WATCHES, WARNINGS, AND ADVISORIES

As you follow the weather, you will notice that certain meteorological conditions may pose a threat to life and property. Under these conditions, the National Weather Service issues advisories, weather watches, and weather warnings. A weather **watch** informs us that current atmospheric conditions are favorable for hazardous weather. When the hazardous weather will soon occur in an area, a warning is issued. Weather watches and warnings are issued for a wide variety of hazardous weather, including tornadoes, hurricanes, severe thunderstorms, winter storms, and flooding. FIGURE 1-20 depicts the weather warnings and watches issued on the day corresponding to Figure 1-18. Notice that many of the watches and warnings are in the vicinity of the clash between warm and cold air over the central United States.

The National Weather Service issues weather watches and warnings under specific weather conditions (TABLE 1-2). It is important to understand the difference between a weather watch, a weather warning, and a weather advisory. The term **watch** implies that you should be aware that a weather hazard may develop in your area. A **warning** is issued when the hazard is developing in your area. You should take immediate action in the event of a warning. An advisory is a less urgent statement issued to bring to the public's attention a situation that may cause some inconvenience or difficulty for travelers or people who have to be outdoors.

Pinpointing the location of hazardous weather in advance is extremely difficult. For this reason, watches are usually issued for large regions, sometimes covering several states (Figure 1-20). Warnings are often issued for smaller areas because they are based on actual observations of hazardous weather.

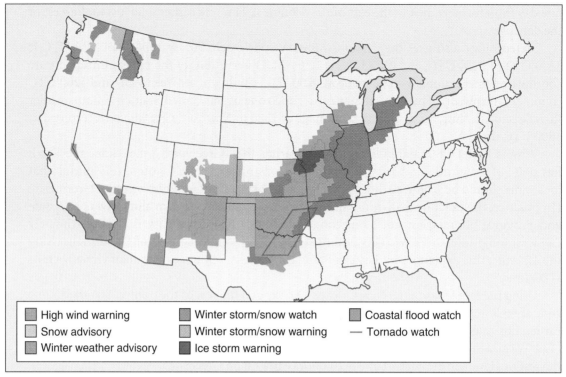

■ High wind warning	■ Winter storm/snow watch	■ Coastal flood watch
■ Snow advisory	■ Winter storm/snow warning	— Tornado watch
■ Winter weather advisory	■ Ice storm warning	

FIGURE 1-20 The National Weather Service's weather watches and warnings for the busy weather day of November 29, 2006 (compare with Figures 1-1, 1-6, and 1-18, which are for the same day and time). Winter storm watches and warnings cover a large part of the central United States from Texas to Michigan, as well as parts of New Mexico and Colorado and the Pacific Northwest. An ice-storm warning has been issued for northwest Missouri, with a flood watch for the Ozark region of southwestern Missouri. A tornado watch box extends through parts of Texas and Oklahoma. A high-wind warning has been posted for parts of Southern California and other scattered regions of the West. (Courtesy of SSEC, University of Wisconsin-Madison.)

You should know in advance what to do if hazardous weather threatens your area. Throughout this text we will discuss many types of weather hazards, the ways in which they kill, and the steps you can take to protect yourself from the awesome, and sometimes awful, pageant of weather.

PUTTING IT ALL TOGETHER

■ Summary

Weather is the state of the atmosphere at a particular time, and meteorology is the study of weather. The state of the atmosphere on longer time scales, including its interactions with oceans, land, and living things, is called climate. As with daily weather, the climate of a region changes with time, although on much longer time scales than the weather.

Earth's atmosphere has been shaped by billions of years of volcanic emissions and plant life. It is composed primarily of nitrogen and oxygen gases. Atmospheric concentrations of water vapor, carbon dioxide, methane, nitrous oxide, ozone, CFCs, and aerosols are very small but also very important to the study of weather and climate. Understanding the concentrations of water vapor and carbon dioxide, among other gases, requires taking into account the cycles of water and carbon from the air to the land and ocean and back to the air. The concentrations of most of these trace gases are increasing as a result of human activities.

The different molecules of the atmosphere are continually moving, exerting pressure in all directions. Atmospheric pressure is related to the weight of the column of air above you. As your altitude increases, the number of molecules above you decreases. For this reason, atmospheric

TABLE 1-2 Typical National Weather Service Criteria for Issuing Selected Weather Watches and Warnings

Weather Hazard	Watch	Warning
Hurricane	Hurricane conditions are possible in a given region within 48 hours.	The arrival of a hurricane is expected within the warned region within 36 hours.
Severe thunderstorm	Conditions are favorable for the development of severe thunderstorms in and close to the watch area.	A severe thunderstorm has been sighted visually or indicated by radar. This is issued when a thunderstorm produces hail 1 inch or larger in diameter and/or winds exceed 58 mph.
Tornado	Conditions are favorable for the development of tornadoes in and close to the watch area.	A tornado has been sighted visually or indicated by radar.
Flash flood	Conditions are forecast that could result in flash flooding in or close to the watch area.	Flash flooding will occur within 6 hours and will threaten life and/or property. Dam breaks or ice jams can also create flash flooding.
Winter storm	Conditions are favorable for the development of hazardous weather such as heavy snow, sleet, or freezing rain.	Hazardous winter weather is imminent or very likely.
Ice storm	Conditions are favorable for a significant and possibly damaging accumulation of ice on exposed objects, such as trees, power lines, and roadways.	Freezing rain/drizzle is occurring with a significant accumulation of ice (more than 0.25 inch) or accumulation of 0.5 inch of sleet.
High wind	Conditions are favorable for high wind speeds developing that may pose a hazard or is life-threatening during the next 24 to 36 hours.	Sustained wind speeds of 40 mph or greater will last for 1 hour or longer, or winds of 58 mph or greater are occurring for any duration of time.

pressure always decreases with distance from the surface. This simple concept is important in understanding the formation of clouds and the transfer of energy in the atmosphere. To compare measurements of atmospheric pressure at locations that are at different altitudes, the measurements must be adjusted to sea level.

Unlike pressure, temperature does not always decrease with increasing distance from the surface. The vertical changes in temperature neatly divide the atmosphere into four distinct layers: the troposphere, stratosphere, mesosphere, and thermosphere. Most weather occurs in the troposphere, which is the focus of this text.

Weather maps show the weather conditions for many locations at the same time. Many variables describe weather at a specific location, including temperature, pressure, and wind. The station model is a graphic way to compress large amounts of weather data for one location into a very small area on a weather map. Weather observations are taken at the same time at locations all over the world. This requires a system of time zones and a universal standard of time, known as UTC and referenced to Greenwich, England.

The National Weather Service issues watches when hazardous weather may occur in a region and issues warnings for more localized areas when hazardous weather has been observed nearby. Warnings require immediate action on your part to protect life and property.

Key Terms

You should understand all of the following terms. Use the glossary and this chapter to improve your understanding of these terms.

Aerosols	Isobar	Source
Anthropogenic	Isopleth	Stationary front
Argon	Isotach	Station model
Atmosphere	Isotherm	Stratopause
Barometer	Mesopause	Stratosphere
Barometric pressure	Mesosphere	Surface chart
Carbon dioxide	Meteorology	Thermosphere
Chlorofluorocarbons	Methane	Time zones
(CFCs)	Micron	Trace gases
Climate	Millibar	Transpiration
Climatology	Molecule	Tropopause
Cloud	Nitrogen	Troposphere
Cold front	Nitrous oxide	Universel Temps Coordonné
Condensation	Occluded front	(UTC)
Density	Oxygen	Variable gases
Diffuse	Ozone	Warm front
Evaporation	Permanent gases	Warning
Front	Pressure	Watch
Gravity	Sea-level pressure	Water vapor
Hydrologic cycle	Sink	Weather

Review Questions

1. What is one difference between weather and climate?
2. Explain why the meteorology of the Northern Hemisphere differs from that of the Southern Hemisphere, with reference to Figure 1-2.
3. Although the production of CFCs has been drastically reduced over the past quarter century, their atmospheric concentrations are only slowly decreasing. Why?
4. If pressure is force per area, then the atmospheric pressure over a large area such as the United States must be tiny compared with the atmospheric pressure over a city. Find the flaw in this reasoning.
5. What aspect of meteorology might make it easier to hit a home run at Coors Field in Denver than at PETCO Park in coastal San Diego, California?
6. Why is there no such field as lunar (Moon) meteorology?
7. The pressure at a weather station 1.6 kilometers above sea level is 900 mb. Based on the example discussed in the chapter, what is the sea level pressure of this station? Is this station experiencing unusually high, low, or normal atmospheric pressure?
8. Why does atmospheric pressure decrease with altitude?
9. If a raft filled with air is thrown into a cold lake, it appears to deflate, even though no air escapes. Explain.
10. Why does air that is rapidly compressed into a small volume get hot?
11. You are halfway between two layers of the atmosphere in which temperature increases with altitude. What is your approximate pressure, and in what layer of the atmosphere are you?
12. Using the station model decoding information in Figure 1-17, completely decode the station model for New York City in Figure 1-18.
13. You are asked to investigate the weather conditions at Boston at 7:00 PM on Monday, February 28, 2012. To get this information, you have to know the right day and time for this situation in coordinated universal time (UTC). What is the correct UTC day, date, and time?
14. Practice contouring weather maps using the Web link to Contouring Weather Maps.
15. Practice decoding the weather station model using the Web link to Decoding the Weather.

▪ Observation Activities

1. Place a balloon over the open end of an empty glass bottle. Run hot water over the bottle (not the balloon). What happens to the balloon? Explain what you observe. Then fill a balloon with air, and place it in a refrigerator. Remove it from the refrigerator after 20 minutes, and describe its appearance. Explain what you observe.

2. Visit http://www.spaceweather.com/aurora/gallery_01nov04.htm to see beautiful photographs from the ground of the November 2004 aurora shown from space in Figure 1-16.

3. Weather reports include pressure observations as well as how the pressure has been changing (increasing, decreasing, or steady). Observe these pressure changes with respect to changes of precipitation. Do you observe any relationship between precipitation and change in pressure?

4. Throughout the course, visit a NOAA web page to keep track of the different types of warnings and watches issued nationally and for your region (see, for example, the graphic on http://www.weather.gov/). How many different types of potentially hazardous weather were occurring when you visit the site?

5. Find a good internet site that plots the station model for your region (e.g., http://profhorn.meteor .wisc.edu/wxwise/station_model/sago.html for Wisconsin). Observe the weather outside, and see how your observations are plotted on a weather map. How does the weather in your region compare with that in neighboring cities?

This rain cloud icon is your clue to go to the *Meteorology* Web site at http://physicalscience.jbpub.com/ackerman/meteorology/. Through animations, quizzes, web exercises, and more, you can explore in further detail many fascinating topics in meteorology.

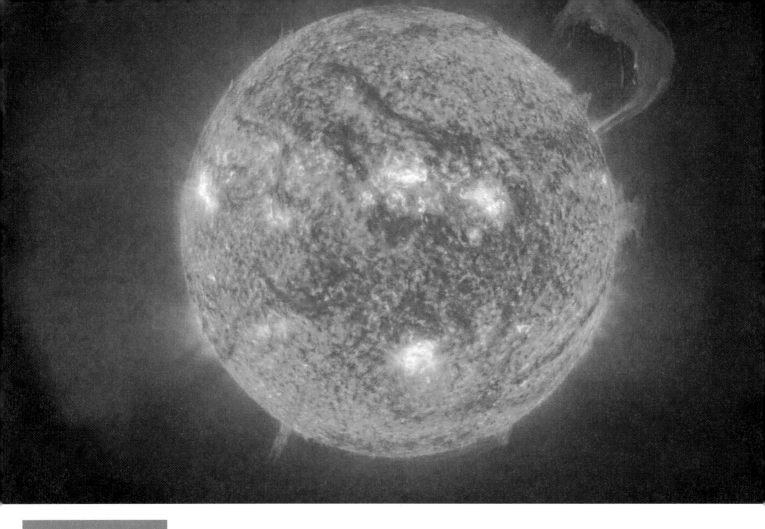

2 The Energy Cycle

AFTER COMPLETING THIS CHAPTER, YOU SHOULD BE ABLE TO
- Describe the different ways in which energy is transferred in the atmosphere
- Determine why very hot objects glow but room-temperature objects do not
- Explain why the Earth has seasons
- Relate the greenhouse effect to the presence of trace gases in the atmosphere

INTRODUCTION

Whether something warms or cools is related to its energy gains and losses. So, as you stand facing an evening bonfire, your front warms because it gains more energy than it loses, while your back cools as it loses more energy to the cooler night air than it gains. If the night is chilly and you are too close to the fire, you become uncomfortable; your front is too hot, and your back is too cold. You can modify your energy imbalance in several ways. You can turn around and place your back to the fire or step further away and put a blanket over your back. In both cases you have changed your energy budget.

The same thing happens to the Earth. At any particular moment half of the Earth is facing the Sun and half is not. Over a year, the tropical regions of the planet experience a net energy gain while the poles have a net energy loss. This global imbalance of energy is largely the result of differences in how high the Sun gets in the sky and the length of day. The atmosphere and oceans respond to this energy imbalance by transporting energy from the equatorial regions toward the poles. Transport of energy is the reason we have weather.

This chapter introduces the methods of energy transfer important for understanding weather and climate. The chapter contains definitions of terms and concepts used throughout the book. If you learn them now, the next chapters will be easier to follow.

FORCE, WORK, AND HEAT

The movement of air is important in defining the weather and climate of a given region. What causes this movement? To explain this, we have to understand a little about **force**, work, and heat.

Put your book down, and give it a push. When you push your book, you exert a force on the book that causes it to move, but your book doesn't keep moving. Why not? How fast and how far the book moves depends on how hard you push it and the force of friction acting between the book and the surface.

The same forces act in the atmosphere. In mathematical terms, the force exerted on an object is the mass of the body multiplied by the acceleration the force causes in the body. **Acceleration** is a change in speed or a change in direction of an object's movement. When you push your book, its speed changes. Similarly, forces acting on the atmosphere cause the wind to speed up or slow down or change direction. (We explore atmospheric forces in detail in Chapter 6.)

When you push your book, you use energy to do work on the book. **Work** is done on an object—whether it is your book or the air—when a force moves it. The amount of work done on your book is the distance traveled times the force in the direction of that displacement. Wind is air in motion; to move air requires work.

Doing work requires energy. **Energy** is the capacity to do work. The amount of energy needed depends on the amount of work to be done. It does not take much energy to push a book or lift a glass of water. The energy required to carry a bucket of water up to the top of the Empire State Building is a different matter!

We will study four different forms of energy: heat energy, electrical energy, kinetic energy, and potential energy. Energy can be converted from one form to another, but the total energy is always conserved.

In meteorology and climatology, we are primarily concerned with kinetic and potential energy. **Kinetic energy** is the work that a body can do by virtue of its motion. A major part of this energy is related to how fast a mass is moving. The kinetic energy of a moving object is also related directly to its mass. A freight train moving at only 16 km/hr (10 mph) has 36 times less kinetic energy than the same train moving at 96 km/hr (60 mph). A sport-utility vehicle on a highway has more kinetic energy than a car moving at the same speed.

Potential energy is the work an object can do as a result of its relative position. You do work as you lift a book off the desk. The higher you lift the book the greater the potential energy becomes. Once off the desk, the book has potential to do work because of gravity. When you let go of the book, it falls and potential energy is converted to kinetic energy. If you drop the book from the 10th floor of a building, it becomes a dangerous missile. If you drop the same book from a height of an inch, it can't do much damage. So, potential energy represents stored energy that can be converted to other forms of energy, such as kinetic energy. The potential energy of the book is represented by its height from the surface of the desk.

In studies of the atmosphere, meteorologists find it useful to refer to an isolated **parcel of air**. A parcel of air is a hypothetical balloon-like bubble of air, flexible but impermeable, and perhaps as large as a parking lot. As this parcel of air moves around in the atmosphere, mass and energy do not cross its imaginary boundary. The air around the parcel is the "environment" or the "surroundings." Although the real atmosphere is not composed of parcels, meteorologists can make sense of the atmosphere by looking at it in this way. If we imagine such a parcel of air, we can move it through the atmosphere as an intact unit and compare its temperature with the temperature of its surrounding environment.

Temperature relates to energy because temperature is a measure of the average kinetic energy of a substance. A thermometer in a glass of ice water records a lower temperature than a thermometer in a pan of boiling water because the molecules of H_2O in the boiling water are much more energetic. That energy is imparted to the thermometer and shows up as a higher temperature.

There are three commonly used scales for measuring the temperature of an object. **Fahrenheit** (named after the German instrument maker G. D. Fahrenheit) is commonly used in the United States to report temperatures near the surface (**FIGURE 2-1**). At sea level, ice melts at 32° F, and water boils at 212° F. (It is common to say that water freezes at 32° F, but we will see in Chapter 4 that water droplets can stay unfrozen at temperatures as low as −25° F!) The **Celsius** (or centigrade, named after the Swedish astronomer A. Celsius) temperature scale is based on the melting and boiling points of water—ice melts at 0° C, and water boils at 100° C. This temperature scale is used every day throughout the world to report the air temperatures near the ground. The **Kelvin** scale (named after one of Britain's foremost scientists, William Thomson,

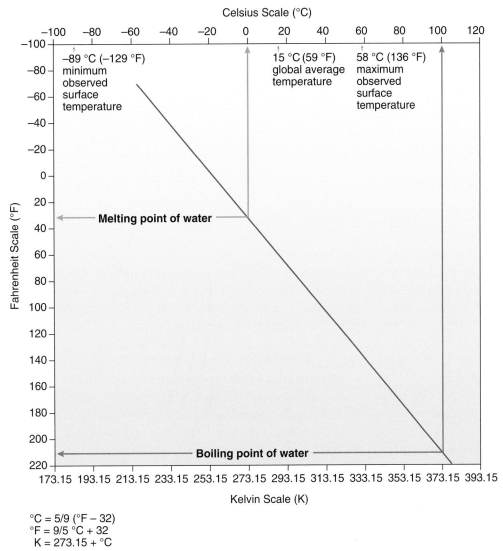

FIGURE 2-1 The three scales of temperature Fahrenheit (°F), Celsius (°C), and Kelvin (K) all represent the temperature of matter. You can convert between scales using the graph or by the arithmetic formulas.

who in 1892 became Lord Kelvin) is an absolute scale in which zero is the lowest possible temperature. The Celsius and Kelvin scales are the same, except that the Kelvin temperature is a little more than 273 degrees higher than the Celsius temperature. All three temperature scales are summarized in Figure 2-1.

With this definition of temperature, we can now define the **calorie** (abbreviated cal) as the unit used to measure amounts of energy. A calorie is the energy needed to raise the temperature of 1 gram of water 1 degree Celsius (from 14.5° C to 15.5° C.) The dietary "Calorie" (with a capital "C") used in quantifying the energy content of foods is actually a kilocalorie or 1,000 calories (with a small "c"). A **Joule** is another unit used to measure amounts of energy. One Joule equals 0.2389 calories.

The term **power** refers to the rate at which energy is transferred, received, or released. The **watt** (W) is a unit of power that represents the transfer of 1 Joule of energy per second. You are probably familiar with the term watt from light bulbs. A 100-watt light bulb indicates the rate at which electric energy (from your outlet) is consumed by the bulb. A 100-watt bulb consumes more electrical energy than a 60-watt bulb and consequently is brighter. In weather and climate, the term watt is also used to indicate the rate of energy change. We are also interested in how much energy flows across an area. This energy flow is expressed in units of watts per square meter of

Click on "Temperature Conversion" to convert Fahrenheit to Celsius

TABLE 2-1 **The Specific Heat of a Substance Is the Amount of Heat Required to Increase the Temperature of 1 Gram of the Substance 1° C**

| Substance | Specific Heat | |
	(cal/g/° C)	(J/kg/° C)
Water	1.0	4,186
Ice	0.50	2,093
Air	0.24	1,005
Sand	0.19	795

area (W/m²). For example, the average amount of solar energy hitting the top of the atmosphere is 1368 W/m².

Heat is the energy produced by the random motions of molecules and atoms; it is the total kinetic energy of a sample of a substance. Both heat and temperature are related to kinetic energy and therefore to one another. Consider the heat of a freshly brewed cup of coffee and that of Lake Erie. The temperature of the coffee is greater than that of Lake Erie; however, the total kinetic energy—the heat contained in Lake Erie—is much greater than the heat contained in a cup of brewed coffee because there are many more molecules that are moving. If the cup of coffee is gently placed into Lake Erie without spilling any coffee, the temperature of the coffee will eventually be the same as that of the lake. For the temperature of the coffee to decrease, the kinetic energy of the molecules must decrease, and because energy cannot be created or destroyed, it must be converted to another form or transferred to the environment. In this case, energy, or heat, was transferred from the cup of coffee to Lake Erie. It may help to think of heat as the energy transferred between objects as a result of the temperature difference between them.

Although the cup of coffee transferred heat to Lake Erie, the temperature of the lake did not increase perceptibly. This is because the temperature change of an object depends on the following:

1. How much heat is being added—a single cup of coffee, although hot, does not contain much heat compared with Lake Erie because the coffee does not have much mass relative to the lake.
2. The amount of matter—the more matter, the more heat is required to change its temperature. Lake Erie contains a lot of water molecules and therefore a lot of matter!
3. The specific heat of the substance—but what is specific heat?

The **specific heat** of a substance is the amount of energy required to increase the temperature of 1 gram of that substance 1 degree Celsius. Because it takes a lot of energy to raise the temperature of water, water has a high specific heat (**TABLE 2-1**). A low specific heat means that a substance warms and cools easily, requiring little energy. Table 2-1 indicates that it takes more than four times as much energy to warm 1 gram of water 1 degree Celsius than it takes to warm 1 gram of air by the same amount.

TRANSFERRING ENERGY IN THE ATMOSPHERE

To change the temperature of a substance, such as air, we need to add or remove energy. Methods of energy transfer important to weather and climate discussed in this chapter are **conduction**, **convection**, **advection**, **latent heating**, and **radiation**. In discussing energy transfer, we concentrate on the direction of the energy transfer and the factors that determine how fast the energy transfer occurs.

Conduction: Requires Touching

Conduction is the process of heat transfer from molecule to molecule; energy transfer by conduction requires contact (**FIGURE 2-2**). An example of energy transfer by conduction is when we

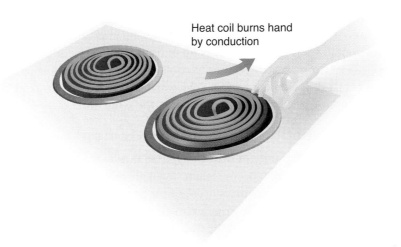

Heat coil burns hand
by conduction

FIGURE 2-2 Touching objects is an example of heat transfer by conduction. When something you touch feels hot, heat is being conducted from the object to your body.

touch an object to feel whether it is warm or cold. Heat is transferred from the warmer object to the colder one. The amount of energy transferred by conduction depends on the temperature difference between the two objects and their **thermal conductivity**. The ability of a substance to conduct heat by molecular motions is defined by its thermal conductivity. The larger the temperature difference and the higher the thermal conductivity, the faster the energy is transferred via conduction.

If you walk into a cool room and touch a piece of wood and a piece of metal, the metal "feels" colder. The two objects are actually at the same temperature, but the metal feels colder because metal has a much higher thermal conductivity than wood. Heat is rapidly conducted from your warm finger to the cooler metal, making your finger cool. Wood has a low heat conductivity, and the amount of heat transferred is smaller; it does not feel as cold as metal even though the wood and metal are at the same temperature.

Water is a good conductor of heat, whereas air is a poor heat conductor when it is not moving. This is why air between two pieces of glass in a storm window keeps a room insulated from the cold in winter. Because air is a poor conductor, conduction is not an efficient mechanism for transferring energy in the atmosphere on a global scale. Conduction is good for transferring energy over small distances, however, and it is an important form of energy transfer between the ground and the air in contact with the ground.

Convection: Hot Air Rises

If the ground is hot, energy is transferred to air molecules in contact with the surface via conduction. The heated parcel of air rises, and cooler air sinks to replace the rising warm air. This results in a net transfer of heat upward, away from the surface. Warmth moves upward, far away from the surface, because of the movement of the fluid (air, in this case), not because the air high up is in direct contact with the ground. This process of transferring energy vertically is called convection (**FIGURE 2-3**).

In the atmosphere, the rate of energy transfer by convection depends on how hot the rising air parcel is and the vertical temperature pattern of the surrounding atmosphere. In certain regions of the globe where the ground is much warmer than the air above it, convection is an important process for moving heat vertically. Convection is strong over deserts during the summer, where energy from the Sun rapidly warms the sand. Convection is an inefficient mode of heat transfer in polar regions, where the surface air is in contact with a surface that is often cooler than is the air above it.

Heat flow

Air flow

FIGURE 2-3 The shape of a flame is the result of convection. Air near the flame heats by conduction, becomes less dense than the surrounding air, and rises. Because of convection, the rising air creates a flow of fresh air to the flame.

Free convection refers to the situation when an air parcel is heated and becomes less dense than the air around it and therefore rises. Strong winds can also cause air to rise through forced convection. Horizontally moving air can break up into small swirls or eddies. These eddies mix the air and transport heat from warmer regions to cooler regions. An example of forced convection is a cloudy and windy day, where heat from the warm ground will be mixed upward into the atmosphere. Free and forced convection can occur at the same time.

Temperature Advection: Horizontal Movement

The horizontal transport of energy in the atmosphere is referred to as temperature advection (**FIGURE 2-4**). Warm air advection occurs when warm air replaces cooler air. On weather maps (see Chapter 1), temperature advection is occurring wherever the wind "flagpoles" are pointed across the isotherms. In winter snowstorms, warm air advection moves warm air poleward while cold air advection brings cold air toward the tropical regions. Advection is an important process throughout the troposphere, especially close to the Earth's surface.

Latent Heating: Changing the Phase of Water

In the atmosphere, only water exists in all three phases: solid, liquid, and gas. Ice is the solid form of water, and water vapor is the gas phase. In everyday language, the liquid form of water is generally referred to as "water." In this section, however, using this common term would lead to some confusion. So, in this section only, water in the liquid phase is referred to explicitly as liquid water.

Changing the phase of a substance either requires or releases energy. Changing the phase of water adds or removes energy from its surroundings. Because the Earth's surface is 70% liquid water, understanding the phase changes of water is important for understanding atmospheric energy. How does a change of phase occur?

Ice cubes melt and puddles evaporate because energy is added to the water, causing a change of phase. **Latent heat** is the heat absorbed or released per unit mass when water changes phase. This change of phase does not necessarily result in an increase in the water temperature. For example, adding heat to an ice cube may result in a change of phase of the water from a solid to a liquid; however, the temperature of a liquid water–ice mixture will not increase until all of the ice is melted. Energy cannot be destroyed, so what happens to the heat energy if it is not used to change the temperature of the water?

Water molecules of ice and liquid water are bound together by molecular forces. In the ice phase, the water molecules have low enough kinetic energies that intermolecular attractions bind them into a highly ordered, crystalline form. To break these intermolecular attractions and transform ice into liquid water, energy must be added. **Latent heat of melting** is the amount of energy absorbed by water to change 1 gram of ice into liquid water, and it is equal to 80 cal for each gram of ice.

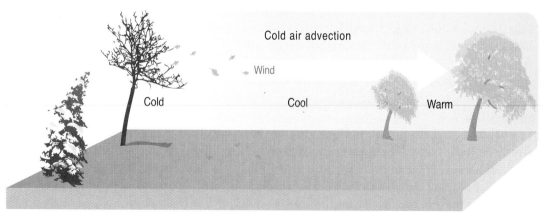

FIGURE 2-4 Cold air advection occurs when colder air replaces warmer air. Warm air advection occurs when warmer air replaces colder air.

If energy must be added to melt ice, the opposite occurs when water freezes. The amount of energy released into the environment when water freezes is also 80 cal per gram of ice and is referred to as the **latent heat of fusion.**

The transition of water from the liquid phase to the gas phase is called vaporization or evaporation. The molecules of a gas are essentially free of one another, having no bonds between them. To convert liquid water to a vapor requires the addition of a sizable amount of energy (more than seven times the energy required to melt ice) to break the binding forces that keep the molecules in a fluid state. The amount of heat required to evaporate 1 gram of liquid water is referred to as the **latent heat of vaporization.** The latent heat of vaporization is a function of water temperature, ranging from 540 cal per gram of water at 100° C to 600 cal per gram at 0° C—that is, it takes slightly more energy to evaporate cold water than to evaporate the same amount of warmer water. (In addition, to raise the temperature of liquid water from 0° C to 100° C also requires an input of energy; this is where the specific heat of water comes in.)

Water vapor condenses to form liquid water. Condensation is the opposite of evaporation. **Latent heat of condensation** represents the amount of energy released when water vapor condenses to a liquid form. The latent heat of condensation is a function of temperature and has the same range as the latent heat of vaporization.

Water vapor may change directly to ice in a process known as **deposition.** Conversely, ice may also directly enter the gas phase without melting (called **sublimation**). The **latent heat of sublimation** equals the **latent heat of deposition.** A total of 680 cal are required to change 1 gram of ice at 0° C (32° F) into vapor. This number is equal to 600 cal + 80 cal. This means that the latent heats of sublimation/deposition are simply the sum of the latent heats of melting/fusion and the latent heats of vaporization/condensation. However, on Earth the sublimation and deposition of water happen less frequently, and on a far smaller scale, than do the processes of evaporation, condensation, melting, and freezing.

Where do we see phase changes of water in everyday life? Changes of phase affect both human beings and the atmosphere by removing or adding energy to the air in which phase changes of water are occurring. For example, when we physically exert ourselves, we sweat. Perspiration is a method for maintaining our body temperature. To evaporate the liquid water on our skin requires energy; some of this energy is taken from our skin. This energy transfer from our skin to the sweat cools the skin. So, evaporation is a cooling process that removes energy from the physical environment—in this case, our bodies. On the other hand, condensation, the opposite of evaporation, is a heating process that supplies energy to the environment. When water vapor changes into the liquid water or ice phase to form clouds, energy is released into the atmosphere. Changes of phase are also important in the energy gains and losses at ground level. The formation of dew (condensation) or frost (deposition) releases heat.

Changing the phase of water is an efficient method of transferring energy globally and provides an energy source for much of our weather. We are interested in some phase changes more than others. The reason is simple: 600 cal per gram is a lot more than 80 cal per gram. In other words, the phase changes of evaporation and condensation have by far the largest latent heats of the common phase changes of water. This obvious fact means that throughout this text we will be very interested in where, when, and how much evaporation and condensation occur. Wherever these two processes occur, there is a large transfer of energy going on.

To appreciate the amount of energy involved in condensation and evaporation, consider how much energy you would use to boil away a pot of water. Condensing an equivalent amount of water vapor would release a similar amount of energy. Now remember the heaviest rainstorm that you have experienced in your life. Imagine how much time and energy that was required to evaporate all the water that eventually condensed to form the rains. All of that rainwater was at one time water vapor and later condensed, thus releasing energy into the atmosphere and supplying fuel for the storm. If just 2 kilograms (4.4 pounds) of atmospheric water vapor are condensed into a liquid, 1,194,230 cal of latent heat are released. If all this energy were used to heat 2 kilograms of air, the air would warm more than 2000° C! The air does not get this hot when storm clouds form, however, because much of the released latent heat energizes the movement of air within

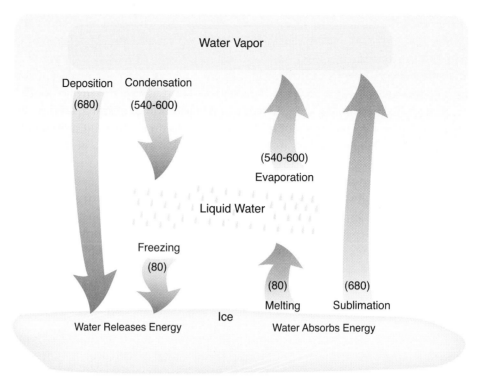

FIGURE 2-5 The phase changes of water. Blue lines represent a phase change that removes energy from the atmosphere. Phase changes that add energy to the environment are represented by red lines. When water vapor changes to liquid water or ice, clouds are formed, and large amounts of energy are released into the atmosphere. The latent heats associated with the phase change are given in parenthesis in units of cal per gram.

the storm. This is a major, yet invisible, energy source for the winds of the atmosphere that we explore later in this text.

FIGURE 2-5 summarizes the phase changes of water vapor and the associated energy changes. To understand much of meteorology, it is very important that you know which processes in Figure 2-5 absorb energy and which ones release energy—so this is a figure to study now and refer to later. We say more about how latent heating and cooling affect cloud development in Chapter 4.

◼ Radiative Heat Transfer: Exchanging Energy with Space

Earth receives energy from the Sun. This solar energy moves through empty space from the Sun to the Earth and is the original energy source for our weather and climate. **Solar radiation** is one form of **radiant energy**, or energy in the form of waves that are not composed of matter. Radiant energy is also called **radiation** and **electromagnetic energy**. In meteorology, "radiation" refers to a broad range of energy types, not just the type associated with nuclear energy. All electromagnetic radiation propagates through matter or empty space in the form of a wave with electric and magnetic fields.

Waves are characterized by two properties: **wavelength**, the distance between wave crests, and **amplitude**, half of the height from the peak of the crest to the lowest point of the wave (**FIGURE 2-6**). The size of the wavelengths of radiation range from ultra-long radio waves to high-energy gamma rays (**FIGURE 2-7**). The amount of energy in the wave increases with smaller wavelengths. The distance between wave crests is measured in terms of a micrometer or micron (Chapter 1; abbreviated μm), which is approximately one one-hundredth the diameter of a human hair. In meteorology, we are concerned with radiant energy with wavelengths between about 0.1 μm and 100 μm.

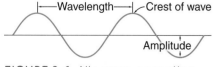

FIGURE 2-6 All waves, no matter what kind, are defined by the distance between crests (the wavelength) and their amplitude.

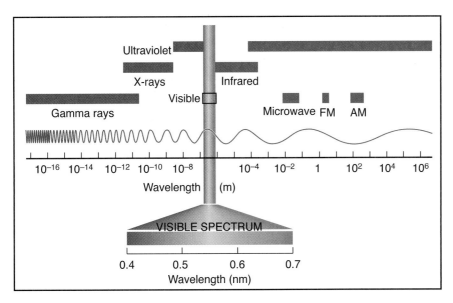

FIGURE 2-7 Electromagnetic energy spans a large spectrum of wavelengths. In this course we are interested primarily in solar (or shortwave) and infrared (or longwave) radiation.

We can make a useful distinction between the types of radiation in this wavelength range. The Sun emits most of its radiant energy with a wavelength between 0.2 μm and 4 μm, on the short end of the 0.1-μm to 100-μm range. For this reason, solar radiation is sometimes referred to as **shortwave radiation**. Solar energy includes UV, visible, and near-infrared radiation.

Within the wavelength range of solar energy lies **ultraviolet (UV) light**, where wavelengths range from 0.2 μm to 0.4 μm. UV radiation is responsible for the tanning of our skin, sunburn, and skin cancer. Our eyes are sensitive to visible radiation, meaning that they detect electromagnetic energy characterized by wavelengths ranging from 0.4 μm to 0.7 μm. The color blue has a wavelength of approximately 0.45 μm, and red has a wavelength of 0.7 μm. Because visible light has wavelengths that are longer than UV light, they are not as energetic. Chapter 5 has more detail on how we see colors in the atmosphere.

Longwave radiation (or **terrestrial radiation**) emitted by the Earth is less energetic than solar radiation and is characterized by much longer wavelengths, primarily between 4 μm and 100 μm, with a maximum near 10 μm.

Why do the Earth and the Sun emit energy in such different wavelength bands? To understand, we must learn a few fundamental laws of radiation.

Radiation Laws

The first law of radiation we need to know is that all objects with a temperature above absolute zero emit radiation. You, this book, a hot cup of coffee, a roaring fire, and a cold cup of coffee all emit radiation. Even cold, nearly empty outer space emits some radiation from the original "Big Bang" billions of years ago.

Just as some substances are better conductors of heat than others, some objects emit and absorb radiation better than others. The emissivity (denoted by the symbol ε) is a measure of an object's ability to emit radiation. Materials are assigned an emissivity value between 0 and 1.0. A **blackbody** has an emissivity of 1 and is an object that absorbs all the electromagnetic energy that falls on the object, no matter what the wavelength of the radiation. A perfect blackbody does not exist, but it is a useful reference for determining how good a body is at emitting and absorbing radiation. Although an object may visually appear black, it does not mean it is a blackbody, as the object may not absorb energy at wavelengths outside the visible region of the electromagnetic spectrum.

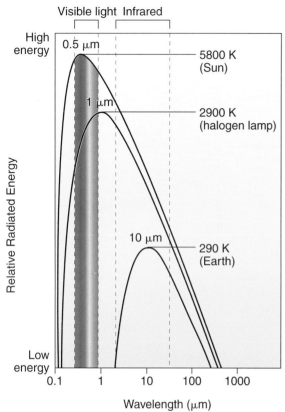

FIGURE 2-8 Energy curves for three objects having temperatures of 5800 K, 2900 K, and 290 K. These temperatures correspond roughly to the temperature of the sun, the temperature of a halogen lamp filament, and the globally averaged temperature of the earth. The hotter the object, the larger is the curve (Stefan-Boltzmann Law). The figure also shows that the hotter the object, the shorter is the wavelength of the peak in the curve (Wien's Law).

"Energy Curves" to explore the relationship between temperature, wavelength, and radiation.

The amount of radiant energy that is emitted by an object depends on its temperature. The **Stefan-Boltzmann Law** makes our first law more precise by stating that the amount of radiative energy that is emitted by an object (watts per square meter) is related to the fourth power of its Kelvin temperature:

$$\text{Emitted Radiation} = \varepsilon \times 5.67 \times 10^{-8} \times T^4$$

Therefore, *as an object warms, it emits more radiation.* The Sun emits about 160,000 times as much energy as the Earth because it is about 20 times hotter.

The wavelength of radiant energy emitted depends on the temperature of the emitting body. German physicist Wilhelm Wien (pronounced "ween") won the 1911 Nobel Prize in physics for this discovery, which is now called **Wien's Law**. His law is simple division:

$$\frac{\text{Wavelength (in microns) of maximum radiation emitted by an object}} = \frac{2900}{\text{Object's temperature in Kelvin}}$$

Let's use this equation to answer our questions about solar and terrestrial radiation. Using Wien's Law, an object with a temperature of 290 K (17° C or 62° F) maximizes its radiant energy output at 10 μm. From Figure 2-7, we see that this maximum occurs in the longwave region of the radiation spectrum; therefore, room-temperature objects emit longwave radiant energy, not visible light or X-rays. The Earth's average temperature is 288 K (15° C or 59° F), and so Earth emits longwave radiation.

Now consider a very hot object with a temperature of about 2900 K, such as the filament in a halogen lamp. By Wien's Law, it has a wavelength of maximum energy output of about 1 μm, which is much shorter than the 10 μm wavelength that characterizes the energy coming from the room the lamp is in. The Sun is about twice as hot as a halogen lamp filament, with a temperature of approximately 5800 K. By Wien's Law, its energy output therefore maximizes at wavelengths that are half as long as in the halogen lamp case, in the shortwave region around 0.5 μm. This explains why halogen lamps and the Sun emit light—but you, this text, and the room around you emit only heat, not light. While the peak energy of a halogen lamp is emitted at 1 μm, lots of energy is also emitted at visible wavelengths, which is why we can see the hot filament. Wien's Law can be summarized as follows: *the hotter the object, the shorter the wavelength of maximum emission of radiation.*

FIGURE 2-8 illustrates both the Stefan-Boltzmann Law and Wien's Law. The curves shown are the amount of energy emitted by a blackbody object as a function of wavelength for the three different temperatures we have discussed: 290 K, 2900 K, and 5800 K. The 2900 K energy curve is higher than the 290 K energy curve at every point, and the 5800 K curve is higher than the 2900 K curve. This is a consequence of the Stefan-Boltzmann Law. Also, the 5800 K curve peaks at the shortest wavelengths, followed closely by the 2900 K curve. The 290 K curve, in contrast, peaks farther to the right in the infrared region and, unlike the 2900 K curve, does not include the visible wavelengths. This shift in peak energy emission is the main point of Wien's Law.

For those who don't like an arithmetic or energy curve, **FIGURE 2-9** tells the same story in a single photograph. It is a photograph of a person holding a lighted match, but the camera is seeing infrared energy, not light. The burning match is the hottest object in the picture and is emitting the most infrared energy, just as it should by the Stefan-Boltzmann Law. It is color coded white. The person's eyeglasses are emitting the least energy and are color coded purple. Wien's Law explains why you can see a hot match glow but you can't see eyeglasses glow.

The Art of Radiative Energy Transfers

If rooms and eyeglasses don't glow, how can we see them? To explain this, we need to learn what happens when emitted radiation interacts with an object.

As with all forms of energy, radiation can change form, but it must be conserved. When radiation interacts with an object, it can be

1. *Absorbed:* Absorption of radiation by a molecule increases the energy of the molecule. If this energy is in the form of kinetic energy, then the temperature of the object or gas increases. The atmosphere is an effective absorber of infrared energy.
2. *Reflected:* Energy reflected by an object is sent back. Mirrors are good reflectors of visible light, as are clouds.
3. *Transmitted:* Energy transmitted through an object passes through the object, although it may change direction. The atmosphere is almost completely transparent to visible radiation.

FIGURE 2-9 An infrared photograph showing the heat coming from a man holding a lighter. White indicates the most infrared energy emission, and purple indicates the least infrared energy emission.

The initials of these three processes are ART, making them easy to remember. (A fourth process, scattering, will be discussed in Chapter 5.)

Most of what we see comes from the reflection of visible light off objects. The **albedo** of an object describes the percentage of light that it reflects. The higher the albedo, the brighter the object. Freshly fallen snow has an albedo of 90%, whereas green grass has an albedo of 25%.

Albedo is a key determiner, along with distance from the sun, of the temperatures of the planets in the solar system. Brighter planets are able to absorb less solar radiation than darker planets. **FIGURE 2-10** shows visible photographs of Venus, Earth, and Mars from U.S. space missions. Bright white Venus has the highest albedo, at about 80%; Earth's albedo is around 30%, and Mars, the darkest of the three, has an albedo close to 20%. Mars is darker than Earth and farther away from the Sun, so it receives much less energy than the Earth does but absorbs a higher percentage of what reaches it. The temperature of Mars is lower than Earth's, on average 40° C cooler. Venus is much whiter than Earth, which should keep it cooler than Earth despite Venus's proximity to the Sun; however, a thick carbon dioxide atmosphere causes Venus to be much hotter than Earth, an effect that we will discuss shortly.

The percentage of energy that is not reflected is either absorbed or transmitted. The structure of a molecule can be altered if it absorbs high-energy radiation. **Photodissociation** occurs when absorption of UV radiation results in breaking of chemical bonds. Photodissociation is an important process in the formation of ozone, O_3 (**BOX 2-1**).

Absorption of electromagnetic energy at wavelengths longer than UV does not disrupt the molecule's structure. Instead, the molecule vibrates or spins after absorbing the radiation and thus increases its kinetic energy. Absorption of radiation with wavelengths greater than 0.2 μm warms the atmosphere and surface.

How much radiation energy the atmosphere or an object absorbs depends on the following:

1. *The radiative properties of the material:* Most objects are selective in the wavelengths that they absorb.
2. *The amount of time the object is exposed to the emitted energy:* The longer it is exposed to radiation, the more energy it can absorb.
3. *The amount of material:* Very thin objects may transmit and not absorb all of the energy reaching them. Increasing its thickness can increase the amount of energy an object absorbs.

"Planetary Effective Temperatures" to explore the relationship between the effective temperature of a planet and its energy gains.

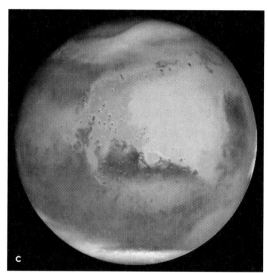

FIGURE 2-10 Photographs: (a) Venus on February 5, 1974. (b) Earth on December 7, 1972. (c) Mars on June 26, 2001. These photographs were taken from the Mariner 10, APOLLO 17, and Hubble Space Telescope missions, respectively. Venus is whitish. Earth is blue and white, and Mars is red. The albedo of Venus is high. Earth's albedo is relatively low, and Mars's albedo is even lower than Earth's.

4. *How close the object is to the source of energy:* The farther an object is from a source of radiation, the less energy that reaches it, and it can absorb less energy.

5. *The angle at which the radiation is striking the object:* Radiation shining directly onto an object in a concentrated beam is absorbed more effectively than is a spread-out beam hitting the object at an angle.

Just as some substances are better conductors of heat than others, some objects emit and absorb radiation better than others. A blackbody is an object that absorbs all the electromagnetic energy that falls on the object, no matter what the wavelength of the radiation. A perfect blackbody does not exist, but it is a useful reference for determining how good a body is at emitting and absorbing radiation. The energy curves drawn in Figure 2-8 are based on the assumption that the objects are perfect blackbodies.

If an object absorbs electromagnetic energy of a certain wavelength, it will also emit energy at that wavelength. This is **Kirchhoff's law**, our last law of radiation: *A good absorber of radiation is also a good emitter of radiation at that same wavelength.* Thus, because our bodies are good emitters of radiant energy with a wavelength of 10 μm, we are also good absorbers of this infrared energy. Remember that the amount of radiation an object emits also depends on its temperature. So, while the black text on this paper absorbs lots of visible light that falls on it, the letters do not glow in the dark because the paper is at room temperature.

Box 2-1 Ozone

Ozone (O_3) is a molecule formed by three O atoms. In the lower troposphere, O_3 is considered a pollutant, as it is a chemically reactive gas and can cause respiratory problems when breathed. Ozone also resides in the stratosphere, where it absorbs UV rays, protecting life on the Earth's surface from these high energy electromagnetic waves. In this box, we explore how ozone is formed and destroyed in the stratosphere.

The amount of ozone in a column of air is measured in Dobson Units (DUs), after G. M. B. Dobson, who was one of the first scientists to study atmospheric ozone. One DU of O_3 is equal to 0.01 mm thickness of the gas if the entire column were compressed down to standard temperature and pressure (stp is 0° C and 1000 mb). Three hundred DU is 3 mm thick, or about the thickness of a dime. That is not a large quantity of O_3.

Ozone is a trace gas that plays a role in myriad radiative processes in the Earth's atmosphere. Ozone is formed by interacting with solar radiation. It is destroyed by solar radiation; it absorbs terrestrial radiation, and recent losses in stratospheric ozone are firmly linked to the presence of CFCs, another trace gas that also absorbs terrestrial radiation.

Stratospheric ozone represents about 90% of the total amount of ozone in the atmosphere. It is produced in the stratosphere by the combination of UV radiation and an oxygen molecule (O_2). The basic steps in its formation are as follows:

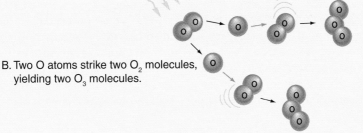

A. UV radiation from the sun strikes an oxygen (O_2) molecule, and splits it into two O atoms.

B. Two O atoms strike two O_2 molecules, yielding two O_3 molecules.

This reaction is written in chemical equations as follows:

$$O_2 + UV \rightarrow O + O$$

$$2O + 2O_2 \rightarrow 2O_3$$

$$\text{Net Reaction: } 3O_2 + UV \rightarrow 2O_3$$

UV radiation is also involved in the destruction of O_3.

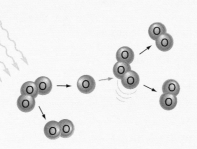

A. UV radiation from the sun strikes an ozone (O_3) molecule and splits off an O atom.

B. The free O atom strikes the O_3, yielding two O_2 molecules.

(continued)

Box 2-1 Ozone, continued

This destruction is expressed as follows:

$$O_3 + UV \rightarrow O + O_2$$

$$O + O_3 \rightarrow 2O_2$$

Net Reaction: $2O_3 + UV \rightarrow 3O_2$

This cycle of creation and destruction of ozone does not cause any long-term loss of ozone; however, in 1970, meteorologist Paul Crutzen proposed the following catalytic reaction that results in the chronic destruction of ozone by molecules such as chlorine:

$$X + O_3 \rightarrow XO + O_2$$

$$O_3 + UV \rightarrow O_2 + O$$

$$O + XO \rightarrow X + O_2$$

Net Reaction: $2O_3 + UV \rightarrow 3O_2$

In this sequence of reactions, X is an atom or molecule that acts as a catalyst to convert O_3 to O_2. Note that X does not change in the net reaction, so it can continue to destroy O_3 molecules.

There is a delicate balance between the production and destruction of ozone, resulting in what is referred to as an ozone shield that protects us from high-energy UV radiation. This natural balance has recently been disrupted by human activities causing ozone depletion (meaning that the destruction of O_3 exceeds the creation of O_3). How has this happened?

One molecule that can serve as the catalyst molecule X is chlorine (Cl). But how does chlorine get into the stratosphere? In the 1930s, CFCs were first produced for use in refrigeration, air conditioning, solvents, aerosol spray cans, and Styrofoam puffing agents. They leaked into the atmosphere as a result of their use. CFCs are very stable in the troposphere, with lifetimes of approximately 100 years. This long lifetime allows CFCs to eventually make it into the stratosphere, where they are split, or dissociated, by UV light to produce chlorine atoms. The destruction of ozone then occurs with the following chemical reactions:

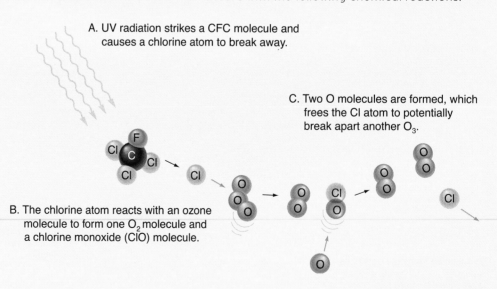

A. UV radiation strikes a CFC molecule and causes a chlorine atom to break away.

C. Two O molecules are formed, which frees the Cl atom to potentially break apart another O_3.

B. The chlorine atom reacts with an ozone molecule to form one O_2 molecule and a chlorine monoxide (ClO) molecule.

The chemical reactions shown here predicted a slow but worrisome decrease in stratospheric ozone. Observations during the 1980s indicated, in contrast, a very rapid and alarming decrease in ozone over Antarctica. Why?

The long winter months over Antarctica have unusual upper atmospheric conditions. A whirlpool of stratospheric winds, called the polar vortex, isolates the upper air over Antarctica from air in the middle latitudes. Because there are no solar energy gains at this time, the trapped air in the vortex gets so cold that clouds form, even though the Antarctic air is extremely dry. The particles composing these polar stratospheric clouds may be water, ice, or nitric acid, depending on the temperature. Complex chemical reactions involving chlorine and other ozone-depleting molecules occur on the surface of these cloud particles that greatly accelerate the depletion of ozone. When the Sun rises in the Southern Hemisphere springtime (e.g., late September), this adds the crucial ingredient of UV radiation to the mix of clouds and chemicals. Ozone is then destroyed rapidly. The depletion of ozone over Antarctica is so pronounced that it has been called the "**ozone hole**" (the blue region in the satellite image of ozone amounts below). The region labeled the "ozone hole" does contain ozone gas, but the amount of O_3 in that area is less than 220 DU. This value was chosen because, prior to 1979, ozone values of less than 220 DU were not observed. A smaller but noticeable Arctic "ozone hole" has also developed at times in the past decade.

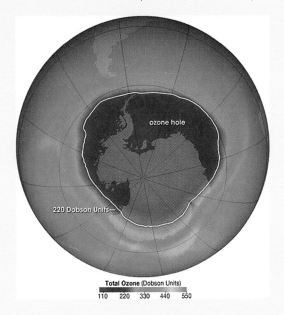

Total Ozone (Dobson Units)
110   220   330   440   550

Fortunately, chlorine alone does not remain for centuries in the stratosphere. Thanks to the cooperation of scientists, governments, and industry, the use of CFCs has declined drastically in just the past two decades. As a result, stratospheric ozone may eventually return to pre-1930 levels—but it will take decades, if not a century, to do so.

In 1995, Drs. Crutzen, Mario J. Molina, and F. Sherwood Rowland won the Nobel Prize in chemistry for their work concerning the formation and decomposition of ozone. Crutzen became the first meteorologist to win the Nobel Prize.

Now that we know how radiation behaves, we need to examine more closely how solar radiation reaches Earth, how it is distributed across the Earth, and how the atmosphere affects the transfer of radiation.

THE SUN AND THE SEASONS

The amount of solar energy reaching Earth at any particular latitude is defined by how the Earth orbits the Sun. In the 17th century, Johannes Kepler discovered that Earth, along with the other planets, orbits the Sun in a path that traces out an ellipse. It takes 1 year (365.24 days) for Earth to make one complete revolution around the Sun. Because of its elliptic path, Earth's distance from the Sun varies with the time of year. Currently, Earth is farthest from the Sun on approximately July 4 (**aphelion**) and is closest to the Sun on approximately January 4 (**perihelion**) (FIGURE 2-11). The difference in the Earth–Sun distance between the time of aphelion and perihelion is so small that it is hardly noticeable in Figure 2-11; it is only about 5 million kilometers (3 million miles), or 3% of the total Earth–Sun distance.

If Earth is closer to the Sun in January than July, then why is it colder over North America in January than July? It is because the tilt of the Earth, not its distance from the Sun, is the primary orbital parameter that affects the Earth's temperature. The Earth's tilt causes the seasons.

How? As the Earth orbits the Sun, its axis of rotation is tilted at an angle of 23°27′ from its orbital plane (Figure 2-11). This tilt is referred to as the **angle of inclination**. Because Earth's axis of spin always points in the same direction—toward the North Star—the orientation of Earth's axis to the Sun is always changing as the Earth orbits around the Sun (Figure 2-11). As this orientation changes throughout the year, so does the distribution of sunlight on Earth's

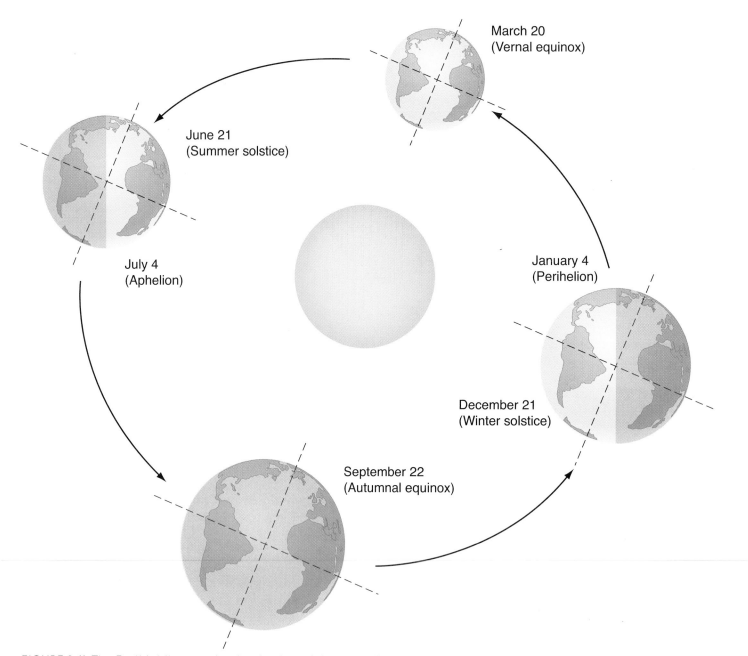

FIGURE 2-11 The Earth's tilt, or angle of inclination, determines the amount of solar energy a given region of Earth receives as it orbits the Sun. The Northern Hemisphere summer occurs when the Earth is tilted toward the Sun. Earth is about 3% further away from the Sun in June than it is in December.

surface at any given latitude. This links the amount of solar energy reaching a location to the time of year and causes some months of the year to always be warmer than others—in other words, the seasons. Later in this chapter, we explain in more detail how the tilt of the Earth affects where the Sun is in the sky, leading to the seasons.

At high noon on the **solstices** (sol, "Sun," and stice, "come to a stop"), the Sun's rays strike the equator at an angle of 23°27′. On these days, the Sun is highest in the mid-latitude summer sky. On approximately June 21, the northern spin axis is tilted 23°27′ toward the Sun. On this day, the Northern Hemisphere summer solstice, latitudes south of approximately 66°33′S remain in complete darkness. This latitude is referred to as the Antarctic Circle. Around December 21, on the Northern Hemisphere winter solstice, the northern spin axis is pointed away from the Sun, and latitudes north of the Arctic Circle (66°33′N) have 24 hours of darkness. On June 21 the Sun is directly overhead the Tropic of Cancer (23°27′N) at noon. The Tropic of Capricorn (23°27′S) is the latitude at which the noon Sun is overhead on December 21.

The **equinoxes** (equi, "equal," and nox, "night") occur when the Sun's rays strike the equator at noon at an angle of 90 degrees. In the Northern Hemisphere, the vernal or spring equinox occurs around March 20, and the autumnal or fall equinox occurs on September 22 or 23. During the equinoxes, all locations on Earth experience 12 hours of daylight and 12 hours of darkness.

The tilt of the Earth's axis is responsible for the seasonal variation in the amount of solar energy distributed at the top of the atmosphere and plays a key role in determining the seasonal variation in surface temperature. The variation of solar energy at the surface by latitude is caused by the following:

1. Changes in the angle that the Sun's rays hit the Earth
2. The number of daylight hours
3. The amount of atmosphere the Sun's rays have to pass through

Let's consider these variations one at a time.

Angle of Incidence of the Sun's Energy

If you shine a flashlight at the ceiling, the region that is illuminated shrinks or grows depending on whether you point it straight up at the ceiling or at an angle. Similarly, the Sun's energy spreads out over differing geographic areas when it reaches Earth's spherical surface. Solar energy is most concentrated in the area where the Sun is overhead at noon, the location of which changes with the season (FIGURE 2-12).

The angle at which the Sun's energy strikes a particular location on Earth is called the **solar zenith angle** (FIGURE 2-13). This angle is equal to 0° when the Sun is directly overhead and increases as the Sun sets, until the Sun is on the horizon and the solar zenith angle equals 90°. The solar zenith angle indicates the concentration of the Sun's rays at a given instant. Maximum intensity of the Sun's radiation occurs when the solar zenith angle is 0°, or directly overhead. FIGURE 2-14 demonstrates how sunlight intensity depends on solar zenith angle. The direct beam of light is less intense when it is aimed obliquely at a surface, as the illuminated area gets larger. In each area in Figure 2-14, the total amount of light is the same, but the light is always less intense when it has spread over a larger area.

The solar zenith angle is a function of time of day, time of year, and latitude. FIGURE 2-15 demonstrates typical paths that the Sun traces from sunrise to sunset at 2 days in the Northern Hemisphere at a latitude of approximately 55°N. At local noon the Sun reaches the highest point on its path. Poleward of 23°27′N and 23°27′S, the Sun can never be seen directly overhead. Poleward of the Arctic and Antarctic Circles and between the spring and autumnal equinoxes, the Sun never sets. Instead, the solar zenith angle remains approximately fixed throughout the day as the Sun circles the horizon. Referring back to Figure 2-12, we see that the Sun's energy spreads out over a larger area when the solar zenith angle is large. This causes the solar energy to be distributed over a larger area. As a result, a given spot on the surface receives less energy.

Length of Daylight

The tilt of the Earth's axis also defines the length of daylight for a given latitude. As Earth orbits the Sun, it spins about its axis approximately once every 24 hours. This spinning explains our

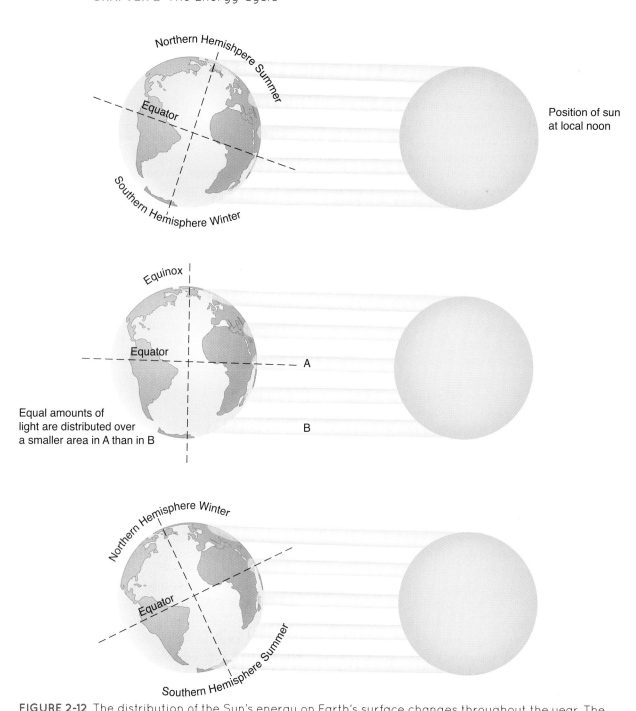

FIGURE 2-12 The distribution of the Sun's energy on Earth's surface changes throughout the year. The Sun's energy is always more concentrated in the equatorial regions than in the polar regions. The top figure represents conditions in Northern Hemisphere summer, the middle figure the equinoxes, and the bottom globe Northern Hemisphere winter. The position of the sun at local noon on each day is represented by the yellow circle.

"Incoming Solar Radiation" to analyze the relationships between latitude, time of year, and the incoming solar energy.

daily cycle of night and day and the resulting daily, or **diurnal**, variations in the amount of solar energy and in temperature. (Diurnal temperature changes are discussed in detail in Chapter 3.) On June 21, the North Pole, because it is facing the Sun, experiences 24 hours of daylight, while the South Pole is in complete darkness. **FIGURE 2-16** shows the number of daylight hours for four latitudes (70°N, 42°N, 30°N, and the equator) as a function of time of year. As previously noted, on the equinoxes, each region of the globe has 12 hours of daylight. The equator always has 12 hours of daylight. Because the axis is tilted at 23°27′ from the plane of the orbit, only latitudes poleward of 66°33′ (or poleward of the Antarctic Circle or Arctic Circle) can experience

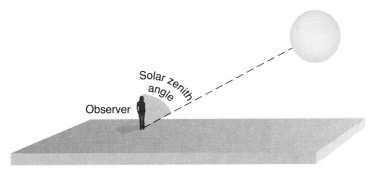

FIGURE 2-13 The solar zenith angle is measured from directly overhead to the position of the Sun. A solar zenith angle of 90 degrees means the sun is on the horizon.

24 hours of daylight. So, only the 70°N line in Figure 2-16 has 24 hours of daylight during the Northern Hemisphere summer. Compare the daylight hours for 42°N and 30°N. The further poleward you travel, the more daylight you have during the summer and the fewer daylight hours you have during winter.

The average amount of solar energy that reaches the outer limits of our atmosphere on a surface that is perpendicular to the solar rays is referred to as the **solar constant**. Its value is about 1368 watts per square meter (or W/m²). This "constant" actually can fluctuate by as much as 0.4% in a week, and it regularly changes by about 0.1% over a regular period of 11 years. We will discuss how small changes in the Sun's energy output might impact global climate in Chapter 14.

Let's put all these ideas together now as we look at the changing amounts of incoming solar energy striking the top of the atmosphere at different latitudes over the course of a year. The length of day and solar zenith angle are both determined by the tilt of the Earth's axis. They act together in complex ways to define the incoming solar energy. **TABLE 2-2** demonstrates the combined effects by listing the solar zenith angle at noon, the daylight hours, and the average incoming solar radiation for four different latitudes at three times of the year. We have chosen 42°N (the latitude of Detroit, Michigan) and 30°N (the latitude of New Orleans, Louisiana) to represent typical conditions of the northern and southern continental United States, respectively. On March 20, all latitudes have 12 hours of daylight, and the incoming solar energy at 30°N is greater than at 42°N. On June 21, the solar zenith angle is larger at a latitude of 42°N than it is at a latitude of 30°N, but the more northern latitude also has more daylight hours during the year, which results in more solar energy at the top of the atmosphere. A latitude of 70°N has 24 hours of daylight and a low sun (high solar zenith), yielding a large supply of solar energy at the top

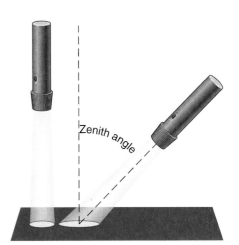

FIGURE 2-14 This figure illustrates how the intensity of sunlight spreads out over a surface when the solar zenith angle is greater than zero.

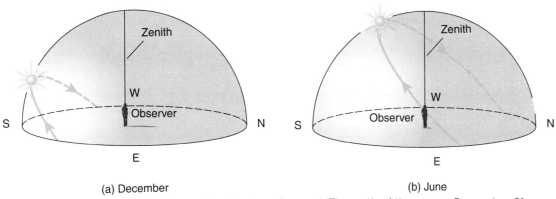

(a) December   (b) June

FIGURE 2-15 The Sun rises in the east and sets in the west. The path of the sun on December 21 and June 21 at about 55°N is shown. The more northern path represents summertime conditions.

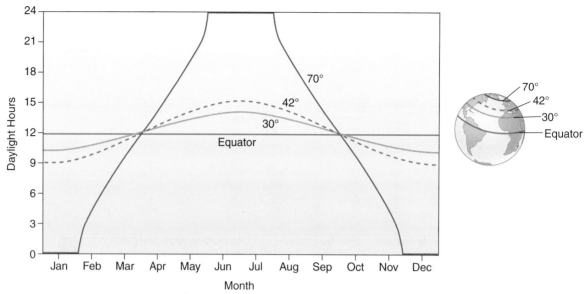

FIGURE 2-16 The total number of daylight hours for four different latitudes (70°N, 42°N, 30°N, and the equator) as a function of time of year.

of the atmosphere. On December 21, at 42°N, the sun is lower in the sky (larger solar zenith) than at 30°N, but now the fewer hours result in less solar energy at the more northern latitude. **FIGURE 2-17** shows how the amount of energy incident at the top of the atmosphere varies with time of year for four latitudes: 70°N, 30°N, the equator, and 70°S.

During the late winter, 70°N receives no solar energy because the Sun is below the horizon (Figure 2-16). The Sun finally appears above the horizon in late January, and the energy input continues to increase until June 21, when the Sun reaches its smallest solar zenith angle. This is also the day that the Sun appears directly overhead the Tropic of Cancer (23°27′N) at noon. At the equator, the amount of solar energy at the top of the atmosphere is greatest at the equinoxes, when the Sun appears directly overhead at noon at the equator.

The shape of the energy distribution at 70°S is opposite to the distribution at 70°N: a minimum occurs in June, and a maximum occurs in December. During their respective summers, the incident

TABLE 2-2 The Solar Zenith Angle at Local Noon, the Number of Daylight Hours, and the Average Incoming Solar Energy for Four Different Latitudes on Different Days of the Year

Latitude/Date	Solar Zenith at Local Noon (Degrees)	Daylight Hours (Hours and Minutes)	Average Incoming Solar Radiation (W/m²)
70°			
March 20	70	12 hr	150
June 21	46.5	24 hr	494
December 21		0 hr	0
42°			
March 20	42	12 hr	326
June 21	18.6	15 h 18 m	485
December 21	65.4	8 h 57 m	143
30°			
March 20	30	12 hr	380
June 21	6.6	13 h 54 m	476
December 21	53.4	10 h 4 m	229
Equator			
March 20	0	12 hr	438
June 21	23.5	12 hr	387
December 21	23.5	12 hr	412

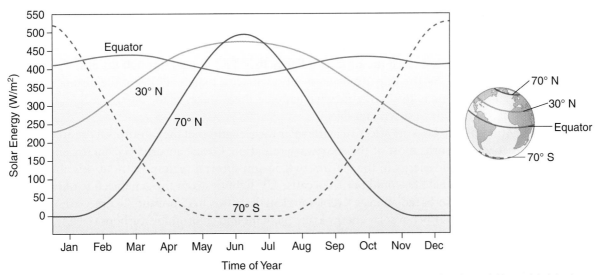

FIGURE 2-17 The amount of solar energy striking the top of the atmosphere for four different latitudes (70°N, 30°N, the equator, and 70°S) as a function of time of year. Zero solar energy means that the Sun never rises.

energy at 70°S is a little greater than the amount at 70°N. This is because of the Earth's elliptical path around the Sun, which brings the Earth a little closer to the Sun in December than in June.

The energy distribution at 30°N has the single summer peak characteristic of locations poleward of the Tropics of Cancer and Capricorn, but its summertime maximum and especially its wintertime minimum are less drastic than at 70°N.

Path of Solar Energy Through the Atmosphere

The incoming solar energy in Figure 2-17 represents the amount of energy at the top of the atmosphere. How much solar energy eventually gets absorbed by the Earth's surface is a function of the cloud amount, the properties of the surface, and how much solar energy the atmosphere absorbs. How much energy the atmosphere absorbs is a function of how much atmosphere the sun's rays pass through. The tilt of the axis and the spherical nature of the Earth influence the amount of atmosphere the sunlight passes through before reaching the surface. FIGURE 2-18 demonstrates that at larger solar zenith angles the Sun's energy must pass through more atmosphere. This means there is more chance for atmospheric absorption of solar energy. The surface, then, receives less energy. The atmosphere is largely transparent to sunlight, but some radiation is absorbed by some atmospheric gases. In the next section, we discuss the absorbing properties of the atmosphere.

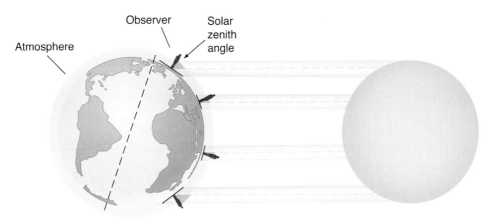

FIGURE 2-18 Sunlight incident at a higher solar zenith angle passes through more of the atmosphere than light incident at a smaller solar zenith angel (more overhead sun).

RADIATIVE PROPERTIES OF THE ATMOSPHERE

Before solar rays can reach the surface of Earth, they have to pass through the atmosphere. The trek through the Earth's thin atmosphere represents less than 0.0001% of the distance the rays travel from the Sun to the Earth, but what happens there makes all the difference to us because it permits human life to survive on Earth.

Atmospheric gases are selective in the solar wavelengths that they absorb. Of the atmospheric gases, ozone absorbs most of the shortwave radiation. Ozone absorbs UV energy and a small amount of visible energy in the 0.4-μm to 0.56-μm spectral region. Without this absorption, human and animal life would perish because UV light destroys the genetic code of life. Water vapor weakly absorbs radiation at several wavelengths between 0.7 μm and 4.0 μm. Carbon dioxide is a very weak absorber of solar energy as are methane and chlorofluorocarbons (CFCs). The left-hand side of FIGURE 2-19 summarizes the wavelength dependence of atmospheric absorption of solar radiation. Nearly all of the visible energy from the Sun is transmitted to the surface, whereas radiation with wavelengths less than 0.28 μm does not reach the surface. The percentage of solar energy absorbed by the atmosphere basically depends on how much ozone and water vapor are present. A 1% decrease in ozone, for example, leads to a 2% increase in the amount of UV light at the surface.

Atmospheric gases are also selective in the wavelengths at which they absorb and emit terrestrial radiation (right-hand side of Figure 2-19). In addition to absorbing solar radiation,

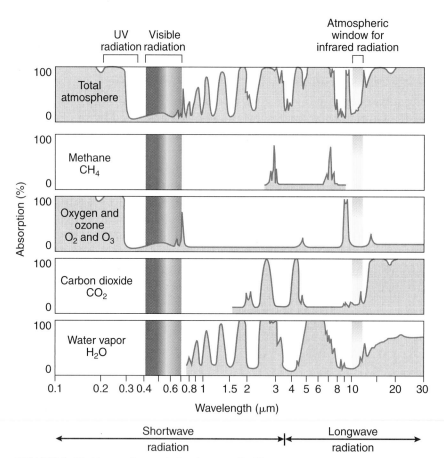

FIGURE 2-19 Absorption of shortwave (left) and longwave (right) radiation by the atmosphere. Low amounts of total absorption (white areas in bottom row) indicate wavelength regions in which solar radiation can reach the Earth's surface and terrestrial radiation can escape to outer space. The visible light spectrum and the infrared atmospheric window stand out in this regard. Notice the prominent roles of water vapor and carbon dioxide in absorbing Earth's longwave radiation.

water vapor absorbs (and, therefore, emits) terrestrial energy at wavelengths between 5 µm and 8 µm and beyond 12 µm. Carbon dioxide and ozone emit and absorb energy at wavelengths near 15 µm and 9.6 µm, respectively.

Atmospheric gases only weakly emit and absorb energy in the 10-µm to 12-µm region. This spectral region is referred to as the **infrared atmospheric window** because the atmosphere is relatively transparent to infrared radiation emitted by the surface at these wavelengths. Another **atmospheric window** exists around 4 µm, but the other atmospheric window is more important for weather and climate because the Earth's peak in emitted energy happens to occur right in this window, at 10 µm. The atmospheric window thus allows the Earth to cool off by emitting energy into outer space.

Clouds are good reflectors of solar energy, and they are good emitters and absorbers of longwave energy. Clouds also emit and absorb radiation in the 10-µm to 12-µm infrared atmospheric window. So, when clouds are present, the window is effectively "shut." **BOX 2-2** discusses how weather satellites make use of the science of radiation to observe weather patterns from space.

The Greenhouse Effect

The selective nature of radiation absorption by atmospheric gases is the fundamental cause of the **greenhouse effect**. Much of the shortwave, or solar, energy passes through the atmosphere and warms the surface. Although the atmosphere is nearly transparent to shortwave radiation, it efficiently absorbs terrestrial or longwave radiation emitted upward by the surface. So, although carbon dioxide and water vapor make up only a very small percentage of the atmospheric gases, they are extremely important because these gases absorb this longwave radiation and emit it throughout the atmosphere.

"Greenhouse Warming" to use an interactive simple climate model to explore some of the causes of the greenhouse effect.

What happens when the atmosphere absorbs longwave radiation emitted by the surface? The atmosphere gains energy through absorption but does not accumulate this energy continually. Instead, the atmosphere loses energy by emitting longwave radiation in all directions. Some of this longwave energy is emitted toward the Earth and absorbed by the surface. The Earth's surface is heated by shortwave and longwave absorption emitted by gases in the atmosphere. If the atmosphere did not absorb and emit longwave radiation, the surface of Earth would be up to 33° C (60° F) cooler than it is today!

Because the greenhouse effect keeps the Earth from freezing into a subzero slab of ice, it is a good thing. A separate issue is whether humans have added to the greenhouse effect and made too much of a good thing.

Greenhouse Warming: The Basics

Greenhouse warming and the **enhanced greenhouse effect** are the terms used to explain the relationship between the observed rise in global temperatures and the observed increase in atmospheric carbon dioxide. In this chapter we have considered one aspect of the enhanced greenhouse effect: absorption and emission of radiation by certain atmospheric gases. How does this play a role in the enhanced greenhouse effect? Let's use carbon dioxide as an example.

Increasing the carbon dioxide concentrations in the atmosphere does not appreciably affect the amount of solar energy that reaches the Earth's surface; however, because carbon dioxide absorbs longwave radiation, then the more carbon dioxide there is, the more infrared energy is absorbed. This increases the temperature of the atmosphere. Then, by the Stefan-Boltzmann Law, the now-warmer atmosphere will emit more longwave energy than it did when there was less carbon dioxide in it. This, in turn, increases the amount of longwave energy striking the Earth's surface, and it too warms up. By this simple argument, an increased concentration of carbon dioxide could result in a warming of Earth's atmosphere and surface.

Gases that are transparent to solar energy while absorbing terrestrial energy will warm the atmosphere because they allow solar energy to reach the surface and inhibit longwave radiation from reaching outer space. These radiatively active gases are called **greenhouse gases**. This

Box 2-2 Satellite Images

Satellite instruments measure electromagnetic energy that the Earth and the atmosphere reflect and emit. The most common satellite instruments measure visible light (0.6 μm), infrared window (10 μm to 12 μm), and an infrared absorption band of water vapor near 6.7 μm—the so-called water vapor channel. Analysis of satellite images allows meteorologists to locate thunderstorms, hurricanes, fronts, and fog. Weather events can be tracked using time sequences of satellite images. This allows weather forecasters to predict their movement over short time periods of 30 minutes or less. Television weather forecasters frequently show animations of time sequences of infrared (IR) images.

The IR radiometers on satellites measure radiation with wavelengths between 10 μm and 12 μm. The information in this "spectral channel" is very different than in the visible image. Instead of light that has been reflected, the IR radiometer measures radiation emitted from the planet out to space. An IR image provides information on the temperature of land, water, and clouds by measuring the IR radiation emitted from surfaces below the satellite. All objects emit radiation proportional to their temperature and also the degree to which they can emit radiation. Because the cloud-free atmosphere only weakly absorbs in this spectral region (Figure 2-18), much of the energy emitted by the Earth's surface is transmitted through the atmosphere to the satellite. Clouds are very effective at absorbing and emitting IR radiation. Thus, clouds block the satellite view of the surface and emit energy at colder temperatures.

In IR radiometric images, cold objects (such as high clouds) are white, and hot surfaces (such as deserts) appear black (see figure below).

nickname makes an analogy with a greenhouse, which lets solar radiation come through its glass ceiling and walls but traps energy inside. In addition to carbon dioxide, other important greenhouse gases are water vapor, ozone, methane, nitrous oxide, and CFCs. Methane and CFCs are important despite their very small concentrations because they absorb terrestrial radiation in the 10-μm to 12-μm infrared atmospheric window.

Water vapor is the most important greenhouse gas because of its relative abundance and its ability to absorb a lot of longwave energy in many different wavelengths. A warmer atmosphere can mean more water vapor in the atmosphere and possibly more clouds; however, clouds affect both shortwave and longwave radiation, complicating the situation considerably. We revisit the enhanced greenhouse effect in Chapter 4, where we cover water vapor and clouds in more detail.

THE GLOBAL AVERAGE ENERGY BUDGET: ENERGY IS TRANSFERRED FROM THE SURFACE TO THE ATMOSPHERE

When you balance your checkbook you are concerned with deposits and withdrawals. Deposits represent a gain to the account, and withdrawals represent a loss. Gains are positive, and losses are negative. The net result of adding all of the gains and losses tells you whether you are getting richer or poorer.

Similarly, when we study energy, it is helpful to make a "budget" of the gains and losses of all types of energy over long periods of time, such as a year. Net gains in energy lead to warmer temperatures, and losses lead to cooler temperatures. This approach works on many different scales, from the entire planet to a single location. The budget for each square meter of Earth is shown in FIGURE 2-20, and now we step through this "bookkeeping."

The globally averaged, total amount of solar energy incident on each square meter of area at the top of the atmosphere is 342 W (top middle portion of Figure 2-20). This is an energy gain for the planet. The annual average albedo of the planet is 30%. So, approximately 30%, or 107 Watts per square meter (W/m^2), of the incident solar energy at the top of the atmosphere is sent back out into space. The solar energy reflected back to space is considered a loss of energy. Seventy percent (235 W for each square meter of area) is absorbed by the atmosphere and surface. Solar radiation absorbed by each square meter of the atmosphere is 67 W. Each square meter of the surface absorbs 168 W. In the atmosphere, water vapor, clouds, aerosols, and ozone absorb solar energy.

"The Energy Budget" to observe how the Earth's energy budget varies throughout the year.

The Earth's surface also gains energy as a result of atmospheric emission of longwave energy (left-hand side of Figure 2-20). At the same time, it loses its own longwave energy by emission to the atmosphere. On average, each square meter of the Earth's surface emits 390 W. Some of the surface-emitted energy escapes to space, and the atmosphere absorbs the rest. Only 40 W/m^2 of the Earth's surface is transmitted through the atmosphere directly to space, much of it through the infrared atmospheric window. The atmosphere absorbs 350 W/m^2. The atmosphere emits radiant energy out to space and toward the surface. On average, each square meter area of the atmosphere emits 195 W to space and 324 W to the surface.

Now it is time to add up the numbers of the energy budget for the Earth and the atmosphere's "joint bank account." The Earth's surface has a net gain of 102 W/m^2 of radiant energy (168 W/m^2 − 390 W/m^2 + 324 W/m^2), and the atmosphere experiences a net loss of 102 W/m^2 of radiant energy (67 W/m^2 + 390 W/m^2 − 40 W/m^2 − 324 W/m^2 − 195 W/m^2). The surface's gains equal the atmosphere's losses. This means that the Earth and its atmosphere, when viewed as one system, are in radiative balance.

A loss of 102 W for each square meter area (102 W/m^2) of the atmosphere is equivalent to the atmosphere cooling more than 200° C over the course of a year! We do not observe this large cooling because energy is transferred from the surface to the atmosphere through ways other than radiation. The transfer of 102 W/m^2 is accomplished by two kinds of heat transfer: sensible and latent heat transfers (center of Figure 2-20). **Sensible heating** represents the combined

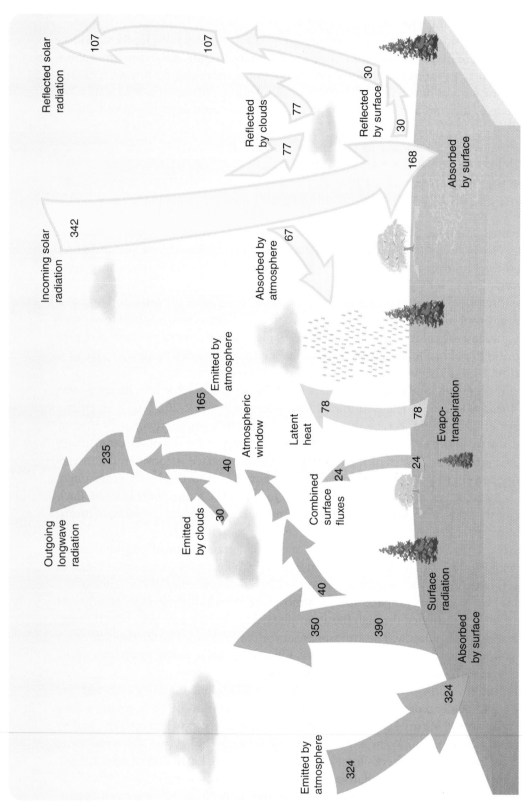

FIGURE 2-20 The annual average energy budget of Earth. The flow of radiative energy is depicted with yellow (solar) and red (infrared) arrows, with shaft widths proportional to the amount of energy. On average, the atmosphere is losing energy by radiative processes while the surface of the Earth has a surplus of radiative energy. Energy transfer from the surface to the atmosphere is one reason why the average temperature in the troposphere decreases with increasing distance from the surface. Convection and conduction fluxes are represented by the orange arrow. Latent heat losses from the surface are represented by the blue arrows.

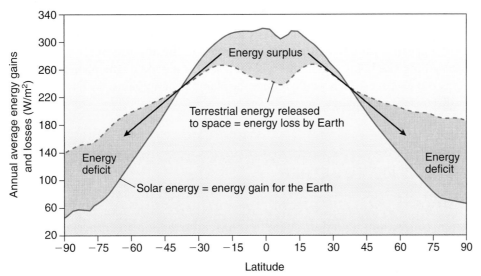

FIGURE 2-21 The radiation budget of the planet as a function of latitude. Atmospheric and oceanic circulations move energy from the latitudes with a surplus of radiation to regions with a deficit (the energy transfer is represented by the black arrows). Because tropical latitudes cover more surface area than polar latitudes on a sphere, the apparently small area of surplus does equal the deficit.

processes of conduction and convection and amounts to a total of 24 W/m² transferred from the surface to the atmosphere. Latent heating transfers 78 W/m² from the surface to the atmosphere. Evaporation from oceans and lakes and sublimation from glaciers cools the surface. Some of the water that evaporates into the atmosphere condenses to form clouds and precipitation, releasing latent heat into the atmosphere.

When globally averaged over a year, Earth's net energy gains balance energy losses, or nearly so, but this is not the case when the radiation gains and losses are averaged in relation to latitude (**FIGURE 2-21**). As we learned earlier, any object with a temperature above absolute zero emits energy. This means that all parts of the globe, including the poles, emit longwave energy at all times. In budget terms, all parts of the globe are always "paying out" energy; however, only the tropics receive enough sunlight all year long to counterbalance these losses and "turn a profit." The poles run deep into energy debt during the dark winter and are not able to make it up during the summer because even 24-hour sunlight is at a perpetually high zenith angle. The end result is that the tropics gain and the poles lose radiant energy on a yearly basis.

Because the tropics are not observed to be continually heating and the polar regions continually cooling, energy must be transported from the tropics to the poles. This transport is accomplished by the atmosphere and ocean currents and is a major reason we have weather.

■ Radiative Forcing

A climate change will alter the energy budget of the Earth system. There are many processes that can result in a change of our climate system—changes in incoming solar energy, changes in cloud amount, or changes in ocean currents. Climate scientists need a useful way to compare and relate these different perturbations to a climate system. The International Panel on Climate Change, or IPCC, has adapted the concept of **radiative forcing** to compare factors that affect climate and alter the energy budget of the Earth. Radiative forcing is defined as the change in the net radiation at the tropopause. This change in the net radiation energy is the difference between the incoming and outgoing radiative energies. The change in the net radiation is with respect to some reference state of the Earth and the units are watts per square meter (W/m²).

A positive radiative forcing means a larger energy gain, which tends to warm the system. A negative forcing indicates an increase in the energy losses, which tends to cool the system. An example of a positive radiative forcing is an increase in the Sun's energy output. With nothing else changing, this would increase the energy gains of the system and result in a warming. An example of a negative radiative forcing is an increase in the amount of sea ice cover over the ocean. Everything else being equal, this will increase the albedo of the planet reflecting more solar energy out to space, which would reduce the planet's energy gains. We'll revisit this concept when we discuss climate change.

PUTTING IT ALL TOGETHER

■ Summary

Conduction, convection, advection, latent heating, and radiation are important processes that transfer energy in the atmosphere. Conduction moves energy by physical contact. Convection and latent heating transfer heat over great distances through vertical motions and phase changes of water. Advection by the wind moves heat horizontally. Radiative processes transfer heat throughout the entire atmosphere and into space. Radiation can be absorbed, reflected, or transmitted.

The Sun powers Earth's weather and climate. Solar energy streams through space as electromagnetic waves. Most of this solar radiation has wavelengths between 0.2 and 4 microns (or micrometers, or μm). The reason the Sun emits radiation with such short wavelengths is because, according to Wien's Law, the Sun is very hot.

Earth intercepts a portion of this solar energy. The total amount intercepted is determined by the tilt of the Earth's axis and how far the Earth is from the Sun. Although Earth is a little closer to the Sun in January than in July, Northern Hemisphere temperatures are warmer in July because Earth's axis of rotation points toward the Sun in July. The Earth's tilt, combined with its orbit around the Sun, causes seasons.

Approximately 50% of the solar energy reaching the planet passes through the atmosphere and is absorbed by the surface of the Earth. The surface albedo is about 30%, meaning that the Earth reflects 30% of the incoming sunlight.

Because any object that has a temperature emits radiation, the Earth emits energy and is constantly losing energy to space as longwave or terrestrial radiation. This energy emitted to space has wavelengths of 4 to 100 microns. Both the surface and atmosphere emit longwave radiation, not shortwave radiation like the Sun, because they are much cooler than the Sun.

Much of the energy that escapes from the Earth to space is in the narrow band of 10 μm to 12 μm, called the infrared atmospheric window. At other wavelengths, the Earth's atmosphere absorbs much of the longwave radiation emitted by the Earth and warms the Earth's surface in return. This is known as the greenhouse effect, and it keeps the Earth from freezing.

The atmosphere is always losing radiant energy, and the Earth's surface has a surplus of radiation energy. Conduction, convection, evaporation, and condensation transfer energy from the surface of the Earth to the atmosphere.

Averaged over many years, the tropical and subtropical regions of the planet gain more solar energy than these regions lose to space by longwave radiation. Polar regions lose more longwave energy than they receive from the Sun over a given year. This energy imbalance, with net energy losses at the pole and net energy gains in the tropics, is the driving force of weather and climate.

■ Key Terms

You should understand all of the following terms. Use the glossary and this chapter to improve your understanding of these terms.

Acceleration	Aphelion	Conduction
Advection	Atmospheric window	Convection
Albedo	Blackbody	Deposition
Amplitude	Calorie	Diurnal
Angle of inclination	Celsius	Electromagnetic energy

Energy	Latent heating	Solar radiation
Enhanced greenhouse effect	Latent heat of condensation,	Solar zenith angle
Equinoxes	deposition, fusion, melting,	Solstices
Fahrenheit	sublimation, and vaporization	Specific heat
Force	Longwave radiation	Stefan-Boltzmann Law
Greenhouse effect	Ozone hole	Sublimation
Greenhouse gases	Parcel of air	Temperature
Greenhouse warming	Perihelion	Terrestrial radiation
Heat	Photodissociation	Thermal conductivity
Infrared atmospheric	Potential energy	Ultraviolet (UV) light
window	Power	Visible radiation
Joule	Radiation/radiant energy	Watt
Kelvin	Radiative forcing	Wavelength
Kinetic energy	Sensible heating	Wien's Law
Kirchhoff's Law	Shortwave radiation	Work
Latent heat	Solar constant	

■ Review Questions

1. What are the major mechanisms for transferring energy in the atmosphere?

2. A documentary filmmaker once interviewed graduating Harvard University seniors on camera and asked them the question: "Why are we warmer in summer than in winter?" The consensus answer from the Harvard graduates was, "Because we're closer to the Sun in summer than in winter." Is this answer correct? If not, what is the correct answer?

3. I was walking through a store one day and came across an advertisement of an object that claimed to defrost a 1-inch steak in less than 10 minutes. Made of some sort of metal, the object was about the size of a cookie sheet and was not electric. When I touched it, it felt very cold. The salesperson demonstrated the "super defroster" by placing an ice cube on it. The ice cube melted before my very eyes in a matter of minutes. How could something that felt so cold melt ice so quickly?

4. To keep warm when sleeping on a cold winter night, some animals burrow into snow. How does this keep them warm?

5. In extremely cold weather, birds fluff their feathers to keep warm. In terms of energy losses, what is the bird doing to keep warm?

6. Explain why "a watched pot of water never boils" in terms of specific and latent heats.

7. Consider a puddle of water that is evaporating. The faster molecules escape, leaving behind those with less energy. Thus, during evaporation, the average energy of the molecules in the water decreases as molecules continually leave. The water temperature should be lowered; however, the temperature of a puddle often is not much different than the surrounding air. Explain why the temperature of the puddle does not continually decrease.

8. You pump up a bicycle tire with a hand pump in which air is compressed in the barrel of the pump and forced into the tire. You observe that the barrel of the pump gets hot. Why?

9. Explain why satellites make observations in the 10-μm to 12-μm spectral region. Do you think satellite observations at 6.7 μm would provide information on the temperature of the Earth's surface?

10. If an object in a fire is heated until it glows "red hot," what is its approximate temperature? Assume that its wavelength of peak energy emission is 3 μm so that the front part of its energy curve will include the red part of the visible spectrum.

11. The leftover radiant energy from the "Big Bang" is measured to be about 2.9 K. What is the wavelength of maximum emission of this energy? What kind of equipment would you need to detect this energy? (Hint: It would be a specialized version of a common electronic device [see Figure 2-7].)

12. Why will covering a soil with a white powder, such as lime, reduce the soil's daytime temperature?

13. The outer dwarf planet, Pluto, and Earth have the same albedo. Why is Pluto much colder than Earth?

14. The figure below shows the absorption of nitrous oxide as a function of wavelength. The shaded areas represent the percentage of radiation absorbed at particular wavelength. Explain why N_2O would be considered a greenhouse gas.

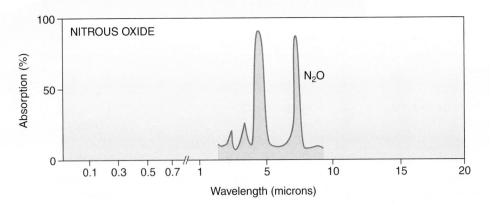

15. How would the length of day vary with latitude if the Earth's angle of inclination were 0 degrees?

16. Why is the greenhouse effect beneficial? If enhanced, how can it become a problem?

17. Our eyes are sensitive to visible light. We cannot see radiation at near-infrared or infrared wavelengths. A halogen lamp emits peak radiant energy at 1 micrometer. Explain why we can see light emitted by these lamps.

18. If you needed to observe the distribution of methane in the atmosphere from a satellite platform, what wavelength range would you use to make these measurements? Explain your selection.

Observation Activities

1. The purpose of this exercise is to apply energy transfer mechanisms to an experiment with two balloons. Blow one balloon up with air, and place it over a lit match or flame. What do you observe? Fill the second balloon with water, and place it over a lit match or flame. What happens?

2. Shine a flashlight on the center of a globe or large ball. Without tilting the flashlight, move it up and down so that the "spot" falls on different parts of the globe. What happens to the flashlight beam as it strikes the globe at different angles? How is this related to seasonal weather?

3. How does a microwave oven warm up food? Think about this in terms of the absorption of electromagnetic energy by a certain molecule found in most foods.

This rain cloud icon is your clue to go to the *Meteorology* Web site at http://physicalscience.jbpub.com/ackerman/meteorology/. Through animations, quizzes, web exercises, and more, you can explore in further detail many fascinating topics in meteorology.

Temperature

3

AFTER COMPLETING THIS CHAPTER, YOU SHOULD BE ABLE TO:
- Describe the various surface temperature cycles
- Interpret temperature cycles in terms of the surface energy budget
- Explain differences in temperature cycles resulting from the effects of location, altitude, and cloud cover
- Understand temperature changes in the vertical and relate this to the concept of static stability
- Relate temperature inversions to the surface energy budget and to stability

INTRODUCTION

Can the weather determine the outcome of wars? The experiences of Napoleon I and Hitler suggest it can. Their armies encountered two bitter Russian winters over a century apart. The rest, as they say, is history.

In the fall of 1812, Napoleon's army—a half-million strong, the largest the world had ever seen—entered the outskirts of Moscow but was forced to retreat. The premature onset of winter turned the retreat into a complete disaster. FIGURE 3-1 depicts Napoleon's army as a dwindling river of soldiers advancing from left to right and then retreating from right to left as temperatures dropped as low as −34° C (−29° F). One of Napoleon's generals reported that it was so cold that birds fell dead to the Earth. So did Napoleon's soldiers. Fewer than 10,000 men fit for combat remained with Napoleon's main force after the retreat. (Napoleon met his Waterloo two and a half years later, where once again weather was his nemesis.)

In October 1941, the German Nazis' Operation Barbarossa blitzkrieg on Moscow bogged down as the ground turned into a quagmire because of heavy rains. The rainy season was followed by one of the coldest winters in two centuries. Adolf Hitler's invading force had expected such a quick victory that they did not even bring winter gear with them. Cold outbreaks that winter plummeted the temperature just west of Moscow to a mind-numbing −53° C (−63° F). The Germans' equipment could not function properly at such bitter temperatures. Desperately needed supplies failed to reach

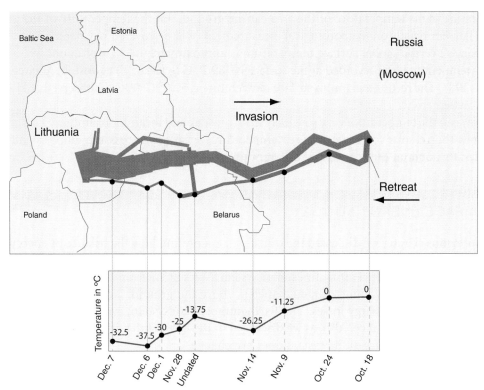

FIGURE 3-1 Cartographer Joseph Minard's famous graphical description of Napoleon's march toward (dark green shading, from left to right) and retreat from (orange shading, from right to left) Moscow in 1812. The size of Napoleon's army during the campaign is depicted proportionately by the width of the shaded lines. The scale at the bottom records the minimum temperatures observed during the retreat. (Adapted from Winters, Harold A. *Battling the Elements: Weather and Terrain in the Conduct of War*. Johns Hopkins University Press. 1998.)

the army because of breakdowns in the rail system brought on by the cold weather. Bombs were muffled by deep snow, and railway bridges had to be constructed of blocks of ice.

Stymied by the weather, the German troops abandoned the attack on December 6, 1941 (the day before Japan bombed Pearl Harbor). Then the Soviet army launched a counteroffensive against the Nazis. By the end of December more German troops died because of exposure to the cold than through the more conventional horrors of war.

Like Napoleon, Hitler was haunted by weather woes. Two and a half years later, the weather forecast for D-Day (see Chapter 13) would help put a permanent end to the Nazi quest for world domination.

Historians can provide many examples of how weather, especially temperature, has controlled battles or even entire wars. In this chapter, we provide examples of temperature observations at different locations to explain what controls the temperature.

SURFACE TEMPERATURE

Temperature, as we learned in the last chapter, represents the average kinetic energy of the air molecules. A change in the air's temperature depends on its net energy budget, its specific heat, and whether phase changes of water have occurred. Before we explore temperature, however, we must be clear on what a "surface temperature" really is.

For consistency's sake, throughout the world, meteorologists measure temperature at the same reference height, 1.5 meters (about 5 feet) above the ground, usually on a grass-covered surface. To avoid solar heating of the thermometer, temperature measurements are made in the shade. Air temperatures at this 1.5-meter height are called **surface temperatures**. The surface temperature is the temperature of the air near the ground, not the temperature of the ground itself. (Further details on meteorological measurements will be covered in Chapter 5.)

Today's average global surface temperature is approximately 15° C (or about 59° F). The coldest temperature ever recorded at the surface is −89.2° C (−128.6° F) at Vostok, Antarctica, on July 21, 1983. The record maximum surface temperature is 57.8° C (136° F) observed at El Azizia, Libya, on September 13, 1922.

How can the temperature vary so widely from one day and location to another? This chapter examines the reasons for the observed temperature variations across the globe, at different latitudes, throughout the year, and at different times of the day.

SURFACE ENERGY BUDGET

In the previous chapter we discussed the transfer of energy and how the balance of energy gains and losses determines an object's temperature. The simple diagram in FIGURE 3-2 summarizes the flow of energy through the Earth–atmosphere system. Any system, from Earth and its atmosphere all the way down to a volume of air 1.5 meters above the ground, is in energy balance if its energy gains equal its energy losses. The Earth–atmosphere system, averaged over a year, is in approximate energy balance. Also, as we discussed in the last chapter, if energy gains and losses are equal, the temperature of the system does not change.

Energy gains do not always balance energy losses over short periods of time and in localized regions. When an energy imbalance exists, energy is stored within or removed from the system. When energy gains exceed losses, the temperature increases. Conversely, when energy losses exceed energy gains, the system will cool.

An energy imbalance is common for many systems with which we are familiar. For example, imagine running barefoot on the blacktop in a parking lot on a hot sunny summer day. You will immediately notice that the blacktop is very hot. Why? The blacktop is exchanging energy with the bottom of your feet. Your feet are naturally warm and therefore emit energy. However, the blacktop is even warmer and emits much more energy. Enough heat is transferred from the blacktop to your feet to overcome their energy loss. As a result, each foot's energy gains are greater than its losses, and the soles of your feet get hot. At the same time, heat is also exchanged with the air above the blacktop, affecting the air temperature.

The temperature of the air near the ground is determined by energy exchanges with the surface. Convection, conduction, latent heating, and radiation all play roles in transferring energy between the air and the ground. In addition, irregular air motions, referred to as **turbulence**, mix heat and moisture from the surface higher up in the atmosphere. We examine the complicated topic of turbulence in depth in Chapter 12.

Conduction, convection, radiation, sensible and latent heat transfers, and turbulence all act at the same time to transfer energy between the atmosphere and its surroundings. This makes it surprisingly difficult to express the relationship between air temperature and surface conditions in a simple formula. Instead, let's turn our attention to observed variations in the air temperature across the world. By comparing the observed temperature patterns, we will reveal the processes that govern temperature across the globe.

Energy gains are equal to energy losses

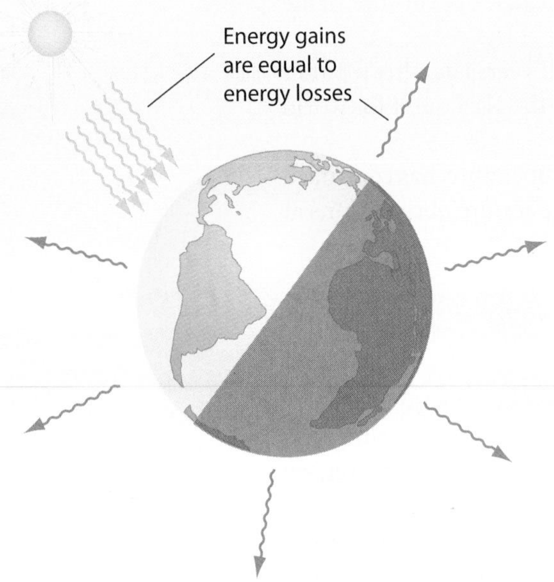

FIGURE 3-2 Earth gains energy from the sun and loses terrestrial radiation to space.

GLOBAL DISTRIBUTION OF TEMPERATURE

FIGURE 3-3 shows a 30-year average surface air temperature across the globe for the months of January and July. These two months are chosen as they represent the seasonal extremes of many geographic

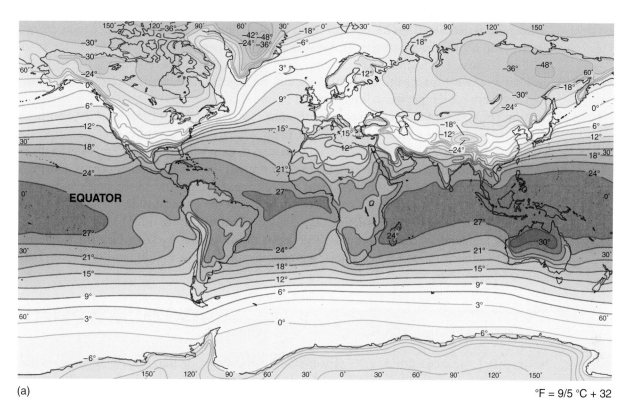

(a)

°F = 9/5 °C + 32

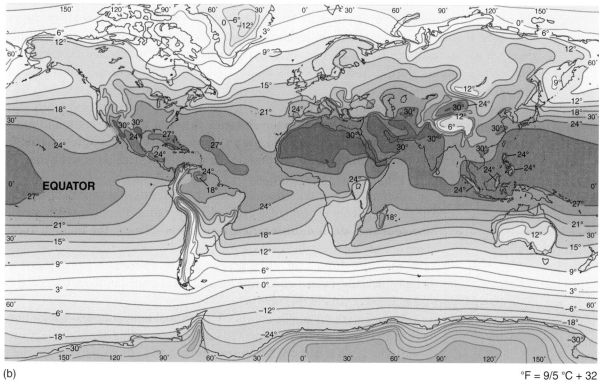

(b)

°F = 9/5 °C + 32

FIGURE 3-3 A 30-year average surface air temperature, in °C, across the globe for the months of (a) January and (b) July. These two months are chosen as they ordinarily represent the seasonal extremes of a geographic region. (*Source*: Data from NOAA.)

regions. Some of the patterns we see in the global distribution of the temperature correlate with the incoming solar energy. For example, the lowest temperatures are over the polar regions during winter, where the annual average of incoming solar energy is at a minimum. Temperatures decrease from the equator to the poles, as does the average incoming solar energy. The tropical regions have a smaller seasonal change in temperature, partly because of the small seasonal changes in incoming solar energy. There are large changes in temperature from the summer and winter in the middle latitudes and polar regions. The observations also show that there are land–sea temperature differences.

Lines of constant temperature, or **isotherms**, are oriented east–west over the Southern Hemisphere, where the surface is mostly ocean. In the winter, the temperatures over the continents are colder than nearby oceans. This difference is seen in each winter hemisphere, where a given isotherm dips further toward the equator over continents than over adjacent oceans. During a summer, the isotherms over the interior of the continents bulge toward the poles, indicating that the land regions are warmer than the ocean bodies at the same latitude.

A **temperature gradient** is defined as a change of temperature divided by the distance over which the temperature change occurs. Large temperature gradients exist where the isotherms are close together. In winter, when a polar region is in darkness, the middle latitude regions have large temperature gradients.

TEMPERATURE CYCLES

In Chapter 1, we studied the cycles of water and carbon dioxide that described the ways that those molecules repeatedly moved from the atmosphere to the land and back again. Repetitive patterns happen in weather and are also called cycles. For example, you have probably noticed that temperatures are usually warmer in the afternoon than during the night, day after day. This repeating pattern of daily temperatures is the **diurnal temperature cycle**. This cycle includes the maximum and minimum daily temperatures and the times of day that they usually occur (**FIGURE 3-4**). In the typical diurnal temperature cycle, the maximum temperature occurs during mid to late afternoon. The minimum temperature is reached around dawn.

One way to measure temperature cycles is by calculating the **temperature range**. For example, the **diurnal temperature range** is the difference between the maximum and minimum temperatures of any given day. The **daily mean temperature** is usually determined by averaging the maximum and minimum temperature for a 24-hour period or sometimes by averaging all 24 hourly temperature measurements.

Everyone knows that temperatures are usually colder in winter and warmer in summer. This very regular cycle of temperature throughout the year is the **seasonal** or **annual temperature cycle**. Plotting the monthly mean temperature as a function of month represents the annual temperature cycle. The **monthly mean** (or monthly average) **temperature** is calculated by adding the daily mean temperature for each day of the month and dividing by the number of days in the month.

Similar statistics can be compiled for annual temperatures. The **annual average temperature** is simply the sum of the monthly mean temperatures divided by 12. The **annual temperature range** is the difference between the warmest and coldest monthly mean temperatures of a given geographic location. It can also be defined as the difference between the highest and lowest daily temperatures observed in a given year or years. **FIGURE 3-5** shows the annual range in temperature as the difference between the average temperatures of January and July. The annual range is relatively small for the oceans because of the high specific heat of water and the mixing that occurs, as discussed in the previous chapter. Ranges in temperature are also not large

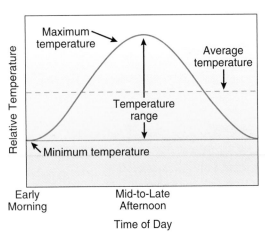

FIGURE 3-4 Temperature cycles are characterized by a maximum, minimum, the range between maximum and minimum, and the average temperature.

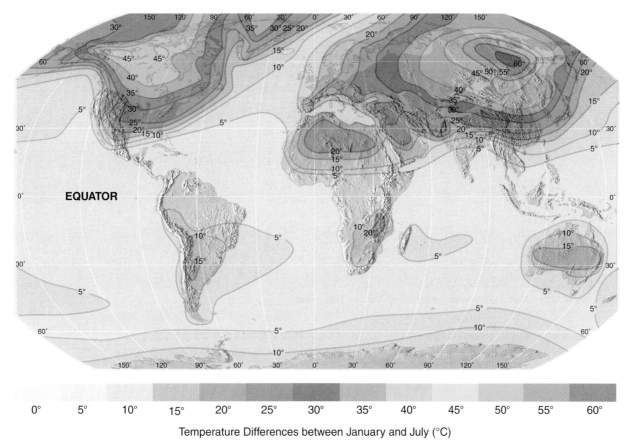

0°   5°   10°   15°   20°   25°   30°   35°   40°   45°   50°   55°   60°

Temperature Differences between January and July (°C)

FIGURE 3-5 The global distribution of the annual range in temperature plotted as the difference between the average temperatures of January and July. These two months are chosen, as they ordinarily represent the seasonal extremes of a geographic region.

near the equator, where the incoming solar energy is relatively constant throughout the year. The largest range occurs over the continents, particularly in the interior of the continents in the Northern Hemisphere. The Sahara Desert also has a large range in the annual temperature.

Temperature cycles of a given region reflect the net energy gains and losses throughout a particular time period. When energy gains exceed losses, the temperature warms, and the temperature cools when energy losses surpass energy gains. We know that temperatures usually rise and fall day after day and from season to season. This means that energy imbalances occur locally on daily and seasonal time scales. What causes these energy imbalances? The major factors are as follows: latitude, surface type, elevation and aspect, relation to large bodies of water, advection, and cloud cover. First we will discuss these factors in terms of how they affect the annual temperature cycle.

Annual Temperature Cycle

The geographic setting influences the temperatures at a given location. Comparing the annual temperature cycles of different locations allows us to observe how latitude, surface type, elevation and aspect, relation to large bodies of water, advection, and cloud cover all influence temperature cycles.

"Annual Temperature Cycle" enhances your understanding of what controls annual temperature cycles.

Latitude

As discussed in Chapter 2, the tilt of Earth's axis—the angle of inclination—affects the amount of incoming solar energy and is the reason for the seasonal cycle in temperature. The amount of incident solar energy at the top of the atmosphere, or **insolation**, is a function of time of the year, time of day, and latitude.

FIGURE 3-6 shows the monthly mean temperatures for each month of the year (which together make up the annual temperature cycle) of New York City, New York, and Miami, Florida. Both

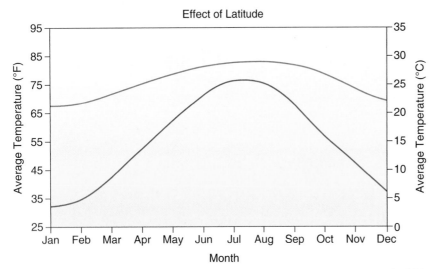

FIGURE 3-6 Comparison of the annual temperature cycle observed at Miami, Florida, and New York City. The difference in latitude between Miami and New York creates large differences in temperature between these two cities, which are otherwise similar in location and altitude.

are large cities on the East Coast of the United States. Both are near a large body of water, and both are at nearly the same low altitudes above sea level. Miami's annual average temperature is 24° C (75° F), whereas New York City's is 11.5° C (53° F). The main reason for this temperature difference is that Miami is closer to the equator. Therefore, the Sun is higher in the sky all year at Miami, and Miami receives a greater amount of insolation than New York City throughout most of the year (FIGURE 3-7).

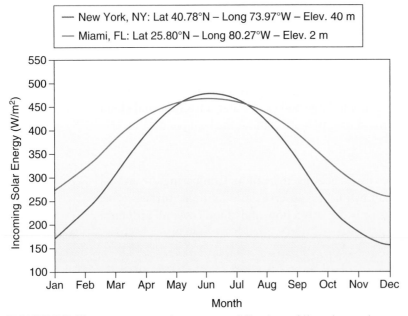

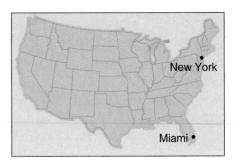

FIGURE 3-7 The incoming solar energy at the top of the atmosphere over Miami, Florida, and New York City. New York City receives more solar energy than Miami near the summer solstice because the length of day is greater at higher latitudes in summer. The greater variation of incoming solar energy over New York City explains the larger amplitude in the annual cycle of temperature (Figure 3-6).

The maximum monthly mean temperatures of both New York City and Miami occur in July, and the minimum in each city occurs in January. The maximum temperature occurs after, or **lags**, the time of maximum solar input, which occurs in June on the summer solstice. However, after the summer solstice, energy gains still exceed energy losses. This means that the atmosphere continues to warm. Not until late July are the energy losses (such as emission of radiation) larger than the energy gains, causing the mean temperature to begin to decrease. Energy losses are greater than the energy gains until late January or February, although the minimum solar energy received occurs at the winter solstice. This explains why the coldest temperatures come in January, not December. These lags exist largely because of the Earth's oceans.

New York's annual range of temperature is much larger than Miami's range, 24° C (44° F) versus 8° C (15° F), respectively. Latitude influences the annual temperature range because it affects the following: (1) the seasonal variation of the insolation, (2) the solar zenith angle, and (3) the length of day. Throughout the year these three factors vary less in Miami than in New York, causing the temperature range to be smaller for Miami.

Surface Type

As discussed in Chapter 2, the surface of the Earth absorbs approximately 50% of the solar energy incident at the top of the atmosphere. So, the surface contains heat that can be transferred to the atmosphere. Because the atmosphere is heated by energy transfers from the Earth's surface, the surface type plays an important role in determining the surface air temperature.

Deserts such as the Sahara in northern Africa have a large annual range in temperature compared with coastal locations with vegetation, such as Miami (**FIGURE 3-8**). Why is this? Deserts gain large amounts of solar energy because of the persistent clear-sky conditions over the desert. But dry sand is a poor conductor of heat and has a low specific heat. Therefore, the dry sand at the surface rapidly heats up during the day and cools down during the evening. Because the surface and the atmosphere transfer energy to each other, the surface air temperature also has a large annual range (Figure 3-4).

On land surfaces, vegetation also modifies the annual range in temperature. Vegetation reduces the temperature range in several ways. Plants transpire and use some of the solar energy that

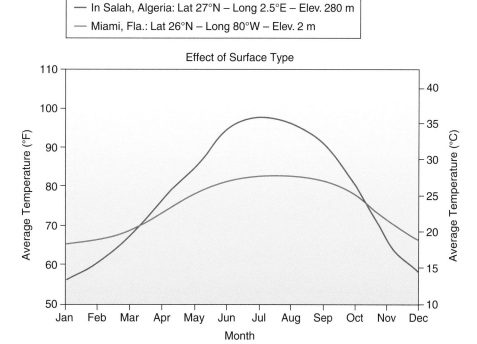

FIGURE 3-8 Comparison of the annual temperature cycle of In Salah, Algeria, in the Sahara Desert with the annual cycle at Miami, Florida, which is at approximately the same latitude. Note the large annual temperature range in the desert.

reaches the surface. Evaporation in the vicinity of plants takes in energy that would otherwise go into the raising of temperature. This prevents vegetated surfaces from having the high surface temperatures observed over the bare, dry soil of deserts.

Elevation and Aspect

When we consider energy exchanges over mountains and hills, the altitude and the direction a slope faces are important. The effect of altitude is demonstrated by comparing the annual temperature cycle of Burlington, Vermont (elevation 92 meters [300 feet]), and Mount Washington, New Hampshire (elevation 1748 meters [5727 feet]) (FIGURE 3-9). The two locations have similar latitudes but very different altitudes. The higher elevation station, Mount Washington, is on average colder than Burlington in every month of the year. At the higher elevation of Mount Washington, the air is less dense, and there are fewer molecules to absorb incoming solar radiation. In addition, terrestrial radiation emitted by the surface can more easily escape to space and therefore does not heat the atmosphere. Mount Washington is also much windier than Burlington (see Box 12-2 in Chapter 12). The high winds and associated turbulence rapidly carry energy away from the surface and mix the energy throughout the lower troposphere.

The **aspect** is also an important influence on the energy budget of a region, particularly the solar-energy side of the ledger. Aspect is the direction that a mountain slope faces. In the Northern Hemisphere, under cloudless skies, a north-facing slope receives less solar energy than a south-facing slope (FIGURE 3-10). Because south-facing slopes receive more solar energy, they are warmer. South-facing slopes are also generally drier. More solar energy results in increased evaporation and reduced moisture in the soil.

The effect of aspect can be seen by comparing the vegetation type of south-facing slopes versus the type of vegetation growing on north-facing slopes (FIGURE 3-11). In many regions of the western United States where the amount of vegetation depends heavily on available moisture, only sparse vegetation grows on south-facing slopes, whereas plants grow densely on the moister north-facing slopes. These differences in vegetation further enhance temperature differences between the north- and south-facing slopes.

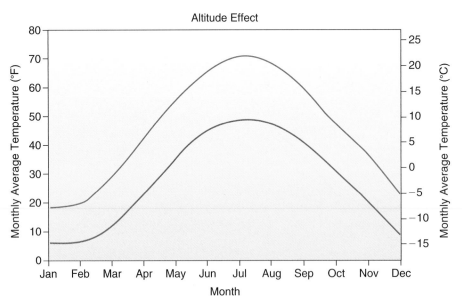

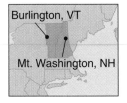

FIGURE 3-9 Comparison of the annual temperature cycle of Burlington, Vermont, and Mt. Washington, New Hampshire, which is at approximately the same latitude but at a much higher altitude than Burlington.

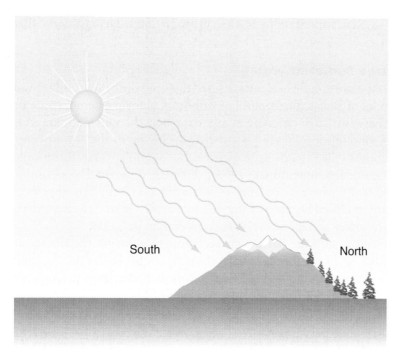

FIGURE 3-10 Schematic illustrating that in the northern hemisphere the Sun's rays are more intense on a south-facing slope than one facing north. As a result, temperature, moisture levels, and vegetation types can vary widely from one side of a mountain to another.

FIGURE 3-11 Photograph confirming the reality of the effect shown schematically in Figure 3-10. More trees are found on the north-facing slope of this mountain than on the south-facing slope.

Because the elevation and aspect of the land affect temperatures and moisture of the ground, they also influence the economic planning of a region. Ski resorts are built on cooler north-facing slopes where melting and evaporation are less and snow cover persists longer. Vineyards and apple growers in New York State plant their fruit on south-facing slopes where the growing season is longer.

Effects of Large Bodies of Water

The seasonal temperature cycle of a city is related to its proximity to a body of water. The annual temperature cycles of Dallas, Texas, and Los Angeles, California, demonstrate this, as shown in FIGURE 3-12. These two large cities are at approximately the same latitude and therefore have about the same amount of solar energy entering at the top of the atmosphere. Los Angeles is located on the shore of the Pacific Ocean, whereas Dallas is inland, far from a large body of water. The two cities' seasonal temperature cycles are distinctly different. Dallas's annual temperature range is 22° C (39° F), whereas Los Angeles' is only 8° C (15° F). Los Angeles's monthly mean temperatures are modified because of the nearby Pacific Ocean. The summertime maximum temperatures are cooler and the winters are warmer than in Dallas.

Energy exchanges with the Earth's surface strongly influence the surface air temperature. Large water bodies act to stabilize thermally the temperature of the surrounding air so that the differences between months are reduced. The factors that contribute to temperature differences between continental and maritime regions are as follows:

1. The specific heat of water is almost three times greater than of land. More heat is therefore required to raise the temperature of water. Water also cools more slowly than land.
2. Evaporation of water reduces temperature extremes over and near lakes and oceans.
3. Solar radiation absorbed by water is distributed throughout a large depth of the water body as a result of mixing and the transparency of water to solar radiation. Over land, the solar radiation is absorbed by the surface, and heat can quickly be transferred to the atmosphere above it.

The temperature of a nearby body of water also plays an important role in modifying a region's temperature, as we see in the next section.

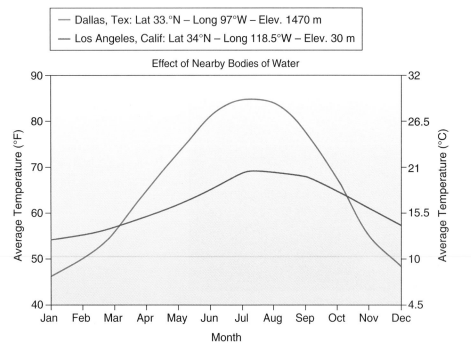

FIGURE 3-12 A comparison of the annual temperature cycles of Dallas, Texas, and Los Angeles, California. Despite the similarity in latitude, the two cities' temperature cycles are quite different because of the effect the Pacific Ocean has on Los Angeles weather.

Advection

The energy gained by the atmosphere through processes such as conduction and radiation can be transferred from one region of the globe to another by advection. For example, southerly winds in the Northern Hemisphere middle latitudes are usually associated with warmer temperatures than winds from the north. Air moving over the ocean can moderate the temperatures of the continent that is downwind. Oceans store heat in the summer and release heat to the atmosphere in winter. The east–west difference in surface air temperature between eastern North America and western Europe (Figure 3-3) during winter is partly a result of this energy exchange and partly because of advection of cold air over eastern North America and advection of warmer air to western Europe.

Consider the cities of St. John's, Canada, and London, England (**FIGURE 3-13**). Both cities are at similar latitudes and altitudes, and both are near large bodies of water. London's annual temperature range is approximately 14° C (24° F) while St. John's is 22° C (39° F). London is influenced by advection of air over the warm Gulf Stream ocean current (see Chapter 8) during winter. During winter, as air moves over the Gulf Stream on its trek to Europe, sensible heating from the ocean warms the air above. The two locations have similar summertime temperatures, but London's winters, unlike St. John's, are made milder by the influence of the relatively warm ocean waters.

Although far from these two cities, the Rocky Mountains also play an important role in determining the annual temperature cycle of these two cities. As we discuss in Chapter 6, large mountain ranges influence large scale flow patterns. During the winter, the Rocky Mountains set up an atmospheric circulation pattern that results in cold air being advected from the Arctic to eastern North America and warmer winds from the south advected toward London and western Europe. These flow patterns are not as prevalent in summer as they are in winter, and so, the summertime temperatures are similar for the two cities.

London's relative warmth in winter was often ascribed to the direct effect of the Gulf Stream in the past. But recent scientific experiments have conclusively shown that it is advection by the persistent southwesterly winds in winter that keeps London warm. These winds

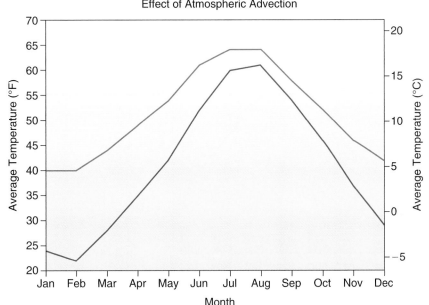

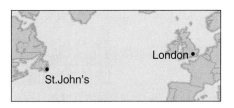

FIGURE 3-13 The annual temperature cycles in London, England, and St. John's, Canada. The warmer winter in London is caused by large-scale advection patterns and London's proximity to a nearby warm ocean current.

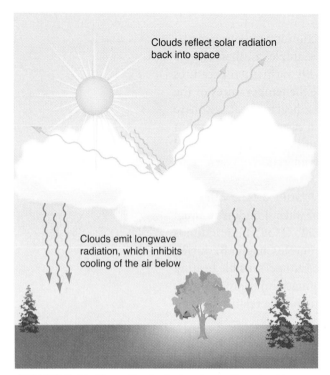

FIGURE 3-14 The effect of clouds on the daytime energy budget of the surface during the day. Clouds reflect solar radiation back to space and reduce energy gains of the air below the cloud and the surface. Clouds emit longwave radiation and inhibit cooling of the air and surface below.

are part of the large-scale Rossby waves that we will discuss in Chapter 7.

Cloud Cover

Clouds have a large impact on the solar and terrestrial energy gains near the surface. Clouds reflect and absorb solar energy. They reduce the amount of solar radiation reaching the surface and cause daytime cooling. The thicker the cloud, the more pronounced the cooling is. Clouds also have a warming effect, however, because they emit longwave radiation toward the surface (**FIGURE 3-14**). This warming effect can be very pronounced at night, keeping minimum temperatures higher than clear-sky conditions.

Consider the annual temperature cycles of Grand Rapids, Michigan, and Madison, Wisconsin (**FIGURE 3-15**). Winds usually blow from west to east across this region in winter. Air blowing eastward over Madison arrives from the dry Great Plains. In contrast, air blowing eastward over Grand Rapids has passed across Lake Michigan and becomes moister as a result. This leads to chronic "lake-effect" clouds in Grand Rapids, but not Madison, from late fall into early spring. The strong warming effect of clouds at night causes the mean temperature in Grand Rapids to be as much as 5° F warmer than in Madison in January. However, in the summertime, the winds are more often from the south, and cloudiness at the two locations is similar (Figure 3-15). Consequently, Madison and Grand Rapids temperatures in July are nearly identical.

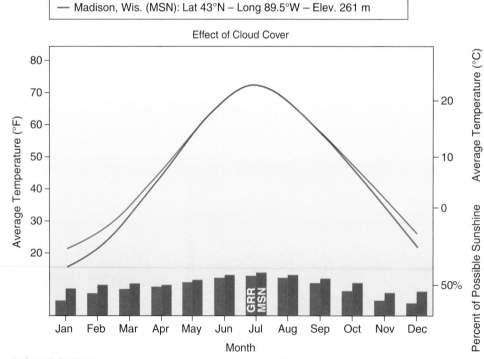

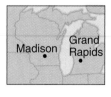

FIGURE 3-15 The annual temperature cycles of Grand Rapids, Michigan, and Madison, Wisconsin. Wintertime cloudiness in Grand Rapids (indicated by low percentage of possible sunshine in bar graphs at bottom) keeps it warmer there than in Madison.

Interannual Temperature Variations

The previous discussion on annual temperature variations was based on temperatures averaged over a few decades. These climatological temperatures are computed by averaging all temperatures over a 30-year period. These averages are called **normal temperatures.**

We know from personal experience, however, that the temperature can be abnormal. Some winters are colder than others, and some summers are hotter than others. Interannual temperature variations are temperature changes from one year to the next. Sometimes these changes are subtle and the causes are unknown, whereas at other times, the changes are dramatic and the causes are known.

Meteorologists study interannual temperature variations by plotting the departure from the climatological normal temperatures by year. Departures for a given year are found by subtracting the normal value from that year's mean value. These departures are called **anomalies.**

FIGURE 3-16 plots the surface temperature anomalies over the world's landmasses (where most observations are made) for the years 1880–2008. The 30-year "normals" from 1951–1980 are used as the reference period in Figure 3-16, which explains why the anomalies are negative before that period and positive after it. The red line in Figure 3-16 represents the anomalies of individual years. It wiggles up and down, indicating year-to-year changes. The blue line reveals longer-term trends by averaging these yearly anomalies over 5 years, which smoothes out the sharp bumps but leaves untouched any trends longer than a few years.

What does Figure 3-16 tell us? In general, temperatures across the globe increased from 1880 until about 1940, after which the temperature trend reveals a slight decrease over the next two decades. The global temperature has steadily increased since 1970. Over the last 120 years, the average global surface temperature has increased approximately 0.6° C (1° F).

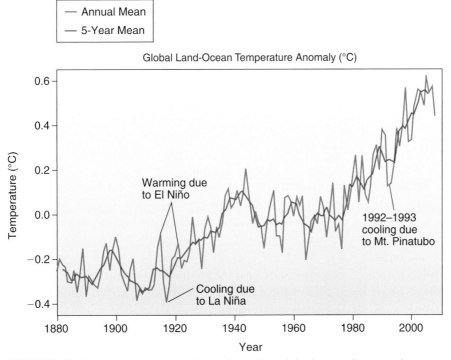

FIGURE 3-16 Temperature departures from the global mean temperature over land since 1880. Base period is from 1951 to 1980. The upward trend is evidence that the average global temperature is increasing. Note the shorter term effects of the ocean temperature phenomena El Niño and La Niña, which have periodically affected global temperatures throughout this period. The episodic cooling effect of the Mt. Pinatubo volcanic explosion is also indicated.

Global temperature changes of 0.6° C near the surface do not sound very significant. To help put this change in perspective, consider that temperatures during the latest ice age (20,000 years ago) were about 5° C colder than today's temperatures. So on a global scale, the change of a degree or so is big news—it is a sizable fraction of the difference between today's climate and a harsher climate for which we are not adapted.

Changes at particular geographic locations may be greater or smaller than the global mean, and year-to-year temperatures may vary considerably. For example, notice in Figure 3-16 that the temperatures of 1992 and 1993 are much cooler than the years before or after. These cool temperatures are the direct result of the eruption of a single volcano, Mt. Pinatubo, in the Philippines, on June 15, 1991. This eruption provided the first direct observational evidence that strong volcanic eruptions tend to cool the Earth (BOX 3-1). During a violent eruption, huge quantities of ash, dust, and sulfur dioxide are ejected into the stratosphere (FIGURE 3-17a). Volcanic ash falls back to Earth quickly, but the other debris stays in the stratosphere for a couple of years. There, its presence modifies the energy balance of the planet. The sulfur injected into the stratosphere by the volcanic eruption is chemically transformed into small droplets of sulfuric acid (Figure 3-17b). These tiny drops absorb solar radiation and, more importantly, increase the amount of solar energy reflected back to space. Less solar energy is transmitted to the surface, resulting in a cooling of the air near the surface.

The oceanic temperature phenomena known as El Niño and La Niña (to be covered in detail in Chapter 8) can also alter the world's temperatures, even over land. El Niño causes unusually warm waters over the eastern Pacific Ocean and, because the ocean affects the atmosphere, leads to warming temperatures across a sizable portion of the globe. La Niña cools the Pacific and other parts of the globe. During the period from 1916 to 1921, a La Niña pattern was sandwiched between two potent El Niño events, leading to the sharp zigzag in global temperatures during that period. The rapid climb in temperatures during the 1990s may also be related to the persistent El Niños during that decade.

In summary, the interannual temperature pattern of the Earth appears to be a persistent upward trend, punctuated by interludes of less warm periods. What can explain this upward trend? Much of this warming is attributed to an increased concentration of greenhouse gases, such as carbon dioxide, chlorofluorocarbons, nitrous oxide, and methane. In Chapter 2, we briefly discussed how increasing the concentrations of these gases could lead to a warming trend. Before revisiting this problem, we must discuss the relationship between air temperature and water vapor content—something we will do in the next chapter.

"The Mount St. Helens Eruption": observe a sequence of satellite images showing this eruption.

▮ Diurnal Temperature Cycle

Closer to home than the question of global warming is the daily cycle of local temperature. If you made temperature measurements every hour of a given day for many years, and averaged the measurements, a temperature cycle would emerge that shows a regular pattern of change according to the time of day. This is the diurnal temperature cycle that is driven by the daily changes in the energy budget near the ground. This variation is driven by the Earth's spin which causes day and night.

The daily variation of air temperature near the ground for a cloud-free day is demonstrated in FIGURE 3-18. We have learned that temperature changes are driven by the difference in incoming solar radiation gains versus outgoing terrestrial energy losses. Therefore, these quantities are also plotted in Figure 3-18.

Figure 3-18 summarizes the following story. After sunrise, the ground warms as a result of absorption of solar energy. As the ground warms, it transfers heat to the atmosphere. Cool air emits less energy than warm air, as we learned in Chapter 2. The rising Sun and warming surface adds more energy to the air than the air is emitting, and therefore, the air temperature increases all morning. Incoming solar energy peaks at noon, as does the Sun's trajectory in the sky. After noon, the solar energy gains are reduced, but the energy gains are still greater than the energy losses and so the surface temperature continues to increase. The energy losses usually

"Diurnal Temperature Cycle" activity explores the controls of the diurnal temperature cycle.

Box 3-1 Volcanoes and Temperature

In this chapter, we discuss the possible links between volcanoes and temperature. The most famous United States volcano of the 20th century, Mount St. Helens in Washington, provided proof of this link on a small scale when it exploded on the morning of May 18, 1980 (blob in center of satellite image below). The cloud of ash from Mount St. Helens spread rapidly eastward and enshrouded Spokane, Washington, for days. However, the ash cloud was narrow enough that Boise, Idaho, less than 450 kilometers (280 miles) south of Spokane, escaped the influence of the volcano.

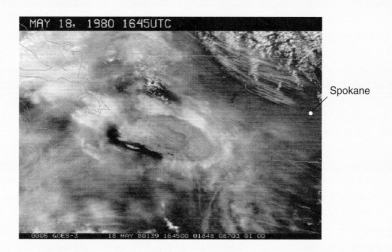

The effect of the Mount St. Helens eruption on temperature at Spokane versus Boise is shown in the figure. The volcanic ash flattened out the usual diurnal cycle of temperature at Spokane as compared with Boise. In fact, right after the arrival of the ash cloud at Spokane (arrow on the graph) the temperature actually decreased during the afternoon.

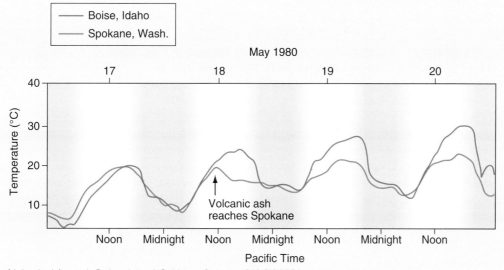

(Adapted from A. Robock and C. Mass, *Science* 216 [1982].)

(continued)

Box 3-1 Volcanoes and Temperature, continued

This strange turn of events can be explained in terms of an energy budget. The amount of solar energy reaching the surface at Spokane was greatly reduced on the day of the explosion and significantly reduced for several days afterward. This kept solar energy gains low for several days. Meanwhile, outgoing terrestrial radiation remained relatively high. The net result of gains minus losses was a small surplus on the days after the explosion and a deficit on the day of the explosion. As a result, the temperature at Spokane rose slowly, or not at all, even during the daytime.

The debris from Mount St. Helens remained mostly in the troposphere and had little global effect. However, much larger volcanoes can eject material into the stratosphere, where it can reflect solar radiation for years. As a result, the Earth's temperature can experience a pronounced, if temporary, cooling—just like Spokane, but on a global scale and for a few years, not just a few days.

Historical evidence supports the premise that large volcanic eruptions cool the Earth. Between 1812 and 1817 there were three major volcanic eruptions: Soufriere on St. Vincent Island in the eastern Caribbean Sea erupted in 1812; Mayon in the Philippines in 1812; and Tambora (the largest) in Sumbawa, Indonesia, in April 1815.

Debris from the eruption of Mt. Tambora (8° south latitude, 118° east longitude) took 1 year to spread globally. The following year, 1816, is known as the "year without a summer" in eastern North America. Although extensive meteorological observations did not exist at this time, people's diaries and weather journals documented the cold weather of the summer of 1816.

In New England snow fell in June and frost occurred in July and August. Although late frost killed a large number of crops, the entire summer was not below freezing. Indeed, on June 5, the day before the snowfall, the temperatures in Vermont were in the low 30s Celsius (upper 80s Fahrenheit). After the early June cold spell in New England, farmers, hoping for a good crop, replanted their crops as temperatures returned to normal. Another cold spell hit in early July, bringing freezing temperatures to the area. Harvests were poor that year and resulted in severe food shortages in parts of New England. The poor harvest had an economic impact throughout the United States. In Philadelphia in May 1817 a bushel of corn cost twice as much as it had in April 1816.

Weather in Europe and other regions of the globe also went haywire in 1816. In Europe, the cold and wet weather contributed to a disastrous harvest as crops rotted in the field. Famine, food riots, grain hoarding, and government embargoes followed. The cold, moist weather patterns may have contributed to the typhus epidemic of 1816 to 1819 in Europe that killed approximately 200,000 people. At about the same time, a cholera outbreak originated in Bengal, India, and spread throughout the world.

Bad "volcano weather" also had an unexpected and lasting impact on the world of literature and the movies. In the summer of 1816, a small group of friends gathered in Switzerland, where the cold wet summer prevented outdoor activities. One of the friends proposed a ghost story contest, a fitting match to the dreary weather. Storytelling came to mind naturally because two of the friends—Lord Byron and Percy Bysshe Shelley—were already among the greats of British Romantic poetry. However, their ghost stories are lost to history. Shelley's 19-year-old wife, Mary, bested them all with a tale that had come to her in a dream during that year without a summer. In 1818, Mary Shelley's ghost story was published, giving birth to a creature as chilling as the volcano that helped create it: Frankenstein's monster.

(a)

(b)

FIGURE 3-17 (a) A schematic illustrating how volcanic explosions affect solar radiation. (b) A photograph showing the volcanic aerosol layer as a result of Mt. Pinatubo. Astronauts on the space shuttle Atlantis took this photograph during mission STS-43, about 3 weeks after the eruption of Mt. Pinatubo. The volcanic aerosol is seen in the stratosphere (the thunderstorm clouds over South America are below the aerosol layer). The volcanic debris was a 20-megaton cloud of sulfur dioxide that reduced the amount of solar energy gains of the planet and resulted in a global cooling in 1992 and 1993 (Figure 3-16).

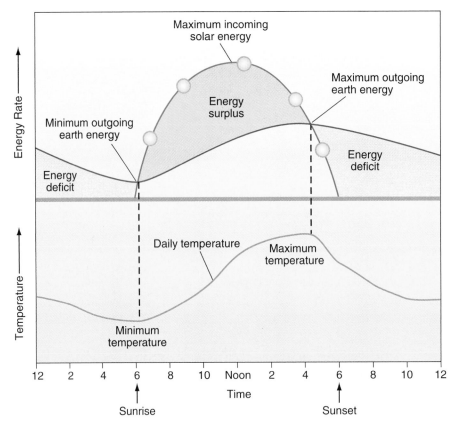

FIGURE 3-18 The diurnal variation in temperature is primarily controlled by energy gains from the sun and energy losses caused by emission of infrared radiation. Incoming solar energy is confined to the daytime. The amount of outgoing infrared radiation is related directly to the temperature (see Chapter 2) and is indicated by the green dashed lines. Temperature increases when the energy gains exceed the energy losses (orange shading), and the temperature decreases when the net energy is negative (purple shading).

exceed the energy gains by 4:00 PM, when the maximum daily temperature is reached. This is when the maximum amount of outgoing energy is emitted. The energy losses exceed gains all night because the source of gains is the Sun and it is below the horizon. Therefore, the temperature decreases all night. Temperatures reach a minimum around sunrise, and then the cycle starts over.

The factors affecting the diurnal temperature changes during a cycle are similar to those that determine the annual temperature cycle: latitude, surface type, elevation and aspect, relationship to large bodies of water, and cloud cover. To review their effect and to see how they affect daily as opposed to monthly temperatures, let's go over them one by one:

1. Latitude: How much solar energy a region gains plays an important role in establishing the daily variation in the region's energy budget. Latitude, in turn, determines the intensity of the Sun's rays and number of daylight hours. In equatorial regions, the position of the Sun in the sky changes dramatically throughout the course of a day, from below the horizon to nearly overhead. Solar energy gains, therefore, vary greatly, resulting in large variations in air temperature during the day. Equatorial regions, in general, have a greater daytime variation of temperature than polar regions because of the larger variations in solar zenith angle in equatorial regions.

2. Surface type: How much the ground, and therefore the air near the ground, warms will depend on the condition and composition of the surface. Solar energy falling on bare, dry,

sandy soil is absorbed within a thin layer near the surface. Because this type of soil has a low specific heat and because the solar energy is all absorbed in a thin layer, the surface quickly warms. The heated surface then transfers its energy to the air via convection and conduction, warming the air. During the night, the soil surface cools quickly as it loses energy in the form of terrestrial radiation. For all these reasons, desert regions experience a large diurnal temperature range. When vegetation is present, some of the solar radiation is converted to chemical energy by photosynthesis. Some heats the plant tissue and never reaches the surface, and some is used to evaporate water from the plant (transpiration). The result is that the maximum temperature is less over the vegetative field than over bare soil.

3. Surface elevation and aspect: Air temperatures are normally warmer at lower elevations. At lower elevations, greater amounts of water vapor absorb the terrestrial radiation emitted by the ground, thus warming the air. By comparison, high-elevation regions have less water vapor and less atmosphere. As a result, these regions tend to have larger diurnal temperature ranges than regions at sea level. This is because the same amount of solar energy warms fewer molecules at high altitudes during the day, making the maximum temperature very high. At night, there is only a small "greenhouse effect" as a result of water vapor at high altitudes, so the Earth cools rapidly, leading to a low minimum temperature. The aspect of a surface affects how long a surface is exposed to direct sunlight and the angle at which the Sun's rays strike the surface. Regions that directly face the Sun tend to have larger daily temperature ranges than surfaces sloped away from the Sun.

4. Relationship to large bodies of water: The diurnal temperature range is usually greater over regions far from large bodies of water than locations surrounded by water. The presence of large bodies of water reduces daytime maximum temperatures and increases nighttime minimum temperatures. The large amounts of energy required to heat and evaporate water are a key reason for this moderating effect; much of the Sun's energy goes into the water, not into heating the air. This explains why islands usually have a small diurnal temperature range.

5. Cloud cover: Clouds reduce the diurnal temperature range by minimizing the range between the daytime maximum and nighttime minimum temperatures. When clouds are present, the maximum daytime temperature is reduced because less solar energy reaches the ground. During the night, clouds increase the minimum temperature by increasing the longwave radiation absorbed by the air and the ground. The amount of reduction in the diurnal temperature range depends on the type of cloud. Thick clouds that are low in the atmosphere have the largest effect. Low clouds are warmer and therefore emit more terrestrial energy than high clouds. Thick clouds also let less solar energy pass through them. High, thin clouds have the smallest impact on an otherwise clear-sky diurnal temperature range.

Now that we understand the controls of daily temperature, we can appreciate the rare combination of circumstances that lead to record high and low temperatures, which are explored in BOX 3-2.

TEMPERATURE VARIATION WITH HEIGHT: LAPSE RATES AND STATIC STABILITY

So far we have limited ourselves to discussing temperature changes just 1.5 meters above the ground. However, temperature also varies with height above the ground. The change of temperature with height is called the **lapse rate**. These vertical variations of temperature depend on several factors. First, we need to understand how the temperature of air changes as a result of changes in pressure.

Box 3-2 Record Cold and Record Heat Across the United States

Records fascinate meteorologists as much as they do sports fans, and when temperatures approach all-time records, even the least weather-conscious person becomes a "weather junkie." Let's examine the patterns of record high and low temperatures across the 50 United States to see what we can learn about temperature from its extremes.

The figures that follow show the location and thermometer reading for the state-record minimum and maximum temperatures. First, record lows tend to occur in mountainous regions. Record highs are more likely in low-lying areas. Thirty of the record lows were observed at or above 1000 feet of elevation, whereas 37 record highs occurred at or below the 1000-foot mark. This is a natural consequence of the adiabatic cooling of air that is above sea level. Mountains thus have a built-in advantage when it comes to cold temperatures. In addition, the relative lack of "greenhouse warming" in the thin, dry atmosphere at high altitudes promotes the rapid loss of energy that must characterize a record-cold night.

Record Lowest Temperature (°F) (through 2009)

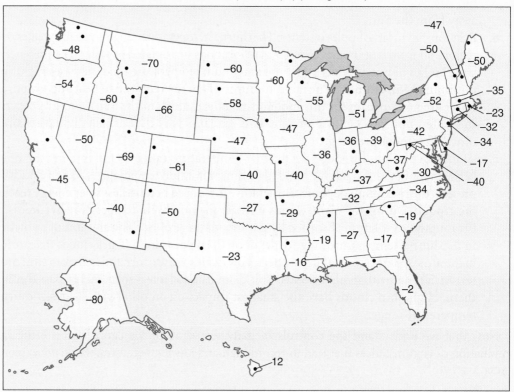

Other interesting differences arise in the numerical and geographical spread of the state records. Every state has a record high of 100° F or higher, but the highest all-time U.S. temperature is 134° F, yielding a range of just 34° F. There is also no obvious dependence of the record highs on latitude. In contrast, the difference between the "warmest" and coldest state-record low temperatures is a whopping 92° F. Latitude strongly affects record lows, with the values steadily dropping from south to north (e.g., from Texas to North Dakota).

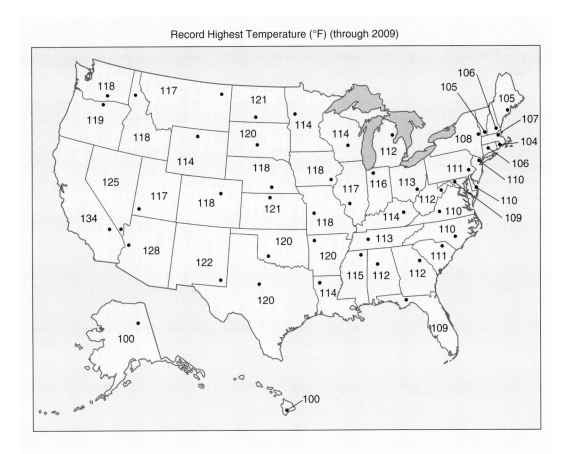

Record Highest Temperature (°F) (through 2009)

Still other oddities arise from a closer look at the records. Utah has the largest spread between its record high and its record low, an amazing 186° F. Remarkably, both of Utah's state temperature records were set in the same year, 1985. East of the Mississippi River, the state with the largest range in record temperatures is, surprisingly, Wisconsin at 169° F. The effect of "continentality"—hot summers and bitter winters far away from the moderating effects of an ocean—gives Wisconsin a more extreme climate than states with high mountains, such as New Hampshire or North Carolina.

When do records occur in a given year? Almost half of all state-record lows across the United States cluster in two periods: January 17 to 23 and February 8 to 14. Both of these periods "lag" the winter solstice by several weeks. This is because the Earth–atmosphere system does not respond instantaneously to the minimum of solar heating. Instead, it takes weeks of longwave cooling over heavy snow cover to create the kind of bitter-cold air that is required for a record low.

What this explanation does not account for is the strange lull in records in late January. This gap may be a manifestation of the "January thaw," a period of warmer than usual weather that is observed in some winters across the eastern United States at around the same time. However, the small number of state-record lows is not enough information on which to base a conclusion.

Record highs are much more spread out across the summer months, but the period of July 10 to 15 holds 15 state records. As with record lows, a lag of a few weeks after the solstice is required for the Earth-atmosphere system to "catch up" with the cycle in solar heating.

Hawaii is an anomaly. Its record high is in April, but its record low was recorded later in the year, in May. We can explain this by recalling what we have learned about solar zenith angles. Hawaii is located just south of the Tropic of Cancer. As a result, it receives the most direct sunlight in spring and in fall, not in summer. A spring record high is therefore logical. Even in winter the Hawaiian sun is high and the Pacific Ocean keeps temperatures moderate, making a May record low possible.

(continued)

Box 3-2 Record Cold and Record Heat Across the United States, continued

A fear concerning global warming is that it will lead to extreme record high temperatures unlike anything that our planet has experienced. Does this trend show up in state-record temperatures? The figure shows all state-record temperatures by decade, and it tells a surprising story: The decade of extreme temperatures was not the 1990s, it was the 1930s! Of all U.S. state-record highs, 70% were set in the 1930s or earlier, despite the overall trend since then toward warmer mean temperatures worldwide (Figure 3-16). The exceptional weather of the 1930s led to the catastrophic "Dust Bowl" of the Texas and Oklahoma Panhandles, which we discuss in later chapters.

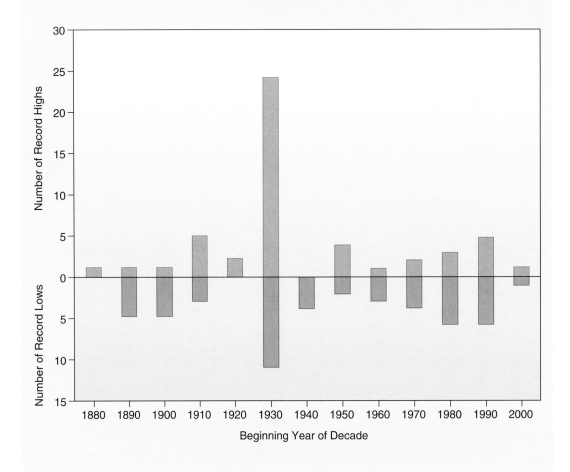

This graphic also reveals that a decade with many record high temperatures usually has a lot of record low temperatures, too. For example, there were more state-record lows than highs set during the 1990s.

■ Adiabatic Cooling and Warming

Air that moves up and down in the atmosphere undergoes temperature changes. To illustrate this concept, we follow a parcel of air as it ascends in the atmosphere.

We start with our parcel of air near the ground. As we learned in Chapter 1, pressure always decreases with altitude. Therefore, as the parcel rises (FIGURE 3-19), the pressure exerted on the outside of the parcel decreases. The molecules inside the rising parcel exert a pressure that is greater than the pressure exerted by the molecules outside the parcel. The internal pressure

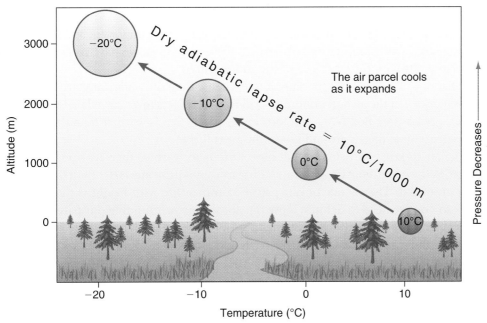

FIGURE 3-19 Adiabatic cooling and warming occurs as a parcel of air moves up or down in the atmosphere. As an air parcel rises, it expands because the surrounding air pressure decreases with altitude. This expansion leads to cooling of the parcel, at a rate of 10° C per kilometer for dry air.

increases the parcel's volume until the internal pressure equals the external pressure of the environment. In other words, *a rising parcel always expands because atmospheric pressure always decreases with altitude.*

Increasing the volume of the parcel requires work, which means energy must be involved. The air molecules are expending energy—in this case, kinetic energy—to do the work required for expansion. Where does this kinetic energy go? Remember that energy cannot be destroyed, but it can be converted to another form. As the parcel rises away from the ground, its potential energy increases. So the air molecules' kinetic energy is being converted to potential energy.

The change in kinetic energy of the parcel affects the parcel's temperature. Recall that a parcel's temperature is the average kinetic energy of its molecules. Because the average kinetic energy is decreasing, the temperature of the parcel must also be decreasing. This leads to a hard-and-fast rule of meteorology: *A rising parcel of air always cools.* How fast the parcel cools depends on gravity, the amount of moisture in the parcel (as we'll discuss in Chapter 4), and the specific heat of the air in the parcel. A dry parcel's temperature will decrease by approximately 10° C for each kilometer it rises.

Now let's consider what would happen if the parcel returned to its original position. As it descended, it would be compressed because atmospheric pressure increases closer to the Earth's surface. This compression increases the average kinetic energy of the molecules as they collide within the collapsing imaginary boundary of the parcel. As the parcel descended, the potential energy of the molecules would be converted back to kinetic energy. When the parcel returned to its original level, its temperature would have returned to its initial temperature. *A descending parcel of air always warms.*

In following our air parcel, we've been tracking an **adiabatic process**. An adiabatic process is a process in which no heat energy is gained or lost by the system in question (in our case, a parcel of air). The 10° C per kilometer rate of cooling resulting from expansion (which is also the rate of warming as a result of compression) is referred to as the **dry adiabatic lapse rate**. We call the parcel "dry" because no clouds formed as it rose; in Chapter 4, we factor in the effect of moisture on the lapse rate.

The Environmental Lapse Rate and Static Stability

Now that we know how fast air cools if it rises, we can ask two very important and basic questions: Will air rise, and will it keep rising? To answer these questions, we have to compare the dry adiabatic lapse rate with the lapse rate of the air around it.

A rubber duck floats in water because it is less dense than the water. Similarly, a parcel of air that is less dense than the air surrounding it will rise. Density is inversely related to temperature by the ideal gas law (see Box 1-2); therefore, an air parcel that is warmer than the air around it is also less dense than the air around it—so the air will rise.

For the air to keep on rising, it will have to remain warmer than its surroundings. This means that, as the parcel goes up, its temperature must not drop any faster than the temperature of the air around it. The change of our parcel's temperature with height is the dry adiabatic lapse rate, so now we need to define a lapse rate for the air surrounding our parcel.

In Chapter 1 we defined layers of the atmosphere by their temperature variations. The troposphere is the atmospheric layer where, on average, the temperature decreases with increasing altitude at a rate of 6.5° C per kilometer (3.6° F per 1000 feet). This is the average lapse rate; on any given day, the lapse rate will vary. The specific change of temperature with altitude at any particular time and location is referred to as the **environmental lapse rate**. The environmental lapse rate changes from day to day and hour to hour. It can be measured by attaching a thermometer to a helium-filled balloon. The temperature measurements are radioed to the surface and recorded. Vertical temperature measurements of the environmental lapse rate occur twice a day at many locations throughout the world.

Will a parcel of air keep rising after it is given a push upward? To find out, all you have to do is compare the dry adiabatic lapse rate with the environmental lapse rate. If the environmental lapse rate is larger than the dry adiabatic lapse rate, the parcel will remain warmer than its surroundings and keep rising. If the environmental lapse rate is smaller than the dry adiabatic lapse rate, the parcel will eventually reach the same temperature as its surroundings as it rises, and it will then stop rising.

Air that keeps rising because it is warmer than its surroundings is called statically unstable. When the environmental lapse rate is greater than 10° C per kilometer, the atmosphere is said to be **absolutely unstable** (FIGURE 3-20). An absolutely unstable atmosphere is very favorable for strong upward motions of air because if a parcel of air is lifted upward, it will accelerate away from its original position.

In contrast, a **statically stable atmosphere** inhibits the vertical movements of air parcels. Stable atmospheres occur in completely dry air when the environmental lapse rate is less than the dry adiabatic lapse rate of 10° C per kilometer. Air parcels that rise in a stable atmosphere rapidly become cooler than their surroundings and sink back to where they came from. Thus, as an example, if the environmental lapse rate is measured to 11° C per kilometer, the atmosphere is absolutely unstable.

Does this seem confusing? If so, don't feel bad. Stability is one of the hardest subjects in all of meteorology for students to understand and for instructors to teach. It's also one of the most important subjects, however, because rising air leads to much of what we call "bad weather." To cement your understanding of stability, we strongly recommend that you visit this textbook's Web site and play with the Introduction to Stability applet to help visualize how lapse rates relate to the vertical changes of temperature and the movement of air parcels.

Because stability is a challenging subject, we've limited the discussion here to dry, cloud-free air, but that's a simplification of the real atmosphere. When we include the effects of moisture in Chapter 4, we have to add in the effect of water's phase changes on the parcel's lapse rate.

"Introduction to Stability" to enhance your understanding of how and when air can rise.

Temperature Inversions

In some regions of the atmosphere, the environmental lapse rate is a negative value. This is the reverse, or the inverse, of what we citizens of the troposphere normally expect. Accordingly,

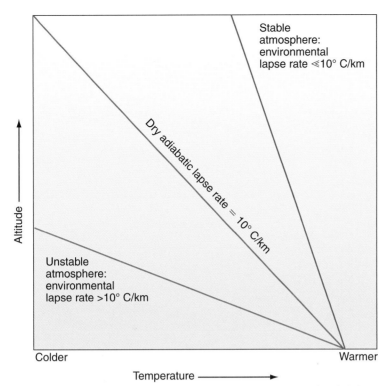

FIGURE 3-20 The vertical temperature profile in an absolutely unstable atmosphere tilts more to the left than the dry adiabatic lapse rate profile (green line). This means that in an absolutely unstable atmosphere the temperature decreases with altitude very rapidly, at a higher rate than 10° C per kilometer. In a stable atmosphere, the temperature profile tilts less to the left than does the dry adiabatic lapse rate. This means that temperature decreases very slowly with altitude.

regions of the atmosphere in which the temperature increases with altitude are called **temperature inversions.**

A temperature inversion is an extreme case of a stable atmosphere. When a temperature inversion is present, parcels that are displaced from their position will soon become much cooler than their surroundings and will quickly sink back to their original positions. One very large example is the stratosphere. Closer to home, temperature inversions also can develop near the ground, most often on calm, clear nights and early mornings. Now, let's examine how a temperature inversion can develop near the surface.

The range of diurnal and annual cycles in temperature depends on altitude, the distance from the ground. **FIGURE 3-21** shows the temperature measured at various times on a cloud-free day in summer. We start at 3:00 PM (Figure 3-21a). At this time, the temperature is highest close to the ground, where solar heating has raised the temperature to near its daily maximum. By 8:00 PM (Figure 3-21b) the Sun is setting, and the temperature below an altitude of about 200 meters (0.2 kilometers) has cooled because energy losses have exceeded gains for several hours. The temperature increases up from the surface to 200 meters, and therefore, a temperature inversion is present.

By 5:00 AM (Figure 3-21c) nighttime cooling of the surface and a lack of sunlight have led to many hours of energy deficits. As a result, the near-surface temperature has cooled significantly. However, the air higher up has cooled less. Therefore, the temperature at 5:00 AM increases sharply in the lowest 600 meters of the atmosphere. The temperature inversion is near its peak in terms of depth and temperature difference between the bottom and the top of the inversion.

At sunrise, solar energy heats the ground, and conduction and convection transfer heat upward. This warms the air near the surface more effectively than higher up. By 10:00 AM

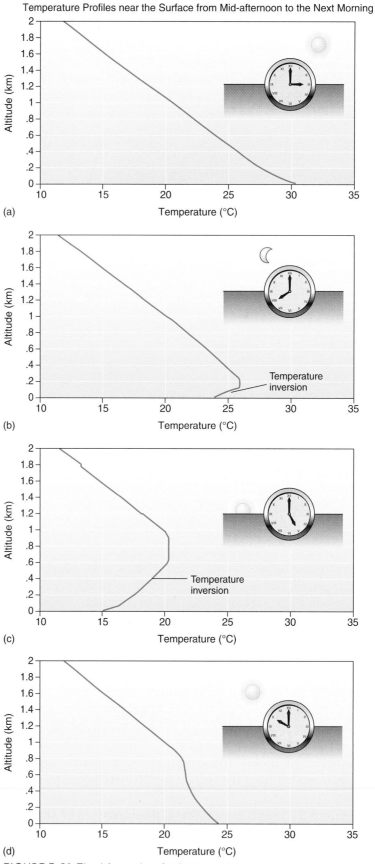

FIGURE 3-21 The life cycle of a temperature inversion. (a) Mid afternoon. (b) Evening. (c) Sunrise. (d) Mid morning. Overnight cooling at the surface as a result of longwave radiation emissions causes a surface temperature inversion to be present in b and c. In an inversion, temperature increases with altitude.

(Figure 3-21d) the temperature near the surface is warmer than above it, and the inversion has dissipated.

A temperature inversion that develops near the ground during the night, as shown in Figure 3-21, is referred to as a **nocturnal inversion** (sometimes also called a **radiation inversion** because of the key role that terrestrial radiation emissions play in its formation). Nocturnal inversions often occur on clear, calm nights and are more prevalent during winter than summer. The primary factors controlling the development of a nocturnal inversion are clouds, wind, length of the night, and the condition of the ground, for the following reasons:

- Clouds inhibit the formation of a nocturnal inversion by emitting terrestrial energy toward the surface, reducing the surface energy losses. The cooling of the ground is suppressed, and development of the surface nocturnal inversion is inhibited.
- Winds are the mixers of the atmosphere. If a temperature inversion exists and the winds suddenly increase, the inversion is destroyed. This is because warm air aloft is mixed with the cooler air below.
- Winter nights are longer than summer nights. The longer the Sun is down, the more time the surface has to cool down and form the nocturnal inversion. Nocturnal inversions are prevalent over the polar caps in winter.
- The condition of the ground is another variable to consider when forecasting nocturnal inversions. For example, snow is a very good emitter of terrestrial energy, but because of the air trapped within it, it is a poor conductor. The top of the snow surface cools rapidly but does not warm from below. As a result, the air just above a deep snow cover cools rapidly, a condition favorable for a radiation inversion and for record cold temperatures.

Why are temperature inversions important? As noted earlier, inversions suppress the upward movement of air. The stratosphere keeps weather below it, in the troposphere. A nocturnal inversion keeps parcels of air from rising very high before they sink back down. Severe-weather experts sometimes refer to a temperature inversion as a "lid" or "cap" because it is difficult for air to move vertically through it. Episodes of dense air pollution are often associated with temperature inversions. The pollutants are released near the surface but cannot escape because vertical mixing is reduced by the existence of the inversion (**FIGURE 3-22**). We explore the implications of inversions for pollution in Chapter 10 and for severe weather in Chapter 11.

FIGURE 3-22 Photograph of pollution over Mexico City. This pollution is often trapped by a temperature inversion over the city, which is located at a high altitude (2239 meters, or 7347 feet) valley ringed by towering mountains as high as 5452 meters (17,887 feet).

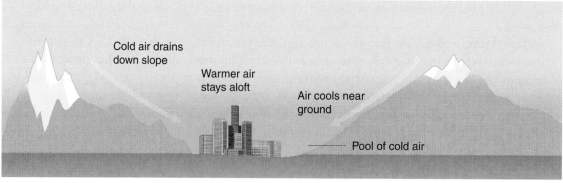

FIGURE 3-23 Cold air drains down mountain slopes into the valley, where pollution can get trapped near the ground as a result of a temperature inversion near the surface.

Temperature inversions also often develop in valleys. Consider the conditions of a valley on a clear calm night in October. During the evening, the air begins to cool by radiation losses. Air that is colder than its environment sinks. The coldest air drains down the hill, settling at the bottom of the valley (**FIGURE 3-23**). By late evening, the air at the bottom of the valley is often much colder than the air above it.

The record low temperature for the state of Utah (Box 3-2) was set in 1985 in a bowl-shaped 2.4-kilometer (8000-foot) high valley descriptively named Peter Sink. Bitterly cold air over snow cover at high altitude sank into the valley and caused the temperature to plummet to an astounding −56° C (−69° F), only 1° shy of the all-time record low for the lower 48 United States.

WIND-CHILL TEMPERATURE

"Wind-Chill Temperature" to calculate the wind-chill temperature for a wide range of temperatures and wind speeds.

On a cold, windy day, you try to keep yourself warm by wearing appropriate clothing or seeking shelter from the wind. It feels colder in the wind because the wind sweeps away heated air in contact with your body and replaces it with colder air. Whereas still air is a poor conductor, moving air is not! The cooling power of the wind is measured by the wind-chill factor. The **wind-chill temperature** describes the increased loss of heat by the movement of the air. The wind chill is relevant to humans and other animals that need to maintain a constant temperature that is higher than their surroundings.

The wind-chill factor cannot be measured with a thermometer; it must be computed. The wind-chill temperature, expressed in degrees, translates your body's heat losses under the current temperature and wind conditions into the air temperature with a 3-knot wind that would produce equivalent heat losses. This is not an easy conversion. The National Weather Service computation of the wind-chill temperature is given in **TABLE 3-1**. From this table we see that a temperature of 5° F with a 40-mph wind has an equivalent heat loss to a temperature of 0° F and a 20-mph wind. The wind-chill temperature for both of these situations is −22° F. Cold temperatures plus wind cause danger to exposed flesh. **BOX 3-3** discusses some of the hazards of extreme temperatures and actions you can take to avoid harm.

TEMPERATURE AND AGRICULTURE

"Frost Prediction" provides some rules-of-thumb for predicting frost conditions.

For both large-scale farming and home gardeners, a cold-air outbreak at the wrong time can be costly. A freeze in Florida or California can reduce harvests and increase the price of citrus or vegetable products. Cold-air outbreaks accompanied by nocturnal inversions are the farmers' worst enemies.

Because of nocturnal inversions in winter, the air temperature measured at the standard height of 1.5 meters may be above freezing (0° C, 32° F), whereas temperatures at the ground near crop plants may fall below freezing, injuring them. To protect the plants, their energy budget must be modified to keep their temperature from falling below freezing.

TABLE 3-1 The Wind-Chill Temperature Chart

								Temperature (°F)										
Calm	**40**	**35**	**30**	**25**	**20**	**15**	**10**	**5**	**0**	**-5**	**-10**	**-15**	**-20**	**-25**	**-30**	**-35**	**-40**	**-45**
5	36	31	25	19	13	7	1	-5	-11	-16	-22	-28	-34	-40	-46	-52	-57	-63
10	34	27	21	15	9	3	-4	-10	-16	-22	-28	-35	-41	-47	-53	-59	-66	-72
15	32	25	19	13	6	0	-7	-13	-19	-26	-32	-39	-45	-51	-58	-64	-71	-77
20	30	24	17	11	4	-2	-9	-15	-22	-29	-35	-42	-48	-55	-61	-68	-74	-81
25	29	23	16	9	3	-4	-11	-17	-24	-31	-37	-44	-51	-58	-64	-71	-78	-84
30	28	22	15	8	1	-5	-12	-19	-26	-33	-39	-46	-53	-60	-67	-73	-80	-87
35	28	21	14	7	0	-7	-14	-21	-27	-34	-41	-48	-55	-62	-69	-76	-82	-89
40	27	20	13	6	-1	-8	-15	-22	-29	-36	-43	-50	-57	-64	-71	-78	-84	-91
45	26	19	12	5	-2	-9	-16	-23	-30	-37	-44	-51	-58	-65	-72	-79	-86	-93
50	26	19	12	4	-3	-10	-17	-24	-31	-38	-45	-52	-60	-67	-74	-81	-88	-95
55	25	18	11	4	-3	-11	-18	-25	-32	-39	-46	-54	-61	-68	-75	-82	-89	-97
60	25	17	10	3	-4	-11	-19	-26	-33	-40	-48	-55	-62	-69	-76	-84	-91	-98

Wind (mph)

Frostbite Times ▢ 30 minutes ▢ 10 minutes ▢ 5 minutes

$$\text{Wind Chill (°F)} = 35.74 + 0.6215T - 35.75(V^{0.16}) + 0.4275T(V^{0.16})$$

Where, T= Air Temperature (°F) V= Wind Speed (mph) *Effective 11/01/01*

Courtesy of NWS/NOAA.

Box 3-3 Temperature and Your Health

Extreme temperatures threaten your well-being in a surprising number of ways. Knowledge of the risks can help keep you from becoming a victim of extreme weather.

Frostbite occurs when your skin and other tissues cool to the point that ice crystals form in your bodily fluids. Cold temperatures combined with strong winds can cause frostbite in a matter of minutes (Table 3-1). Frostbite can also occur as a result of long exposure to less severe, but subfreezing, conditions. Frostbite usually strikes your body's extremities first: fingers, toes, and ears. Your face is also at risk because it is often exposed to the cold. Numbness in the skin is followed by the death of cells as the water in your body freezes. The simple way to prevent frostbite is to avoid direct exposure of your skin to cold and wind in winter.

Hypothermia occurs when your body temperature drops dangerously low. It occurs when the energy budget of your body runs at a deficit, which leads to a temperature decrease. Hypothermia is possible when your body is immersed in cold for long periods—for example, by floating in a cold lake or ocean. Body temperature can drop to a point where metabolic activities and organs, such as the brain, cannot operate normally. The first signs of hypothermia are confusion and a loss of judgment, followed by stupor and possibly death, as depicted graphically in the movie *Titanic*. Again, insulation from the cold is essential for survival.

Warm temperatures can also threaten your health. During summer, there are many regions in the United States and the world where the temperature reaches 38° C (100° F) or higher. The environmental temperature in this case is greater, not less, than your body temperature. In this case, your body's energy gains exceed energy losses, and your temperature can rise to dangerous levels. We focus on what to do in these situations when we discuss heat waves in Chapter 9.

One method of frost protection is putting certain plastic coverings over the plants. The plastics used do not transmit longwave radiation well, so they trap heat. This approach is similar to adding cloud cover, in the sense that certain plastics decrease the overall longwave radiative energy losses. This slows the plants' temperature decrease.

Another method of frost protection is to supply heat to the area using orchard heaters, large heaters that heat the air directly. In addition to supplying heat to the air around the plants, the rising hot air mixes the air throughout the inversion. This mixing inhibits the development of a near-surface temperature inversion in the same way that a strong wind does.

Another frost protection strategy is direct mechanical mixing of the air. Large fans are used to stir the air surrounding the crops, and the turbulence generated by the fans mixes warm air from above down into cold surface air. Some desperate farmers even resort to using a helicopter, flying at very low altitudes, to mix warmer air from the nocturnal inversion down to the surface.

A popular strategy to protect plants from freezing is, ironically, to cover them with ice. Many plants are not damaged if the temperature is at 0° C (32° F). Moderate to severe damage by frost generally occurs at approximately −2° C (28° F). If the temperature of the plants is at freezing, spraying them with water will cause the water to freeze. As we learned in Chapter 2, a phase change of 1 gram of water from liquid to ice releases 80 calories of heat to its environment. This small amount of heat can be just enough to keep plants from falling to −2° C, saving them from frost damage.

The never-ending battle between agriculture and weather is a testament to the importance of temperature to society. The strategies for frost protection, all of which involve principles we have covered in the first three chapters, illustrate how an understanding of meteorology can have very practical and useful applications.

Growing Degree Days

The **growing degree-days** (**GDD**s) is a heat index that is related to plant development. It can be used to predict when a crop will reach maturity. Each day's GDDs are calculated by subtracting a reference temperature, which varies with plant species, from the daily mean temperature, setting values less than zero to zero. The reference temperature for a given plant is the temperature below which development for that plant either slows or stops. For example, cool season plants, like peas, have a reference temperature of 40° F while warm season plants, like sweet corn and soybeans, have a reference temperature of 50° F.

The sum of all the GDDs over the growing season is related to plant development. The development of plants depends on the accumulation of heat. Plants that like cooler weather naturally experience more GDDs than plants that require warmer temperatures. When plants are not overly stressed by drought or pests, the summation of the GDDs can be used to measure the accumulation of heat and thus predict when a crop will reach maturity. Corn requires 1360 GDDs to reach maturity. If a corn region has a cooler than normal summer, it falls behind in GDDs, and it takes longer for the crop to mature.

GDDs can be computed using climatic temperatures of an area. With that computation, we can estimate good crops to grow in a given region, similar to plant hardy zones you may have seen on seed packages.

HEATING AND COOLING DEGREE DAYS

Other types of degree-days include the heating degree-day and cooling degree-day, used by engineers to estimate fuel requirements for heating and cooling buildings.

Engineers determined that when the mean outdoor temperature drops below 65° F, most buildings require heating to maintain an indoor temperature of 70° F. So, in an effort to make weather data easy to use in planning, they developed the **heating degree-days** (**HDD**) to estimate fuel-consumption needs. One heating degree-day is defined as each degree that the mean temperature is below 65° F. So, an average daily temperature of 55° F equates to 10 heating degree-days. Fuel distributors often use this index to schedule home deliveries. The amount of heat required to maintain a building's temperature is proportional to the accumulated heating degree-days. Electrical and natural gas utilities predict power demands adding up this index over time. So, heating degree-day (HDD) totals are usually reported each day, as well as the total for the season. This allows us to judge quickly whether the season is above, below, or near normal.

Another quantitative index designed to reflect energy demands is the **cooling degree-days** (**CDD**s). The amount of energy used to cool a building is related to the CDDs. Most U.S. buildings require air conditioning to maintain a 70° F temperature inside when average outdoor temperature is 65° F or more. If the average daily temperature is above 65° F, the CDDs are then computed by subtracting 65° F from the average daily temperature.

Of course, the amount of energy required to maintain a given indoor temperature depends on more than the outdoor temperature. It is also a function of how well the building is insulated, the amount of solar exposure, the wind conditions, and the number of electrical appliances running. Still, when engineers are familiar with a particular building's construction, the heating degree-days and CDDs can provide a baseline number to predict energy usage.

PUTTING IT ALL TOGETHER

Summary

Temperature near the surface is governed by the energy gains and losses at ground level. The ground gains energy during the day from the Sun and emits longwave radiation both day and night. An imbalance between gains and losses causes a change in temperature. The surface exchanges energy with the air in a variety of ways. Conduction between the ground and the atmosphere is slow and only affects the air temperature close to the ground. Convection, turbulent mixing by the winds, and radiative and latent heat fluxes exchange most of the energy between the ground and the rest of the air.

If energy gains and losses ebb and flow in a regular pattern over time, then the surface temperature will change in a cyclical pattern. The surface temperature has cycles on daily, annual, and interannual time scales. The annual and diurnal temperature cycles are primarily driven by the periodic nature of solar energy gains at the ground. The diurnal and annual temperature cycles are related to the following:

1. Solar energy input, and therefore latitude, the time of day, and the time of year
2. The surface type, including proximity to bodies of water, which determines albedo and specific heat and influences evaporation
3. Advection, a steady wind direction, which can advect warm or cold air into a region
4. Cloud cover, which suppresses temperature changes
5. Altitude and aspect, which cause generally cooler temperatures at higher altitudes and on north-facing slopes in the Northern Hemisphere

Temperature changes with altitude above the Earth's surface. The dry adiabatic lapse rate of 10° C per kilometer describes how the temperature of an air parcel changes as it goes up or down. The environmental lapse rate describes how the temperature varies with altitude at any given time and location. Comparing these two lapse rates allows us to know if an air parcel will keep rising. If the environmental lapse rate is bigger than the dry adiabatic lapse rate, the parcel will keep rising. This situation is called "absolutely unstable."

If the environmental lapse rate is less than the dry adiabatic lapse rate of 10° C per kilometer, then the atmosphere is stable. An extreme case of this is a temperature inversion, when warmer air overlies colder air. Vertical motions of air parcels are inhibited by these "stable" temperature inversions. This is an example of an absolutely stable atmosphere.

Conditions favorable for the formation of a near-surface temperature inversion are clear skies, long nights, calm or light winds, and little vegetation. These conditions promote radiative energy losses that are greatest near the surface. Drainage of cold air into valleys can also form inversions near the ground. Temperature inversions play a role in air pollution episodes and agriculture.

During winter in the northern United States and Canada, nightly weather reports often include the wind-chill equivalent temperature. The wind-chill temperature indicates how cold it feels outside as a result of the combination of surface temperature and wind.

Temperature indices are used to help predict agricultural and energy consumption needs. Growing degree-days are used to predict when a given crop will reach maturity. Heating and cooling degree-days are indicators of a building energy demand for heating and cooling, respectively.

■ Key Terms

You should understand all of the following terms. Use the glossary and this chapter to improve your understanding of these terms.

Absolutely unstable	Dry adiabatic lapse rate	Radiation inversion
Adiabatic process	Environmental lapse rate	Seasonal temperature cycle
Annual average temperature	Growing degree-day (GDD)	Statically stable atmosphere
Annual temperature cycle	Heating degree-day (HDD)	Surface temperature
Annual temperature range	Insolation	Temperature inversion
Anomalies	Isotherm	Temperature gradient
Aspect	Lag	Temperature range
Cooling degree-day (CDD)	Lapse rate	Turbulence
Daily mean temperature	Monthly mean temperature	Wind-chill temperature
Diurnal temperature cycle	Nocturnal inversion	
Diurnal temperature range	Normal temperatures	

■ Review Questions

1. Why is the energy budget at the ground a critical factor in determining the air temperature above the ground?

2. The temperature on a cloudy, snowy winter day is $0°$ C from midnight to 11:00 PM. Then, during the last hour of the day, the clouds break up, and the temperature drops to a reading of $-6°$ C. What is the daily mean temperature as calculated by the average of the daily maximum and minimum temperature? What is the daily mean temperature if you average the sum of each hourly temperature?

3. Why does the maximum daily temperature lag several hours behind the peak in incoming solar energy from the Sun?

4. Take three pans of water, and fill one with cold water, one with hot water and the final one with warm water. Place one hand in the cold water while the second hand is in the hot water. Hold your hands in the water for 2 minutes. Remove your hands and place them both in the third pan of water. Does the water feel warm or cold? Explain your observations.

5. *Leucochroa candidissima* is a species of desert snail. During the day, they climb up the branches of plants. Why do you suppose the snails do this?

6. When I was a child, I was envious of my friend during the winter. We lived in houses that were very similar, although he lived across the street from me on the north side. After a winter storm passed and the sky cleared, he often had much less work to do in shoveling his sidewalk than I! Why do you think this was the case?

7. If you were building a house in the Northern Hemisphere, why would you want to put your large windows on the south-facing side? Would your answer be the same if you lived in Sydney, Australia?

8. Why do cities located on islands or near the coast have a smaller annual temperature range than cities located in the interior of continents?

9. Observations show that the diurnal temperature range is wider over dry sand than over wet sand. Why would this be?

10. Place a thermometer on a windowsill in direct sunlight and another thermometer on the floor below the window next to the wall. Come back in 1 hour and write the temperatures measured in the two locations. Explain your temperature observations in terms of energy gains and losses.

11. Name two factors that appear to control interannual changes in global temperature.

12. Explain why the diurnal cycle of temperature of a city in Maine is related in part to the time of year.

13. Why are nocturnal inversions in the middle latitudes more common during winter than summer?

14. Let's say that the temperature at the surface in winter is 0° C and the temperature at 1 kilometer above the surface is −5° C. What is the environmental lapse rate for this situation? Is this situation absolutely unstable or stable? Is there a temperature inversion present?

15. If a dry parcel of air at the surface has a temperature of 20° C and rises to an altitude of 3 kilometers above the surface, what will its temperature be? If it sinks back to the surface, what will its temperature be at that point?

16. The best way to prevent frost damage is to plan the location of your crop. Based on our discussions of the factors controlling temperature, can you explain why it is better to plant frost-sensitive crops on a hillside rather than at the bottom of a valley?

17. Calculate what the temperature of a dry air parcel from near the tropopause (temperature, 200 K; altitude, 10 kilometers) would be if it were brought to the surface and it warmed adiabatically all the way down. Compare this temperature to the "freezing point" of water. Why don't farmers simply pump air from the tropopause down onto their crops when frost threatens?

■ Observation Activities

1. Temperature data for a city near you can be found on several state climatology Web sites (e.g., http://www.ncdc.noaa.gov/oa/climate/regionalclimatecenters.html). Plot the daily average, maximum, and minimum temperature over a 2-week period. How do these compare with the climatological observations?

2. On a sunny afternoon, take a thermometer, and measure the temperature at various sites across campus. Plot your observations and the time of the observations on a campus map. Describe the temperature pattern you observed, and explain this pattern in terms of energy gains and losses. What would be the advantage of a group of fellow students all making temperature observations at different locations across campus, but at the same time?

This rain cloud icon is your clue to go to the *Meteorology* Web site at http://physicalscience.jbpub.com/ackerman/meteorology/. Through animations, quizzes, web exercises, and more, you can explore in further detail many fascinating topics in meteorology.

Water in the Atmosphere

AFTER COMPLETING THIS CHAPTER, YOU SHOULD BE ABLE TO:
- Define saturation and its importance in the atmosphere
- Explain why there can be more water vapor in warm air than cold air and how this affects the atmosphere
- Describe how clouds and precipitation form
- Identify the major cloud and precipitation types, and explain the significant differences among them

INTRODUCTION

On January 26–28, 2009, a layer of ice encased the central United States from Texas to West Virginia. The precipitation fell mostly as rain, but the temperatures at the ground were cold enough for the rain to freeze on contact with anything it touched (see figures below). The storm coated many areas with more than an inch (2.5 cm) of ice, killing more than 60 people—35 in Kentucky alone. Most of the deaths were due to traffic accidents, extreme cold, and carbon monoxide poisoning (caused by power generators or kerosene heaters being used indoors without proper ventilation). Trees fell and power lines snapped under the weight of the ice, leaving more than 1.3 million people without electricity. It seemed as if Mother Nature had declared an icy war on the region—and as if to confirm this impression, the entire Kentucky Army National Guard was mobilized to help with the many problems left in the wake of this storm.

In this chapter, we explore water in the atmosphere in all its phases: water vapor, liquid water, and ice. We will explain how fog, clouds, and precipitation form. We will also learn how slightly different temperature conditions can turn a cold rain into pellets of ice or into a destructive freezing rain that can paralyze half a nation.

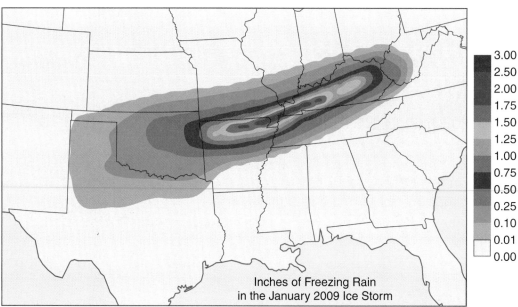

Inches of Freezing Rain
in the January 2009 Ice Storm

Source: NOAA/NWS.

EVAPORATION: THE SOURCE OF ATMOSPHERIC WATER

Water is, near the surface, the atmosphere's most abundant trace gas. How does water enter the atmosphere? Evaporation puts it there. As we learned in Chapter 1, **evaporation** is the process by which water is converted from liquid form into its gaseous state, water vapor. Evaporation occurs constantly over the surface of the Earth. When water molecules at the surface of liquid water gain enough energy to escape as vapor into the air above, evaporation results. As we learned in Chapter 2, it takes a lot of latent heat energy to change liquid water to vapor. Evaporation therefore occurs more rapidly over warmer surfaces, which supply water molecules with enough energy to escape into the atmosphere. Evaporation is also greater when the atmospheric pressure is low, the wind speed is high, and there is relatively little water vapor already in the air.

To understand evaporation better, let's consider the following example. Put some liquid water in a closed container. Keep the container at a constant temperature and pressure. Initially the container has only liquid water in it (**FIGURE 4-1a**). Some individual molecules in the liquid water will have more (and some will have less) kinetic energy than the average. For instance, a water molecule in the liquid phase might gain kinetic energy considerably above the average because of several rapid collisions with neighboring molecules. Now imagine such a molecule at the liquid's surface, the boundary between the water and the air. If it has enough kinetic energy to overcome the attractive force of nearby molecules and is moving toward the air, it may escape from the liquid (Figure 4-1b).

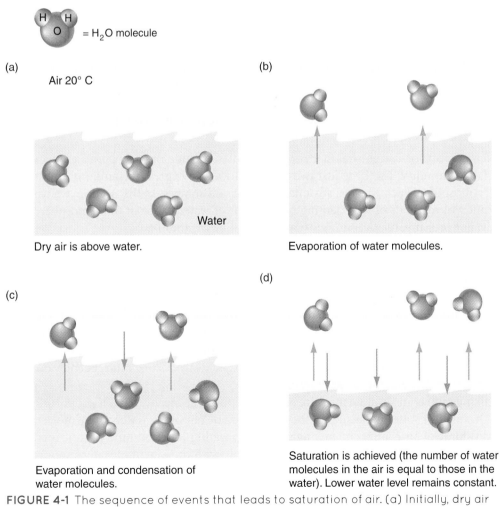

= H₂O molecule

(a)

Air 20° C

Water

Dry air is above water.

(b)

Evaporation of water molecules.

(c)

Evaporation and condensation of water molecules.

(d)

Saturation is achieved (the number of water molecules in the air is equal to those in the water). Lower water level remains constant.

FIGURE 4-1 The sequence of events that leads to saturation of air. (a) Initially, dry air lies above water at 20° C. (b) Evaporation into the air begins. (c) Condensation back to the water occurs; however, evaporation exceeds condensation and the number of water molecules in the air increases. The air is saturated in (d), where the number of evaporating and condensing water molecules is equal.

As time goes on, some water molecules at the liquid surface will be escaping and evaporating. Other molecules in the air will be captured (Figure 4-1c). Eventually, because the container is sealed, the number of molecules leaving the surface of the liquid will be the same as the number entering. This means there will be no net change in the number of molecules in the liquid or the vapor phase (Figure 4-1d).

A situation in which there is no net change is described as being in equilibrium. When the number of molecules leaving the liquid is in equilibrium with the number condensing, the air above the surface is **saturated**—that is, the rate of return of water molecules is exactly equal to the rate of escape of molecules from the water. As we will see, the concept of saturation is central to understanding the formation of clouds and precipitation.

Counting the number of molecules in the container above the water is one way to measure the amount of water. There are several methods of measuring the amount of water vapor in the atmosphere that do not require counting molecules. We look at these different methods in the following sections.

MEASURING WATER VAPOR IN THE AIR

Why do we need to measure the amount of water vapor in the atmosphere? There are several reasons:

1. The change of phase of water is an important energy source for storms, atmospheric circulation patterns, and cloud and precipitation formation.
2. Water vapor is the source of all clouds and precipitation. The potential for cloud formation and dissipation depends on the amount of water vapor in the atmosphere.
3. The amount of water in the atmosphere determines the rate of evaporation. Rates of evaporation are important to weather and many forms of plant and animal life, including humans.
4. Water vapor is a principal absorber of longwave radiant energy. It is the most important greenhouse gas.

News reports of current weather conditions often include the dew point temperature and the relative humidity. These are just two of several ways to express the amount of water vapor in the atmosphere. Each way has advantages and disadvantages. In this section, we will discuss four different methods of representing the amount of water vapor in the atmosphere: mixing ratio, vapor pressure, relative humidity, and dew point/frost point.

Mixing Ratio

One way of expressing the amount of water vapor in the atmosphere is the ratio of the weight of water vapor to the weight of the other molecules in a given volume of air. This is the **mixing ratio**. The unit of mixing ratio is grams of water vapor per kilogram of dry air. Typical values of the mixing ratio near the surface of the Earth range between less than 1 gram per kilogram in polar regions to more than 15 grams per kilogram in the tropical regions. Because the surface of the Earth is the source of water vapor for the atmosphere, the mixing ratio generally decreases as you get farther above the surface (**FIGURE 4-2**).

Mixing ratio is an absolute measure of water vapor. This means that it is proportional to the actual number of water molecules in the air. Adding or removing water vapor molecules from a fixed volume of air changes its mixing ratio. Evaporating water into the volume increases the mixing ratio. Because the mixing ratio has to do only with the weight of water vapor relative to the total weight of an air mass and because the total mass and total number of molecules remain unchanged, cooling the air or expanding the air has no effect on the value of the mixing ratio.

Vapor Pressure

Gas molecules exert a pressure when they collide with objects. The atmosphere is a mixture of gas molecules, and each type of gas contributes its part of the total atmospheric pressure. The

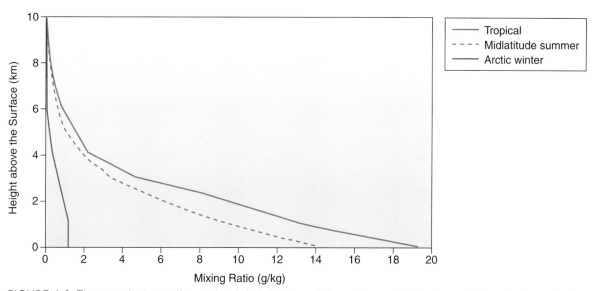

FIGURE 4-2 This graph shows the vertical distribution of the mixing ratio for three different atmospheric conditions. Atmospheric water vapor comes from the surface of Earth. This explains why values of mixing ratio usually decrease with altitude even at different regions of the world and during different seasons.

pressure the water molecules exert is another useful method of representing the amount of water vapor in the atmosphere. The pressure caused by these water vapor molecules is called the **vapor pressure**. Atmospheric vapor pressure is expressed in millibars (mb).

As we learned in Chapter 1, water vapor is at most only 4% of the total atmosphere. The average surface pressure as a result of all atmospheric gases is approximately 1000 mb. Therefore, the vapor pressure attributable to water vapor alone is never more than about 4% of 1000 mb, or 40 mb.

A variety of factors can change the vapor pressure. Increasing the air temperature will increase the vapor pressure. Changing the air temperature changes the average kinetic energy of the molecules and therefore the pressure exerted by the molecules. Increasing the number of water vapor molecules in a specific volume of air will also raise the vapor pressure. When water evaporates into a volume of air, both the vapor pressure and the mixing ratio increase. However, if the air cools, the vapor pressure decreases along with the total air pressure, but the mixing ratio remains constant. Atmospheric scientists often use vapor pressure to express the amount of water in the atmosphere when they discuss the formation of clouds.

When air is saturated (as in Figure 4-1d), the pressure exerted by the water vapor molecules is called the **saturation vapor pressure**. Saturation vapor pressure in the atmosphere is reached whenever the atmospheric water vapor exerts a pressure equal to what the saturation vapor pressure would be at that particular temperature in a closed container.

FIGURE 4-3 reveals several facts about vapor pressure and saturation vapor pressure. The data in Figure 4-3 represent more than 6 years of hourly observations of vapor pressure and temperature on a ridge top 80 kilometers (50 miles) north of New York City. What does this graph tell us? The lack of observations in the top left part of the graph implies that vapor pressure cannot exceed a certain value for a given temperature. The maximum value at each temperature is the saturation vapor pressure for that temperature. Following the maximum (pink) values from left to right, you see a curve that swoops upward rapidly as the temperature increases.

This last point is the most important fact about saturation vapor pressure: It increases rapidly as the temperature increases. Why? As the temperature of water increases, the number of molecules with enough kinetic energy to evaporate from the water surface increases. Increasing the temperature also increases the number and speed of the water molecules in the vapor phase. As a result, more molecules move at greater speed and exert a higher pressure.

It is often said that "warm air holds more water vapor than cold air," but this is a misleading simplification. This saying implies that warm air expands and has more room for water vapor, which is incorrect. Instead, the amount of water vapor in the air is, as we have discussed here, in

"Moisture Graph" to learn how moisture affects static stability.

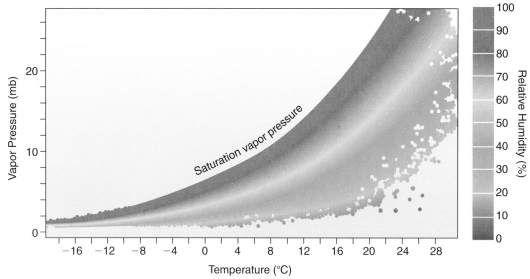

FIGURE 4-3 Observations of vapor pressure as a function of temperature on a ridge top at Black Rock Forest along the lower Hudson River in New York. The observations were made hourly from December 1994 through mid April 2001. Relative humidity is indicated by the color coding. Notice how the highest observed values of vapor pressure at each temperature form an arc that curves upward from left to right. This upper limit on vapor pressure at each temperature is the saturation vapor pressure. (*Source*: Black Rock Forest Consortium.)

equilibrium between the air and the surface beneath it. It is more accurate to say that a saturated parcel of warm air will contain many more water vapor molecules than a saturated parcel of cooler air.

Relative Humidity

Neither the vapor pressure nor the mixing ratio tells us how close the air is to being saturated. The ratio of the actual vapor pressure exerted by molecules of water vapor versus the saturation vapor pressure at the same temperature indicates just how close the air is to saturation. This ratio is called the saturation ratio. Multiplying the saturation ratio by 100% yields the **relative humidity**.

Relative humidity describes how far the air is from saturation. Saturated air has a relative humidity of 100% because the vapor pressure equals the saturation vapor pressure. In Figure 4-3, the pink colors indicate relative humidities near 100%, and they are also the maximum values of vapor pressure for a particular temperature. This implies that when the relative humidity is close to 100%, the vapor pressure is very close to the saturation vapor pressure—which is exactly what the formula tells us. A relative humidity of 50% (light green in Figure 4-3) tells us that the vapor pressure is half that required for saturation.

The amount of relative humidity also affects the rate of evaporation. At the same pressure and temperature, water evaporates more slowly in air that has a high relative humidity and more quickly in air that has a low relative humidity. This fact is of prime importance to the public because high humidity makes perspiration an inefficient way of removing heat by evaporation. This can lead to uncomfortable and even life-threatening conditions (BOX 4-1). Relative humidity is more generally an important indicator of the rate of moisture and heat loss by plants and animals.

Relative humidity can change in response to a wide range of circumstances. For example, in the case of a constant volume of air at a constant temperature, changing the vapor pressure changes the relative humidity. Why does the relative humidity change in this case? Adding water molecules to a fixed volume of air increases the vapor pressure but has no effect on the saturation vapor pressure because the temperature has not changed. The saturation ratio is changed and so is the relative humidity.

Box 4-1 Atmospheric Moisture and Your Health

Water in the atmosphere can be a killer in all of its phases: vapor (humidity), liquid (fog and rain), and ice (freezing rain and ice storms). The victim can be a National Football League star—or you.

On August 1, 2001, 27-year-old Minnesota Vikings football player Korey Stringer died just hours after practicing with his team. How could such a physically fit young athlete, who was selected by his peers to play in the Pro Bowl the previous year, die so suddenly?

When our bodies get hot we cool down by sweating. It is not the sweating that cools our bodies; it is the evaporation of the sweat. If the air has a high vapor pressure, then the rate of evaporation is reduced. This hampers the body's ability to maintain a nearly constant internal body temperature. This is why we are uncomfortable on hot, muggy days. Like an engine without proper ventilation, we are overheating.

When athletes practice for hours in summertime heat and humidity, their bodies can overheat to the point that organs can fail and death can occur. This is called heatstroke. For example, Stringer's temperature when he reached the hospital was above 108.8° F, more than 10 degrees above normal.

The apparent temperature index or heat index (see the accompanying table) indicates how hot it feels. It is expressed as a function of air temperature and the relative humidity. R. G. Steadman developed this index in 1979. When the temperature is high but the relative humidity is low, the heat index is less than the actual temperature. This is because cooling by evaporation of sweat is very efficient in these situations. However, high relative humidities prevent evaporation and make it seem hotter than it really is. In these cases, the heat index is greater than the actual temperature.

Heat Index Values (°F)

Relative Humidity (%)		0	10	20	30	40	50	60	70	80	90	100
Air Temperature (°F)	70	64	65	66	67	68	69	70	70	71	71	72
	75	69	70	72	73	74	75	76	77	78	79	80
	80	73	75	77	78	79	81	82	85	86	88	91
	85	78	80	82	84	86	88	90	93	97	102	108
	90	83	85	87	90	93	96	100	106	113	122	
	95	87	90	93	96	101	107	114	124	136		
	100	91	95	99	104	110	120	132	144			
	105	95	100	105	113	123	135	149				
	110	99	105	112	123	137	150					
	115	103	111	120	135	151						
	120	107	116	130	148							

- ◼ Great risk to health, heatstroke imminent.
- ◻ Risk of heatstroke.
- ◻ Prolonged exposure and physical activity could lead to heatstroke.
- ◼ Prolonged exposure and physical activity may lead to fatigue.

A combination of high temperature and high humidity leads to extreme heat indices as much as 40° F above the actual air temperature. In these situations, exercising outside

(continued)

Box 4-1 Atmospheric Moisture and Your Health, continued

can be fatal. For example, the Vikings practiced in Mankato, Minnesota, in the morning to avoid daytime maximum temperatures. The air temperature at noon in Mankato on the day of Stringer's last practice was 89° F, which does not sound particularly unusual for summertime; however, the dew point was a sultry 80° F, giving a relative humidity of 75%. The high humidity in Mankato, combined with the temperature, yielded a heat index of 106° F. As the table shows, heatstroke was indeed a possibility at the Vikings' practice that morning.

Athletes are not the only ones at risk from high heat index values. Prolonged periods of very high temperatures in association with high humidities can be extremely dangerous even to those who are not exercising at the time, as discussed in Chapter 9.

Fog can be a killer, too. Each year 680 fatalities occur in the United States as a result of traffic accidents during which fog is present. A combination of high speed and low visibility is often to blame. On December 27, 1996, fog caused a 54-vehicle series of pileups on the Sunshine Skyway Bridge over Tampa Bay, Florida. The wrecks killed one person and snarled traffic for 7 hours. Fog also played a key role in the United States' worst modern passenger train wreck. A towboat lost in fog bumped a railroad bridge near Mobile, Alabama, early on September 22, 1993. The bridge was pushed out of alignment, causing an Amtrak train to derail as it crossed the bridge a few minutes later. The train plunged into the water, killing 47 people.

Rain also kills. Up to 20% of all fatal highway accidents occur on wet pavement. A little rain after a dry spell, combined with the built-up residue of oil on roads, turns concrete and asphalt into a slick and oily mess. Cars can "hydroplane" on the wet surface and skid out of control because of reduced friction between the tires and road surface.

Ice storms consisting of freezing rain and sleet are perhaps the most dangerous of all. Driving in an ice storm is life threatening and is much more uncontrollable than in rain. Even off the roadways, you can easily fall and break bones on slippery steps or a sidewalk. Overall, in the United States each year, approximately 7000 highway deaths and 450,000 injuries are associated with poor-weather–related driving conditions, according to Congressional testimony by Dr. Richard Anthes, President of the University Corporation for Atmospheric Research in Boulder, CO. This means that weather plays a role in about 28% of all crashes and 19% of all highway fatalities. Whether exercising in hot humid weather or driving in fog, you should give careful consideration to possible threats to your safety related to the water in your environment.

Changing the saturation vapor pressure also changes the relative humidity. As shown in Figure 4-3, the saturation vapor pressure decreases rapidly when the temperature of the air decreases. Therefore, a decrease in temperature results in an increase in the relative humidity, and increasing the temperature decreases the relative humidity.

The effect of temperature on relative humidity is illustrated in FIGURE 4-4, which is based on observations at a national monument in New Mexico. For every season of the year, relative humidity peaks around sunrise and is at its lowest in mid afternoon. Why? The saturation vapor pressure is lowest at sunrise, when the temperature is lowest. Therefore, without a change in the actual moisture content of the air, the relative humidity will be highest at sunrise. Conversely, daytime heating raises the temperature and the saturation vapor pressure, reducing the relative humidity. This seesaw pattern of temperature and relative humidity is seen during mostly clear, relatively calm, and precipitation-free conditions, which exist most of the time in New Mexico.

The effect of temperature on relative humidity explains why regions with cold winters may have very low indoor relative humidities. As cold outside air finds its way into a building, it is eventually heated, and this greatly decreases the air's relative humidity. Explore this for yourself using our Exploring Humidity learning applet.

To summarize, adding water vapor, cooling the air, or both increases relative humidity; removing water vapor, warming the air, or both decreases relative humidity.

"Exploring Humidity" to explore why, for example, homes in cold climates are so dry in winter.

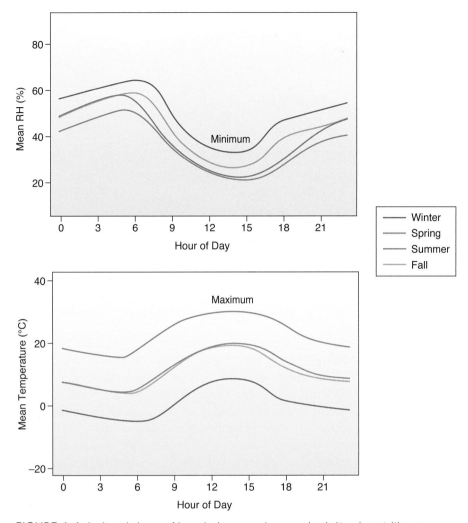

FIGURE 4-4 A climatology of hourly temperature and relative humidity, as observed at Bandelier National Monument, New Mexico, from October 1988 through May 1999. In the usually dry, clear climate of New Mexico, the daily cycle of relative humidity is exactly the opposite of the temperature cycle, with a peak at sunrise and a minimum at mid afternoon. The same daily cycle of relative humidity is observed at any location during a mostly calm, clear, dry day. (Data from Diurnal Cycle. Retrieved December 10, 2010, from http://vista.cira.colostate.edu/improve/data/graphicviewer/diurnal.htm.)

▓ Dew Point/Frost Point

When air is saturated, the vapor pressure is equal to the saturation vapor pressure. The air cannot contain any more moisture. If the vapor pressure is greater than the saturation vapor pressure, the relative humidity exceeds 100%. Ordinarily, this is not possible in the atmosphere. The excess moisture must condense out of the air until the relative humidity is once again reduced to 100%. This condensed water is called **dew** (FIGURE 4-5).

Dew forms when moisture is added to the air, when the air is cooled, or when a combination of both moistening and cooling occurs. The most commonplace occurrences of dew are caused by cooling. An everyday example of dew is the moisture that forms on the outside of a glass of an ice-cold drink. Overnight cooling of the air near the ground causes morning dew on grass, car windshields, and spider webs.

The temperature to which air must be cooled to become saturated without changing the pressure is called the **dew point**. The dew point temperature is determined by keeping the pressure fixed because changing the pressure affects the vapor pressure and therefore the temperature at which saturation occurs.

FIGURE 4-5 When the temperature of the air around this web cooled to the dew point temperature, dew formed, making the web more visible.

To know how close the air is to saturation, we need to know the dew point and the air temperature. The closer the dew point is to the air temperature, the closer the air is to saturation. When the dew point equals the air temperature, the air is saturated, so the dew point temperature cannot be greater than the air temperature. The temperature difference between the air and the dew point is called the **dew point depression**. TABLE 4-1 compares all the different measures of water vapor in the atmosphere for two different temperatures and two different relative humidities.

If temperatures are below freezing, saturation or moistening of cooled air will lead to deposition (see Chapter 2) rather than condensation. The ice crystals that form are called **frost**. The temperature to which air must be cooled at a constant pressure to cause frost to form (normally below 0° C [32° F]) is called the **frost point**. Dew may form and then freeze if the temperature falls below freezing, forming **frozen dew**. Frozen dew is different than frost. Frozen dew first condenses as liquid water before freezing, instead of becoming ice via deposition, as in the case of frost. Yet another frozen water type is **rime**, which is a white, opaque deposit of ice formed by the rapid freezing of water drops as they collide with an object at or below freezing.

We can use the energy budget concepts from Chapters 2 and 3 to explain dew and frost. Dew and frost form on objects in air close to the ground, such as blades of grass. Whether a blade of grass cools below the dew or frost point is a function of its energy gains and losses. At night, a blade of grass loses energy by emission of longwave radiation while gaining energy by absorbing the longwave radiation emitted from surrounding objects. Under clear nighttime skies, objects near the ground emit more radiation than they receive from the sky, and so a blade of grass cools. If the temperature of a grass blade falls below the dew or frost point, dew or frost will form on the grass.

TABLE 4-1 Various Humidity Quantities for Two Air Temperatures and Two Relative Humidities for an Atmospheric Pressure of 1000 mb

Temperature	−10° C (14° F)		20° C (68° F)	
Relative humidity	25%	75%	25%	75%
Mixing ratio (g/kg)	0.45	1.35	3.67	11.15
Vapor pressure (mb)	0.72	2.16	5.87	17.60
Saturation vapor pressure (mb)	2.88	2.88	23.47	23.47
Dew point temperature	−26.2° C (−15.2° F)	−13.5° C (7.8° F)	−0.5° C (31.3° F)	15.6° C (60.1° F)
Dew point depression	16.2° C	3.5° C	20.5° C	4.4° C

FIGURE 4-6 Frost, ice crystals formed by deposition of water vapor on subfreezing surfaces, will form in open fields before forming under a tree. In the background, steam fog is forming over the water.

This explains why frost forms in an open field but not under a tree (**FIGURE 4-6**). Trees emit more radiation toward the ground than does the clear sky. Energy gains of the grass in the open field are less than those of the grass under the tree. The grass in the open field cools faster and reaches the frost point before the grass blades under the tree.

The dew point is useful in forecasting minimum temperatures. On a clear, calm night, the temperature will often drop to near the dew point. This is because condensation releases energy, and this energy release counteracts cooling below the dew point.

Formation of frost and dew are examples of phase transitions between the gas phase of water and its solid and liquid states. These transitions are vital to understanding how clouds form, which we now examine in detail.

CONDENSATION AND DEPOSITION: CLOUD FORMATION

Clouds are the atmosphere's equivalent of movie stars, instantly recognizable worldwide and the subject of endless curiosity. Children and poets look for shapes in the clouds. Parents struggle to explain how something so large can keep from falling out of the sky but then can disappear in a matter of minutes.

The key to understanding clouds is water. Clouds, from the fair-weather wisps to the mightiest thunderstorms, are composed of nothing more than tiny 20-micron–sized particles of liquid water called **cloud droplets** and particles of ice called **ice crystals**. However, the formation and growth of these particles is one of the most complicated aspects of weather and climate. Before we can look at clouds, we must first examine in some detail how they are made.

Solute and Curvature Effects

As a volume of unsaturated air cools, its relative humidity increases. If the air is sufficiently cooled, the temperature equals the dew point and the relative humidity equals 100%. Based on what we have learned so far, condensation should occur at this point, forming a cloud. But cloud droplets can actually form at relative humidities other than 100%. Why?

Over the oceans and seas, waves and winds add salt to the mix of atmospheric components. When salt dissolves in water, the salt particles become dispersed among the water molecules. Salt particles dissolved in water attract water molecules even more strongly than neighboring water molecules. The greater the concentration of salt, the more the rate of evaporation is reduced, all other things being equal.

The ability of dissolved salt to hold onto water molecules is called the **solute effect**. By suppressing evaporation, the solute effect enhances the growth of droplets by condensation, thereby allowing cloud formation at relative humidities much less than 100%. As the droplet

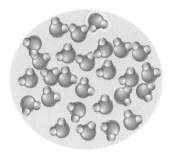

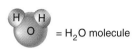

= H_2O molecule

FIGURE 4-7 The smaller the drop, the more curved the surface, reducing the number of neighbors for each water molecule at the surface. This curvature effect makes it easier for small drops to evaporate.

grows, however, the solution becomes more dilute and the solute effect decreases.

In our previous discussion of evaporation, we discussed a single molecule near the edge of a flat surface of still water—but cloud droplets are not flat surfaces. A molecule on any surface feels attracted to its neighbors, which attempt to keep it part of the water. A molecule on a curved surface such as a cloud droplet has fewer neighbors to attract it (FIGURE 4-7) and can, therefore, escape the fluid more easily.

As a result, even if air is saturated with respect to a flat surface of water, it may be unsaturated with respect to a curved surface. This is called the **curvature effect**. This effect opposes the formation of small droplets by condensation. As a result, the relative humidity must be higher than 100%—a condition known as **supersaturation**—for cloud formation to occur.

It is surprisingly difficult to form a water droplet out of air that contains only water vapor. It takes a relative humidity of more than 200% for water vapor molecules to form a tiny cloud droplet of pure water. This is because a tiny droplet has a strongly curved surface, but relative humidities this high are never observed in the atmosphere. So how do liquid droplets form from pure vapor? This question leads us to the subject of nucleation.

NUCLEATION

Droplets form around particles. The initial formation of a cloud droplet around any type of particle is called **nucleation**. There are two types of nucleation: homogeneous and heterogeneous nucleation. In **homogeneous nucleation**, the droplet is formed only by water molecules. Homogeneous nucleation requires that enough water molecules bond together to form a cluster, or particle, that then acts as a nucleus for further condensation. Water-only bonding only works if the water molecules have low kinetic energy. If the kinetic energy of the molecules is too high, the cluster cannot form. For this reason, homogeneous nucleation only occurs at temperatures colder than −40° C (−40° F).

You can see homogeneous nucleation for yourself when you open a chilled bottled beverage that has very clean air in the bottle's neck. Brewers sterilize and clean the bottles to keep the beverage from going bad. By removing the cap, you allow the air inside the neck to expand adiabatically and cool rapidly, but temporarily, to −40° C. The smoky cloud in the neck of the bottle is the result of homogeneous nucleation.

We learned in Chapter 1 that temperatures are as low as −40° C only in the upper troposphere, close to the stratosphere. Clouds usually form in much warmer air. Therefore, most clouds must develop through a different process. **Heterogeneous nucleation** occurs when small, nonwater particles serve as sites for cloud droplet formation. The particles are usually aerosols such as those we studied in Chapter 1. The aerosols that assist in forming liquid droplets are called **condensation nuclei**.

In the next sections, we first consider the formation of liquid droplets around condensation nuclei and then address how ice crystals form around ice nuclei.

Condensation Nuclei

There are two types of condensation nuclei: hygroscopic and hydrophobic. **Hygroscopic nuclei** dissolve in water, and **hydrophobic nuclei** do not. Nucleation is more favorable on hygroscopic ("water-seeking") nuclei. Droplet formation can occur on hygroscopic nuclei even when the relative humidity is below 100% because the solute effect reduces the rate of evaporation. Hydrophobic ("water-repelling") nuclei resist condensation but can form droplets when relative humidities are near 100%.

There are plenty of condensation nuclei in the atmosphere in the form of dust, salt, pollen, and other small particles. The surface of the Earth is the major source of aerosols. The concentration of condensation nuclei is therefore usually greatest near the surface and decreases with altitude. In general, there is no lack of condensation nuclei for forming water droplets. Polluted cities have more condensation nuclei than wilderness environments. Over the oceans, the air has fewer condensation nuclei than over land. Many of the nuclei over the oceans also contain salt thrown from waves, making them hygroscopic nuclei.

Ice Nuclei

When ice crystals form, water molecules cannot deposit onto the crystal haphazardly, as they can when condensing onto an existing water droplet. The molecules must fit into the shape of the crystal. **Ice nuclei**, the particles around which the ice crystals form, are important in the beginning stages of ice crystal formation. The ice nuclei make it easier for deposition to occur. Ice particles can form in four ways: deposition nucleation, freezing nucleation, immersion nucleation, and contact nucleation.

In **deposition nucleation**, ice forms from vapor by deposition onto the ice nucleus when the air is supersaturated with respect to ice. This happens most often on particles, such as clay, that have a molecular geometry resembling the molecular structure of ice. This geometry helps water molecules in the surrounding air to align in the proper molecular structure for forming ice when they deposit on the surface of the nuclei.

Liquid water at a temperature below 0° C is referred to as **supercooled water. Freezing nucleation** is the process by which a supercooled drop freezes without the aid of a nonwater particle.

The existence of supercooled water requires some explanation. There is a big difference between freezing a small water droplet and freezing a larger body of water. The freezing point of a large body of water (such as the water in an ice tray) is 0° C at standard pressure; however, a 1-millimeter diameter droplet will generally not freeze until the temperature falls below −11° C (12.2° F). A tiny droplet, but not a large body of water, can be supercooled.

How can this happen? For ice to form, all the water molecules must align in the proper crystal structure. First a few molecules align, and then the rest quickly follow, turning the liquid into a block of ice. The larger the volume of water, the greater the chances that a few of the molecules will align in the proper manner to form ice when the temperature falls below freezing. In a small volume of water, the chance that some of the molecules will align in the correct structure is reduced, simply because there are fewer molecules. For this reason, 0° C is more accurately called the melting point, not the freezing point, of water.

In **immersion nucleation**, the nucleus is submerged in a liquid drop. After the drop reaches a given temperature, the immersed nucleus allows the supercooled liquid to rapidly align in the crystalline structure of ice, causing the drop to freeze.

Ice nuclei may also collide with supercooled drops. The drop freezes immediately on contact with the ice. This is referred to as **contact nucleation**.

Cloud Particle Growth by Condensation and Deposition

After a cloud particle forms, it can grow if the air around it is saturated. If the particle is a liquid droplet and if the vapor pressure of the air is greater than the vapor pressure just above the surface of the particle, it will continue to grow by condensation. If the particle is ice and the air is saturated, it grows by deposition.

Growth by condensation and deposition produces droplets, but they are small. It takes a long time to create droplets large enough to fall as precipitation. We look at precipitation particle growth a little later in this chapter. First, we examine how growth by condensation produces fog, a cloud at the ground.

FOG FORMATION

The air in contact with the ground can become saturated if it cools or when water from the surface evaporates into it. Water vapor then condenses on cloud condensation nuclei to form a suspension of tiny water drops. This is a cloud at the ground, which we call **fog**.

FIGURE 4-8 Fog consists of tiny water droplets that can reduce visibility to less than 1 kilometer (0.6 miles).

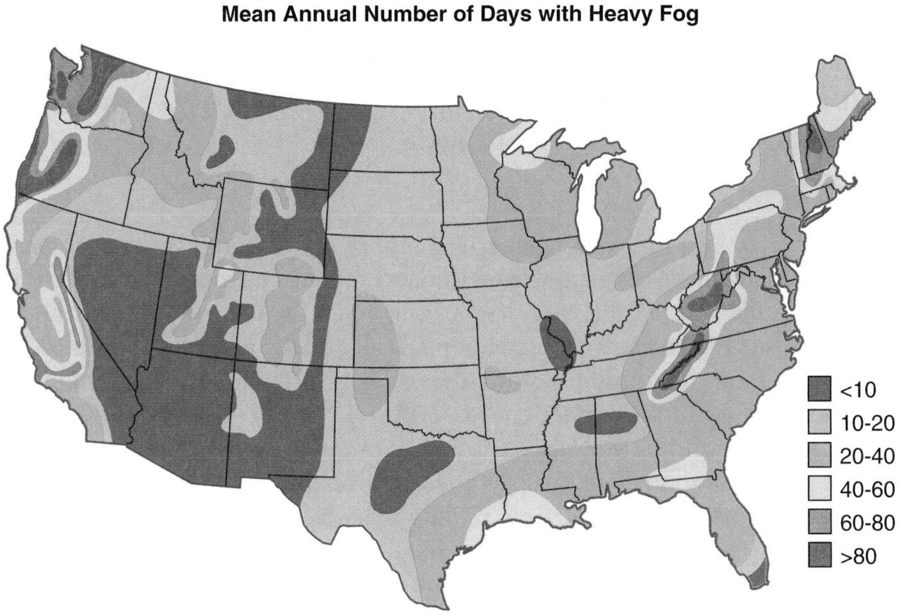

Mean Annual Number of Days with Heavy Fog

Legend:
- <10
- 10-20
- 20-40
- 40-60
- 60-80
- >80

FIGURE 4-9 The mean annual number of dense fog days with visibility less than 300 meters across the United States.

The formation of heavy fog often reduces visibility to the point where certain modes of transportation become hazardous (**FIGURE 4-8**). The appearance of fog on highways can trigger chain-reaction accidents involving scores of vehicles. Fog also contributed to the famous collision between the *Titanic* and an iceberg. In early December 1952, a fog in London became so thick (partly because of pollution) that people walked into canals and rivers because they could not see the ground.

The distribution of heavy fog over the continental United States is shown in **FIGURE 4-9**. Heavy fog in Alaska, Hawaii, and Puerto Rico occurs on fewer than 10 days per year. Fog is most common in the Appalachian Mountains and near bodies of water, especially along the northwest and northeast coasts.

Fogs are named for the ways in which they form. We explore four different types of fog below: radiation fog, advection fog, evaporation fog, and upslope fog.

Radiation Fog

Radiation fogs form in the same way that dew does. On clear, long nights, the ground rapidly cools by radiation, and the air just above the ground cools by conduction and radiation. As the temperature of the air drops, the relative humidity increases. Radiation fogs tend to develop on clear nights, when radiative cooling near the ground is more rapid. Light winds are also required because they can gently mix moist air near the ground. Winds that are too strong mix the air near the ground with the drier, warmer air above, keeping the air near the surface from saturating.

Radiation fogs are common in autumn in river valleys and small depressions. The cold air sinks to the bottom of the valley, providing the cool air. Rivers and streams provide the water vapor needed to increase the relative humidity via evaporation. These fogs are often called valley fogs (**FIGURE 4-10**).

There are some rules for forecasting a radiation fog. If the dew point temperature is approximately 8° C (14° F) below the air temperature at sunset and if the winds are predicted to be less than 9 kilometers per hour (5 knots), there is a good chance that a radiation fog will form during the night.

Advection Fog

When warm air is advected (blown horizontally) over a cold surface, the air near the ground cools because of energy exchanges with the surface. The relative humidity increases, and an **advection fog** may form (**FIGURE 4-11**).

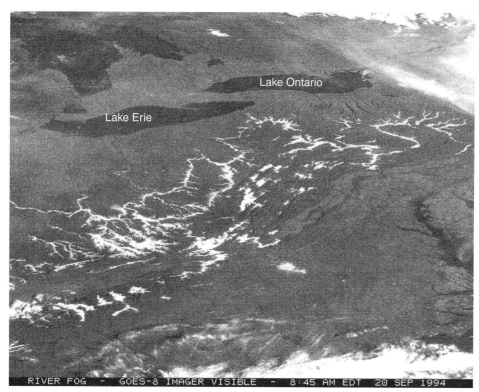

FIGURE 4-10 Satellite photograph on the morning of September 20, 1994, showing radiation fog (narrow white areas) in the Ohio River valley and its tributaries.

Advection fog is common off the coast of California as warm moist air over the Pacific is advected over the cold coastal waters. Off the East Coast, warm air over the warm Gulf Stream current may be advected over the colder coastal waters, forming a fog. Another foggy region is off the coast of Japan, where the cold water of the Oyoshio current meets the warm Kuroshio current. These fogs form at all times of the year and can last for more than a week.

Advection fogs can also occur when warm air flows from over the water to cooler land. Fog is common along the coast of the Gulf of Mexico during fall and winter. During these times, saturation of the air occurs when warm moist air flows from the Gulf of Mexico over the cooler land. These types of fog are also common in New England and give London its reputation for fog.

FIGURE 4-11 Advection fog is common off the coast of California near San Francisco as warm moist air over the Pacific is advected over the cold coastal waters.

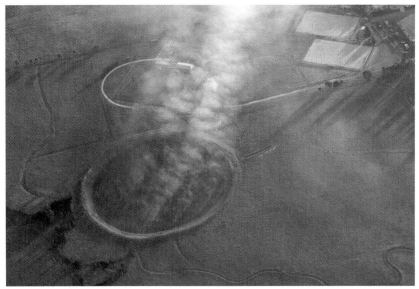

FIGURE 4-12 An aerial view of steam fog rising and moving downwind from a small lake in Australia. Using our knowledge of relative humidity plus the shadows in this photograph, can you determine what time of day this picture was taken?

Evaporation Fog

If you take a long, hot shower, you may "fog up" the bathroom. Some of the warm water from the shower evaporates into the cooler bathroom air, moistening it to saturation and forming a fog. **Evaporation fogs** also occur in the vicinity of warm fronts and are sometimes called **frontal fogs**. These fogs form when water evaporates from rain that falls from warmer air above the ground into cold air near the surface. Frontal fogs form only after it has been raining for hours because it takes time for the evaporating drops to saturate the air. Similarly, it is difficult to fog up the bathroom by taking a short shower.

Evaporation fogs also form over lakes when much colder air moves over warmer water. The vapor pressure of the cold air is less than that of the air over the water. As a result, evaporation is rapid. This rapid evaporation saturates the air above the surface. The condensation further warms the air. This warmed air rises and mixes with the cold air above it, reaching saturation and causing more fog to form.

Evaporation fog over a lake gives the appearance of steam rising out of the water and is sometimes referred to as a **steam fog** (FIGURE 4-12). It is common over lakes during late autumn or early winter in the more northern midlatitude regions of the globe. Steam fog is common when very cold air rushes over unfrozen waters.

Upslope Fog

Consider air rising over a mountain barrier. As the air rises, it expands and cools, and the relative humidity rises. If the air becomes saturated, an **upslope fog** forms. Upslope fog is common in moist mountainous regions such as the Appalachian Highlands. This type of fog forms best when the air near the ground, before flowing upslope, is cool and moist. Therefore, it does not require much lifting before saturation occurs.

LIFTING MECHANISMS THAT FORM CLOUDS

"Cloud Base Altitude" to explore the relationship between temperature, dew point, and the base of a cumulus cloud.

Upslope fog occurs when air moves along a rising ground surface. In general, most clouds form when air cools to the dew point as a parcel of air rises vertically.

FIGURE 4-13 depicts four mechanisms that cause air to ascend. Air is lifted as it moves against a mountain range (Figure 4-13a). The air cannot go through the mountain, and so it flows over the mountain. This is **orographic lifting**.

Other lifting mechanisms can also cause clouds to form. For example, at the same pressure, cold air is denser than warm air. Fronts represent the boundaries between these air masses of different densities. As fronts move, **frontal lifting** occurs when less dense warm air is forced to rise over the cooler, denser air (Figure 4-13b). Frontal lifting is common in winter. We study fronts in detail in Chapters 9 and 10.

During summer, **convection** is an important lifting mechanism. In summertime convection, solar energy passes through the atmosphere and heats the surface. The air near the surface warms, becomes less dense than the air around it, and rises (Figure 4-13c).

The final mechanism, **convergence**, occurs when air near the surface flows together from different directions. When the air near the ground converges, or is squeezed together, it causes upward motion (Figure 4-13d). The opposite of convergence is **divergence**, which is the horizontal spreading out of air.

In each of these examples of lifting, the rising air creates an **updraft**. The updraft keeps the cloud particles suspended in mid air despite the force of gravity that acts to bring them to the ground.

Certain atmospheric conditions are less favorable for cloud development than others. In the following section, we examine the role of moisture and stability in cloud development.

STATIC STABILITY AND CLOUD DEVELOPMENT

We introduced the concept of static stability in Chapter 3. The basic question regarding stability is as follows: Will a rising air parcel keep on rising? To answer this question, we learned to compare the lapse rate of the environment with the dry adiabatic lapse rate.

A quick glance out the window during a thunderstorm, however, tells you that the atmosphere is not dry (i.e., it is not unsaturated). Therefore, we have to modify our understanding of static stability to take phase changes of water into account. This leads to two key new definitions: (1) a definition for a parcel's lapse rate that incorporates latent heating as a result of the phase changes of water and (2) a definition for stability that includes this latent heating.

The Saturated Adiabatic Lapse Rate

A saturated parcel of air is one in which the air contains the maximum amount of water vapor possible; its relative humidity is therefore 100%. In saturated air, water molecules are changing phase from vapor to liquid or ice. As shown in Chapter 2, a phase change of water vapor to liquid water or ice releases energy, warming the parcel through latent heating.

This means that for an ascending moist parcel of air, two processes are going on at once. Expansion is cooling the parcel, while condensation (or deposition) is warming the parcel by latent heating. The cooling process from expansion is always larger than the latent heating, so the parcel temperature decreases. The rate that the rising saturated air parcel cools is called the **saturated adiabatic lapse rate.**

Because heat is being added by the phase change of water vapor, the cooling rate of a rising saturated parcel is always less than the

Orographic lifting

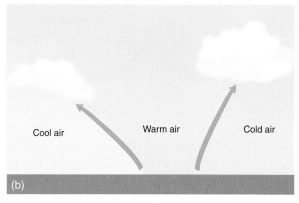

Frontal lifting

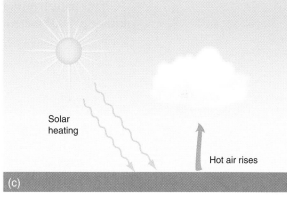

Convection

Convergence of air at surface

FIGURE 4-13 Depiction of the four mechanisms that cause air to ascend and form a cloud.

dry adiabatic lapse rate. The exact saturated adiabatic lapse rate for a given parcel depends on whether liquid or ice particles form and how much water vapor changes phase and, as a result, differs from one place and time to another. To simplify our discussion, in this text, we will use the commonly assumed saturated adiabatic lapse rate of 6° C per kilometer.

To understand the saturated adiabatic lapse rate better, let's consider a rising parcel at the ground with an initial temperature of 10° C (50° F) (**FIGURE 4-14**). This is the same example we used in Chapter 3 to illustrate static stability with a completely dry parcel of air. This time, however, we will assume that our parcel becomes saturated at an altitude of 1000 meters.

As the parcel rises up to 1000 meters, it cools at the dry adiabatic lapse rate. When it reaches 1000 meters it has a temperature of 0° C (32° F). At this point, the figure indicates that the water vapor in the parcel is condensing; in other words, the temperature and the dew point are equal at that altitude, and the relative humidity is 100%. This altitude is called the **lifting condensation level** (**LCL**) because it is the height at which water vapor in a rising parcel of air starts condensing. The bottoms of puffy clouds on sunny days are at the altitude of the LCL. The LCL varies with temperature and dew point, as you can demonstrate for yourself using the Cloud Base Altitude learning applet.

If the parcel is warmer than its environment or if it is being forced orographically up a mountain or front, it will continue rising. However, because the parcel is now saturated, as it rises above 1000 meters it will cool at the saturated adiabatic lapse rate. This is crucial! Up to now, our parcel of air has behaved just like the example in Chapter 3. But from here onward (and upward), there are big differences.

The parcel then rises to 2000 meters, where it has a temperature of −6° C (21° F), not −10° C as in the case of an unsaturated air parcel. The saturated air parcel is warmer than would be the case if it were unsaturated. Where does the added warmth come from? It comes from latent heating that is released as the water vapor condenses.

If the parcel continues at the saturated adiabatic lapse rate, what will its temperature be at an altitude of 3000 meters? Because the parcel is still saturated, it will cool to 6° less than −6° C, or

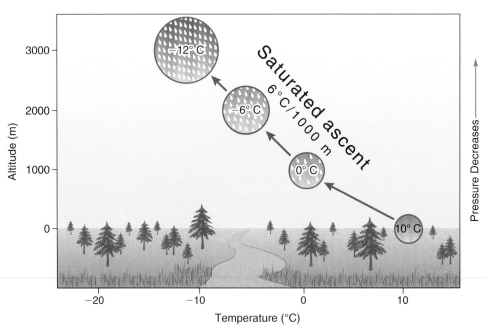

FIGURE 4-14 The ascent of a parcel of air that becomes saturated after traveling upward 1 km. The rate of cooling during ascent is less than the dry adiabatic lapse rate because of the warming as a result of latent heating, as water vapor molecules condense. A typical saturated adiabatic lapse rate is 6° C per kilometer, but the exact value varies. Compare this figure with Figure 3-19, which depicts the ascent of an unsaturated air parcel. How do the two differ?

−12° C (10° F). This is 8° C warmer than an unsaturated parcel that started at the ground with the same temperature.

The essential point is that ascending parcels that are saturated cool less quickly than do unsaturated parcels. This is the same thing as saying that the saturated adiabatic lapse rate is less than the dry adiabatic lapse rate. Next, how does this add to our understanding of static stability?

◼ Conditionally Unstable Environments

In Chapter 3, we determined that an environment was either absolutely stable or absolutely unstable by comparing the environmental lapse rate to the dry adiabatic lapse rate. Now there is a third lapse rate to consider: the saturated adiabatic lapse rate. This creates a third possibility: that air parcels might be stable if they are "dry" (i.e., unsaturated) but unstable if they are saturated.

This third possibility depends on the condition of the air parcel—is it saturated or not? Therefore, the case where saturated air parcels are unstable, but unsaturated air parcels are stable, is called a **conditionally unstable environment.**

A conditionally unstable environment exists when its lapse rate is in between the saturated adiabatic lapse rate of about 6° C per kilometer and the dry adiabatic lapse rate of 10° C per kilometer. In this situation, a dry air parcel will rise, become colder than its environment, and sink back down. Because the parcel returns to the altitude where it started, it is a stable situation. However, a rising saturated air parcel will become progressively warmer than its environment as it rises and therefore will keep rising. In such unstable situations, tall clouds can form, especially thunderstorms.

FIGURE 4-15 depicts our expanded understanding of static stability, including conditional instability. If you prefer words to figures, **TABLE 4-2** summarizes the same information. Either way, the bottom line is that condensed moisture improves the ability of air to rise, form clouds, and cause "bad weather." In Chapter 11, we explore how severe thunderstorms can be understood and predicted using these same concepts of saturation and stability.

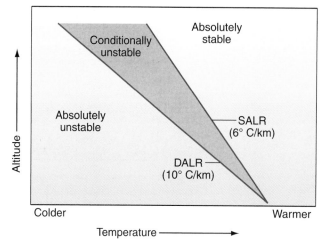

FIGURE 4-15 An environment can be described in three different ways in terms of lapse rates and stability: (1) Absolutely unstable, when the environmental lapse rate is greater than the dry adiabatic lapse rate (DALR; left); (2) conditionally unstable, when the environmental lapse rate is in between the DALR and the saturated adiabatic lapse rate (SALR; middle wedge); and (3) absolutely stable, when the environmental lapse rate is less than the SALR (right). In a conditionally unstable environment, saturated parcels are unstable and will keep on rising, leading to the development of tall clouds.

TABLE 4-2 Atmospheric Stability Summary

Environmental Lapse Rate Is . . .	Environment Is . . .	Means What?
Less than saturated adiabatic lapse rate	Absolutely stable	No parcels keep rising
Greater than dry adiabatic lapse rate	Absolutely unstable	All parcels keep rising
Less than dry adiabatic lapse rate and greater than saturated	Conditionally unstable	Only saturated parcels keep rising

CLOUD CLASSIFICATION

Earlier we learned that fogs are named for the process that caused the air to become saturated. In 1803, British pharmacist and chemist Luke Howard devised a different kind of classification system for clouds above the ground. It has proved so successful that meteorologists have used Howard's system ever since, with minor modifications. According to his system, clouds are given Latin names corresponding to their appearance—layered or convective—and their altitude. Clouds are also categorized based on whether or not they are precipitating.

TABLE 4-3 Common Cloud Types

Cloud Type	Layered Cloud	Convective Cloud	Mixed/ Neither	Typical Altitudes	
				Feet	Kilometers
High	Cirrostratus	Cirrocumulus	Cirrus	20,000–40,000	6–12
Middle	Altostratus	Altocumulus		6500–20,000	2–6
Precipitating	Nimbostratus			Surface–10,000	0–3
		Cumulonimbus		Surface–50,000	0–15
Low	Stratus	Cumulus	Stratocumulus	Surface–6500	0–2

Layered clouds are much wider than they are tall. They generally have flat bases and tops and can extend from horizon to horizon. The Latin word *stratus* describes the layered cloud category, just as "stratosphere" describes a layered region of the atmosphere. Stratus-type clouds form in relatively stable air that is forced to rise.

Convective clouds are as tall, or taller, than they are wide. These clouds look lumpy and piled up, like a cauliflower. Convective cloud types are indicated by the root word *cumulo,* which means "heap" in Latin. Convective clouds may become very tall and are rounded on top. They generally form in unstable air.

Clouds are also be classified by their altitude and their ability to create precipitation. The root word *cirro* (meaning "curl") describes a high cloud that is usually composed of wispy ice crystals. The Latin word *alto* ("high") is used to indicate a cloud in the middle of the troposphere that is below the high cirro-type clouds (just as altos in a choir sing lower notes than sopranos, but higher notes than basses). The prefix or suffix *nimbus* ("rain") denotes a cloud that is causing precipitation.

Using the combination of appearance, altitude, and ability to make precipitation, a wide range of cloud types can be identified. **TABLE 4-3** classifies ten common cloud types and their typical heights above the ground, and **FIGURE 4-16** depicts them pictorially. Now we will examine each cloud type, from the ground up.

"Name That Cloud" to quiz yourself on the cloud types discussed in this section.

Low Clouds

Stratus

Stratus clouds, abbreviated St, are fog that hovers just above (rather than on) the ground. They are fuzzy and featureless in appearance (**FIGURE 4-17**). From the ground, these clouds appear light to dark gray in color and cover the sky. They are common along coastlines and in valleys. Early morning fogs may lift and form a stratus cloud. Stratus clouds may also originate when moist, cold air is advected at low altitudes over a region. No precipitation normally occurs with stratus clouds, although a fine mist is sometimes visible on car windshields during thick stratus.

Stratocumulus

Stratocumulus (Sc) clouds are low-lying clouds that cover the sky and appear white to gray in color (**FIGURE 4-18**). They are a combination of layered and convective cloud types. This is because stratocumulus clouds often occur in a shallow layer of unstable air near the surface that is overlain by stable air. Unlike featureless stratus clouds, stratocumulus clouds often appear in rows or patches. You can distinguish stratocumulus from stratus by looking for more variations in color and a lumpier appearance.

Stratocumulus clouds are common in certain regions, such as coastlines and in valleys. Marine stratocumulus layers are very persistent off the California and South American coastlines. In those regions, moist air flows over cooler waters and becomes saturated. Stratocumulus clouds are also associated with fronts. When accompanying a large weather system, stratocumulus clouds are often the last clouds to appear before the skies clear completely.

Precipitation typically does not occur with stratocumulus. However, if the unstable surface air grows deeper (e.g., as a result of daytime heating), the stratocumulus can grow taller and develop into convective clouds that produce rain or snow.

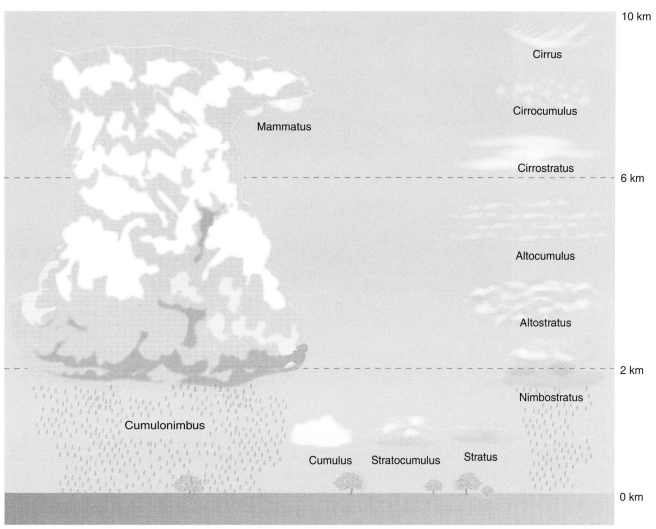

FIGURE 4-16 The major cloud types arranged by their typical altitude.

FIGURE 4-17 Stratus (St) are low-altitude layered clouds.

FIGURE 4-18 Stratocumulus (Sc) are low-lying clouds with both layered and convective aspects. Stratocumulus are distinguished from stratus clouds by variations in color across the sky.

Cumulus

Cumulus (Cu) clouds generally have well-defined, flat bases and intricately contoured domed tops resembling cauliflower. The edges of the cloud are distinct. The bases are generally dark gray and the sunlit sides are bright white.

These clouds form whenever fairly humid air rises, usually by convection. The height of the bottom of the cloud (the cloud base) is related to the temperature and the dew point of the rising air. Cumulus clouds in the dry southwestern United States generally have much higher bases than those in the Southeast. Cumulus clouds may also form over mountains or large hills if the air is unstable. These orographically forced clouds appear stationary, although they continually form and dissipate.

The two basic forms of cumulus clouds are fair-weather cumulus and cumulus congestus. Fair-weather cumulus clouds symbolize pleasant weather conditions all over the world (**FIGURE 4-19**). They have a height similar to their width. These clouds are common in summer when solar heating of the surface triggers convection. During autumn and winter, cumulus clouds often form in cold air over large open lakes that are still warm. Fair-weather cumulus are not deep enough to cause rain, although some may grow into large storms.

Cumulus congestus, or towering cumulus, are tall relative to their width. For these clouds to form, the atmosphere must have a deep unstable layer, deeper than is required for the formation of the fair-weather cumulus. These towering clouds are common in summer and may have light rain falling from them. In regions such as Florida, cumulus congestus may produce heavy rains for a few minutes. When cumulus congestus form in the morning it is a good indicator that storms may form later in the day. If the cloud tops appear fuzzy, ice is forming, and the cloud may be developing into a cumulonimbus.

Precipitating Clouds

Nimbostratus

Nimbostratus (Ns) are deep clouds that bring precipitation and appear dark gray to pale blue in color (**FIGURE 4-20**). The cloud base is difficult to see because precipitation is falling from

FIGURE 4-19 Cumulus (Cu) clouds are often observed on summer days.

the cloud. For this reason, nimbostratus sometimes look similar to stratus, stratocumulus, or altostratus clouds. Nimbostratus clouds often precede warm fronts.

The precipitation that falls from nimbostratus clouds is usually continuous and light to moderate in intensity. More episodic and intense precipitation is associated with cumulonimbus clouds.

Cumulonimbus

Cumulonimbus (Cb) are thunderstorm clouds. They extend upward to high altitudes, often to the tropopause and sometimes into the lower stratosphere. Cumulonimbus clouds produce large amounts of precipitation, severe weather, and even tornadoes (**FIGURE 4-21**).

FIGURE 4-20 Nimbostratus (Ns) are deep layered clouds that bring precipitation and appear dark gray to pale blue in color.

FIGURE 4-21 Cumulonimbus (Cb) over Lake Wingra, Madison, Wisconsin.

A distinguishing feature of cumulonimbus is the flattened **anvil** shape of the top of the cloud. The anvil develops when the updraft slows and spreads outward horizontally as it encounters the very stable air in the stratosphere. Underneath the anvil, sinking air may create pouches called **mammatus** (FIGURE 4-22, and see this book's cover). Although mammatus clouds are not severe weather, they can form under the anvils of strong thunderstorms.

Cumulonimbus clouds develop in unstable, moist atmospheres and are fairly common in the United States in spring and summer. They often occur ahead of cold fronts. In summer they can form over mountains because of orographic lifting in combination with solar heating. Cumulonimbus clouds can be isolated or organized in groups. When cumulonimbus clouds

FIGURE 4-22 Pouchy mammatus clouds (top half of photograph) sometimes form on the underside of cumulonimbus anvils.

develop into an organized system, the chance of severe weather often increases, as we will see in Chapter 11.

Middle Clouds

Altostratus

Altostratus (As) are layered clouds made up mostly of liquid water droplets. They are gray to pale blue in appearance (**FIGURE 4-23**). Altostratus form when the middle layers of the atmosphere are moist and slowly lifted. If the Sun appears through these clouds, it has a "watery" appearance, whereas stratus clouds normally obscure the Sun. Altostratus clouds are often observed ahead of a warm front, before the nimbostratus.

Altocumulus

The appearance of **altocumulus** (Ac) clouds varies considerably. They can be thin or thick, white or gray, and organized in lines or randomly distributed. They occur in the middle levels of the atmosphere when the air is moist, not too stable, and is being lifted. They are similar in appearance to stratocumulus, although with a higher cloud base (**FIGURE 4-24**). Altocumulus clouds often appear ahead of a warm front, prior to altostratus. If other cloud types accompany altocumulus, a storm is probably approaching.

You can distinguish between various types of cumulus clouds using the "fist-thumb-pinkytip" rule. Because of distance, clouds that are higher up appear smaller to your eye than those closer to the ground. If you extend your arm on a line from your eye to the cloud, cumulus clouds are generally about as big as your fist. Altocumulus clouds, in contrast, are only as big as your thumb. If the lumps of cumulus are even smaller, as small as the tip of your little finger, then the cloud is probably cirrocumulus (described in the next section).

High Clouds

Cirrocumulus

Cirrocumulus (Cc) clouds are thin, white clouds that often appear in ripples arranged in a regular formation (**FIGURE 4-25**). The smaller size of the individual cumulus lumps in cirrocumulus

FIGURE 4-23 Altostratus clouds (As) are layered clouds that exist in the middle layers of the troposphere and give the Sun or Moon a "watery" appearance.

FIGURE 4-24 Altocumulus (Ac) occur in the middle levels of the atmosphere when the air is moist.

FIGURE 4-25 Cirrocumulus (Cc) are thin, white clouds that appear high in the troposphere.

clouds distinguish this cloud type from altocumulus. Cirrocumulus clouds are composed of ice crystals and occur high in the atmosphere in regions that are relatively moist and unstable.

Although these clouds occur year-round, they are not very common and are usually present with other cloud types. Their tiny, delicately shaped features make cirrocumulus among the most beautiful of clouds. A "mackerel sky" is one that contains cirrocumulus clouds (or small altocumulus clouds) in a pattern that resembles fish scales.

Cirrocumulus clouds appear in association with large precipitation-causing weather systems, especially warm fronts. This cloud type usually follows cirrus and precedes altocumulus as a precipitation-causing warm front approaches. For this reason, the saying "Mackerel sky, not three days dry" became a popular piece of weather folklore in the days before modern weather forecasting.

Cirrostratus

Cirrostratus (Cs) clouds can cover part or all of the sky. They are uniform in appearance and can be thin or thick and white or light gray in color (**FIGURE 4-26**). Sometimes cirrostratus clouds are almost invisible and the Sun shines through easily, unlike the "watery sky" of altostratus. They occur high in the atmosphere and are composed of ice crystals.

Cirrostratus clouds are common during winter in association with large-scale weather systems. If the cirrostratus cloud thickens into altostratus, an approaching storm is indicated. They may also appear far out in advance of a tropical or subtropical weather disturbance. Cirrostratus clouds are most famous for the optical effects that occur when the Sun or Moon shine through them. Halos, bright arcs, and brilliant spots form when light passes through the ice crystals composing the cirrostratus. We examine these optical effects in the next chapter.

Cirrus

Cirrus (Ci) are wispy, fibrous, white clouds that are made of ice crystals. They often occur as wisps here and there across the sky and are aligned in the same direction as the upper-level winds (**FIGURE 4-27**). They are a very common cloud type associated with all weather systems, including fair-weather high-pressure areas. Cirrus clouds precede warm fronts and accompany

FIGURE 4-26 Cirrostratus (Cs) are layered clouds that are sometimes observed in connection with optical effects, such as halos and sundogs.

FIGURE 4-27 Cirrus (Ci) are wispy, fibrous, white clouds that are composed of ice crystals.

jet streams. Mountains can also generate cirrus clouds when air is forced over high peaks. When isolated cirrus occur, they do not indicate approaching bad weather. Mares' tails are cirrus clouds that are long and flowing, like a horse's tail.

There are many other types of clouds, including some that are associated with specific weather and climate patterns such as the ozone hole and mountain wind circulations. We will examine those clouds in later chapters when we discuss the phenomena associated with them.

CLOUDS AND THE GREENHOUSE EFFECT

Before we delve into the details of cloud composition, let's discuss their crucial role in the global warming debate. As we learned in Chapter 2, greenhouse gases such as water vapor and carbon dioxide warm the atmosphere by absorbing the longwave radiation emitted from the surface. Water vapor is an important greenhouse gas because it absorbs longwave energy effectively. Absorption of that energy warms the atmosphere. Increases in greenhouse gases over time can result in a climate change because the atmosphere becomes more effective at absorbing longwave energy emitted by the surface.

As the atmosphere warms, initially the relative humidity should decrease. Evaporation depends on relative humidity. With a lower relative humidity, more evaporation occurs, which adds more water molecules to the atmosphere and enhances the greenhouse warming. Increases in the temperature of the atmosphere would affect the dynamics of weather and climate.

Greenhouse gases are not the whole story, however. Clouds have a large impact on the energy gains of the atmosphere. Clouds reflect solar energy into space, away from the air beneath them, and clouds reduce the amount of solar radiation reaching the surface. Because of this, clouds tend to cool the Earth (**FIGURE 4-28**). The thicker the cloud, the more energy reflected back to space and the less solar energy available to warm the surface and atmosphere below the cloud. By reflecting solar energy back to space, clouds tend to cool the planet.

Clouds also have a warming effect on atmosphere below them because they are very good emitters and absorbers of terrestrial radiation. Clouds block the emission of longwave radiation to space and inhibit the ability of the planet to emit its absorbed solar energy to space in the

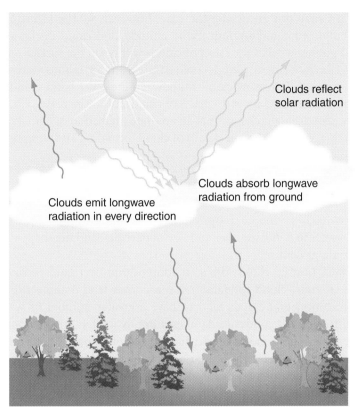

Clouds reflect solar radiation

Clouds absorb longwave radiation from ground

Clouds emit longwave radiation in every direction

FIGURE 4-28 In the solar spectrum, clouds tend to cool Earth. In the longwave spectrum, they tend to warm the planet.

form of longwave radiation (Figure 4-28). Thus, in the longwave, clouds act to warm the planet, much like the greenhouse gases do.

To complicate matters, the altitude of a given cloud is important in determining how much it warms the planet. Cirrus are cold clouds. Thick cirrus clouds emit very little energy out to space because of their cold temperature, according to the Stefan-Boltzmann Law from Chapter 2. At the same time, cirrus clouds are effective at absorbing the surface-emitted heat, which keep that energy from being lost to space. Thus, with respect to longwave radiation losses to space, cirrus clouds tend to warm the planet. The longwave effect dominates, and cirrus clouds, in general, tend to warm the planet in comparison to clear-sky conditions.

Stratus clouds over water tend to cool the planet. This is because stratus clouds are very effective at reflecting solar energy out to space reducing the net energy gain. Stratus are low in the atmosphere and therefore have temperatures that are close to the surface temperatures, so adding them to clear sky conditions does not change the outgoing longwave energy. The shortwave effect dominates, and maritime stratus clouds tend to cool the planet.

To complicate matters still further, a cloud's effectiveness at reflecting sunlight is related to how large the cloud droplets or cloud ice crystals are. We will investigate the reasoning behind this in Chapter 5, but it is easily demonstrated with ice in a familiar form. If you look at a glass filled with crushed ice next to a glass filled with ice cubes, you can see that the crushed ice is brighter (whiter) than the glass of ice cubes. The crushed ice particles are smaller than the cubes. Similarly, clouds consisting of small droplets are brighter than clouds consisting of large particles. Clouds composed of small particles therefore have a higher albedo and reflect more solar radiation back into space, causing more cooling than clouds with large particles.

In summary, clouds can act to cool or warm the planet, depending on how much of the Earth they cover, how thick they are, how high they are, and how big the cloud particles are. Measurements by NASA indicate that, on average, the reflection of sunlight by clouds more than compensates for the clouds' greenhouse warming. Thus, today's distribution of clouds tends to cool the planet.

Hexagonal plate
0 to -5° C
-10 to -12° C
-16 to -25° C

Needle
-5 to -10° C

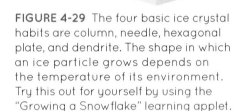

Dendrite
-12 to -16° C

Column
-5 to -10° C
-25 to -50° C

FIGURE 4-29 The four basic ice crystal habits are column, needle, hexagonal plate, and dendrite. The shape in which an ice particle grows depends on the temperature of its environment. Try this out for yourself by using the "Growing a Snowflake" learning applet.

"Growing a Snowflake" make your own beautiful and complex ice crystals.

"Precipitation Formation" explore how cloud particles become precipitation size.

This may not always be the case, however. As the atmosphere warms, the distribution of cloud amount, cloud altitude, and cloud thickness all may change. We do not know what the effect of clouds will be on the surface temperatures as the global climate changes. Clouds could dampen any greenhouse warming by increasing cloud cover or decreasing cloud altitude. On the other hand, clouds could increase a warming if the cloud cover decreases or cloud altitude increases. Climate is so sensitive to how clouds might change that an accurate prediction of climate change hinges on correctly predicting the fine details of cloud formation and composition.

CLOUD COMPOSITION

Every cloud has a unique composition. The composition of a cloud includes the phase of the water in it, the number and size of particles, and the shape of any ice particles, if they are present.

Water-bearing clouds differ in composition over land versus over the oceans. There are more cloud condensation nuclei over land than over the oceans. For this reason, continental clouds tend to have a greater number of water droplets than maritime clouds. Clouds over land have approximately 500 million to 1 billion cloud droplets per cubic meter of air; maritime clouds have about one tenth as many. Because the water content of maritime and continental clouds are similar, however, the drops in continental clouds are usually smaller and more numerous than the maritime counterparts. Maritime clouds have large, soluble, heterogeneous nuclei that favor the formation of large droplets.

Ice-containing clouds vary greatly in terms of the number, size, and shape of the ice crystals in them. The size and shape of a crystal is called its **crystal habit**. Temperature determines the particular crystal habit of ice. FIGURE 4-29 shows the four basic shapes of ice crystals, each of which occur preferentially in the following temperature ranges: the **hexagonal plate** (0° C to −5° C; −10° C to −12° C; −16° C to −25° C), the **needle** (−5° C to −10° C), the **column** (−5° C to −10° C; −25° C to −50° C), and the **dendrite** (−12° C to −16° C). The dendrites are hexagonal with elongated branches, or fingers, of ice. They most closely resemble what we think of as snowflakes. We will soon learn that this is because ice crystals grow fastest around −15° C, the range in which dendrite formation is preferred.

PRECIPITATION

Precipitation is any liquid or solid water particle that falls from the atmosphere and reaches the ground. Precipitation can be long lasting and steady, or it may fall as a brief and intense **shower.** Because precipitation is formed from water vapor, it removes water vapor from the atmosphere, returning it to the Earth's surface. Rain, snow, sleet, freezing rain, and hail are all forms of precipitation.

Dew and frost also remove water vapor through condensation or deposition onto surfaces on the ground. Dew and frost are not precipitation because they do not fall from a cloud under the force of gravity. In precipitation, water vapor condenses onto a particle that eventually grows large enough to fall out of a cloud and to the surface. These growth processes are dependent on the cloud temperature. **Warm clouds** are those that have temperature greater than freezing throughout the cloud. **Cold clouds** have temperatures that are below freezing. After discussing how particles grow into precipitation, we examine precipitation types.

Precipitation Growth in Warm Clouds

Rainmaking, natural or artificial (BOX 4-2), is not easy. Cloud particles are usually 10 microns (μm) in size. (For comparison, the period at the end of this sentence is about 500 μm in diameter.)

Box 4-2 Controlling the Weather

It is an age-old question: Can humans control the weather? In the past, people rang bells or fired cannons to prevent lightning or cause rain—producing sound and fury, but nothing in the way of success.

The scientific era of weather modification began in the 1940s and 1950s with the advent of cloud-seeding experiments. In cloud seeding, airplanes drop particles of dry ice or silver iodide into cold clouds. These particles act as extremely effective ice nuclei, potentially increasing the amount of rainfall or snowfall. However, progress in cloud seeding has been slow. Today, scientists agree that the seeding of orographic clouds can enhance precipitation by a modest 10% on a seasonal basis. How this actually happens is still poorly understood.

Farmers have long sought for a way to suppress hail. One hailstorm can destroy a year's crops in a few minutes. Cloud seeding during the early stages of cumulonimbus development can reduce hail damage by keeping hailstone sizes small, but the results so far have been mixed. Research on hail suppression continues.

Another active area of weather modification is fog dispersal. Fog can shut down an airport for hours, causing delays and cost overruns. Like clouds, fog can be seeded with materials that cause the water droplets in fog to turn into ice, which cleanses the air of fog. This is now a routine practice at many airports worldwide. However, it is impractical over large regions.

At one time or another you may have wondered, "Why can't meteorologists make tornadoes and hurricanes go away?" Perhaps a nuclear explosion or, less violently, cloud seeding could disrupt the circulation of a tornado or a hurricane. Alternatively, oceans could be covered with water-impermeable chemicals, cutting off a hurricane's fuel source.

Severe weather is here to stay, however. Hurricane seeding in the 1950s and 1960s produced few firm results. Since then, research on this subject has been curtailed. Tornadoes are even less well understood than hurricanes, and currently no anti-tornado research is in progress. Legal and ethical dilemmas arise whenever severe weather modification is proposed. The atmosphere cannot be put into a test tube, and so experiments must be conducted on actual storms. However, what if a cloud-seeded hurricane hits New York City instead of moving out to sea harmlessly? Who is to blame? How would the environment respond if the Gulf of Mexico were covered with a film of chemicals? Obviously, the consequences of exploding nuclear bombs in storms would be horrific. The difficulty of weather modification, combined with the risks and dilemmas associated with it, has kept work on this subject to a minimum. We will revisit some of these same concerns in Chapter 16 when we discuss geoengineering as a "fix" for climate change.

As we will see throughout this text, humans have changed the atmosphere in many ways. These changes usually come about as unintended by-products of modern civilization. So far, humans have been more effective at modifying weather and climate by mistake than by design.

Small raindrops are usually 1000 μm in diameter. About 1 million cloud droplets have to combine to form a single raindrop. How does this happen? Just as with cloud particle growth, the formation of precipitation is complicated by the different properties of water at different temperatures.

The simple, but incorrect, explanation for precipitation is growth by condensation. Why can't a cloud droplet grow into a raindrop by condensing water onto its surface? Because this process does not work fast enough to produce the precipitation particles we see in real life. It would take several days for a full-sized raindrop to form by condensational growth. Precipitation forms much more quickly than that. So, while condensational growth is an important beginning step, there must be another mechanism for the relatively rapid growth of precipitation-sized particles.

Collision–Coalescence

One process that could produce a larger drop quickly would be to combine many smaller particles. To do this the cloud particles have to bump into each other and merge together, or coalesce. This is called the **collision-coalescence** process.

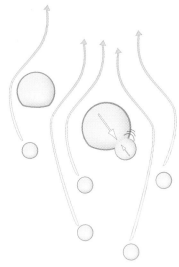

FIGURE 4-30 Large water drops fall faster than smaller ones. Because of the different fall speeds, water drops sometimes collide and coalesce. Very small droplets may flow around the larger drops and avoid colliding with them.

To explain how this process works, let's consider water droplets in an updraft. Water droplets in clouds with different sizes move at different speeds, as gravity and vertical motions act on them. The difference in speed increases the chance of collisions, just as the combination of fast trucks and slow cars increases the chance of collisions on a highway.

Just because two droplets approach one another does not guarantee they will collide. A large droplet in motion creates an air current around it that can be strong enough to force tiny droplets to flow around it. (For the same reason, it is rare for cars on a highway to hit flying birds because the air current around the car carries the bird around, not into, the car.) This reduces the likelihood that the two drops will collide, although initially they may be headed straight for one another (**FIGURE 4-30**). The percentage of collisions, termed the collision efficiency, between large drops and very tiny droplets is low. The collision efficiency is also low for two drops of the same size, as they will likely be falling at the same speed and therefore will not collide. Turbulence in the cloud can increase the number of collisions.

Even if two droplets collide, they don't always stick together or coalesce. The two drops can bounce off one another. This is not too common, however. If the cloud particles are charged, the coalescence process is enhanced.

After a drop grows to a size where the downward force of gravity exceeds the updraft force of air currents, it falls downward through the cloud. As the drop falls through the cloud, it can sweep up smaller droplets in its path, collecting them and growing by collision-coalescence.

The process of combining particles through collision-coalescence is an important mechanism for forming precipitation in clouds composed solely of liquid water droplets. Therefore, it is most effective in "warm" clouds in the tropics. Outside of the tropics, clouds contain ice particles, even in the summertime. The next section discusses how precipitation forms in cold clouds.

Precipitation Growth in Cold Clouds

Most clouds outside the tropical regions have temperatures that are below freezing. Our personal experiences confirm that these clouds can produce precipitation. This section explores the processes that enable droplets and crystals in cold clouds to reach precipitation size.

Accretion and Aggregation

Collision also helps create precipitation in cold clouds. When an ice crystal falls through a cloud it may collide with and collect supercooled water drops. This process of ice crystal growth by sweeping up supercooled water drops is called **accretion**, which can be thought of as a riming of the crystals. When ice crystals collide with supercooled drops, the drops freeze almost instantly. Accretion thus provides a mechanism for the particle to grow quickly.

An ice particle produced by the accretion process that has a size between 1 and 5 millimeters (0.04 to 0.2 inches) and no discernible crystal habit is called **graupel** (pl. graupeln). On collision and freezing, the supercooled water often traps air bubbles. Because of this trapped air, the density of a graupel is low, and it can easily be crushed, unlike a hailstone.

Aggregation is the process by which ice crystals collide and form a single larger ice particle (**FIGURE 4-31**). The probability that two crystals will stick together depends on the shape of the crystals. If two dendrites collide, it is likely that their branches (arrows in Figure 4-31) will become entangled and that the two crystals will stick together. When two plates collide, there is a good chance that they will simply bounce off one another.

Temperature also plays a role in aggregation. If the temperature of one crystal is slightly above freezing, it may be encased in a thin film of liquid water as it is melting. If this particle collides with another crystal, the thin film of water may freeze at the point of contact and bond the two particles into one. (This is why you should never lick a cold metal flagpole!)

A **snowflake** is an individual ice crystal or an aggregate of ice crystals. Snowfalls do not consist of single crystals. More commonly, they are composed of flakes that are aggregates of ice crystals. Snowflakes composed of aggregates can sometimes reach 7.5 to 10 cm (3 or 4 inches) in size.

In both warm and cold clouds, how big a droplet or crystal grows by collision processes depends on how long it stays in the cloud. The longer a particle is in the cloud, the more particles it can collect and the larger it grows. The strength of the updrafts and the thickness of the cloud determine how long it stays in the cloud. This is why only tall clouds with strong updrafts, such as cumulonimbus clouds, produce large precipitation particles. In contrast, stratus clouds are shallow in depth and have much weaker updrafts than cumulonimbus. Particles usually do not stay in a stratus cloud long, and large particles rarely form.

The Bergeron-Wegener Process

The presence of both water and ice in a cloud gives it a unique precipitation-making ability. To understand it, let's consider two sealed containers connected by a tube with a valve that can be opened (**FIGURE 4-32**). One container holds supercooled water at a temperature of −5° C (23° F), and the second contains ice at the same temperature.

The bonding forces in ice are much stronger than those in water, so fewer molecules have the energy to escape the ice than the number leaving the water. This means there will be fewer molecules of water in the vapor phase in the container with the ice than in the container that has the water—that is, the vapor pressure over water is greater than the vapor pressure over the ice. Because vapor pressure is proportional to the number of water molecules in the air, this also means that at a given subfreezing temperature the saturation vapor pressure is greater over water than over ice. If the valve were opened between the containers, water molecules in the vapor would flow toward the region of lower vapor pressure over the ice.

Now consider an ice crystal surrounded by supercooled droplets (**FIGURE 4-33**). If the air is saturated (100% relative humidity) with respect to the water droplets, it is supersaturated with respect to the ice crystals. Water vapor molecules will deposit onto the crystal, lowering the relative humidity of the air. In response, water molecules evaporate from the water droplet, supplying more water molecules to the air that then deposit onto the crystal.

Put simply, ice crystals grow at the expense of water droplets in a cloud that has both. This ice crystal growth process is called the **Bergeron-Wegener process**. It was first proposed by meteorologist Alfred Wegener (who also famously proposed the theory of

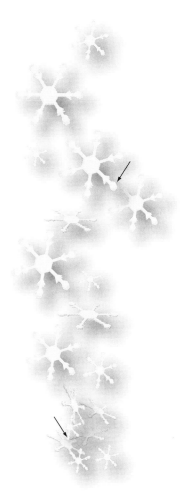

FIGURE 4-31 Ice crystals of different sizes or different shapes may collide and stick together, or aggregate.

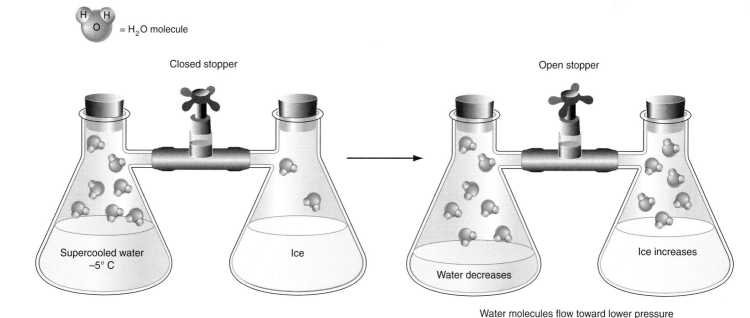

FIGURE 4-32 At a given temperature, the saturation vapor pressure over ice is less than the saturation vapor pressure over water. As a result, water vapor is preferentially attracted to ice versus water.

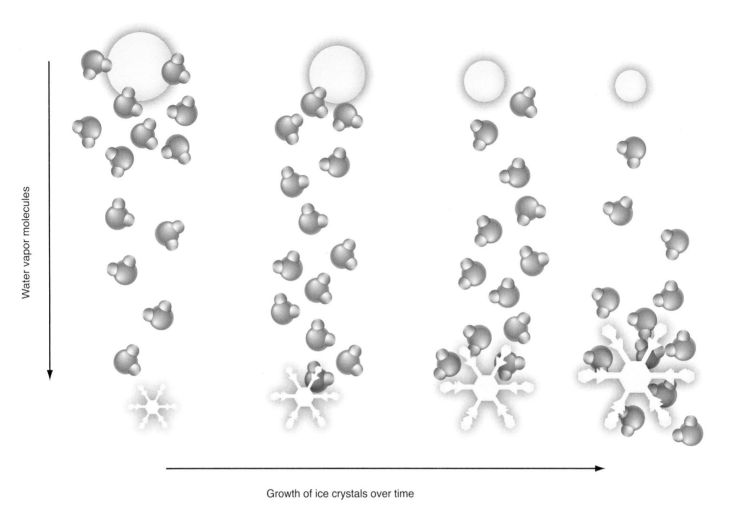

Growth of ice crystals over time

= H₂O molecule

FIGURE 4-33 The larger saturation vapor pressure over a liquid water surface than over an ice surface causes the ice crystal to grow and the supercooled drops to evaporate.

continental drift) in 1911 and explained more extensively by Tor Bergeron, a member of the renowned Bergen School of Norwegian meteorology. This process contributes to the rapid growth of ice crystals and, therefore, to the ability of a cloud to form precipitation.

This ice crystal growth process is fastest when the difference in the saturation vapor pressure between water and ice is large. As shown in **FIGURE 4-34**, these differences maximize when the air temperature is between −12° C and −17° C (10.4° F and 1.4° F). Based on what we have already learned about crystal shapes, this means that dendrites grow the fastest by the Bergeron-Wegener growth process. This explains why snowflakes look more like asterisks than plates, columns, or needles.

◼ Precipitation Types

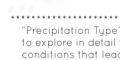

"Precipitation Type" to explore in detail the conditions that lead to various precipitation types.

What happens when a particle falls out the base of a cloud? It is not officially precipitation until it reaches the ground. In some cases this never happens. For instance, if the atmosphere below the cloud is very dry, the particle may evaporate in between the cloud base and the ground. **Virga** is rain that evaporates before reaching the surface.

Similarly, falling ice crystals may sublimate in dry air while being carried horizontally by the strong winds aloft. These **fallstreaks** (**FIGURE 4-35**) often produce visually striking patterns.

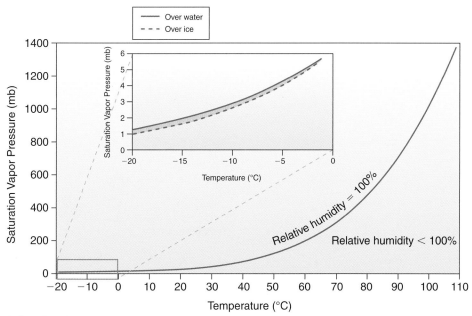

FIGURE 4-34 The differences between the saturation vapor pressure over ice and over water. The larger graph shows the saturation vapor pressure over water, which is the upper boundary of the observations in Figure 4-3. The inset compares the saturation vapor pressure over ice and over water. Because the line for water lies above the line representing ice, when the air is saturated (100% relative humidity) with respect to liquid water, it is supersaturated with respect to ice.

FIGURE 4-35 Fallstreaks are wisps of ice particles that fall out of a cloud but evaporate before reaching the surface. The bright white part of the clouds is composed of supercooled droplets that freeze and form ice crystals, which grow and fall. Fallstreaks usually exhibit a hooked form produced by changes in wind speed with height.

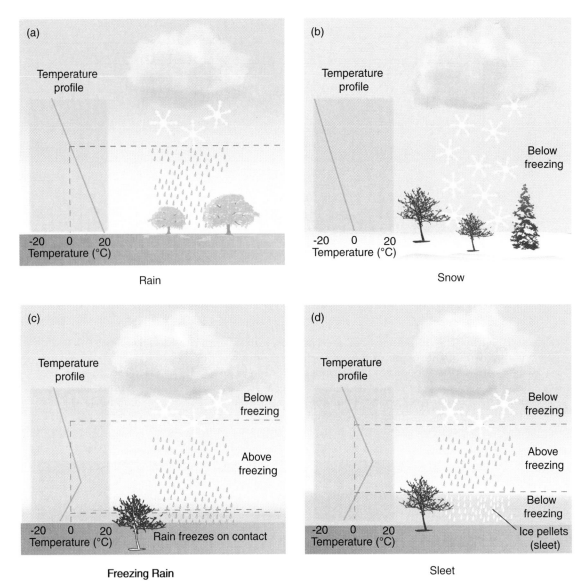

FIGURE 4-36 The vertical variation of temperature will determine whether precipitation falls as rain (a), snow (b), freezing rain (c), or sleet (d). In this example, all of the precipitation particles are initially ice crystals. Particles that fall into the melting layer become liquid drops. For freezing rain or sleet, a temperature inversion is required.

When particles reach the surface as precipitation, they do so primarily as rain, snow, freezing rain, or sleet. Why are there different types of precipitation? Again, the ability of water to change phase is the key. In midlatitude regions, precipitation usually begins as ice particles. **FIGURE 4-36** shows the temperature conditions below the cloud base that lead to rain, snow, freezing rain, and sleet. The dashed line in Figure 4-36 represents the melting line, the altitude at which the temperatures are 0° C (32° F). The main difference between the different types of precipitation is whether an inversion exists near the ground—and if so, how thick the inversion is. Now, let's examine each precipitation type in detail, with reference to Figure 4-36.

Rain

If the temperature remains above 0° C (32° F) from the cloud base to the surface, precipitation particles melt into liquid droplets called **rain** (Figure 4-36a). Raindrops are not teardrop shaped; they are spheres flattened by the pressure of the wind as they fall. Raindrops are at least 500 µm (0.5 millimeters or 0.02 inches) in diameter. Precipitation drops smaller than this are collectively called **drizzle**, which is often associated with stratus clouds.

Average annual rainfall as estimated by satellite measurements is shown in **FIGURE 4-37**. The data in this figure mesh well with what we have learned about water in the atmosphere.

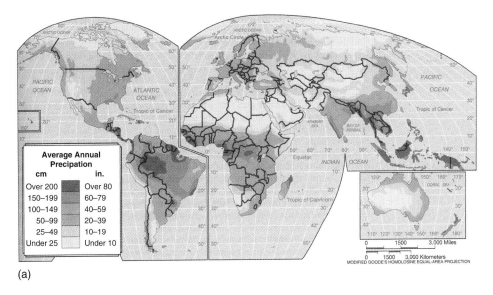

(a)

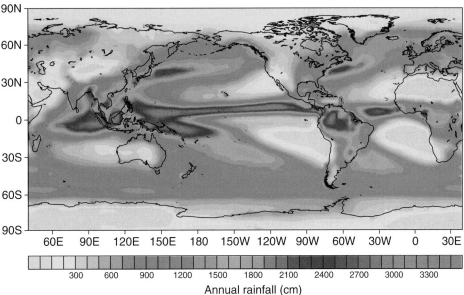

(b)

FIGURE 4-37 (a) Climatology of annual precipitation over the land regions of the world. (b) Climatology of annual rainfall across the world, including the oceans, based on satellite observations. (Courtesy of University of Washington, Joint Institute for the Study of the Atmosphere and Ocean [JISAO].)

For example, the warm tropical regions of the globe receive the most rainfall. This is partly a consequence of the fact that there is more water vapor in warm, moist air than in cooler air (Figure 4-3). But why is there a narrow band of high rainfall near the equator? Also, why are the regions offshore of California and western South America so dry? To explain both of these enigmas, we will need to understand global-scale wind patterns, the topic of Chapter 7.

Rain intensity is classified by the volume of rain that falls in an hour, according to the following scale:

Intensity	Hourly Rainfall
Light	0.25–2.5 millimeters (0.01–0.10 inches)
Moderate	2.5–7.6 millimeters (0.11–0.30 inches)
Heavy	More than 7.6 millimeters (0.30 inches)

Extreme rainfall rates are possible. When this happens, severe flooding usually follows. On the Fourth of July in 1956, Unionville, Maryland, was drenched with a world-record

31.24 millimeters (1.23 inches) of rain in only 1 minute! A sequence of storms over the upper Midwestern United States in late May and early June of 2008 caused severe flooding in Iowa. Two weeks of rain, more than 400 millimeters (16 inches) in some places, led to water levels that exceeded the 100- and 500-year flood levels in several parts of the state. The flooding of the capital city of Des Moines inundated portions of the campuses of the universities and colleges located in the city. The University of Iowa had to replace buildings that housed the School of Art and Art History and the School of Music.

Snow

If the temperature underneath a cloud stays below freezing all the way to the ground, the snow-flakes never melt and **snow** falls (Figure 4-36b). As with rain, snow intensity is recorded in three categories. This is done according to volume of snowfall or, because snow reflects and scatters light effectively, according to the reduction in visibility that results from the snowfall. The categories are as follows:

Intensity	Hourly Snowfall	Visibility
Light	Less than 0.5 centimeters (0.2 inches)	0.8 kilometers (0.5 miles) or more
Moderate	0.5–4 centimeters (0.2–1.5 inches)	0.4–0.8 kilometers (0.25–0.5 miles)
Heavy	Greater than 4 centimeters (1.5 inches)	Less than 0.4 kilometers (0.25 miles)

Water expands when it freezes, and snowflakes trap air between them when they clump together on the ground. For these reasons, an inch of snow and an inch of rain are not the same thing. A general rule is that 10 inches (25 centimeters) of new snow has the same water content as just 1 inch (2.5 centimeters) of rain (Chapter 5 explores this ratio in more detail in the discussion of measuring snowfall). Cold, dry snows may be closer to a 20-to-1 ratio. However, wet snows with relatively warm temperatures can contain much more water, as much as 4 inches of water per 10 inches of snow. Several factors contribute to snow-to-liquid ratio, including crystal habit and size and the amount of sublimation and melting below the cloud base.

These statistics can help us interpret the annual average snowfall across the United States, as shown in **FIGURE 4-38**. The northern Great Plains receive up to 5 feet of snow annually.

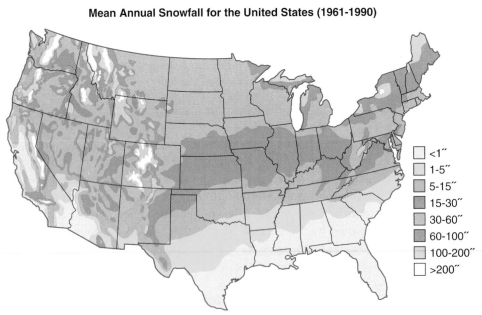

Mean Annual Snowfall for the United States (1961-1990)

☐	<1″
☐	1-5″
☐	5-15″
☐	15-30″
☐	30-60″
☐	60-100″
☐	100-200″
☐	>200″

FIGURE 4-38 Climatology of the average annual snowfall across the lower 48 United States. Latitude and proximity to water strongly affect snowfall amounts over the eastern states, whereas elevation dominates in the western states. (Prepared by Colorado Climate Center, Colorado State University, copyright © 1997.)

When this thick snow cover melts, however, it is equivalent to just a few inches of rainfall. This is one reason why the Great Plains are dry. Elsewhere in Figure 4-38, it is apparent that the snowiest regions are in the Rocky Mountains, where cold temperatures and orographic lifting enhance snowfall.

Like rainfall, snowfall rates can be extreme. In Silver Lake, Colorado, on April 14 and 15, 1921, a whopping 195.6 centimeters (76 inches, or more than 6 feet) of snow fell in just 24 hours, a North American record.

Sleet and Freezing Rain

Ice storms occur when precipitation particles melt and then fall through a layer of cold air near the ground. The two precipitation types most common during ice storms are freezing rain and sleet.

Freezing rain forms when a thin layer of cold air near the surface causes melted precipitation to become supercooled (Figure 4-36c). It then freezes on contact with exposed objects. Freezing rain covers everything in a sheet of ice, creating shimmering landscapes. However, even a little freezing rain causes treacherous road conditions and tree and power line damage. It is also responsible for aircraft icing, which is a cause of fatal plane accidents. Rime on the plane wings is also dangerous for flight.

FIGURE 4-39 shows a 30-year climatology of the number of hours of freezing precipitation (rain and drizzle) across the North American continent. Freezing rain is common across most

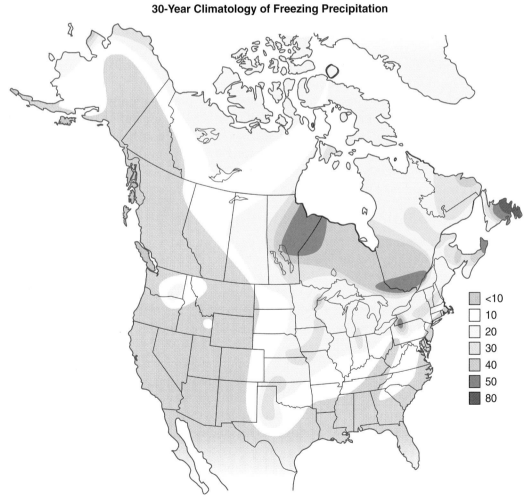

30-Year Climatology of Freezing Precipitation

	<10
	10
	20
	30
	40
	50
	80

FIGURE 4-39 A 30-year climatology of freezing precipitation (rain and drizzle) across North American continent, shown in terms of the number of hours of freezing rain per year. (Adapted from Cortinas, V. J., et al., *Monthly Weather Review*, April 2004.)

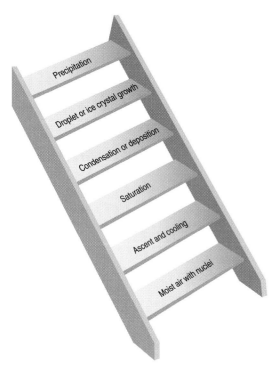

FIGURE 4-40 The steps by which clouds and precipitation are made, starting at the bottom and climbing to the top.

of central and eastern Canada and most of Alaska. A region of freezing precipitation for more than 20 hours a year extends from the western high plains through the Great Lakes region into eastern Canada and New England. Newfoundland has the greatest number of hours of freezing rain, but locations along the southern Appalachians as far south as northeast Georgia can have at least 10 hours of freezing rain per year. These are regions where cold air sinks or can be trapped at the surface (discussed in greater detail in Chapter 12). Meanwhile, warmer precipitating air moves above the cold air, creating an inversion. This thin layer of below-freezing air at the surface is a key ingredient for freezing rain.

Sleet consists of translucent balls of ice that are frozen raindrops. It occurs when the layer of subfreezing air at the surface is deep enough for the raindrop to freeze (Figure 4-36d). Therefore, the difference between sleet formation and freezing rain formation is quite small, although the two precipitation types do not look at all alike. When sleet hits the surface, it bounces and does not coat objects with a sheet of ice, as freezing rain does. Instead, it covers flat surfaces such as roads and driveways like millions of icy ball bearings.

Keep in mind that the different scenarios in Figure 4-36 are idealized. Snow and sleet can occur at temperatures above 0° C when the air underneath the cloud is dry. In these cases, the precipitation particle partly evaporates as it falls. The evaporation cools the particle enough to keep it frozen all of the way to the ground. Sleet seems particularly resistant to melting. One of us has seen sleet three separate times with surface temperatures at or above 10° C (50° F)! You can explore how and under what conditions different forms of precipitation occur using the Precipitation Types learning applet on the text's Web site.

In addition to rain, snow, freezing rain, and sleet, other precipitation types exist. The most notable is **hail**, which is precipitation in the form of large balls or lumps of ice that look like sleet on steroids. The formation of hail is quite different than sleet formation, however. Hail develops in the complex air motions inside a towering cumulonimbus cloud. For that reason, we will discuss hail in connection with thunderstorms in Chapter 11. The "ladder" in **FIGURE 4-40** summarizes the steps in the formation of the precipitation types we have discussed in this chapter.

Clouds, Lapse Rates, and Precipitation Near Mountains

In closing, let's investigate how concepts in this chapter integrate with concepts presented in previous chapters by analyzing airflow over a mountain (**FIGURE 4-41**). Our goal is to explain why semiarid **rain shadow** regions exist downwind (the "lee side") of a large mountain range.

When air rises up a mountain, it expands and cools at the dry adiabatic lapse rate. As the temperature decreases, the relative humidity increases, and the temperature approaches the dew point temperature. Eventually, the temperature will equal the dew point temperature, and a cloud will form, marking the cloud base.

If the air continues to rise, water vapor will continually condense to form cloud droplets. A phase change of water vapor to a liquid releases energy, warming the parcel through latent heating. This causes the parcel to cool more slowly, at the saturated adiabatic lapse rate. As the parcel continues to be orographically lifted, cloud droplets grow by condensation and by collision and coalescence. As the moist parcel continues to rise and cool, some of the liquid drops freeze. The cloud particles can then grow by the Bergeron-Wegener process, accretion, and aggregation.

The precipitation particles continue to grow and eventually fall to the ground on the windward side of the mountain as precipitation. Water molecules leave the air parcel as precipitation. The precipitating water cannot further affect the parcel because the water molecules are now on the ground.

At the top of the mountain, the relative humidity of the parcel is 100%. As the air sinks on the leeward side of the mountain, it warms, so the relative humidity decreases. The cloud

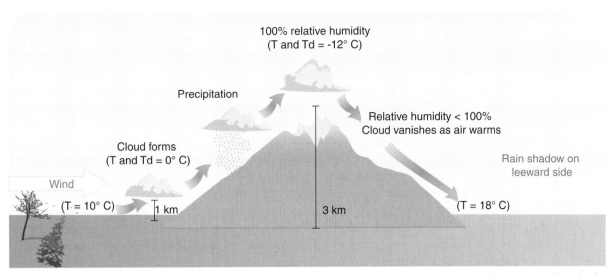

FIGURE 4-41 As moist air flows over a mountain, the temperature decreases, the air becomes saturated, clouds form, and precipitation falls on the upwind side. Sinking motions on the leeward side of the mountain cause the air to become warmer and drier than before. This generates a rain shadow downwind of the mountain.

particles evaporate as the air sinks and warms. The cloud disappears. With no cloud in the parcel, the parcel now warms all of the way down the mountain at the dry adiabatic lapse rate. This is a crucial difference because on the way up the mountain the parcel cools for part of its journey at the (smaller) saturated adiabatic rate. As a result, the parcel ends up on the lee side warmer than it was at the very beginning.

In addition, there are fewer water molecules in the parcel because of the precipitation on the windward side. Increasing the temperature and removing water vapor from the air both act to lower the relative humidity. This is why regions located downwind of a mountain range are both warmer and drier than their windward counterparts. These are the rain-shadow regions. The rain-shadow effect is strongest when the wind is nearly perpendicular to the mountain range. In these cases, the windward side will have more cloud cover than the leeward side. The increased cloud cover will in turn reduce the annual temperature range (see Chapter 3) on the windward side compared with the leeward side. This is supported by weather observations on each side of the Cascade Mountains of Washington, which run north to south in a region of west-to-east winds (TABLE 4-4).

These observations demonstrate the influence of the topography on water in the atmosphere. They also illustrate the far-reaching consequences of the principles we have studied in the first four chapters of this text. Observations are central to the study of weather and climate—so important, in fact, that we devote the next chapter to how we sense the atmosphere.

TABLE 4-4 Difference in Temperature, Cloud Cover, and Precipitation on the Windward and Leeward Sides of the Cascade Mountains in Washington State

	Windward (West) Side (Seattle-Tacoma)	Leeward (East) Side (Yakima)
Mean winter temperature	41° F	32° F
Mean summer temperature	64° F	68° F
Mean annual temperature range (warmest month's average minus the coldest month's average)	25° F	40° F
Number of mostly cloudy days per year	226	164
Average annual precipitation	37.2 inches	8.0 inches

Data from the Western Regional Climate Center, http://www.wrcc.dri.edu/summary/lcd.html.

PUTTING IT ALL TOGETHER

■ Summary

Water in the atmosphere exists as water vapor, clouds, and precipitation. We can measure water vapor in a variety of ways. Mixing ratio, vapor pressure, relative humidity, and dew point temperature are the most common "yardsticks" of water vapor concentrations. The vapor pressure of saturated (100% relative humidity) air increases rapidly as the air is warmed. Also, changing the amount of water vapor or the air temperature changes the relative humidity. In clear, calm, precipitation-free conditions, the relative humidity is usually highest at sunrise and lowest during the mid afternoon.

A cloud is a suspension of water droplets, ice crystals, or both. Updrafts in the cloud keep particles aloft. The formation of a cloud requires water vapor, saturated air, and nuclei onto which the vapor can condense or deposit. These nuclei can be water droplets or aerosol particles. Clouds over land have more, and smaller, cloud droplets than clouds over the oceans.

Fog is a cloud at ground level. Fogs develop through two processes that lead to saturation. Air can cool to the dew point and become saturated, producing radiation fog, advection fog, or upslope fog. Air can also saturate via evaporation of water into the air. This produces steam fog or evaporation fog.

Most other clouds are formed when air cools to saturation as it is lifted. The four primary mechanisms for lifting air are orographic lifting, frontal lifting, convection, and convergence near the surface.

Moisture affects how a parcel's temperature changes as it rises. The saturated adiabatic lapse rate of about 6° C per kilometer is less than the dry adiabatic lapse rate because of latent heating resulting from condensation. This means that saturated air parcels cool less quickly than dry parcels as they rise. As a result, an environment with a lapse rate in between the dry and saturated adiabatic lapse rates can be unstable but only for saturated air parcels. This is called a "conditionally unstable" environment and promotes the growth of clouds and thunderstorms.

Clouds can be classified as layered (strato-) or convective (cumulo-) and also as low, middle (alto-), high (cirro-), or precipitating (nimbo-). The 10 basic cloud types are cirrus, cirrostratus, cirrocumulus, altostratus, altocumulus, cumulus, stratus, stratocumulus, nimbostratus, and cumulonimbus (the thunderstorm cloud).

Clouds play a major role in the greenhouse effect. They warm the surface but also reflect sunlight, cooling the surface. In today's climate, clouds tend to cool the planet.

Precipitation processes differ in warm and cold clouds. In warm water-only clouds, precipitation-sized particles grow by collision and coalescence of large and small water droplets. In cold clouds with ice crystals, accretion, aggregation, and the Bergeron-Wegener process cause precipitation. In the latter process, ice crystals attract water vapor more strongly than do liquid water drops in a temperature range commonly found in midlatitude clouds.

The most common forms of precipitation are rain, snow, freezing rain, and sleet. Freezing rain and sleet form when there is a temperature inversion near the surface. The location of mountains and valleys affect the type and amount of both clouds and precipitation. In particular, regions downwind of a mountain range are drier and sunnier than the upwind slopes.

■ Key Terms

You should understand all of the following terms. Use the glossary and this chapter to improve your understanding of these terms.

Accretion	Bergeron-Wegener process	Collision-coalescence
Advection fog	Cirrocumulus	Column
Aggregation	Cirrostratus	Condensation nuclei
Altocumulus	Cirrus	Conditionally unstable
Altostratus	Cloud droplet	environment
Anvil	Cold clouds	Contact nucleation

Convection
Convective clouds
Convergence
Crystal habit
Cumulonimbus
Cumulus
Curvature effect
Dendrites
Deposition nucleation
Dew
Dew point
Dew point depression
Divergence
Drizzle
Evaporation
Evaporation fog
Fallstreaks
Fog
Freezing nucleation
Freezing rain
Frontal fogs
Frontal lifting
Frost

Frost point
Frozen dew
Graupel
Hail
Heterogeneous nucleation
Hexagonal plates
Homogeneous nucleation
Hydrophobic nuclei
Hygroscopic nuclei
Ice crystals
Ice nuclei
Immersion nucleation
Layered clouds
Lifting condensation level
   (LCL)
Mammatus
Mixing ratio
Needle
Nimbostratus
Nucleation
Orographic lifting
Precipitation
Radiation fog

Rain
Rain shadow
Relative humidity
Rime
Saturated adiabatic lapse rate
Saturation
Saturation vapor pressure
Shower
Sleet
Snow
Snowflake
Solute effect
Steam fog
Stratocumulus
Stratus
Supercooled water
Supersaturation
Updraft
Upslope fog
Vapor pressure
Virga
Warm clouds

◼ Review Questions

1. Name two ways that you can cause a parcel of air to become saturated.

2. One day, the dew point is 20° C. The next day, at the same location, the dew point is 10° C. On which day are there more water vapor molecules in the air? On which day is the relative humidity higher? (Hint: You may not have enough information to answer both questions.)

3. Why does the daily cycle of relative humidity look like the reverse of the daily cycle of temperature, with a maximum when the temperature is at a minimum? (Hint: What's the definition of relative humidity, and how does one of the variables in the definition relate to temperature?)

4. Do you think that relative humidity will reach its daily maximum at sunrise and be at its daily minimum at mid afternoon on a day that is cloudy with rain in the afternoon?

5. On November 28, 2001, the University of Virginia played Michigan State University in a basketball game held at the Richmond Coliseum. It was an unusually warm, muggy evening. Underneath the basketball court was a sheet of ice used for hockey games. The game was halted because of slippery floor conditions. Using your understanding of water in the atmosphere, can you explain why the floor became slippery?

6. On a cold night when frost is predicted, you park your car underneath a tree instead of out in the open. Will frost form on your windshield? Explain your prediction.

7. Sometimes a fog will appear over a roadway after a summer rain shower. What type of fog is this?

8. A parcel of air at sea level has a temperature of 20° C and a dew point of 0° C. Assume that the dew point does not change as the unsaturated parcel rises. What will be the altitude of the LCL where the parcel becomes saturated and the bottom of a puffy cloud forms?

9. Continuing from the previous question, what will the temperature of the air parcel be if it keeps rising to an altitude of 3 kilometers? What will the dew point be?

10. If an environment has a lapse rate of 8° C per kilometer, is it absolutely stable, absolutely unstable, or conditionally unstable? Could thunderstorms be a possibility in this situation? Is a temperature inversion present?

11. Why is there no such thing as a cirronimbus cloud?

12. Name the cloud type associated with each of the following: optical effects, thunderstorms, and fair weather.

13. Walking outside, you hold up your hand to the sky. You see a lumpy cloud that has features as big as your thumb. What is the name of this cloud?

14. Tor Bergeron observed that if a fog formed in a forest and the temperature was above 0° C (32° F), the fog extended down to the ground. If the temperature was below −5° C (23° F), the fog would not reach the forest floor. Explain his observation.

15. Explain how clouds help warm the ground. How do they help cool the ground? How does the altitude of the cloud affect its ability to warm or cool the ground?

16. An ice crystal grows for 5 minutes in a supersaturated environment with a temperature of −1° C. The crystal is carried to a different part of the cloud where the temperature is −14° C and the environment is still supersaturated. The crystal stays in this region of the cloud for another 5 minutes. Draw a picture of what the ice crystal might look like. Compare your picture with what you get using the Growing a Snowflake learning applet.

17. Many public restrooms have automatic hand dryers. Why do they use hot air instead of cold air? The instructions say to place your hands in the airflow and gently rub them together. Explain how this dries your hands more rapidly than just holding them motionless in the air.

18. What temperature pattern must be present to cause freezing rain or sleet?

19. Can you easily dry your wet laundry outside when the temperature is below freezing? Explain your conclusion.

20. You are driving when very large raindrops suddenly splash onto your windshield. Is the updraft in the cloud above you strong or weak? Which cloud type is probably above you: nimbostratus or cumulonimbus?

21. What factors determine the growth of an ice crystal? How are these factors different than the factors that affect cloud droplet growth?

22. It starts snowing in very dry air that is above 0° C. Do you think the air temperature will rise, fall, or remain the same? (Hint: What will happen to the snow, and how will that cause an energy gain or loss by the atmosphere? Test your answer by using the Precipitation Types learning applet.)

■ Observation Activities

1. Throughout the course, either take photographs or keep a written log of the variety of cloud types you observe. Note the day, time, and general weather conditions at the time of your observation. Which cloud types were the hardest to identify or photograph, and why?

2. The purpose of this exercise is to relate the formation of bubbles in beer or soda (pop) to the formation of cloud droplets. Beer, which can certainly be nonalcoholic, tends to form better bubbles than soda. Pour the beer or soda into a clear glass. Where do the bubbles tend to form and why? What happens to the bubbles after they form? Pour some salt into the glass (sand or sugar can be used if salt is not handy). What happens when you pour the salt in? Why does this happen?

3. This experiment requires a can of compressed air, the type used to clean computer keyboards. Spray out the compressed air while someone measures the temperature of the can with an infrared thermometer or a thermometer used in aquariums. You need to do this in a well-ventilated place and keep the can upright. Explain changes in the can's temperature using the concept of adiabatic expansion of air.

This rain cloud icon is your clue to go to the *Meteorology* Web site at http://physicalscience.jbpub.com/ackerman/meteorology/. Through animations, quizzes, web exercises, and more, you can explore in further detail many fascinating topics in meteorology.

Observing the Atmosphere

AFTER COMPLETING THIS CHAPTER, YOU SHOULD BE ABLE TO
- Explain how surface and upper-air weather observations are made
- Compare and contrast how and where weather satellites, radar, and wind profilers sense the atmosphere
- Interpret different kinds of weather satellite imagery
- Shed light on why there are blue skies and red sunrises and sunsets
- Understand why, when, and where rainbows, halos, and mirages may be seen

INTRODUCTION

You have probably heard the saying that "no two snowflakes are alike." We know that this is true because a farmer from Vermont with a passion for the weather saw it with his own eyes, proved it with photographs, and told the world.

Wilson "Snowflake" Bentley grew up in Jericho, Vermont. When he was just 19 years old he began photographing snow crystals in a barn on his farm using a microscope attached to a camera. Over the next 46 years Bentley photographed more than 5000 snow crystals with the same equipment. No two were exactly alike.

The complex steps in ice-crystal formation that we studied in Chapter 4 virtually guarantee the uniqueness of a snowflake by the time it reaches the ground. But seeing is believing in meteorology. Bentley's photographs are still renowned worldwide, more than 70 years after his death. While capturing nature's artwork, Bentley became a scientific pioneer in photomicrography and the study of precipitation.

So far in this text we have focused on the essential facts of weather and climate. In this chapter, we will also emphasize how we know these facts. As in the case of Bentley's snowflakes, we learn about the atmosphere by observing it.

In many cases, scientists observe the atmosphere using advanced equipment and techniques, such as automated instruments, orbiting satellites, and Doppler radar. We will examine these approaches in some detail. Your eyes are as finely crafted as any scientific instrument, and they too can reveal the atmosphere's secrets. So we will also discuss the awe-inspiring world of rainbows, halos, and mirages. Following Snowflake Bentley's example, enjoy the beauty of the photographs in this chapter—and learn from them, too.

METEOROLOGICAL OBSERVATIONS

Meteorology, like every other science, relies on careful and precise measurement of its subject. Meteorologists observe the atmosphere using two basic approaches. **Direct methods**, also called *in situ* for "in place," measure the properties of the air that are in contact with the instrument being used. **Indirect methods**, also referred to as remote sensing, obtain information without coming into physical contact with the region of the atmosphere being measured.

Our skin directly senses the temperature of objects we touch. A thermometer directly measures the temperature of the air it touches. Our eyes measure temperature indirectly when they observe steam above a cup of coffee or the different colors in a flame. Satellites measure temperature indirectly by sensing radiation coming from the surfaces below.

DIRECT MEASUREMENTS OF SURFACE CONDITIONS

How do you think local weather observations are made? Some people imagine that a meteorologist goes outside every hour to make weather observations. This was indeed the routine worldwide many decades ago. Today in the United States, however, most weather observations are made automatically by a combination of electronic sensors called the **Automated Surface Observing System (ASOS)**. ASOS (pronounced "A-sauce") is the United States' primary surface weather observing network. It is a joint effort of the National Weather Service, the Federal Aviation Administration, and the Department of Defense. This observing system measures cloud height, visibility, precipitation, pressure, temperature, dew point, wind direction and speed, and rainfall accumulation (**FIGURE 5-1**). Measuring atmospheric conditions frequently and in many locations

FIGURE 5-1 The Automated Surface Observing System (ASOS) makes continuous observations of the atmosphere using both direct and indirect methods. ASOS measures cloud height, visibility, precipitation, pressure, temperature, dew point, wind direction and speed, and rainfall accumulation. Observations can be made automatically in relatively remote locations.

is the key to improving forecasts. ASOS provides official weather observations up to 12 times each hour (and up to once a minute), 24 hours a day for every day of the year. In many other parts of the world, meteorologists use a combination of automatic and manual methods.

Meteorologists often combine and graphically depict observations of the atmosphere to make them more understandable. The station model (see Chapter 1) is one way to do this with surface data. The meteogram, a chart of one or more weather variables at a given location over a given period of time, is another (**BOX 5-1**). Observations from these and other sources

Box 5-1 The Meteogram

As we learned in Chapter 1, meteorologists use the station model to condense many atmospheric observations made at the same time. The meteogram allows a meteorologist to see how weather variables vary in time at a single location.

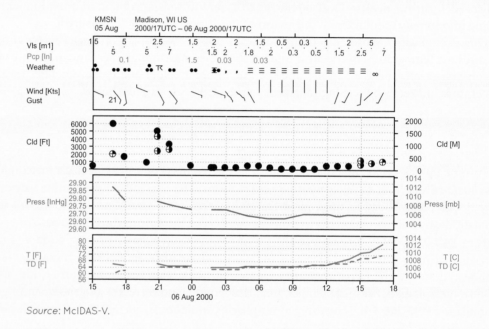

Source: McIDAS-V.

The accompanying figure is an example of a meteogram for Madison, Wisconsin, on August 5 and 6, 2000. Time, in UTC (Universel Temps Coordonné; 5 hours ahead of local time), runs along the x-axis. The meteogram has different weather parameters plotted as functions of time. The top portion of the figure lists visibility in miles. Below the visibility observations are precipitation values in inches and current weather conditions, using the symbols discussed in Chapter 1. The meteogram also includes observations of wind speed and direction and peak wind gusts. The second panel from the top includes cloud base altitude and cloud coverage. The third panel is a graph of station pressure. The bottom panel plots temperature and dew point temperature.

What does this meteogram tell us? Clouds were present over Madison during most of this time period. Precipitation was occurring through most of the evening on August 5, with a thunderstorm at 2100 UTC. By 0000 UTC on August 6, it rained 1.5 inches. The rain gave way to drizzle at 0300 UTC on August 6, and fog set in by 0500 UTC.

Many of these meteogram observations can be explained using what we have learned so far in this text. The winds were very light during periods of fog. The fog reduced visibility by scattering light. The dew point and temperature were equal during rainy and foggy periods. Near sunrise on August 6 at 1200 UTC, the temperature increased as the sun warmed the Earth and lower atmosphere. The dew point temperature also increased, probably because dew was evaporating into the air. The fog lifted by late morning as surface heating reduced the relative humidity and promoted mixing of the air at the and above the surface.

are then analyzed to determine weather patterns and create forecasts, as we will learn in Chapter 13.

In the following sections, we examine how ASOS and human observers measure the primary weather variables.

Temperature

Liquid-filled and metallic thermometers measure temperature by measuring how much the liquid and metal expand or contract when they are heated or cooled by their surroundings. The mercury thermometer is the best-known example. However, mercury freezes at −40° C (−40° F). In bitter-cold situations, observers use alcohol thermometers. Today, however, most thermometers are electronic.

ASOS uses an electronic **resistance thermometer**. Resistance thermometers measure the electrical resistance of a wire made of a metal, usually platinum. As the temperature of the wire increases, its resistance increases. The amount of resistance can be translated to specific temperatures. The temperature of a wire is the same as that of the surrounding air, so the air temperature can be determined by measuring the wire's resistance to the flow of electricity. The data are then automatically reported to the National Weather Service.

To measure the air temperature accurately, thermometers are shielded from sources of energy other than the air. A thermometer in direct sunlight reports a higher temperature than one in the shade, even when the air temperature is the same around both thermometers. FIGURE 5-2 shows the typical "Stevenson screen" shelter used to shield thermometers at non-ASOS facilities. These standard instrument shelters are painted white so that they reflect nearly all sunlight, thereby minimizing the effect of direct sunlight on temperature inside the shelter. Because thermometers measure air temperature directly, shelters are well ventilated. The louvered sides of the shelter permit air to flow in and out so that the air inside is essentially the same as the air outside.

Humidity

ASOS measures dew point using a **dew point hygrometer**. It is based on the simple observation that a bathroom mirror will "fog up" when the temperature equals the dew point during a bath or shower.

A dew point hygrometer uses a beam of light focused on a mirror. The light reflects off the mirror onto an instrument that measures the intensity of the reflected light (FIGURE 5-3A). The mirror surface is then chilled. When the mirror surface cools to the dew point temperature, dew forms on the mirror. The water droplets (or ice crystals for the frost point) block light from reaching the detector (Figure 5-3b). This is interpreted to mean that the mirror is at the dew point temperature. The mirror's temperature is measured with a platinum wire, similar to that used for ambient temperature. The mirror is then warmed above the dew point to evaporate the dew and then cooled again to make a new measurement.

A weakness of this approach is that substances other than dew can cover the mirror and interfere with the laser beam. This fools the hygrometer into "thinking" that the dew point has been reached.

A **psychrometer** is an alternative instrument for measuring relative humidity. A psychrometer consists of two ventilated mercury thermometers, one of which has a wet wick around

FIGURE 5-2 A standard instrument shelter, known as the "Stevenson Screen," houses and protects temperature, pressure, and relative humidity instruments. The front door is open to show the interior, where the instruments are located. The shelter is a white, wooden box with louvered sides; it stands 5 feet above the ground.

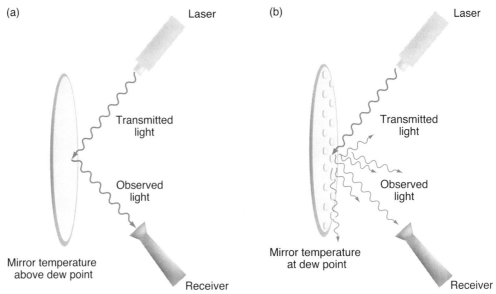

FIGURE 5-3 (a) A laser beam and a mirror are used by ASOS to measure dew point temperatures. (b) When dew forms, less light reaches the receiver because the dew drops scatter light into all directions, as do clouds.

its bulb and is called the wet-bulb thermometer. Evaporation of water off the wick removes heat from the thermometer. The temperature of the wet-bulb thermometer, but not the other thermometer, drops according to the rate of evaporation. To operate correctly, the thermometers have to be ventilated by either whirling the instrument around (**sling psychrometer**) or using a fan (**aspirated psychrometer**). After a few minutes, the temperature of the wet bulb will stabilize at a particular temperature, referred to as the wet-bulb temperature. A table is then used to convert the temperature difference between the two thermometers into relative humidity.

Water vapor is the engine of the weather. The large latent energy associated with water's phase changes significantly affects the energy balance of the atmosphere and the evolution of storm systems. Hence, the distribution of water vapor plays a crucial role in weather and global climate. It is important to measure the total amount of water vapor in the atmosphere. One way to make this measurement as a function of altitude is to attach psychrometers to weather balloons. Another approach is to make use of the **Global Positioning System** (**GPS**), as described in **BOX 5-2**.

> "Wet-Bulb Temperatures and Humidity" to convert wet-bulb and dry-bulb temperatures to relative humidity.

Pressure

Galileo's student Evangelista Torricelli invented the **mercury barometer** in 1643. The mercury barometer consists of a long tube open at one end. Air is removed from the tube, and the open end is immersed in a dish of mercury. As explained in Chapter 1, the weight of the air above will then balance a column of mercury in the tube. The height of the mercury is therefore a measure of the atmospheric pressure.

The **aneroid barometer** (**FIGURE 5-4**) is a flexible metal box, called a cell, which is tightly sealed after air is partially removed. Changes in external air pressure cause the cell to contract or expand. The size of the measured cell is thus related to the atmospheric pressure. A pointer and a dial convert the cell's size to atmospheric pressure.

The aneroid barometer is smaller and more durable than the mercury barometer, and unlike mercury, it is not poisonous. For these reasons, airplanes use aneroid barometers called altimeters that convert pressure to altitude. However, aneroid barometers are also less accurate than mercury barometers.

ASOS uses electronic barometers that combine durability with accuracy and quick response to changing pressure. They are the most reliable ASOS sensors. Atmospheric pressure is an

Box 5-2 GPS and Water Vapor

The same technology used in cars to tell you where you are as you drive around also is used to tell how much water vapor is in the atmosphere and to improve weather forecasts.

Measuring atmospheric water vapor is important in weather forecasting and climate monitoring. Short-term cloud and precipitation forecasts require timely and accurate moisture data. As you learned in Chapter 4, water vapor releases huge amounts of energy as it condenses to form clouds and precipitation, providing an energy source for storms. Water vapor is a greenhouse gas that plays a critical role in the global climate system.

Weather balloons measure the vertical profile of water vapor, which can be used to determine the total amount of water vapor in a column of air. These balloon-borne water vapor sensors, however, are expensive and do not provide the frequent measurements needed for accurate weather forecasting. Remote sensing is an approach to reduce costs and decrease the time between measurements.

The GPS tracking network was established to provide high-precision navigation. The system consists of satellites and a ground-based network of support stations that receive data. Because the GPS uses radio waves, it works in every type of weather—clear, cloudy, or rainy conditions—which is important in weather forecasting. A constellation of GPS satellites transmits radio signals to receivers on Earth. These GPS satellite radio signals are slowed by the atmosphere, which results in a delay in the arrival of a signal as compared with proposition in space. Data recorded by GPS receivers at fixed locations will show delays caused by a variety of effects, one of them being the amount of water vapor in the atmosphere. The measured delay is used to determine the amount of water vapor in the atmosphere.

important weather element, and it is a key measurement for aircraft flight operations. The surface pressure is used in determining the height of an airplane above the surface. Because accurate pressure is critical, three pressure sensors are used at towered airport locations. The ASOS reports a pressure only when two of the observations agree within a defined tolerance.

Wind

Wind has both speed and direction. **Anemometers** measure wind speed, and **wind vanes** measure wind direction (**FIGURE 5-5**). A typical wind vane has a pointer in front and fins in back. When the wind direction changes, the force on the upwind side of the fin is greater than on the downwind side, and so the vane rotates until the forces are balanced. When the wind is blowing, the wind vane points into the wind. For example, in a north wind, the wind vane points northward.

A **cup anemometer** measures wind speed. The cups catch the wind and produce pressure difference inside and outside the cup. The pressure difference, along with the force of the wind, causes the cups to rotate. Electric switches measure the speed of the rotation, which is proportional to the wind speed.

The ASOS wind vane and anemometer are mounted on a tower 10 meters high. This is done around the world to minimize the influence of the ground on the wind observations. Also, the tower's location is chosen so that there are no nearby trees or buildings that would affect the wind at the tower level.

At wind speeds below about 5 kilometers per hour (3 mph) the cup anemometer is prone to error because friction keeps the cups from

FIGURE 5-4 An aneroid barometer converts the size of a partial-vacuum container to atmospheric pressure. The blue pointer indicates the pressure, and the gold pointer can be set by the user so that the change in pressure can be determined at a glance in inches of mercury (outer numbers) and millibars (inner numbers). Because weather is closely related to air pressure, many barometers include short weather forecasts; however, these forecasts apply only to pressure that has been adjusted to sea level.

FIGURE 5-5 The wind vane (left) and cup anemometer (right) are used to measure wind direction and wind speed.

turning. At wind speeds above 160 kilometers per hour (100 mph), cup anemometers often blow away or give unreliable measurements. In freezing rain, the anemometer can literally freeze up and stop turning.

Other devices also are used in wind measurement. **Propellers** are used to measure wind speed. The propeller blades rotate at a rate proportional to the wind speed. A **windsock** is often used at airports. A windsock is a cone-shaped bag with an opening at both ends. When it is limp, winds are light; when it is stretched out, winds are strong. Pilots can quickly determine the wind direction and speed along a runway just by observing the shape and direction of a windsock.

The **sonic anemometers** use ultrasonic sound waves (sound waves humans cannot hear) to measure wind speed and direction. The instrument determines the wind velocity by measuring the time between when a sonic pulse is sent by the instrument and when it is received. Sonic anemometers do not have moving parts, which makes them useful in regions with lots of aerosols, such as deserts and oceans. In those environments, measurements by the cup and vane anemometers can be adversely affected by salt and dust getting into moving parts.

Precipitation

The key to precipitation measurement is to catch the falling precipitation and then to record its amount and intensity. Although it sounds simple, precipitation is difficult to measure accurately.

The precipitation sensor used by ASOS is the **rain gauge**. It consists of a funnel-like receptacle above a bucket. Precipitation falls into the upper portion of the rain gauge, which is called the collector. The collector is heated to melt any frozen precipitation, such as snow or hail. The collected water is funneled into a tipping bucket. The tipping bucket measures water depth in increments of 0.01 inch. It is called a tipping bucket because as water is collected, the tipping bucket fills to the point where it tips over and empties, indicating that 0.01 inch of water was measured.

Although the rain gauge is commonly used, it is subject to many errors. For example, the gauge can leak. During heavy rainstorms, rain can splash out of the gauge. Winds can blow precipitation across, rather than into, the gauge. A wind shield is placed around the collector to reduce this error.

Measuring Snow

ASOS does not measure snowfall. Snowfall depth varies over short distances, and so the National Weather Service relies on trained volunteers to provide snowfall measurements. Accurate measurements of snowfall are surprisingly difficult. The standard method uses a snowboard— a 16″ × 16″ piece of white painted wood. The snowfall accumulated on the snowboard is measured with a ruler in inches and tenths at least once a day, for the amount of new snow since the last snowfall observation. It is important that the snowfall measurement be in locations where the effects of blowing and drifting snow are minimized.

Snowfall is also reported as liquid-equivalent of frozen precipitation. After the snow is collected, it is carefully melted, and the accumulated amount of water is measured in inches and tenths. The general rule that 1 inch of rain is equivalent to 10 inches of snow does not generally hold as the ratio is dependent on many factors. FIGURE 5-6 is a climatology of snow-to-liquid ratio of the contiguous United States. The 30-year climatology indicates that the values generally range from less than 8 to more than 16, with maybe 13 being the most representative value.

Mean Snow-to-Liquid Ratio Values During 1971–2000

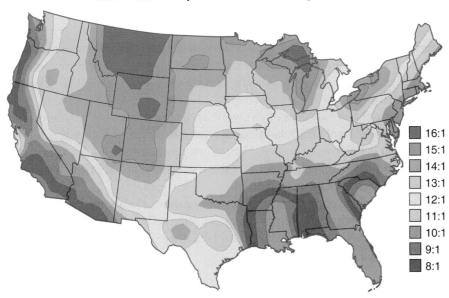

■	16:1
■	15:1
■	14:1
■	13:1
□	12:1
■	11:1
■	10:1
■	9:1
■	8:1

FIGURE 5-6 A 30-year (1971–2000) climatology of the snow-to-liquid ratio for the contiguous United States. (*Source:* M. A. Baxter. C. E. Graves, and J. T. Moore, Climatology of snow-to-liquid ratio for the contiguous United States, *Weather and Forecasting* 20: 729–744).

DIRECT MEASUREMENTS OF UPPER-AIR WEATHER OBSERVATIONS

Meteorologists monitor the atmosphere above the surface by using a radio-equipped meteorological instrument package carried aloft by a helium-filled "weather balloon." These instrument packages are called **radiosondes** (*sonde* is French for *probe*) (**FIGURE 5-7**).

Radiosondes measure vertical profiles of air temperature, relative humidity, and pressure from the ground all the way up to about 30 kilometers (19 miles or 10 millibars [mb]). Temperature

FIGURE 5-7 The vertical distribution of temperature, humidity, pressure, and winds are obtained with an instrument called a rawinsonde. A lighter-than-air balloon carries a small instrument box beneath it. A parachute is in the center of the string between the balloon and the instrument package, and it opens when the balloon bursts in the stratosphere. Measurements are transmitted back to the surface for analysis.

and relative humidity are measured electronically; a small aneroid barometer measures pressure. At low air pressures in the stratosphere, the balloon expands so much that it explodes and the radiosonde drifts back to the ground underneath a small parachute.

Wind speed and direction can also be determined by tracking the position of the balloon. When winds are also measured, the observation is called a **rawinsonde**. Rawinsonde measurements are made worldwide at several hundred locations twice each day at 0000 UTC (Universel Temps Coordonné) and 1200 UTC.

Rawinsondes are the workhorses of the weather data network above the ground; however, they are usually launched only from land-based weather stations, which leaves out the 70% of the atmosphere that lies above the oceans. Also, strong upper-level winds can carry the balloons far downstream, and data can be lost. Because these situations are of prime interest to weather forecasters, this is a drawback of using balloon-borne instruments.

The vertical distribution of rawinsonde observations above a location is known as a **sounding**. Meteorologists plot individual soundings on special thermodynamic diagrams. Although these diagrams are beyond the scope of our discussion here, they rely on concepts covered in this text and are covered in detail on the text's Web site.

"Thermodynamic Diagrams" to explore how meteorologists use these special charts to plot temperature soundings.

INDIRECT METHODS OF OBSERVING WEATHER

There are two basic types of indirect methods of sensing the atmosphere: active sensors and passive sensors. **Active sensors** emit energy, such as a radio wave or beam of light, into the atmosphere and then measure the energy that returns. **Passive sensors** measure radiation emitted by the atmosphere, the Earth's surface, or the Sun.

For example, if you yell in an auditorium and hear an echo, you are doing active sensing. You emit sound waves from your mouth and use your ears to capture the energy that returns. If you make no sound and simply listen to the sounds around you, however, you are doing passive sensing. The eye is also an excellent example of a passive sensor.

Much of what we observe about the atmosphere using indirect measurement techniques deals with how light interacts with molecules or objects, such as water drops, suspended in the atmosphere. To understand the ways in which indirect methods work, we need to discuss some basic laws that govern how light interacts with objects. This is important for explaining satellite and radar observations as well as our visual observations.

■ Laws of Reflection and Refraction

Suppose light traveling through air encounters a pool of water. When the light rays strike the boundary between the air and water, several things can happen. Some rays are turned back in the direction from which they came, or are reflected (**FIGURE 5-8**). Other rays are transmitted into the water. Some of the transmitted rays change direction, or are refracted (Figure 5-8, bottom).

Reflection

Stand directly in front of a mirror, and shine a flashlight at your mirror image. Where does the reflected beam of light go? It comes right back at you. This is called **reflection**. Now stand a little to the side of the mirror and shine the flashlight at the mirror at an angle. Where does the flashlight beam go now? The angle at which the light strikes the surface always equals the angle of reflection (Figure 5-8 top). This simple law of reflection describes how a single ray reflects off a surface, and the law holds whenever light is reflected. As we saw in Figure 5-3, for example, ASOS hygrometers use the reflection of laser light to measure dew point.

Refraction

If you shine a beam of light on a container of water or a block of glass and make careful observations, you will note the following (as illustrated in Figure 5-8 bottom):

1. When a ray of light enters water at an angle other than 90 degrees, it bends toward the line that is perpendicular to the surface of the water or glass, which is called the **normal**.

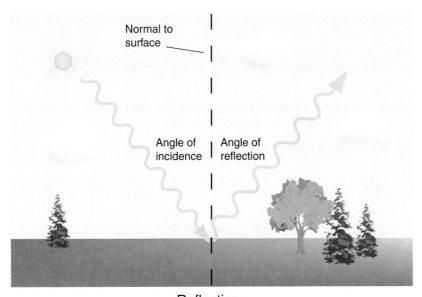

Reflection
Angle of incidence equals angle of reflection

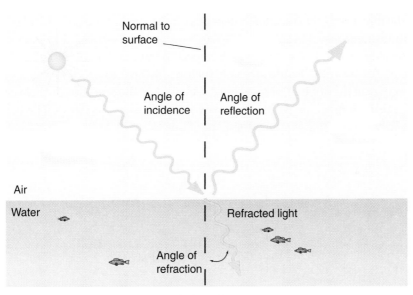

Refraction
Light ray bends toward the normal when entering the water

FIGURE 5-8 Reflection (top) and refraction (bottom). The angles of the reflected and refracted rays depend on the angle of incidences and are measured with respect to the normal, an imaginary line drawn perpendicular to the air–water interface.

2. When a ray of light leaves water and enters air at an angle other than 90 degrees, it bends away from the normal.
3. When a ray of light enters or leaves the water at a right angle (parallel to the normal), the ray does not change direction.

These three observations summarize the laws of **refraction**. Refraction explains many everyday phenomena. For example, refraction causes objects partly immersed in water (such as a person standing in a shallow pool or a straw in a glass) to look bent or broken into two.

Refraction depends on changes in optical properties along a light ray's path to your eye. The ratio of the speed of light in a vacuum to the speed of light in a substance is defined as the **index of refraction** of that substance. The index of refraction is a measure of the optical density of the substance. The higher the index of refraction, the more optically dense the substance. The index of refraction of water is approximately 1.33, whereas that of glass is 1.5. Air has an index of refraction of slightly greater than one, which is a function of temperature. Because the indices of refraction for air and water are different, this means that light moving from air into water, or from water into air, will be bent.

When light is traveling from water into air at a slanted angle, it bends away from the normal. At the **critical angle** the ray exiting the water travels along the air–water surface (**FIGURE 5-9**). When the ray makes an angle with the normal that exceeds the critical angle, the ray cannot pass through the interface and reflects back into the water. This condition, called **total internal reflection**, occurs only when light encounters a medium with a lower index of refraction. This is why light can reflect off the back of the raindrop as the light moves from water to air. As we will see, this is important in the formation of optical effects such as the rainbow. Refraction also explains why stars twinkle (**BOX 5-3**).

■ Scattering

In Chapter 2, we said that radiation could be absorbed, reflected, or transmitted. In addition, light rays can change direction when they encounter small particles. This is called **scattering**. The interaction of light with a sphere was solved mathematically by Gustav Mie in 1908. His complex solution is known as *Mie theory* and explains many optical phenomena. His theory proves that the direction light is scattered depends on the size of the particle and the wavelength of the electromagnetic wave.

Scattering by particles that are small with respect to the wavelength of the incident radiation (such as molecules) is called **Rayleigh scattering**. In these cases, the particle scatters equal amounts of energy in the original, or forward, direction as in the opposite, or backward, direction. **Geometric scattering** occurs with a large particle, such as a cloud drop. The particle scatters more energy forward than backward in these situations.

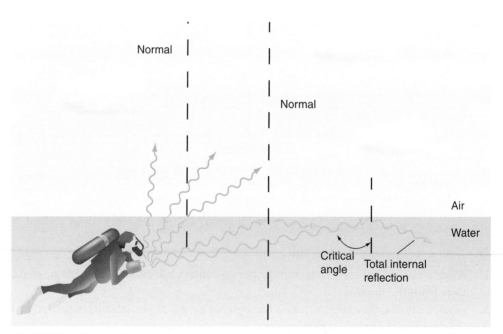

FIGURE 5-9 The critical angle occurs when light travels from water into air. At the critical angle, the refracted ray travels along the surface of the water. An angle of incidence that is greater than the critical angle will be totally reflected and will not enter the air.

Box 5-3 Twinkle, Twinkle, Little Star ...

On a still cloudless night, it's easy to assume that the atmosphere is uniform in structure. In reality, it is always changing. This explains why stars twinkle.

As starlight traverses the atmosphere it continually undergoes small deviations in its direction of travel because of refraction. The refraction is caused by differences in the temperature, density, and moisture content of the air through which the star light travels. The small refraction the light undergoes causes small, rapid changes in the apparent position or brightness of the star. We call these changes of position and brightness "twinkling."

We can explain twinkling using the diagram at right. Small differences in atmospheric structure are often caused by turbulence, which usually exists between the surface and the top of the atmosphere. Now consider a parcel of turbulent air with an index of refraction different from that of its environment that moves into our view of a star. Because of refraction, when the ray encounters the air parcel, the star light appears to come from a different position instead of its true position. As the parcel moves out of our line of sight, the next ray of starlight may encounter another parcel with a different index of refraction, changing the apparent position of the star. If the refraction is strong, the star may fade or even disappear for an instant.

The twinkling of stars is referred to as astronomical scintillation. It is most apparent on clear, cold, and windy nights. The effect is greatest for stars near the horizon because the light from them passes through more atmosphere and is, therefore, more likely to encounter a larger number of air parcels with different indices of refraction.

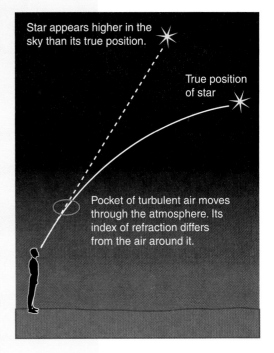

Star appears higher in the sky than its true position.

True position of star

Pocket of turbulent air moves through the atmosphere. Its index of refraction differs from the air around it.

Scattering answers some of the most obvious questions in all of meteorology. For example, why is the sky blue? Until scientific genius Lord Rayleigh described scattering by small particles in 1869, this was a mystery. Although all colors are scattered in all directions by air molecules, as demonstrated by Lord Rayleigh, violet and blue are scattered the most, up to 16 times more than red light. This scattered light reaches our eyes from all skyward directions on a clear day. As a result, the sky looks blue. We do not perceive it as violet because our eyes are more sensitive to blue than to violet.

Another weather enigma explained by scattering is as follows: Why are sunrises and sunsets red? Sunlight passes through more air at sunset and sunrise than during the day when the Sun is higher in the sky. More atmosphere means there are more molecules to scatter the violet and blue light away from your eyes. If the path is long enough, all of the blue and violet light scatters out of your line of sight. The other colors continue on their way to your eyes. This is why sunsets are often yellow, orange, and red (FIGURE 5-10). Volcanic particles such as those present in the sky in this figure accentuate the scattering and the colors.

Have you ever seen a morning or evening sky that looked like brilliant beams of light? This phenomenon, known as **crepuscular rays**, is also caused by scattering (FIGURE 5-11). Irregular objects between you and the Sun, such as a cloud or mountain near the horizon, block some but not all of the light. Where the light peeks through, scattering illuminates its path from the Sun to your eyes. This creates beams in the sky. These beams appear to converge at the horizon; however, this is an illusion, similar to the impression that the rails on train tracks appear to come together in the distance.

The repeated scattering of light, called multiple scattering, causes whitish light because enough light of all colors is scattered to your eye. This explains why piles of salt, sugar, and snow crystals appear white, although the individual crystals are clear. Multiple scattering also explains

FIGURE 5-10 The sky in this photograph (taken in the summer of 1993 in Madison, Wisconsin) is a mixture of yellows and reds near the horizon because the blue end of the spectrum has been scattered away from the beams of the setting Sun by the atmosphere. The colors were accentuated by the presence of volcanic particles from the Mount Pinatubo eruption in 1991.

the blue-white appearance of distant mountains and why haze near the horizon causes the sky to appear whitish. **Haze** is a suspension of small particles in the air, which can reduce visibility by scattering light. Both high and low clouds in sunshine look white as a result of scattering, either by water droplets or ice crystals. Scattered light by clouds also is important in climate change studies (BOX 5-4).

The bottoms of even the whitest clouds appear grayish, sometimes ominously so. Is a terrible storm coming? Not usually. The cloud base is dark not because of absorption of light

FIGURE 5-11 Crepuscular rays shine between objects such as clouds or mountains. Scattering causes the beams to be visible by diverting light toward your eyes.

Box 5-4 Multiple Scattering and Climate Change

Climate change ultimately depends on changes in the global energy gains and losses. Clouds modify the energy budget of the atmosphere and the Earth's surface. The effect of clouds on climate depends on how the amount of cloud cover and on the size of the particles composing the cloud. For this reason, the scattering of light by clouds can be a critical aspect of climate change.

If you smash a clear ice cube, the pile of small pieces appears brighter and whiter than the whole cube. This is because the shards scatter light more extensively than one whole cube. This kind of multiple scattering also happens in clouds. If the amount of water in two clouds is the same but one cloud contains very large drops and the other contains very small drops, the cloud containing the small particles will appear brighter.

Multiple scattering has implications for climate and climate change. If the average particle size of a cloud were to become smaller, the cloud would become brighter, and more solar energy would leave the top of the cloud and be lost from the atmosphere to outer space. That energy would not be available to warm the ground or ocean.

How can the particle sizes in clouds decrease? The average size of cloud particles is reduced if the number of cloud condensation nuclei increases. An increase of nuclei, with a fixed amount of water in the cloud, leads to more, but smaller, particles.

Where would these additional cloud condensation nuclei come from? Human activities are a likely source. If everything else remained the same, changing the number of particles existing in the atmosphere would modify the global energy budget and create the possibility of climate change. This possibility is discussed in Chapter 15.

by a tall severe storm. Instead, it is dark because of multiple scattering that occurs above the cloud base. Multiple scattering redirects the light out the tops and the sides of the cloud. This allows very little light to be transmitted out of the cloud base. The base therefore appears dark (**FIGURE 5-12**).

Now that we understand scattering, we can grasp how ASOS measures two more important weather parameters: cloud ceiling and visibility.

ASOS Indirect Sensors

Visibility Sensor

Visibility is simply the horizontal distance a person with normal vision can see and identify specified objects. We can estimate visibility if we know the distances between objects and ourselves. Visibility is reduced when particles between the object and us scatter or absorb light. If an object a quarter of a mile away cannot be seen, then the visibility must be less than a quarter of a mile.

ASOS uses an active remote sensing method to measure visibility. The visibility instrument uses a flash of light over a very short distance. Particles suspended in the atmosphere, such as fog particles, scatter the light. The greater the number of particles, the more light is scattered and the lower the visibility. The light flash that is scattered by the atmosphere is measured by a receiver and converted into a visibility value. This is tricky because human eyes do not work exactly the same as light sensors. Also, human visibility reports contain information in all directions along the horizon, unlike the ASOS sensor, which is in a single direction.

Cloud Ceiling Sensor

Cloud ceiling is the height of the lowest widespread cloud base. Pilots determine cloud height as they fly through a cloud layer. Trained observers on the ground can estimate cloud ceilings visually. ASOS uses a **ceilometer** (pronounced "seal-OMM-it-er") to measure cloud ceiling.

The ceilometer is an active remote sensing instrument. It uses a beam of radiation—in the case of ASOS, a laser beam—to detect cloud heights. The ceilometer sends pulses of radiation

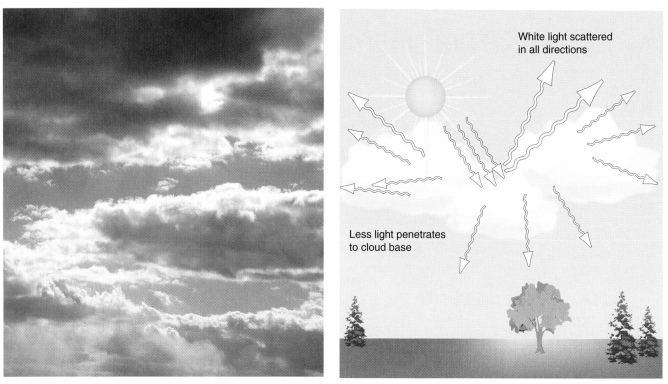

FIGURE 5-12 The bottom of thick clouds appears gray (left) because very little light exits the base. Most of the sunlight is scattered out the sides and top of the clouds (right).

upward (**FIGURE 5-13**). If a cloud is present, part of this beam is scattered by the cloud and sent back to the ceilometer. The time interval between when the pulse is transmitted and when it is received back at the instrument is a measure of the cloud height.

The ASOS ceilometer was designed with the needs of aviation takeoffs and landings in mind. Thin clouds, and all clouds above 3.66 kilometers (12,000 feet), are invisible to ASOS because too little energy returns from the laser beam to be detected. This means that ASOS cannot "see"

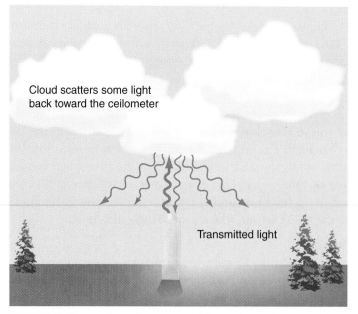

FIGURE 5-13 A ceilometer uses light scattered by clouds to determine the altitude of the cloud base.

clouds such as cirrus and altocumulus even when they are plainly visible to human observers—a distinct disadvantage. Clouds at the horizon are also not "seen" by ASOS. High clouds are monitored using instruments onboard satellites.

Meteorological Satellite Observations

Satellite instruments provide extensive observations of use to meteorologists. They fill in the gaps of surface and weather balloon observations, and they also provide information that cannot be obtained by any other method.

Weather satellites fly around the Earth in two basic orbits: a geostationary Earth orbit (abbreviated GEO) and a low Earth orbit (also called **LEO**) (FIGURE 5-14). Geostationary satellites orbit the Earth as fast as the Earth spins. Therefore, they hover over a single point above the Earth at an altitude of about 36,000 kilometers (22,300 miles). **LEO satellites** orbit often travel over the Earth's polar regions, flying at an altitude of 850 kilometers (530 miles), a little higher than the altitude of the orbiting Space Shuttle. To maintain its position, a **GEO satellite** must be located over the equator. In contrast, polar satellites go around from pole to pole as the Earth rotates beneath the satellite; each orbit is slightly to the west of the previous one.

GEO and LEO satellites have definite advantages and disadvantages relative to each other. A GEO satellite has a continuous high-quality view of the tropics and mid-latitudes. A GEO satellite has a poor view of the polar regions, which the LEO satellite excels at periodically as it passes over the poles. A LEO satellite flies over midlatitude and tropical regions only twice a day. For this reason, its coverage of weather is hit and miss. When the timing is just right, however, the low altitude of a LEO satellite allows it to obtain excellent detailed snapshots of weather events, as we see in images of a severe thunderstorm in Chapter 11.

Meteorologists, therefore, use both GEO and LEO satellites; the choice depends on the specific need. The satellite pictures on your favorite television weather broadcast are from a GEO satellite. This is because a GEO satellite tracks storm systems continuously. Weather events can be tracked using time sequences of satellite images. This allows weather forecasters to predict their movement over short time periods of 30 minutes or less. Horizontal wind speed and direction can also be determined by tracking individual cloud features in a time sequence of GEO satellite images. For those studying global weather, a LEO satellite in a Sun-following orbit views all regions of the Earth in a single day and is a valuable tool.

The United States usually operates two geostationary satellites called GOES (Geostationary Operational Environment Satellite). One has a good view of the East Coast; the other is focused on the West Coast. Another satellite in geostationary orbit is the European METEOSAT (METEOrological SATellite), which views the eastern Atlantic Ocean, Africa, and Europe. The Japanese GMS (Geostationary Meteorological Satellite) has a good view of Asia, Australia, and the western Pacific Ocean.

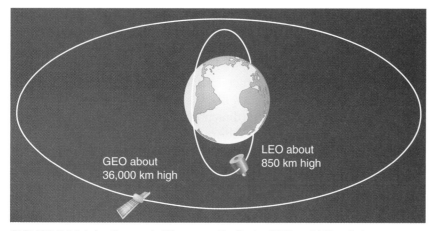

FIGURE 5-14 Weather satellites usually fly in GEO or LEO orbits.

Box 5-5 Next Generation of Weather Satellites

The National Oceanic and Atmospheric Administration (NOAA) is developing the Advanced Baseline Imager (ABI) as the weather monitor for the next generation Geostationary Operational Environmental Satellite (GOES) series. Scheduled to be launched in 2015, the ABI will be used for a wide range of applications that will provide numerous improvements for monitoring weather, oceans, and climate.

The ABI makes measurements at 16 wavelengths, while the current GOES imager has measurements at only five wavelengths. This new capability will allow forecasters to distinguish snow from clouds, detect thin cirrus clouds, identify clouds with super cooled liquid water, and identify aviation hazardous weather.

The ABI will have a better temporal resolution than the current instrument. It will be able to scan the atmosphere faster and acquire images more frequently. The current GOES imager takes approximately 25 minutes to scan the hemisphere. In 15 minutes, the ABI will not only scan the entire hemisphere, it will also scan the continental United States three times, plus it will cover a selectable 1000 km × 1000 km geographic region every 30 seconds.

The ABI will also have better spatial resolution than the current GOES, which means that it can see finer scale weather structures. What does this mean for weather forecasting? For one thing, with the improved spatial and temporal measurements, the ABI will make air travel safer by enabling the detection of upper tropospheric turbulence that impacts aviation. The improved measurements will also enable better determination of winds derived from tracking cloud features, improved fire monitoring, and better detection of fog.

The National Oceanic and Atmospheric Administration typically maintains two polar-orbiting satellites. One views the United States at approximately 2:00 PM and 2:00 AM local time, and the second views regions of the United States around 10:00 AM and 10:00 PM local time. The U.S. Defense Meteorological Satellite Program has used LEO satellites for decades, as we will see in Chapter 10.

Satellites can observe or infer an amazing range of weather variables—winds and chemicals in the stratosphere, rainfall, ocean temperatures, and severe weather potential. Satellites can do it all and do it on a global basis. In the following, we will focus on the most common satellite images: the ones you see on television weather programs. BOX 5-5 gives a preview of what you can expect with the next generation weather satellites.

Interpreting Satellite Images

Satellite instruments measure electromagnetic energy that the Earth and the atmosphere reflect, scatter, transmit, and emit. These passive remote sensing instruments are called **radiometers**. Two common types of radiometers are used in satellite meteorology. One type measures the amount of visible light from the Sun that is reflected back to space by the Earth's surface or by clouds. The second measures the amount of radiation emitted by the surface or clouds.

The radiometers on satellites are not cameras, although they do produce images. Radiometers use moving mirrors to scan the Earth much like the way we read lines of text in a book. The instrument begins at a starting point in one direction and then scans across the Earth below line by line, making observations as it proceeds and creating an image by combining these scan lines. From a geostationary orbit, images are normally updated every 15 to 30 minutes.

As discussed in Box 2-2 in Chapter 2, the most common wavelengths, or spectral channels, observed by radiometers are **visible** (0.6 microns), **infrared** (IR) (10 to 12 microns), and a special channel near one of the IR absorption bands of H_2O that is called "**water vapor**." Better detail or resolution is obtained at smaller wavelengths. Currently, the resolution of visible satellite images from geostationary orbit is 1 kilometer, whereas IR and water vapor images usually have resolutions between 4 and 8 kilometers. This means that weather features on the scale of 1 kilometer, such as individual cumulus clouds, can be seen in visible images but not in IR or water vapor images.

Visible Imagery

A **visible satellite image** represents sunlight scattered or reflected by objects on Earth. Differences in the albedo of clouds, water, land, and vegetation allow us to distinguish these features in the imagery. Dark areas in a visible satellite image represent geographic regions where only small amounts of visible light from the Sun are reflected back to space. The oceans are usually dark, whereas snow and thick clouds are bright (**FIGURE 5-15**).

The brightness of a cloud in a visible satellite image depends on the number of water drops or ice crystals in it. Stratus clouds have lots of particles that scatter solar radiation, so stratus clouds appear white in a visible image. For the same reason, fog is also very easy to see on visible satellite images. Thin cirrus are difficult to see because there are fewer ice particles available to scatter solar radiation. Analysis of visible satellite images allows meteorologists to locate thunderstorms, hurricanes, fronts, and fog.

IR Imagery

The IR radiometers on satellites measure radiation with wavelengths between 10 and 12 microns. The information in this "channel" is very different than in the visible image. Instead of light that has been reflected or scattered, the IR radiometer measures heat. In IR radiometric images, cold objects are white, and hot surfaces appear black (**FIGURE 5-16**). An advantage of the **IR satellite image** over the visible image is that it gives useful information both day and night. (As we saw in Chapter 2, visible images at night are not very informative.) Television weathercasters frequently show animations of time sequences of IR images.

"Interpreting Satellite Visible Images" to explore the appearance of clouds and surfaces in a visible image.

"Interpreting Satellite IR Images" to explore how changes in cloud altitude and surface temperature affect their appearance in satellite IR images.

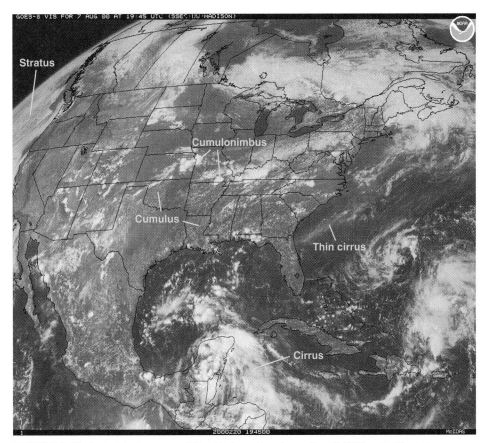

FIGURE 5-15 A visible image of North America from the GOES satellite taken on August 7, 2000, at 1945 UTC. Notice the bright white cirrus and cumulonimbus clouds, as well as the stratus cloud deck at the far left off the Pacific Northwest coast.

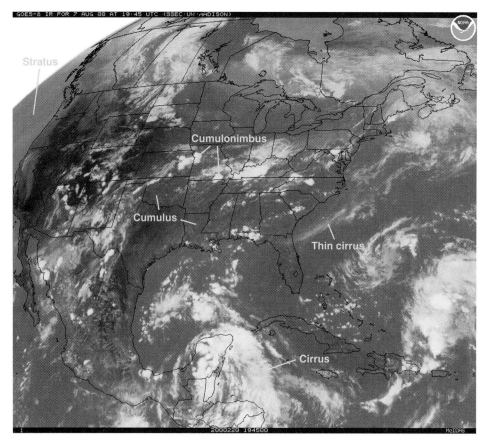

FIGURE 5-16 An IR image of North America from the GOES satellite taken on August 7, 2000, at 1945 UTC. Compare this image with Figure 5-15, which is for the same time and region. High clouds show up more prominently in this image than in the visible image, but low clouds such as cumulus and stratus are hard to distinguish from the surface. This is because IR images depict heat output, and low clouds have similar temperatures as the Earth's surface.

As we learned in Chapter 2, all objects emit radiation proportional to their temperature and also the degree to which they can emit radiation. An IR instrument provides information on the temperature of land, water, and clouds by measuring the IR radiation emitted from surfaces below the satellite. The radiant energy measured by IR radiometers is converted to a temperature.

IR imagery can be used to distinguish low clouds from high clouds. Low clouds are relatively warm and appear gray in satellite IR images. Thick, cold clouds, like the tops of thunderstorms, appear bright white.

Differences in the solar and IR properties of different clouds allow us to use both visible and IR imagery to distinguish among cloud types (**FIGURE 5-17**). For example, stratus can be easily distinguished from thin cirrus. Stratus are gray in the IR image and bright white in the visible image, whereas thin cirrus are white in the IR image and gray in the visible image.

A drawback of both visible and IR satellite images is that they rely on the presence of clouds. What if we want to know what the atmosphere is doing in clear regions? Water vapor imagery can provide this information and much more.

Water Vapor Imagery

Water vapor in the atmosphere is transparent to radiation at visible and 10- to 12-micron wavelengths. This is why we use visible and IR satellite imagery to observe surface features and clouds. Water vapor absorbs and emits radiation very efficiently in wavelengths between 6.5 and 6.9 microns. Satellite radiometers that measure the amount of radiation emitted by the atmosphere at these wavelengths can be used to detect the amount of water vapor in the atmosphere.

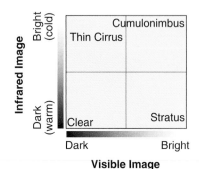

FIGURE 5-17 A matrix for how different types of clouds appear in visible versus IR satellite images. For example, cumulonimbus clouds are bright in both images, but stratus clouds are bright only in the visible image.

As we mentioned in Chapter 1, the amount of water vapor peaks at the surface and decreases higher up in the atmosphere. Satellites look down on the atmosphere, and so a water vapor image "sees" the atmosphere where water vapor increases in concentration. This level is usually between 300 and 600 mb. As a result, water vapor images tell us about the upper and middle troposphere, which is a key region for storm development and growth. Also, water vapor, unlike clouds, is always present in the atmosphere. A water vapor image gives us detailed information everywhere, not just in cloudy regions.

In a water vapor image (FIGURE 5-18), black indicates low amounts of water vapor, and milky white indicates high concentrations. Bright-white regions correspond to thunderstorms. Dark regions are dry air with low relative humidity, such as stratospheric air. In the mid-latitude regions, zones with strong contrast in the amount of water vapor can indicate the presence of a jet stream.

Satellite observations view the world from the top down. The next generation weather satellites are described in Box 5-5. In the next section, we explore a type of indirect observation that sees the atmosphere from the bottom up: radar.

"Interpreting Satellite Water Vapor Images" to explore how changes in cloud altitude and surface temperature affect their appearance in satellite water vapor images.

Radar Observations

The acronym **radar** stands for RAdio Detection and Ranging, a World War II invention. Radar was supposed to detect aircraft in flight, but precipitation frequently got in the way. The military's loss was meteorology's gain. Today, weather radar tracks the development, direction, and movement of storms and estimates precipitation amounts. Radar is becoming the modern, indirect version of the rain gauge.

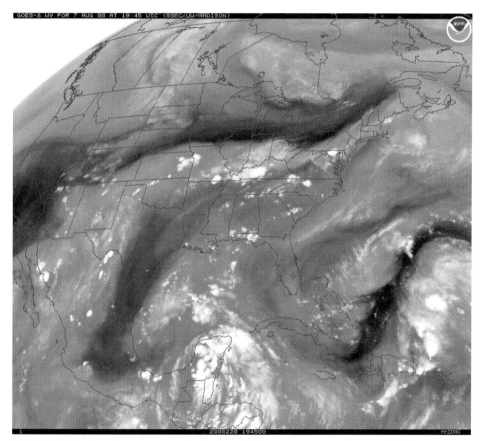

FIGURE 5-18 A water vapor image of North America made from a GOES satellite on August 7, 2000, at 1945 UTC. Light areas are more humid than dark regions. Bright-white regions indicate high clouds. Compare with Figures 5-15 and 5-16, which were made simultaneously. Water vapor images give information in all regions, even those that lack clouds.

FIGURE 5-19 The round, white dome of a national weather service Doppler radar (center in photograph) is silhouetted against an approaching thunderstorm gust front.

A radar transmitter housed in a golf-ball–shaped dome (**FIGURE 5-19**) sends out pulses of radio waves more than 1000 times each second on a path inclined several degrees above the horizon. Precipitation-sized particles scatter these radio waves (**FIGURE 5-20**). Some of the radar waves are scattered back to the transmitting point, where they are detected. The received signal is called the **radar echo**.

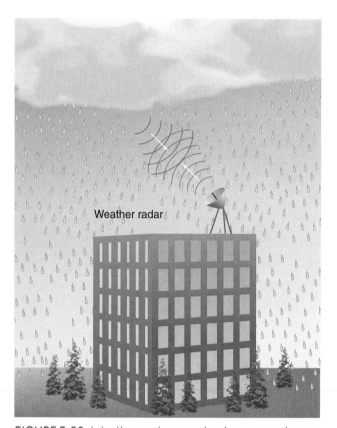

FIGURE 5-20 Weather radars send out a narrow-beam radio wave that is scattered off precipitation. Some of the scattered energy returns to the weather radar, and forecasters interpret it to locate regions of precipitation and the intensity of precipitation. The purple lines represent the transmitted pulse, and the red lines represent the reflected signal.

The radar echo indicates the location and intensity of precipitation. The time the radar signal takes to go out to the precipitation and return (on the order of milliseconds) depends on how far away the precipitation is. Precipitation more than about 240 kilometers (150 miles) away cannot be detected because at such long ranges the radar beam is too high and weak to "see" the precipitating portions of clouds. The direction of the precipitation is simply determined from the direction that the radar is pointing when it receives the echo.

Radars scan 360 degrees around the site and at different elevation angles. Radar scans can be displayed in a horizontal plan or a vertically cross-section through a precipitating storm. We typically see examples of the horizontal scan, referred to as the Plan Position Indicator. A vertical cross-section displays the radar data as a vertical cut along a specific direction and provides valuable information on the structure of the storm, as we will see in Chapter 11.

The intensity of the radar echo indicates the intensity of the precipitation (e.g., millimeters or inches of rain per hour). Relatively high amounts of returned energy indicate high rainfall rates. The amount of energy scattered is proportional to the size and the concentration of particles. Intense radar echoes imply large particles in high concentrations. In these situations, rainfall is extremely heavy, and hail is a possibility.

The radar echo is displayed in a color image (FIGURE 5-21). The colors represent the amount of energy scattered back to the radar site, or the **radar reflectivity.** Radar reflectivity is measured

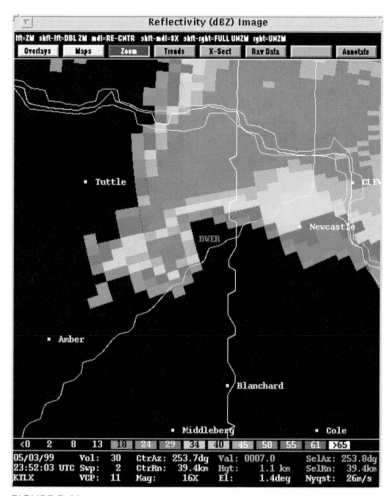

FIGURE 5-21 An example of a weather radar image collected during the May 3, 1999, Oklahoma tornado outbreak. Colors represent the amount of energy scattered back to the radar site, or the reflectivity, in decibels (dBZ; see the colored legend at the bottom). High reflectivities are colored red and indicate high precipitation rates.

in decibels (dBZ). A reflectivity of 20 dBZ usually indicates rain. High reflectivities are colored red and indicate high precipitation rates and, above 55 dBZ, hail. The image is updated roughly every 5 minutes, even more frequently than satellite images. Composites of individual radar images can be used to give regional and national views of precipitation, as we will see in Chapter 11.

Doppler Radar

Today's **Doppler radar** can measure the location and intensity of precipitation and also the speed of the wind in precipitating regions. To explain how it works, we can use an everyday example of traffic noise.

Sound travels as a wave, just like radio waves. As a car with a loud stereo or a siren or a train approaches you and then passes by, the sound you hear changes. The pitch is higher (i.e., the frequency increases) as the sound-maker approaches you, and the pitch goes down as the source of the sound moves away (**FIGURE 5-22**). This phenomenon is the **Doppler Effect**, named after the 19th-century Austrian physicist Johann Christian Doppler who first explained it.

To prove the Doppler Effect to skeptics, a Dutch meteorologist named Buys Ballot devised the following experiment. A group of trumpeters stood on a moving train and played a single note together. The observers on the ground heard the note go up and then go down as the train approached and then passed by (Figure 5-22a).

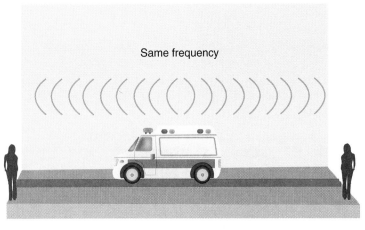

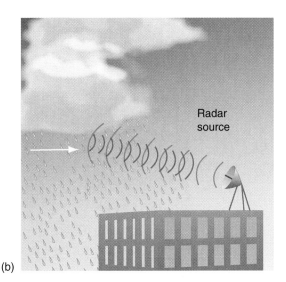

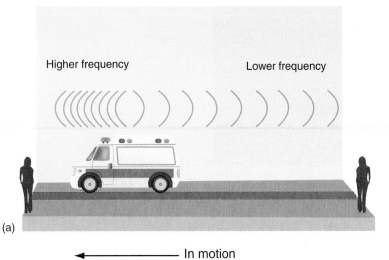

FIGURE 5-22 (a) The frequency of sound waves change when the source or the observer is moving. If the source and observer are converging, the frequency is higher; if they are moving apart, the frequency is lower. (b) This same effect is used by Doppler radar to detect wind speeds in precipitation storms.

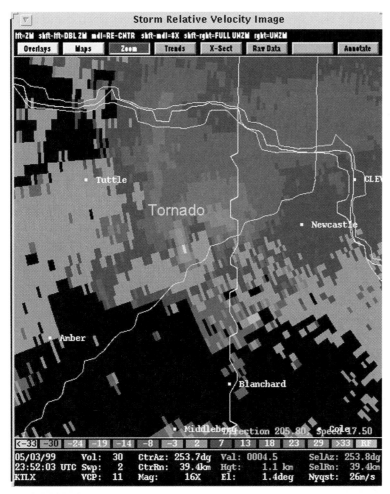

FIGURE 5-23 A radar display in Doppler mode (also known as velocity mode) of a thunderstorm that spawned a tornado and caused destruction in Oklahoma City on May 3, 1999. Greens represent motion toward the radar and the red indicate motion away from the radar (see the color legend at the bottom of the figure). The tornado is located in the vicinity of the red-green couplet in the center. Compare this image with Figure 5-20, which is the reflectivity from the same radar at the same time.

The Doppler Effect works the same with radio waves. Just like the observer on the left in Figure 5-22a, the radar receiver "hears" waves of a higher frequency than the waves that were sent out, if precipitation particles are moving toward the source (Figure 5-22b).

The relative motions of particles measured by a Doppler radar are color coded and displayed in an image for quick analysis by a forecaster. Usually the cool colors (typically greens) represent motion toward the radar. The warm colors (reds and purples) indicate motion away from the radar. Warm colors next to cool colors indicate spinning motion within the storm, which may lead to a funnel cloud or a tornado.

FIGURE 5-23 is a radar display depicting Doppler velocity observations of a thunderstorm that spawned a devastating tornado in and near Oklahoma City on May 3, 1999. The region likely to have a tornado is located where the winds are traveling toward the radar and then away from the radar over a small distance. We will discuss the radar identification of tornadoes in more detail in Chapter 11.

Dual-Polarization Radar

The next generation of weather radars, which are currently being installed throughout the United States, will improve observations of the interior of storm systems. These radars are called **dual-polarization radars** (they are also referred to as *dual-polarimetric radars.*)

A radio wave is an electromagnetic wave and therefore has electric and magnetic fields that are oriented perpendicular to one another. The orientation of these oscillations is referred to as polarization, which is perpendicular to the direction the wave is traveling.

A polarizing filter for a camera, or polarizing sunglasses, can be used to observe the effects of polarization of light in a cloud-free sky. Rotate the filter, or glasses, while looking through them at a portion of the sky away from the sun—at a certain orientation, the intensity of the sky-light will be reduced. This is because the filter is removing polarized light that is not aligned with the filter.

Typical weather radars transmit and receive radio waves with a single orientation of the electric field. Dual-polarization radars emit radio waves that alter their transmitted pulse between horizontal and vertical polarizations. Therefore, these polarization radars can measure information about both the horizontal and vertical dimensions of precipitation sized particles, as these particles will scatter the horizontal and vertical beams differently.

The additional information on polarization improves the precipitation rate measurement as well as the determining the type of precipitation (snow, rain, freezing rain, and possibly hail). FIGURE 5-24 shows a vertical cross-section through a tornadic thunderstorm in Idalia, Colorado, on June 29, 2000, with a dual-polarization radar. The top figure shows the reflectivity in dBZ and the bottom figure the cloud particle type. The identification of cloud particle types is not possible with current radar systems. As we will discuss in Chapter 11, the vertical structure in this storm indicates the potential of severe weather.

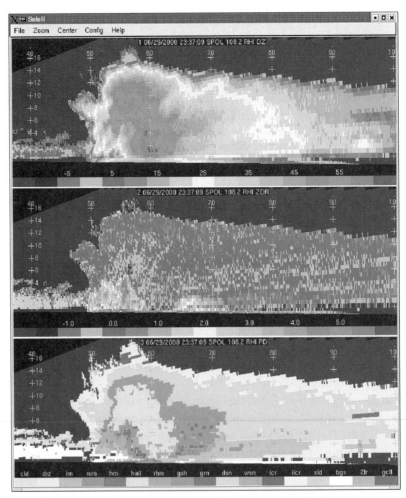

FIGURE 5-24 A vertical cross-section through a tornadic thunderstorm in Idalia, CO, on June 29, 2000, with a dual-polarization radar. The top figure shows the reflectivity in dBZ and the bottom figure the cloud particle type.

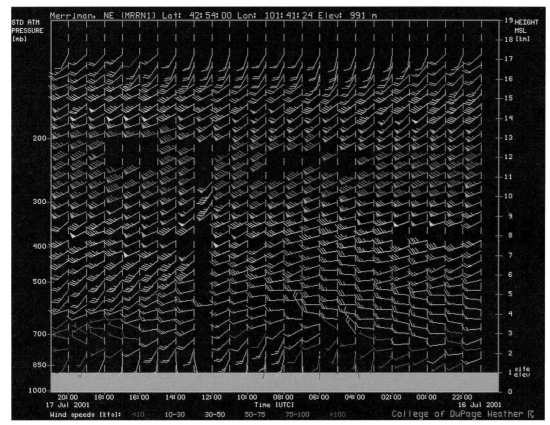

FIGURE 5-25 Hourly wind profiler measurements made at Merriman, Nebraska, during July 16 and 17, 2001. Time elapses are from right to left in this image. Winds are color coded (see the key at the bottom, in knots) and are also represented using the "flagpole" approach in Chapter 1. The strongest winds are from 200 to 300 mb and are from the west-southwest.

Wind Profiler

A recent application of Doppler technology is the **wind profiler**. Wind profilers are Doppler radars pointed skyward. By measuring the change in the frequency of the returning radar beam, a wind profiler can determine the wind speed even when precipitation is not present. By alternately pointing the transmitted radio beam in different directions, the three-dimensional wind field can be determined, providing a continuous measurement of winds at a given location.

FIGURE 5-25 shows a day's worth of hourly wind observations from the ground to the stratosphere at a location in Nebraska, with time elapsing from right to left. The strongest (red) winds are located between 200 and 300 mb, where the jet stream is located. This type of detailed hourly wind information greatly augments radiosonde observations, which are made only twice a day. However, wind profilers are more expensive than radiosondes, and their coverage is mostly limited to the central United States.

ATMOSPHERIC OPTICS

As we have seen throughout this chapter, scientific observations of the atmosphere cannot be divorced from the human senses. The best example is vision. Our eyes are passive remote sensing instruments sensitive to light with wavelengths between 0.39 and 0.78 microns. Our brains then process the collected data. This section focuses on some of the beautiful and fascinating atmospheric phenomena that you can observe with your own eyes.

Mirages

Have you ever seen the surface of a highway look watery on a hot, dry summer day? This is a classic example of a **mirage**. Mirages are not illusions; they are images formed by the refraction of light.

FIGURE 5-26 An inferior mirage in the desert. There is no water in this photograph; instead, refraction causes light from the sky to bend, making it appear that the trees are being reflected in water.

Refraction occurs when light moves through substances with different indices of refraction. Because the index of refraction for air depends on temperature, light can refract if it passes through layers of air that have different temperatures.

On a summer day, the air just above a highway is much hotter than the air a few meters higher up. The strong temperature gradient causes light from above to be bent downward and then upward toward your eye. "Puddles" on dry highways and in deserts (**FIGURE 5-26**) are the result. You are seeing the sky, not the highway or the sand. The refraction is even more obvious when a recognizable object, such as a palm tree (**FIGURE 5-27**), is seen in a mirage. It appears upside down! For this reason, we call this an **inferior** (meaning "below") **mirage**.

Mirages also develop when the surface is much colder than the air above it, such as over snow, ice, or a cold lake or ocean. In these temperature inversions, light rays passing upward through the warm air in the inversion refract downward toward the surface. A mirage image thus appears above, not below, the true position of the object (**FIGURE 5-28a**). This is a **superior mirage**. Superior mirages can bring into view objects that are below the horizon (Figure 5-28b). A special type of superior mirage, the **Fata Morgana**, creates the false appearance of cliffs or castles. In reality, refraction is distorting and vertically stretching small features on the horizon like a funhouse mirror.

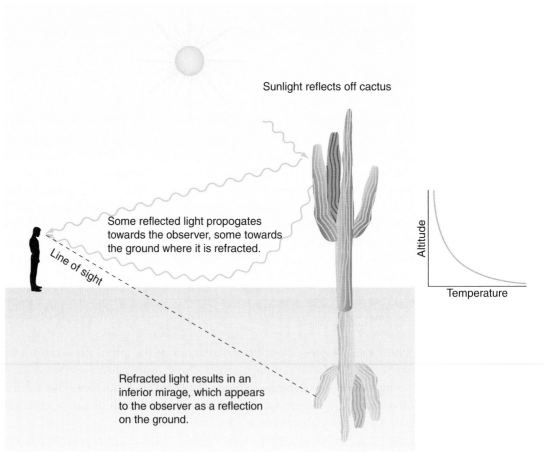

Sunlight reflects off cactus

Some reflected light propogates towards the observer, some towards the ground where it is refracted.

Line of sight

Refracted light results in an inferior mirage, which appears to the observer as a reflection on the ground.

Altitude

Temperature

FIGURE 5-27 An inferior mirage, where the refracted image appears below the object's true position, occurs because light is refracted downward and then upward from the ground toward the observer. This requires a strong temperature gradient near the ground.

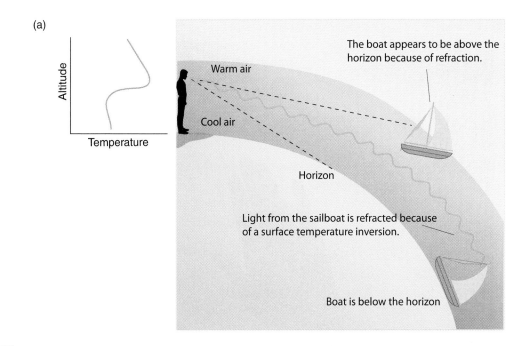

(a)

(b)

FIGURE 5-28 (a) A superior mirage occurs when air is refracted through a temperature inversion. (b) A superior mirage causes objects to appear to hover in midair above the horizon, as seen along the horizon in this photograph.

Halos

A **halo** is a whitish ring that encircles but does not touch the Sun or Moon. It is an optical phenomenon that owes its existence to refraction of light by ice crystals. Because the light must shine through a fairly uniform layer of ice crystals that are thin enough to let light through, halos are usually associated with cirrostratus clouds.

Different crystal habits, crystal orientations, and solar zenith angles can produce a variety of halos. The most commonly observed halo is the 22° halo (**FIGURE 5-29**). The 22° halo encircles the Sun at about a hand's width from the center of the Sun, if your arm is fully extended. On rare occasions, pencil-shaped crystals can cause halos inside or around the 22° halo, a remarkable sight!

Small column-like ice crystals form the 22° halo. Light rays enter a crystal, refract, and then refract again as they exit the crystal (**FIGURE 5-30**). Because the crystals are randomly oriented in space, there are many different directions from which light rays can enter the crystals. More light rays are refracted at this 22° angle than at any other, producing the concentration of light known as the halo.

FIGURE 5-29 The 22° halo, the hand in the middle is used to shield the camera from the direct light of the sun; otherwise the image would be overexposed.

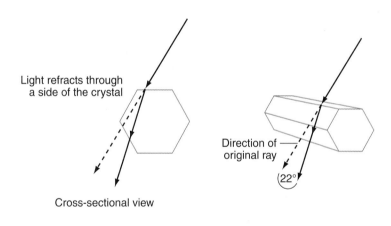

Light refracts through a side of the crystal

Cross-sectional view

Direction of original ray

22°

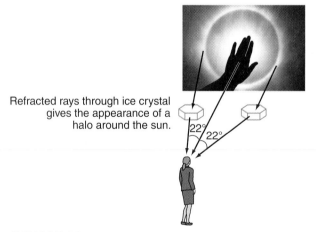

Refracted rays through ice crystal gives the appearance of a halo around the sun.

22° 22°

FIGURE 5-30 Diagram showing how light is refracted through a crystal to form a 22° halo.

FIGURE 5-31 An example of how refraction causes a dispersion of light through a prism.

Dispersion of Light

When sunlight passes through a triangular glass prism, it separates into the colors of the rainbow (**FIGURE 5-31**). This separation happens because different colors, defined by their wavelength, refract by different amounts. The shortest (blue and violet) wavelengths refract the most; red light refracts the least. The separation of colors is referred to as **dispersion**. Not only prisms, but also the atmosphere, water drops, and ice crystals can cause dispersion. This causes a wide range of breathtaking visual effects in the atmosphere.

Green Flash

When the Sun is finally setting or just rising over the horizon, some people have reported seeing a brief, brilliant flash of green from the top part of the Sun's disk. Often discounted as myth, this **green flash** is real and is a product of refraction and dispersion.

As the Sun sets, refraction bends and disperses its light. Red bends least and disappears first. Blue and violet should be the last seen, but multiple scattering keeps this light from reaching your eyes. The color with the next-shortest wavelength is green. Green light is refracted more than red, but it is not scattered as effectively as blue. Therefore, a sliver of green appears for a second or two as the Sun disappears over the horizon (**FIGURE 5-32**). The best conditions for seeing a green flash are a nearly flat horizon and a lengthy sunset or sunrise.

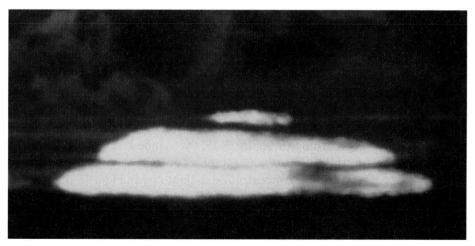

FIGURE 5-32 The green flash, as captured in this photograph. The green flash is the sliver of green light refracted just above the top rim of the sun at the top.

Sundogs

On sunny mornings and afternoons with high clouds, you are likely to see shiny, colored regions at either side of the Sun. These are **sundogs**, another optical effect caused by refraction and dispersion. Sundogs appear because hexagonal ice crystals in the high clouds tend to drift downward with their flat bases parallel to the ground. The sunlight passing through the crystal refracts sideways. If the Sun is low enough in the sky, you see spots of bright light instead of a complete halo. Therefore, sundogs are usually 22° away from the Sun. They can be on one or both sides of the Sun, depending on where the clouds are (FIGURE 5-33). Refraction causes blue light to be bent more than red light, and so sundogs show the spectrum of colors with red nearest the Sun. Both cirrus and cirrostratus clouds produce sundogs.

Sun Pillar

A setting Sun behind high clouds will sometimes project a narrow column of reddish light straight above it. This is a **sun pillar**. Although it sounds very similar to sundogs, the Sun pillar is caused by reflection, not refraction, of sunlight. Flat-plate ice crystals floating in the cloud reflect the Sun's reddened light to your eyes in a line above, or sometimes below, the Sun. Pillars can also appear over streetlights when ice crystals are in the air.

Now that we are familiar with both reflection and refraction with dispersion, we can investigate the most famous optical phenomenon of all: the rainbow.

Rainbows

When did you last see a rainbow? What was the time of day? What were the weather conditions? The chances are that you saw it in the late afternoon or early evening, in the eastern sky, and that it was or had just been raining. Our ability to make these good guesses is based on the science behind the rainbow.

The classic **rainbow** is a single, bright, colored arc. Red is the outermost color of this arc, and violet is always the innermost color. On occasion, you may have seen two rainbows at once. The lower rainbow is the **primary rainbow**, and the higher, fainter, colored arc is the **secondary rainbow** (FIGURE 5-34). The color sequence of the secondary rainbow is opposite to the primary; red is on the inside of the arc, and violet is on the outside. The rainbows usually stand out against a fairly dark sky. To explain these observations, we need to trace rays of light as they enter and leave large drops of water.

Raindrops act as prisms, refracting and reflecting light just like ice crystals. FIGURE 5-35 shows how different rays of light are refracted at the surface of the drop. This happens because the indices of refraction for water and air are different. Then a total internal reflection occurs at the back of the drop, and the light refracts again as it exits the drop. Because of this, sunlight that shines through water drops tends to brighten the region of the sky opposite from the Sun and especially at an angle of about 40°. The shape of the rainbow is a circle, but we only see the arc above the ground.

According to Figure 5-35, the rainbow is located opposite to the Sun; this explains why rainbows are not seen at noon with the Sun overhead. The concentration of light rays in Figure 5-35 reveals why the rainbow seems so bright compared with the sky around it—but we have not yet explained its color.

Raindrops refract sunlight, dispersing white light into its spectrum of colors (FIGURE 5-36). Blue light is bent at a 40° angle, and red is bent at 42°, and the other colors are bent at angles intermediate between these

FIGURE 5-33 Sundogs appear to either side of the sun and in high clouds; the colors are caused by dispersion. The sundog in this photograph appears in the upper part.

FIGURE 5-34 A double rainbow over a valley in Arizona. The primary rainbow is the brighter arc on the inside of the fainter secondary rainbow. Notice that the order of colors is reversed in the two rainbows. Why do we look in the opposite direction from the Sun, as in this picture, to see rainbows?

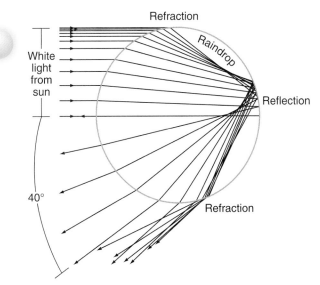

FIGURE 5-35 How light rays are refracted (surface of raindrop, left) and reflected (back of raindrop, right) to form a rainbow. The cluster of light rays at an angle of about 40 degrees explains the location, shape, and brightness of the rainbow. (Adapted from Greenler, R. *Rainbows, Halos and Glories.* Cambridge University Press, 1980.)

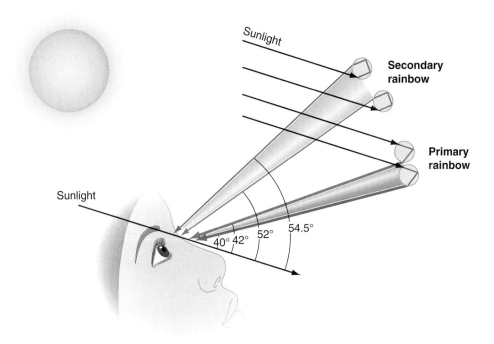

FIGURE 5-36 Colors of the rainbow arise from drops at different altitudes in the sky. Distinct sections of a rainbow are produced by different drops suspended in the atmosphere. When viewing a rainbow, the Sun must always be behind you.

two values. These are the colors of the primary rainbow. If the light is reflected twice instead of once, the bending is at different angles: Blue is bent at a 54.5° angle, and red is bent at 52°. This second region of color is the secondary rainbow.

Figure 5-36 also explains why the blue colors are on the inside of the primary rainbow but on the outside in the secondary rainbow. The rainbow is an assembly of different rays of light leaving millions of falling raindrops. If you looked at only one drop, you could change the color you see by changing the height of your line of sight. If you stood straight up, you would see the drop as red. If you squatted, you would come to a height where the drop looked violet. When you see a rainbow, you are looking at drops at different heights. The top drops look red, just as if you were standing straight up while looking at the single drop. Also, the bottom drops look purple, as though you were squatting. This explains why the outside of the primary rainbow arc is red and the inner portions are violet. Red light appears below violet in the secondary rainbows because of the second reflection. This explains why the color sequence of the secondary rainbow is a mirror image of the colors in the primary rainbow.

Using the concepts of reflection, refraction, and dispersion, we have explained most aspects of the rainbow. Now test your knowledge using the rainbow applet on the text's Web site.

"Rainbows" to explore the appearance of rainbows.

Coronas

Hold the underside of a compact disc up to a light at an angle. Tilt it slightly back and forth. Do you see streaks of color? They resemble a rainbow, but the color stems from a different optical effect: **diffraction**. Diffraction occurs when light is bent around small objects, such as the tiny lines of data in your compact disc.

Cloud droplets also diffract light. When the droplets are of different sizes, the colors blend, and the light is a diffuse white. A **corona** forms when thin water clouds obscure the Sun or Moon and cause a whitish region to form around the Sun or Moon. The key visual difference between the corona and a halo is that the corona is not a ring around the Sun or Moon; it is instead a broader, bright region that appears to touch the Sun or Moon.

The color produced by diffraction is called **iridescence**. Clouds become iridescent frequently in many parts of the world. The only ingredients needed are light and water clouds.

Glories

If you are flying above water clouds with clear skies above the plane, you may notice concentric colored circles on the cloud below around the airplane shadow. This is an optical phenomenon known as a **glory**. Glories always form directly opposite the sun, centered where you would find your shadow, or the airplane shadow.

Glories have a bright center and colored rings with blue on the inside, with reds and purple on the outside. You will often see three or four of these concentric rings. While we can simulate the glory structure using computer solutions to Mie theory, we cannot yet simply explain the cause of a glory. We do know that their formation involves liquid droplets all of a similar size and the processes of refraction and reflection, as in the rainbow, but also diffraction and complex **surface waves**. When a sun-ray grazes the surface of a drop, the ray can travel along the drop surface. Refraction and reflection of these surface waves are important in forming the glory.

Glories can also be seen when hiking along mountain ridges, such as along the top of Brocken, the highest peak in the Harz Mountains of Northern Germany. When a glory is observed from a mountain (**FIGURE 5-37**), it is often called a **Brocken Bow**. In Figure 5-37, the photographer's body casts a shadow on the fog below, and the concentric rings of color are the Brocken Bow.

TABLE 5-1 classifies all of the optical phenomena we have examined in this chapter by the mechanisms that form them.

FIGURE 5-37 A glorious Brocken Bow around the shadow of the photographer in Ha Ling Peak, Alberta, Canada. The rings of color are caused by diffraction of reflected light by the water droplets in the fog.

TABLE 5-1 **Atmospheric Optical Phenomena Categorized by Cause**

By	Air	Ice Crystals	Liquid Water
Refraction	Mirages Green flash Twinkling stars	Halos Sundogs	R a i n
Reflection	—	Sun pillar	b o w s
Scattering	Blue skies Blue mountains Red sunrises/sunsets Crepuscular rays	White clouds	White clouds with dark bases
Diffraction	—	—	Corona Iridescence
Refraction, reflection, diffraction, surface waves			Glory Brocken bow

PUTTING IT ALL TOGETHER

Summary

We can observe the atmosphere in several ways. Direct methods measure the properties of the air in contact with the instrument that is sensing it. Indirect methods obtain data from afar. Indirect methods can be either active or passive. Active sensors emit pulses of energy and observe what comes back to the sensor. A passive sensor "sees" whatever energy naturally comes to it. Our eyes passively sense visible light, whereas meteorological instruments can observe in many different regions of the electromagnetic spectrum.

Observations of the atmosphere, both direct and indirect, contribute greatly to our understanding of weather and climate. The ASOS in the United States observes and records surface temperature, humidity, pressure, wind, precipitation, cloud ceiling, and visibility data as often as once a minute. Rawinsondes provide twice-a-day observations of temperature, humidity, and wind from the surface into the stratosphere.

Satellites observe the global atmosphere by indirect methods. The three most common weather satellite images are the visible, IR, and water vapor images from geostationary satellites. Visible images provide detailed views of clouds but only during daytime. IR satellite images reveal storm systems via their cloud patterns 24 hours a day. Water vapor images provide meteorologists with information in both cloudy and cloud-free regions of the upper and middle troposphere.

Radar actively senses radio waves scattered by large water and ice particles and converts them into information about precipitation. Doppler radar reveals wind patterns within a storm by tracking the relative motions of precipitation particles. Wind profilers provide detailed wind information from the surface to the lower stratosphere.

Light can be reflected, bent (or refracted), scattered, and diffracted, and its colors can be dispersed. One or more of these processes can cause beautiful optical effects in the sky. The most famous, the rainbow, is caused by reflection, refraction, and dispersion of light in raindrops. Refraction of light by ice crystals causes halos and sundogs. Refraction of light by layers of air of different temperatures produces mirages. Scattering of light by air molecules causes blue skies, red sunsets, and white clouds with dark bases. Diffraction of light creates white fuzzy coronas.

> The deeper one enters into the study of Nature, the further one ventures into and along the by-paths that, like a mystic maze, thread Nature's realm in every direction, the broader and grander becomes the vista opened up to the view.
>
> —Wilson "Snowflake" Bentley

Key Terms

You should understand all the following terms. Use the glossary and this chapter to improve your understanding of these terms.

Active sensors	Glory	Rawinsonde
Anemometer	Green flash	Rayleigh scattering
Aneroid barometer	Halo	Reflection
ASOS	Haze	Refraction
Aspirated psychrometer	Index of refraction	Resistance thermometer
Brocken bow	Indirect methods	Scattering
Ceilometer	Inferior mirage	Secondary rainbow
Cloud ceiling	Infrared satellite image	Sling psychrometer
Corona	Iridescence	Sonic anemometers
Crepuscular rays	LEO satellite	Sounding
Critical angle	Mercury barometer	Sundogs
Cup anemometer	Mirage	Sun pillar
Dew point hygrometer	Normal	Superior mirage
Diffraction	Passive sensors	Surface waves
Direct methods	Primary rainbow	Total internal reflection
Dispersion	Propellers	Visibility
Doppler Effect	Psychrometer	Visible satellite image
Doppler radar	Radar	Water vapor satellite image
Dual-polarization radars	Radar echo	Wind profiler
Fata Morgana	Radar reflectivity	Windsock
GEO satellite	Radiometers	Wind vane
Geometric scattering	Radiosonde	
Global Positioning System (GPS)	Rainbow	
	Rain gauge	

Review Questions

1. Classify each of the following as a direct or indirect method of observation: hearing, seeing, and touching. Which of these methods are active sensing? Which are passive sensing?
2. Why should temperature measurement be made in a shaded and ventilated location?
3. A bird manages to build a nest over an ASOS dew point sensor. How could this affect the accuracy of the observations?
4. Why can't radiosondes observe the mesosphere?
5. Describe reflection and refraction.
6. Explain how refraction changes the length of daylight.
7. Which of the following phenomena is not caused by scattering: red sunrises, blue skies, a hot torch with a blue flame, the "blue" in the Blue Ridge Mountains?
8. Explain why clouds are white even though they are composed of water drops, which are transparent.
9. Why are there both LEO and GEO weather satellites? Why not use just one type of orbit?
10. You want to look at a satellite image of a hurricane over the Gulf of Mexico at 0600 UTC. Which type of image do you choose: visible, IR, or water vapor? If you want to see the wind patterns in the upper troposphere that guide the hurricane's direction, which type of image do you choose?

11. If you wanted to launch a satellite to observe the ozone hole, which orbit would you want to place the satellite in: LEO or GEO?

12. Why are weather radars located every few hundred miles across the United States, whereas one GEO satellite can see the entire country?

13. Police radars that measure the speed of cars on a highway are based on what scientific principle?

14. Compare and contrast mirages with images in a mirror.

15. You are in Minnesota during a snowy winter. On a bitterly cold, clear morning, a lone cow stands in a snowy field on the horizon. What kind of optical effect do you expect to see? Will you see an upside-down cow, or a cow looming over the horizon?

16. Visit an art gallery to determine if rainbows, halos, or other optical phenomena are correctly drawn. For example, are the colors in the correct order?

17. Explain how you can see colored "bows" near waterfalls or in the spray of a lawn sprinkler. Can you also expect to see halos in these situations? Explain why or why not.

18. Can a totally dry atmosphere have a corona? Can it produce a green flash?

19. Based on the information in Box 5-2, do you think that astronauts in orbit hundreds of kilometers above the Earth see the stars twinkle? Why or why not?

Observation Activities

1. Put a penny in an opaque bowl. Place it half way between the center of the bowl and the side of the bowl closest to you. Lower your head until you can no longer see the penny. Slowly fill the bowl with water, pouring slowly so that the penny does not move. Explain why the penny appears to move.

2. Carefully observe and describe the colors of a sunset in relation to the position of the sun.

3. Anemometers make measurements of the wind. What visual indicators can you use to estimate the direction and speed of the wind?

4. If you want to see a Brocken Bow from a mountain ridge, what time(s) of day would be best to time to hike?

This rain cloud icon is your clue to go to the *Meteorology* Web site at http://physicalscience.jbpub.com/ackerman/meteorology/. Through animations, quizzes, web exercises, and more, you can explore in further detail many fascinating topics in meteorology.

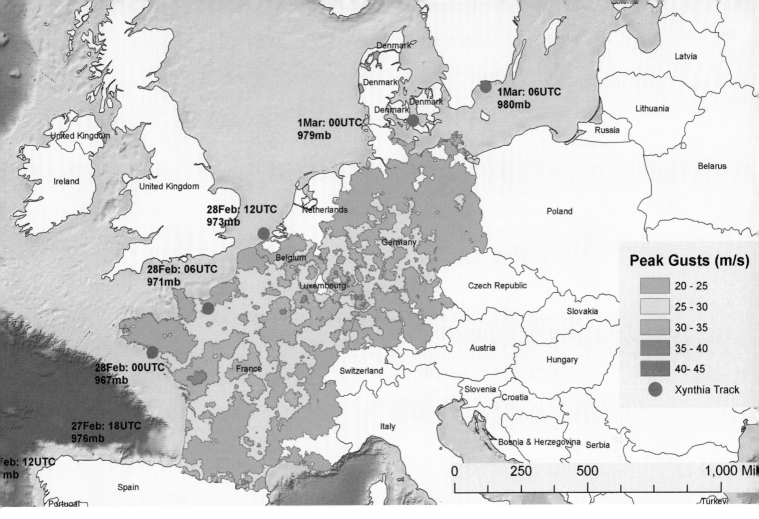

Atmospheric Forces and Wind

6

AFTER COMPLETING THIS CHAPTER, YOU SHOULD BE ABLE TO
- List and describe the forces that act on the atmosphere
- Define geostrophic balance and use it to explain features found on real-life weather maps
- Define other force-balances and relate them to the real-life horizontal wind patterns that they help us understand
- Use force-balance concepts to explain why air rises vertically in a low-pressure system but sinks in a high-pressure system and how this creates different types of weather in highs versus lows

INTRODUCTION

Why does the wind blow? That question was asked across the nation of France at the end of February 2010. An intense low-pressure system with a pressure of 966 millibars blasted much of northern France with winds over 120 kilometers per hour (75 mph), or hurricane force (map on page 181). The winds pushed ocean water against a centuries-old sea wall in the French coastal town of L'Aiguillon-sur-Mer, and the sea wall collapsed (photo on page 181). The resulting flooding killed dozens of people in a nearby mobile home park. The French prime minister declared a national emergency in the storm's wake. Across western Europe, the windstorm caused at least 62 fatalities and at least $2 billion in damage.

Wind is air in motion that arises from a combination of forces. Violent destructive winds, as well as gentle summer breezes, result from a complex interplay of different forces. In this chapter, we explain the forces of nature that affect the atmosphere, how they combine to create wind, and how to use weather maps to interpret wind direction and speed.

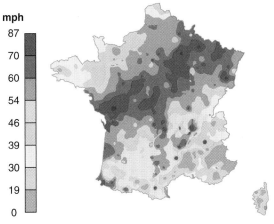

mph

| 87 |
| 70 |
| 60 |
| 54 |
| 46 |
| 39 |
| 30 |
| 19 |
| 0 |

(Data from Météo-France.)

WIND BASICS

Before we delve into the physical laws describing this motion, let's review some basics regarding wind. Simply put, wind is moving air. As we saw in Chapter 1, weather reports include wind speed and direction. Wind speed is reported on U.S. weather maps in **knots**—one nautical mile per hour (equivalent to about 1.15 miles per hour, or 0.5 meters per second). If the wind speed is strong (greater than 15 knots) and highly variable, the weather report will include the **wind gust**, which is the maximum observed wind speed.

Wind direction is the direction from which the wind is blowing. A north wind blows from the north toward the south. It is reported with respect to compass directions (**FIGURE 6-1**). The

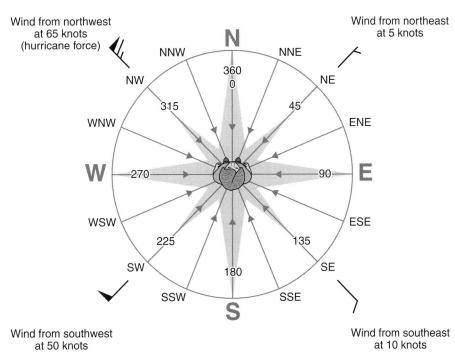

FIGURE 6-1 Wind direction is expressed as degrees around a circle (clockwise from the top, or due north), or as compass points. Outside the circle, wind direction and speed are depicted using "flagpoles" with flags attached, as explained in Chapter 1.

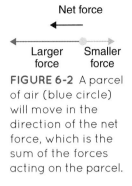

FIGURE 6-2 A parcel of air (blue circle) will move in the direction of the net force, which is the sum of the forces acting on the parcel.

FIGURE 6-3 The graphical method of determining the resultant of two forces acting at an angle to one another on a parcel of air.

prevailing wind direction of a region is the most frequently observed wind direction during a given period. Finally, winds are often associated with pressure systems—for example, low-pressure centers (**cyclones**) and high-pressure systems (**anticyclones**).

PHYSICS OF MOTION BASICS

Fundamental laws of physics explain atmospheric motions. In the 17th century, Sir Isaac Newton—who also first explained the colors of the rainbow—established physical laws that help explain the movement of galaxies, planets, cars, baseballs, and parcels of air.

To understand these laws, let's review some basic facts of physics. **Forces** are characterized by direction and magnitude (or strength). Two or more forces can act on the same point or object. The overall force that results from interacting forces can be expressed as a single force, called the net force or the resultant. If the two forces act in opposite directions and with different magnitudes, the net force will be in the same direction as the stronger force, and its magnitude will be the difference between the two forces (**FIGURE 6-2**). If two forces act at an angle to each other, the resultant force is along a diagonal and away from where the two forces are applied (**FIGURE 6-3**).

A force applied to an object, in the absence of other forces, causes an acceleration and results in movement. An object's **velocity** is the magnitude and direction of its motion. The **speed** of the object, the distance traveled in a given amount of time, is the magnitude of the motion. A change in an object's velocity—magnitude, direction, or both—is its **acceleration**.

NEWTON'S SECOND LAW OF MOTION

Now that we have covered some basics of wind and physics, we can examine the law that explains how forces create wind. **Newton's Second Law of Motion** states that a force exerted on an object (or a parcel of air) of a given mass causes the object to accelerate in the direction of the applied force. The relationship between acceleration, the force, and the mass is expressed mathematically as follows:

$$Force = Mass \times Acceleration$$

or, as it is often written in meteorology,

$$Acceleration = Force \div Mass$$

Newton's Second Law also relates the acceleration of a parcel of air to the sum of *all* of the forces acting on it. As in Figures 6-2 and 6-3, there can be more than one force affecting a parcel of air simultaneously. Sometimes these forces push in the same direction; sometimes they oppose each other, and sometimes they push at right angles to each other.

This law, and most equations in meteorology that involve acceleration, also goes by the name of the **Law of Momentum**. The **momentum** of an object is its mass multiplied by its velocity. For an object whose mass does not change, mass multiplied by acceleration equals the change in momentum. Because force also equals mass times acceleration, this means that applying a force to an object changes its momentum.

The momentum of the wind can propel a sailboat, blow over a truck, uproot large trees, and topple buildings. We now discuss which forces cause the air to move.

FORCES THAT MOVE THE AIR

Five different forces combine to move air: the gravitational force, the **pressure gradient force** (**PGF**), the centrifugal force, the Coriolis force, and the frictional force. Some of these forces are real, meaning that they are observable no matter what your perspective is. Other forces are apparent, meaning that you may observe them in one frame of reference (e.g., on a rotating Earth) but not in others. We must examine these forces to determine where and how hard the wind will blow.

Gravitational Force

The familiar **gravitational force** (**GF**) is directed downward perpendicular to the ground and is approximately equal to the mass times the gravitational acceleration, 9.8 meters per second per second. The "per second per second" is not a typographical error. Remember that acceleration is the change of velocity over time, and the units of velocity are meters per second.

If the gravitational force were the only force acting on the atmosphere, Chicken Little would have been right: The sky would indeed be falling! The gravitational force, however, is almost equally opposed everywhere by the vertical portion of the force we will examine next: the PGF.

Pressure Gradient Force (PGF)

Spray-paint cans have labels warning that the contents are under pressure. When the nozzle is squeezed or the sides are punctured, the large pressure difference over the small distance between the air and the inside of the can forces the contents out of the can. The force that results from pressure differences over distances in a fluid, such as our atmosphere, is called the PGF.

A **pressure gradient** is simply a *change in pressure over a distance*. Atmospheric pressure differs from place to place for many reasons, all of which can usually be traced back to the Sun's uneven heating of the Earth's surface. Generating pressure differences is also important in flying and sailing (BOX 6-1).

The PGF always pushes from higher pressure toward lower pressure (FIGURE 6-4). Its magnitude is equal to the change in pressure over some distance, divided by the air density. Mathematically, it is written per unit mass as follows:

$$PGF = -\frac{1}{\text{Air density}} \times \frac{\text{Change in pressure}}{\text{Distance}}$$

The minus sign is a reminder that the force is pointed "downslope" toward lower pressure. When pressure changes rapidly over a small distance, the PGF is large. Strong winds almost always result from large pressure gradients. However, the reverse isn't always true: Large pressure gradients don't always cause strong winds, as we will now see for ourselves.

The largest pressure gradients in the atmosphere are nearly always right over our heads. As we learned in Chapter 1, typical atmospheric pressure decreases from more than 1000 millibars (mb) at the surface to only half that value only 5 kilometers overhead. This amazingly large pressure gradient is directed upward because the highest pressure is at the surface. But we don't feel sucked or pushed upward off the ground. When we discuss hydrostatic balance later in this chapter, we will learn why this vertical pressure gradient does not cause an enormous upward rush of air.

The pressure gradients that most often cause wind are in the horizontal direction. On the surface weather map, atmospheric pressure measured at the surface is converted to sea-level pressure, as we discussed in Chapter 1, to remove pressure differences caused by altitude variations. This allows meteorologists to isolate the wind-causing horizontal pressure gradients on the surface weather map. Then the sea-level pressure patterns are analyzed. Isobars of constant sea-level pressure are drawn at 4-mb intervals on either side of 1000 mb (FIGURE 6-5). Because the pressure difference between any two isobars is fixed, the closer the isobars, the larger the pressure gradient, the stronger the PGF, and in general the greater the wind speed.

The surface map plots atmospheric pressure adjusted to the altitude of sea level. Another type of weather map reverses this approach and plots the *altitude* of a given *pressure* surface. This map is commonly used when analyzing the weather above the surface and is called a *constant-pressure chart* or **isobaric chart**. Constant-pressure charts are commonly drawn for 850, 700, 500, 300, 250, and 200 mb. FIGURE 6-6 is an example of a 500-mb isobaric chart. The units of altitude are called *geopotential meters* and are nearly equivalent to geometric meters measured

Box 6-1 Planes and Pressure Differences

Pressure differences and wind explain how planes fly. If you chopped an airplane wing in half from front to back and looked at it from the side, you would see that it has a distinctively curved shape. The wings are designed this way so that they can lift the plane off the ground. The forces responsible for lifting the plane off the ground are produced by pressure differences between the wings' top and bottom resulting from the wind flow over the wing.

You can demonstrate the basic concept of airplane flight for yourself by shaping a piece of cardboard like a wing (dashed blue line in figure), securing it to a table, and blowing air over it with a hair dryer. The cardboard will lift off the table, as depicted by the solid blue line.

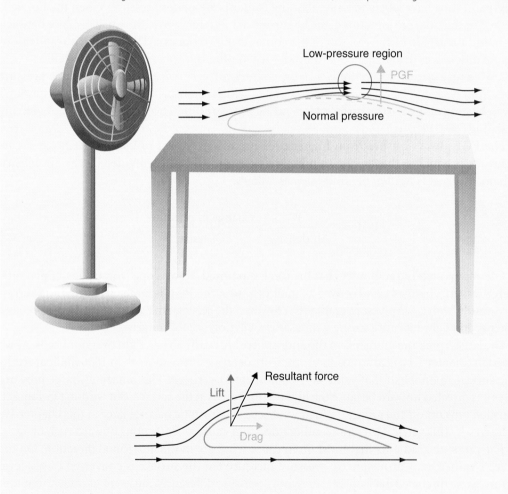

Why does the cardboard lift upward? Scientists still debate the origins of aerodynamic lift. One explanation is based on Bernoulli's Principle (named after the Swiss scientist Daniel Bernoulli). This principle states that velocity and pressure are inversely related. The fast-moving air over the top of the wing is therefore associated with low pressure. The result is an upward pressure gradient, causing lift. There is also a drag force acting to slow down a moving wing. The net effect of lift and drag, if the wing is designed correctly and drag is smaller than lift, forces the wing and the plane upward off the ground—takeoff!

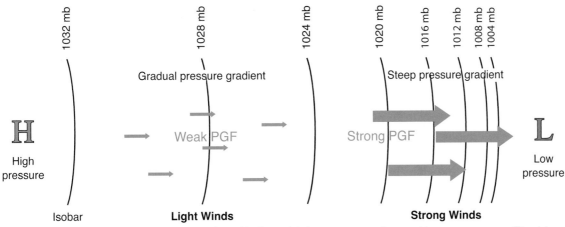

FIGURE 6-4 The PGF always pushes directly from higher pressure toward lower pressure. The bigger the change in pressure over the same distance, the larger is the PGF (e.g., the PGF is larger for the low than for the high in the diagrams above).

with a meterstick. Thus, the altitude at 500 mb across North America as shown in Figure 6-6 is between 5000 and 6000 meters, or very roughly 18,000 feet.

Isobaric maps are useful for portraying horizontal pressure gradients above the ground. The spacing between the lines of constant height (isoheight) is proportional to the PGF. **FIGURE 6-7** illustrates this relationship between the spacing of isoheights and isobars. The colored surface (Figure 6-7a) is a constant-pressure surface of 850 mb. Anywhere on this surface the pressure is 850 mb. The altitude of this 850-mb surface varies—for example, its altitude is higher at point X than at point Y. Figure 6-7b illustrates height lines of the 850-mb pressure surface. The magnitude of the PGF is proportional to the spacing of the contour lines. This is shown in Figure 6-7c. The

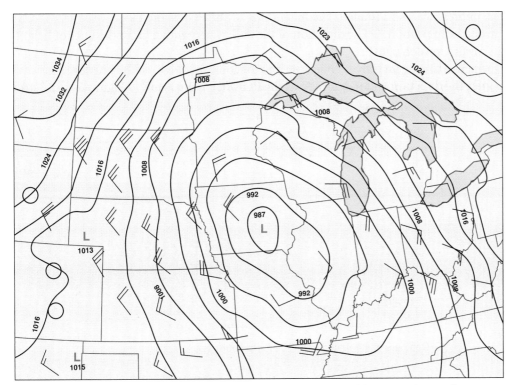

FIGURE 6-5 The surface weather map for the Christmas 2009 blizzard over the Midwest United States, showing the isobars (lines) and wind direction and speed (wind barbs). (From Plymouth State University Weather Center, [http://vortex.plymouth.edu/make.html]. Accessed June 10, 2010.)

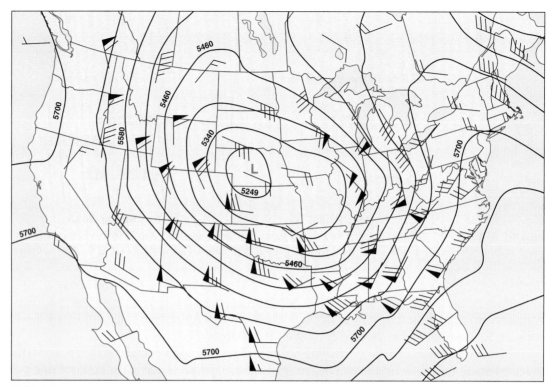

FIGURE 6-6 The 500-mb isobaric chart for the Christmas 2009 blizzard. Note that the isobars are replaced with lines of constant height, or isoheights. Also notice that the highest winds (shaded pennants = 50 knots) occur where the isoheights are closest together. (From Plymouth State University Weather Center, [http://vortex.plymouth.edu/make.html]. Accessed June 10, 2010.)

steeper the slope of the pressure surfaces, the greater the PGF because the pressure gradient is the change in pressure over distance. The magnitude of the PGF is therefore weak near location Y and strong near location X. The direction of the PGF is perpendicular to the lines of height and points toward lower heights.

To summarize, on an isobaric map, the lines of constant height of the pressure surface allow us to infer the direction and magnitude of the PGF. Referring again to Figure 6-6, the stronger

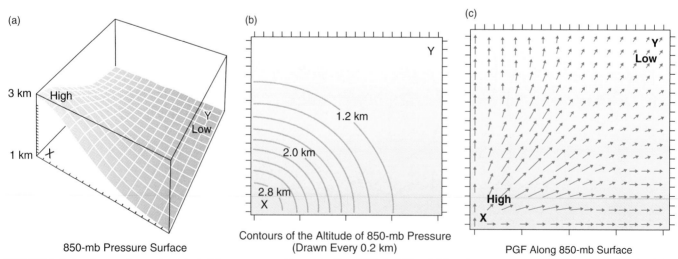

FIGURE 6-7 Isolines of constant height are proportional to the PGF. (a) The 850-mb pressure surface is represented by the colored diagram. The 850-mb pressure is higher over location X than over location Y. (b) Lines of constant height of the 850-mb pressure show the relationship between high isoheights and high pressure. (c) The PGF, represented by arrows, occurs where the lines of constant altitude are closer together. The PGF acts perpendicular to the lines of constant altitude of the pressure shown in B.

winds are located in regions where the spacing of the height lines is at a minimum, as we would expect, because this is where the PGF is the largest. The winds are *not* in the direction of the PGF! The winds, in general, are blowing perpendicular to the PGF and are nearly parallel to the height lines. This indicates that there must be other forces acting on the winds that counteract the PGF.

Centrifugal Force/Centripetal Acceleration

When an object changes its direction of motion, it is accelerating even if its speed does not change. This acceleration is called **centripetal acceleration** (*centripetal* means "pushed toward a center"). You experience this acceleration when you are riding in a car that makes a sudden turn and you are forced away from the direction of the turn. From a nonmoving observer's point of view, an acceleration has been applied to cause a turn, and your momentum has carried you straight ahead. From your point of view, however, a force has been exerted on you—we call this apparent force the **centrifugal force**. (*Centrifugal* means "pushed or fleeing outward from a center"; the ending *fugal* has the same basic meaning as the word *fugitive*). Depending on the context and point of view, we refer to this one effect by either name.

The direction of the centripetal acceleration is always toward the center of the curve, perpendicular to the direction of motion. The centripetal acceleration (CENTF) per unit mass can be written mathematically as follows:

$$CENTF = V^2 \div R$$

where V is the wind speed and R is the radius of curvature of the curved path.

The faster the speed and the tighter the curve of the path traveled (i.e., smaller R), the larger is the centripetal acceleration. In the atmosphere, we define R to be positive when the direction of the curving wind is in the same sense as the Earth's rotation, or counterclockwise in the Northern Hemisphere. This occurs with cyclones. In anticyclones, R is negative. Because V^2 is never negative, the sign of centripetal acceleration is positive for cyclones and negative for anticyclones. This implies that the centrifugal force pushes in the opposite direction of the PGF in lows but in the same direction as the PGF in highs. All of these aspects of the centrifugal force are illustrated in **FIGURE 6-8**.

From the equation, we see that the centrifugal force will give air parcels a strong push only during high winds that curve sharply (i.e., those that have a small "turning radius" R). As a result, this force is most important in hurricanes and tornadoes, although this force plays a significant role in curving jet streams as well. On the largest scales, this force also affects air on the rotating Earth, as we'll see below.

Coriolis Force

There is another important force governing wind, but its effect is best seen in isolation when other forces are small or absent. For example, after a windstorm on a Northern Hemisphere ocean calms, buoys that have been blown by the wind will trace out looping near-circles as the buoys move ever more to the right of their direction of motion (**FIGURE 6-9a**). On rare occasions, you can see the same behavior in the atmosphere, too. When winds from the March 1993 "Storm of the Century" (see Chapter 13) blew through a mountain pass in central America, they emerged out in the Pacific in a region where the PGF was very small. Then the wind and associated clouds began turning to the right of their initial direction (Figure 6-9b). These clues tell us that our next force, the **Coriolis force**, is characterized by a turning or deflection of the direction of moving air.

The Coriolis force is named for the 19th-century French engineer Gaspard Gustave Coriolis who explained it mathematically. But a physical understanding of it turns out to be one of the hardest concepts to grasp in all of meteorology, even

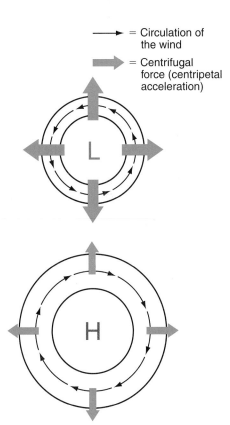

FIGURE 6-8 The centrifugal force CENTF (centripetal acceleration) always pushes outward from the center of circulation and is greater for stronger winds and more tightly curved wind patterns. CENTF does not reverse directions for low-versus high-pressure centers in contrast to the PGF (Figure 6-4).

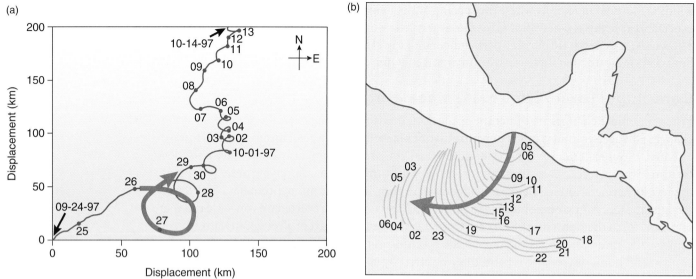

FIGURE 6-9 (a) The motion of a drifting ocean buoy during three weeks in the Bay of Bengal near India in September 24 to October 14, 1997. In the absence of strong PGF or CENTF, the buoy made several near circles (one is highlighted in red), always turning to the right as it slowly moved to the northeast. This clockwise rightward turning is caused by the Coriolis force. (b) Cloud motions over time (blue lines, labeled by hour) in a region of small PGF during the Storm of the Century, March 13, 1993, over the Pacific Ocean near Central America. The red line indicates the clockwise turning to the right of the cloud pattern from 0500 UTC to 2300 UTC. (Note that the wind is turning to the right of its initial southward direction of motion, causing it to go toward the *left* of the graphic.) This clockwise rightward turning is due to the Coriolis force. (Part a adapted from Joseph et al., *Current Science*, 92 [2007]. Part b modified from Steenburgh et al., *Mon. Wea. Rev.*, 126 [1998]: 2679–2712.)

by professionals and professors. In the words of one expert, there have been "four centuries of conflict between common sense and mathematics" on the subject of the Coriolis force. The math is right, but our common-sense notions usually are not. Here we will discuss the basic physics, link it to the observations above, and ultimately rely on the math to give us simple rules for how the Coriolis force works.[1]

The Coriolis force is essentially an extension of the centrifugal force in the case of an object—or air—that is *moving* with respect to the rotating reference frame. All air on the rotating Earth "feels" a centrifugal force that is perpendicular to the Earth's axis of rotation. (It's actually part of what we call gravity.) But if the air is also moving so that it changes *either* the total speed of rotation (west-east motion) *or* the distance from the Earth's axis (north-south motion), or both, then the centrifugal force is changed—because the centrifugal force is equal to V^2/R. This change in the total centrifugal force appears to observers on the rotating Earth as a new force that deflects motion horizontally, perpendicular to the motion (**FIGURE 6-10**). This deflecting force is the Coriolis force.

Which way does the Coriolis force deflect motion? As we learned from the examples in the ocean and atmosphere in Figure 6-9, *the Coriolis force deflects movement to the right in the Northern Hemisphere.* In the Southern Hemisphere, this rule is reversed because the Earth's rotation is clockwise as viewed at the South Pole. (Spin a globe and look at it from below to prove this to yourself.) *The Coriolis force deflects movement to the left in the Southern Hemisphere.*

The magnitude of the Coriolis force per unit mass (CF) is proportional to the distance from the equator and the speed of the wind. This can be written mathematically as follows:

$$\mathrm{CF} = \pm f V$$

---

[1]The common textbook explanation of Coriolis deflection using the increase of west-to-east speed as air moves poleward from the Equator to higher, slower-rotating latitudes is physically incorrect and its use should be avoided (e.g., A. Persson, "How Do We Understand the Coriolis Force?" *Bulletin of the American Meteorological Society,* July 1998, pp. 1373–1385). Thanks to Anders Persson for his suggestions regarding the following discussion.

in which *V* is the wind speed, and *f* is the "Coriolis parameter," which is defined as follows:

$$f = 2 \times \text{Earth's rotation rate} \times \text{the sine of the latitude}$$

In the equation for determining the magnitude of the Coriolis force, the plus sign is for winds in the Northern Hemisphere from the south or east directions, indicating a Coriolis force pushing toward the east for southerly winds or pushing north for easterly winds, respectively. The minus sign is for Northern Hemisphere winds from the north or west directions, representing a Coriolis force pushing toward the west for northerly winds or pushing south for westerly winds, respectively.

Some important properties of the Coriolis force are shown in **FIGURE 6-11**. Motions that are perpendicular to the axis of rotation are strongly affected by the Coriolis force. As a consequence, motions that are parallel to the Earth's axis do not experience a Coriolis force. And so as seen in Figure 6-11a, the Coriolis force has maximum effect for horizontal motions at the North Pole, but is zero at the equator. (This follows from the sine of the latitude in the Coriolis parameter, as the sine of zero degrees is zero.) In between, in the mid latitudes, the Coriolis force is strong; this has huge implications for how we understand midlatitude weather.

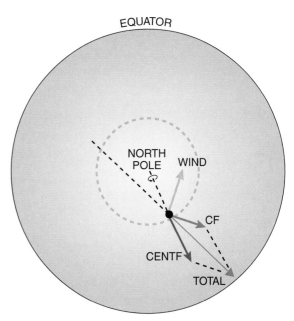

FIGURE 6-10 Diagram illustrating the centrifugal (CENTF) and Coriolis forces acting on an air parcel moving with respect to the rotating Earth. (Modified from A. Persson, *Bull. Amer. Meteor. Soc.,* 79 [1998]: 1378.)

Figure 6-11a does not include west–east motions, but we can use the relationship between the Coriolis and centrifugal forces to demonstrate how west–east motion leads to a Coriolis force. In the mid latitudes, the surface of the Earth rotates at a speed of about 200 m/s at a distance of 4000 km from the axis. This yields a centrifugal force per unit mass of about $(200 \text{ m/s})^2 / 4{,}000{,}000 \text{ m} = 1/100 \text{ m/s}^2$. Air moving eastward with a speed of 10 m/s causes the centrifugal force per unit mass to increase to $(210 \text{ m/s})^2 / 4{,}000{,}000 \text{ m} = 1.10/100 \text{ m/s}^2$, for a surplus of $0.10/100$ or 0.001 m/s^2. This surplus is what we observe as the Coriolis force, and it's the same value you get for the product fV for typical situations in the mid latitudes (Figure 6-11b).

"Coriolis force."

The Coriolis force is important only when dealing with objects or winds that travel over time periods of at least a few hours, that is, a not-too-small fraction of the period of the Earth's rotation (24 hours). In **BOX 6-2** we address a famous question: Does the Coriolis force affect water spiraling down the drain of a sink or bathtub? The answer may surprise you!

Frictional Force

The **frictional force** in the atmosphere is, for our current purposes, caused by the flow of wind over the roughness of the Earth's surface. In these cases, friction opposes, or decelerates, the wind in the same way that a rough road surface slows down a car (**FIGURE 6-12**).

Mathematically, the frictional force per unit mass (FF) can be written very simplistically as

$$\text{FF} = -kV$$

in which *k* is a parameter describing the roughness of the Earth's surface and *V* is the wind speed. The minus sign emphasizes that friction opposes the wind regardless of its direction.

The roughness of the surface and the speed of the wind determine the magnitude of frictional force. The force of friction over smooth, still water is smaller than it is over trees in a forest or a rough, craggy mountain peak, and it is nearly zero above the lowest kilometer or two of the troposphere. The magnitude of the frictional force increases with increasing wind speed.

Through turbulence, friction plays a crucial role in determining the winds near the Earth's surface—so crucial, in fact, that we return to this subject in more detail at the beginning of Chapter 12. The "Friction and Fly Balls" learning applet on the text's Web site explores the role of friction in the flight of a home run in baseball stadiums at different altitudes.

"Friction and fly balls."

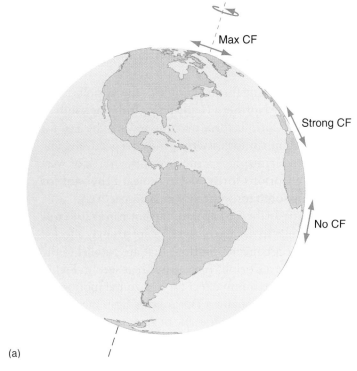

(a)

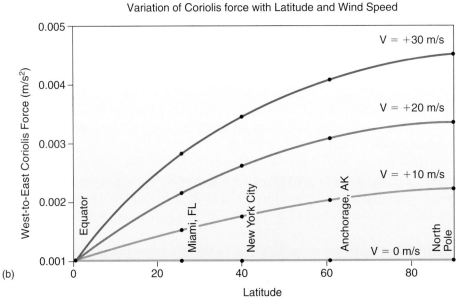

(b)

FIGURE 6-11 (a) The Coriolis force from a physical perspective, as a function of latitude. Motions (arrows) that are perpendicular to the Earth's axis feel the most Coriolis force; the Coriolis force is strong in the mid latitudes, but it is zero at the Equator. (b) The strength of the Coriolis force increases with both latitude and wind speed but is zero at the equator and for calm winds.

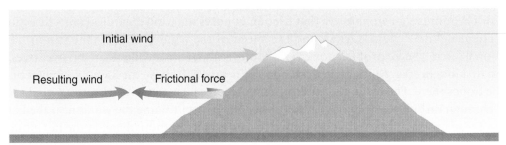

FIGURE 6-12 The frictional force opposes the direction of the wind but usually is not strong enough to stop the wind altogether.

Box 6-2 Going Down the Drain with the Rossby Number—Clockwise or Counterclockwise?

Most people, including cartoonists (see cartoon), seem to think that water draining in a sink or bathtub spins in opposite directions in the Northern and Southern Hemispheres.

 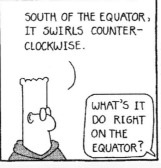 

(DILBERT: © Scott Adams/Dist. by United Feature Syndicate, Inc.)

For this to be the case, the Coriolis force would have to be a major influence on the draining water. However, in this chapter, we state that for the Coriolis force to be important, the motion (of air or water, it makes no difference) must persist for a length of time that is a significant fraction of a day—a few hours, at least. Does this sound like your sink or bathtub? Probably not.

Therefore, the short answer to the age-old question is as follows: The water draining from a tub or sink spins in either direction in either hemisphere because it is influenced more strongly by other forces, such as the centrifugal force, than by the Coriolis force. It does the same at the equator, too! *Water goes down the drain too quickly to "feel" the Earth's rotation via the Coriolis force, no matter where it is.*

We can make this point even more dramatically by using some simple mathematics. Meteorologists quantify which forces are important in a weather situation by calculating their ratios. The ratio of the centrifugal force to the Coriolis force is a special quantity called the *Rossby number,* abbreviated *Ro* in the equation that follows and named after the famous pioneer of research into atmospheric motions, Carl-Gustaf Rossby (photo).

(continued)

Box 6-2 Going Down the Drain with the Rossby Number—Clockwise or Counterclockwise?, continued

$$Ro = \frac{CENTF}{CF} = \frac{\frac{V^2}{R}}{fV} = \frac{V}{fR}$$

When the Rossby number is very small, the Coriolis force is much bigger than the centrifugal force. In these cases, wind and water motions change direction in the Northern versus Southern Hemispheres. In the case of the drain, the spiral is an area of low pressure and—if the Coriolis force mattered—would spin counterclockwise in the Northern Hemisphere (so the "Dilbert" cartoon is wrong, regardless).

When the Rossby number is very large, the Coriolis force is much smaller than the centrifugal force. In these cases, wind and water motions do not "feel" the Coriolis force and spin in whichever direction other forces impart on them.

To summarize, low Ro means flip-flopping flow directions across the equator, whereas high Ro means that water down the drain spins in any old direction, heedless of the Coriolis force.

Now let's do the key calculation: What is the Rossby number for your sink or bathtub? The size of this number will tell us if water spiraling down the sink spins in different directions across the equator (large = no; near zero = yes).

To make this calculation, all we have to do is come up with some ballpark estimates for the size and speed of the water in your sink. Let's assume you are at the latitude of Madison, Wisconsin, where the Coriolis parameter f is equal to .0001 per second. We'll be generous and say that the size of the spiral down the drain is large, around 1 meter. We'll be equally generous and say that the speed of the spiraling flow is fast, about 1 meter per second. Using these numbers, we find the following:

$$Ro = \frac{V}{fR} = \frac{1 m/s}{(0.0001/s)(1m)} = 10,000$$

The Rossby number for your sink is about 10,000! Clearly this is a large value—much, much larger than zero. This estimate tells us with mathematical certainty that the Coriolis force is nowhere close to being a factor in the water spinning in your sink, so water in a sink can spin in either direction in either the Northern or Southern Hemisphere.

PUTTING FORCES TOGETHER: ATMOSPHERIC FORCE-BALANCES

Before we proceed, let's look back at the forces we have discussed. Gravity is the strong, silent type that constantly affects air, but as we'll see, it does not usually lead to wind. The PGF arises from pressure gradients, which ultimately trace their origins to solar heating. This gets the atmosphere moving horizontally and causes wind. Ultimately, the Sun makes the wind blow.

Only when the PGF has caused wind can the other forces play roles. This is easily demonstrated—just look at the equations for the Coriolis, centrifugal, and frictional forces. All of them contain V, the wind speed. Therefore, zero wind speed means no Coriolis, no centrifugal, and no frictional forces. For this reason, it's accurate to say that the PGF acts to cause the wind, whereas the Coriolis, centrifugal, and frictional forces react to the wind to change its speed, direction, or both.

To understand the wind, however, it is not enough to study forces. It is the balance of these forces that usually determines the strength and direction of the wind. Newton's Second Law is more precisely written as follows:

Sum of Forces = Mass × Acceleration

TABLE 6-1 Some Atmospheric Force-Balances

Force-Balance	Explains Why	Gravitational Force	Vertical PGF	Horizontal PGF	Centrifugal Force	Coriolis Force	Frictional Force
		Forces That Balance Each Other (Checkmarks)					
Hydrostatic	• The sky is not falling (or rising)	X	X				
Geostrophic	• Wind blows parallel to, not across, isoheights on isobaric charts • Wind is stronger when isobars or isoheights are closer together • Wind is clockwise around highs, counter-clockwise around lows in the Northern Hemisphere			X		X	
Gradient	• Winds around a high are supergeostrophic, winds around a low are subgeostrophic			X	X	X	
Guldberg–Mohn	• Wind blows across isobars toward lower pressure at/near the Earth's surface • Lows are cloudy and wet, highs are sunny and dry			X		X	X

We need, therefore, to put together all of the forces that act on an air parcel or a region of the atmosphere. When they add up to zero, in terms of both magnitude and direction, a balance has been achieved.

According to Newton's Second Law, a balance of forces equals zero acceleration. This does not necessarily mean zero wind because acceleration is the change of wind, not the wind itself.

Next we examine a few different ways that these five forces balance each other and create wind, as outlined in **TABLE 6-1**. As we study force-balances, remember what we have learned about the forces: A strong PGF means a strong wind; the Coriolis force acts to change the direction of the wind, but has no effect on the wind speed and is zero for calm conditions or at the equator. The centrifugal force acts only on curving winds, and friction slows the wind regardless of its direction.

Hydrostatic Balance

Near the Earth's surface the pressure decreases by about 10 mb for every 100-meter increase in altitude. This is a much stronger pressure gradient than in a hurricane! Why don't we observe strong vertical winds as a result of this strong vertical pressure gradient?

The explanation is that motion results from an imbalance of forces, and in the atmosphere, the vertical pressure gradient is usually in balance with gravity. When these two forces are equal and push in opposite directions, **hydrostatic balance** exists (**FIGURE 6-13**).

In the case of hydrostatic balance, not only are accelerations zero but vertical motions themselves are small—usually only about 1 centimeter per second. Much stronger vertical winds usually occur only in and near thunderstorms, which disrupt the balance between the gravitational and the vertical PGFs.

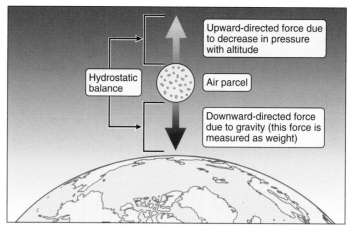

FIGURE 6-13 An air parcel in hydrostatic balance does not accelerate vertically because the vertical pressure gradient and gravitational forces acting on the parcel are balanced. The parcel is able to move vertically, although in reality these motions are tiny. The parcel can still be accelerated horizontally. (Reproduced from Lester, P., *Aviation Weather, Second edition.* With permission of Jeppsen Sanderson, Inc. Not for Navigation Use. Copyright © 2010 Jeppesen Sanderson, Inc.)

We can express hydrostatic balance per unit mass as an equation by adding the expressions for the vertical pressure gradient and the gravitational acceleration and setting them equal to zero (i.e., they are balanced, equal, and opposite to each other):

$$PGF + GF = 0$$

or

$$-\frac{1}{\text{Air density}} \times \frac{\text{Change in pressure in vertical}}{\text{Vertical distance}} - 9.8\,m/s^2 = 0$$

In this equation, if air density decreases, the vertical change in pressure can increase to compensate and maintain hydrostatic balance. Because the density of hot air is less than the density of cold air, this equation implies that the pressure must decrease with altitude more rapidly in cold air than in hot air. We use this fact later when we discuss the thermal wind and the sea breeze.

Geostrophic Balance, the Geostrophic Wind, and Buys Ballot's Law

The most fundamental *horizontal* force-balance arises when the PGF is counterbalanced by the Coriolis force. This is called **geostrophic balance** (the Greek *geo* means "Earth" and *strophic* means "turning"). In other words, this balance exists because the Earth turns and causes the Coriolis force. It can be written as follows:

$$PGF + CF = 0$$

or (limiting ourselves to the Northern Hemisphere for simplicity)

$$-\frac{1}{\text{Air density}} \times \frac{\text{Change in pressure in horizontal}}{\text{Horizontal distance}} + fV = 0$$

Why is geostrophic balance so important? To a good first approximation, it allows us to explain both the direction and the strength of the wind on most weather maps, particularly on isobaric charts representing weather above the Earth's surface.

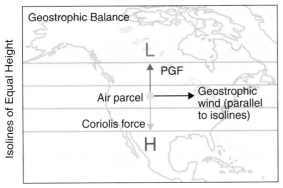
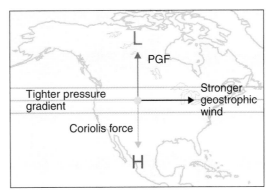

FIGURE 6-14 The geostrophic wind represents a balance between the horizontal pressure gradient and Coriolis forces. The straight lines are lines of equal altitude of the 500-mb pressure surface, or isoheights. The geostrophic wind always is directed parallel to these lines, with lower pressure to its left in the Northern Hemisphere. The closer the isoheights are to each other, the stronger the geostrophic wind.

Here's how geostrophic balance explains horizontal winds. In **FIGURE 6-14**, a typical pressure pattern aloft is depicted. The PGF always pushes from higher toward lower pressure. The Coriolis force opposes it. However, for the Coriolis force to do this, the wind direction must be perpendicular, not parallel, to the PGF. Because the Coriolis force always pushes to the right of the wind in the Northern Hemisphere, this requires a wind in geostrophic balance—which we call the **geostrophic wind**—to blow with lower pressure to its left.

This "low pressure lies to the left of the wind" rule is called **Buys Ballot's Law** after the Dutch meteorologist who devised it. Pilots in the Northern Hemisphere use this rule during flight to avoid bad weather associated with low-pressure systems.

Buys Ballot's Law also explains a fundamental feature of wind in the atmosphere. Because air moves with low pressure on its left in the Northern Hemisphere, *the wind must blow clockwise around highs and counterclockwise around lows* (**FIGURE 6-15**). In the Southern Hemisphere this pattern is reversed, with counterclockwise highs and clockwise lows. (Strictly speaking, we should include the centrifugal force because highs and lows are curved, but the result is the same with respect to wind direction.)

We can explain the strength of the wind using the geostrophic balance equation shown earlier. Remember that balance requires that both terms add up to zero. Therefore, the larger the pressure gradient, then the larger that *V*, the wind speed, must be. In other words, *winds on a weather map are strong where the isobars or isoheights are close together, and winds are weak where the isobars or isoheights are far apart.* This is an excellent rule that works even in cases where geostrophic balance is not such a good approximation.

Like any approximation to reality, geostrophic balance is only as good as its assumptions. It's often a poor explanation of the winds near the equator because there the Coriolis force is nearly zero and does not balance the PGF by itself. More generally, any situation in which other forces besides the pressure gradient and Coriolis forces are acting on the atmosphere will not be in geostrophic balance. Nevertheless, geostrophic balance is a very good first step toward explaining the wind and is probably the single most important and useful concept to learn from this chapter. To understand it better, try out the Balance of Forces learning applet on the text's Web site.

"Balance of forces."

Gradient Balance and the Gradient Wind

Geostrophic balance requires that the wind blow in a straight line. However, this is rarely the case. More often, winds undulate in curvy patterns. This requires the inclusion of a third force, the centrifugal force. The three-way balance of horizontal pressure gradient, Coriolis force, and centrifugal force is called **gradient balance** (**FIGURE 6-16**), and the wind that results from this balance is called the **gradient wind**.

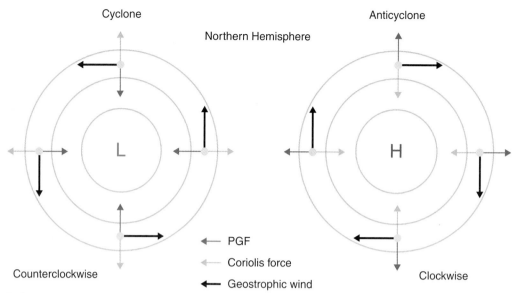

FIGURE 6-15 A force diagram representing the forces acting on an air parcel moving around high pressure (left) and low pressure (right) in the Northern Hemisphere. Because of geostrophic balance, the air rotates clockwise around high pressure and counterclockwise around low pressure. This is called the geostrophic wind. (Here we ignore the centrifugal force, which does not change the direction of the wind.)

The gradient wind is an excellent approximation to the actual wind observed above the Earth's surface, especially at middle latitudes and in the jet stream. Gradient balance can be written mathematically as:

$$CENTF + CF + PGF = 0$$

or

$$\frac{V^2}{R} + fV - \frac{1}{\text{Air density}} \times \frac{\text{Change in pressure in vertical}}{\text{Vertical distance}} = 0$$

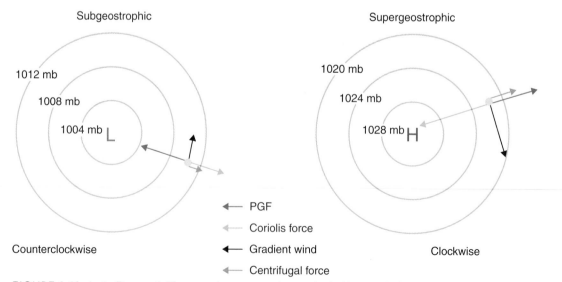

FIGURE 6-16 As in Figure 6-15, except now we also include the centrifugal force, leading to gradient balance. Compared with the geostrophic wind in Figure 6-15, the gradient wind arrows are longer (supergeostrophic) in highs and shorter (subgeostrophic) in lows.

This is a quadratic equation, and the quadratic formula from high school math can be used to solve for the magnitude of the gradient wind as a function of the geostrophic wind, Coriolis parameter, and radius of curvature.

Because the gradient wind depends on the radius of curvature, high- and low-pressure areas differ not only in the direction but also the *strength* of the wind. Let's now explain why physically. Remember that the centrifugal force always pushes toward the outside of curves and is proportional to the square of the wind speed. Therefore, in the case of a curved low-pressure area, the centrifugal force pushes in the same direction as the Coriolis force. Because both of these forces are related to wind speed, the wind does not need to be as strong in this case as in the case of uncurved flow to achieve balance with the PGF. The wind is therefore slower in this situation than in a purely geostrophic case and is called **subgeostrophic flow**.

Conversely, in the case of a curved high-pressure area, the centrifugal force teams up with the PGF, both pushing outward from the high. The Coriolis force attempts to balance this double-team of forces. To accomplish this, the wind speed—to which the Coriolis force is proportional—must be considerably higher than in the purely geostrophic case. This is called **supergeostrophic flow**.

Therefore, *for high- and low-pressure areas that have the same spacing of isobars or isoheights, winds in gradient balance around a high will be stronger than winds around a low.* This is summarized in Figure 6-16. In reality, however, the pressure gradients around lows are stronger than those around highs.

For an extreme example in which the Coriolis force is not important, see the "Cyclostrophic Wind in a Tornado" applet on the text's Web site.

Adjustment to Balance

You may wonder: Does the wind magically develop in perfect geostrophic balance? The answer is no. As with a new pair of shoes, a new car, or a new roommate, a continual process of adjustment leads toward a "perfect fit."

This adjustment to balance by the wind is illustrated in **FIGURE 6-17**. When a pressure gradient arises, initially there is an *imbalance* of forces. An air parcel in the region is therefore pushed toward lower pressure. The Coriolis force responds as soon as there is motion, turning the wind toward the right (in the Northern Hemisphere). The air parcel wiggles a little as it oscillates toward a balance between the pressure gradient and Coriolis forces. Although the net force acting on the parcel of air tends to be zero at this stage, the parcel continues to move. Geostrophic balance is reached, or nearly reached.

In real life, however, the atmosphere is always being jostled around, and even jostles itself. This means that its adjustment to geostrophic—or another—balance is imperfect and temporary at best. Small-scale oscillations result, which we'll explore in more detail in Chapter 12. In this context, it's remarkable how closely the real atmosphere resembles our force-balance approximations. The reason is that the atmosphere adjusts to imbalances rapidly, during a time span of minutes to a few

"Gradient winds."

"Cyclostrophic Wind in a Tornado."

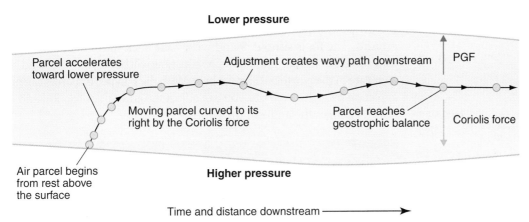

FIGURE 6-17 The adjustment to geostrophic balance of a parcel of air that is suddenly placed in a horizontal pressure gradient.

hours. Twice-per-day radiosonde observations rarely catch the atmosphere red-handed in the act of adjusting.

Like a cat rebounding from an unexpected fall, the atmosphere quickly adjusts to imbalance (licking its wounds, as it were), purrs "I meant to do that," and returns toward the state of balance that we normally see on weather maps.

Guldberg–Mohn Balance and Buys Ballot's Law Revisited

Near the Earth's surface, the frictional force plays a pivotal role in wind. The adjustment to balance in these situations is profoundly affected by the ability of friction to slow the wind and therefore weaken the Coriolis force. This three-way balance can be written as follows:

$$PGF + CF + FF = 0$$

or

$$-\frac{1}{\text{Air density}} \times \frac{\text{Change in pressure in horizontal}}{\text{Horizontal distance}} + fV - kV = 0$$

and is sometimes called **Guldberg–Mohn balance** in honor of its 19th-century discoverers.

FIGURE 6-18 depicts the idealized, stepwise evolution of this balance. As in Figure 6-17, the creation of a PGF creates a wind from higher toward lower pressure; however, near the surface this wind triggers the Coriolis force and the frictional force. Friction opposes the wind, so the wind slows down, which in turn makes the Coriolis force weaker.

As the situation evolves, the PGF wins the tug-of-war. This is because of the balance of the three forces. Without the influence of the Coriolis force, the wind would blow straight into lower pressure. With the Coriolis force at full strength opposing the PGF, the wind would blow parallel to the isobars as in geostrophic balance. In this case, however, friction weakens the wind and the Coriolis force, and the air parcel crosses the isobars toward lower pressure. However, it does so at an *angle*, not straight across from higher to lower pressure.

Therefore, Guldberg-Mohn balance is characterized by near-surface winds that blow across the isobars on a slant toward lower pressure. The angle at which the winds cross the isobars depends on the type of surface and the latitude. Over open water, where friction is low, the winds typically cross the isobars at an angle between 15° and 30°. Over land the angle is usually between 25° and 50°. Friction also damps out the wiggles in the adjustment to balance (**FIGURE 6-19**). For a variety of reasons, the effect of friction is important at the boundary between air and water (**BOX 6-3**).

Because the direction of the surface wind is not the same as the geostrophic wind, we must revise Buys Ballot's Law for use near the ground. The fix is simple: Stand with your back to the wind, and then turn about 30° to your right. Low pressure now will be on your left-hand side if you are in the Northern Hemisphere. With the Balance of Forces learning applet on the text's Web site, you can adjust the amount of friction and see how it changes the angle of the wind.

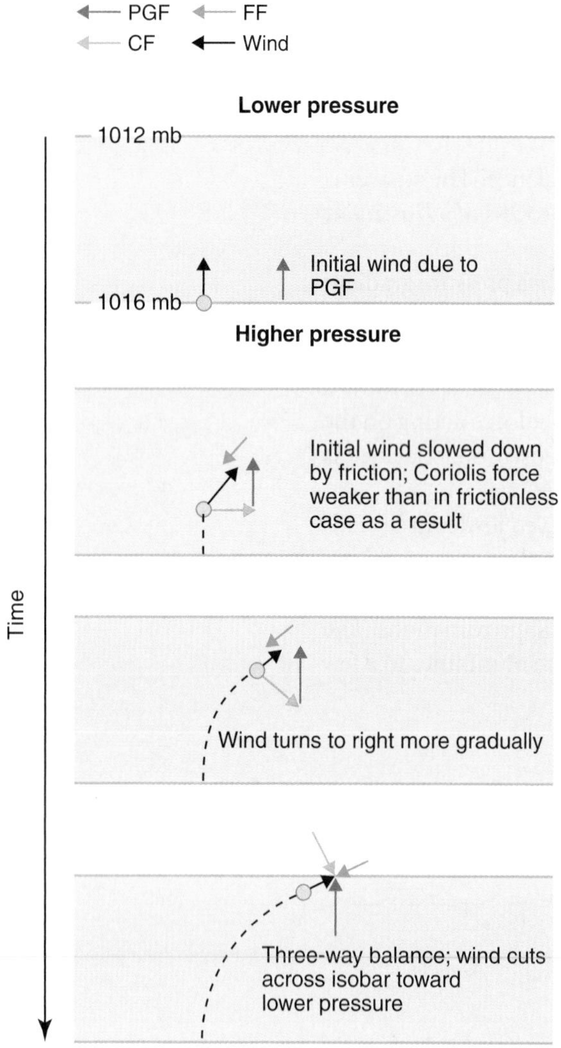

FIGURE 6-18 The adjustment to Guldberg–Mohn balance of a parcel of air that is near the Earth's surface. The parcel crosses the isobar toward lower pressure under the combined influence of the horizontal pressure gradient, Coriolis, and frictional forces.

OBSERVATIONS OF UPPER-LEVEL AND SURFACE WIND

So far we have discussed force-balances individually. However, the atmosphere is a three-ring circus of air motions, and a single weather chart or map can reveal several force-balances simultaneously.

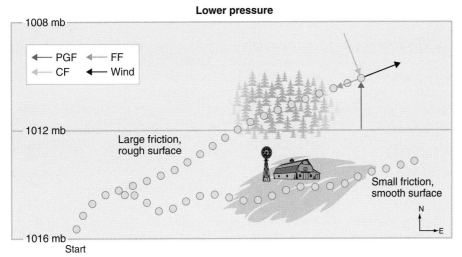

FIGURE 6-19 A numerical simulation of how varying amounts of friction affect the adjustment to Guldberg–Mohn balance. At the surface, friction causes the wind to cross the isobars and converge toward regions of low pressure. The amount of surface friction determines the angle at which surface winds cross the isobars. Over smoother surfaces the angle is smaller than over rougher surfaces. When friction is weak, the paths of parcels are wavy, as in Figure 6-17. (Modified from Knox, J., and Borenstein, S., *J. Geoscience Ed.*, 46 [1998]: 190–192.)

Box 6-3 Winds and Waves

In this chapter we have emphasized how friction at the ground modifies the wind. The wind also modifies the condition of the surface. We see this when wind ripples through trees or across a wheat field. Solid objects remain in place even in a stiff wind, but fluids such as water can be moved by the wind. As a result, winds can have a profound effect on the water conditions at the surface of oceans and lakes.

A good example of this is wind-generated water waves. Waves form as the wind's energy is transferred to the surface of water.

The size of a wind-generated wave depends on the following:

1. The wind speed: The stronger the winds, the larger the force, and thus the bigger the wave. The wind must also be constant, not just a wind gust here or there.
2. The duration of the winds: The longer the wind blows over the open water, the larger are the waves.
3. The fetch: This is the distance of open water over which the wind blows. The longer the fetch, the larger are the waves.

The curve in the figure shows how wind speed (horizontal axis) affects the maximum possible height of waves on an infinitely wide ocean (vertical axis). This curve is not an "ivory tower" result; it is based on real-world observations.

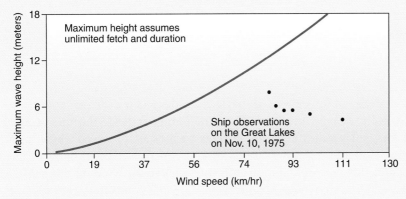

(continued)

Box 6-3 Winds and Waves, continued

The real-world observations plotted with crosses on this figure are for a tragic event in American history. On November 10, 1975, the Great Lakes iron ore freighter *Edmund Fitzgerald* sank in a windstorm on eastern Lake Superior. We will study this shipwreck in great detail in Chapter 10. For now, note how the high winds relate to the wave heights. In this particular storm, not only were the wind speeds high, but the fetch was also long for a lake. Both of these factors increased the size of the waves on Lake Superior during the *Fitzgerald*'s final hours.

The figure below, from a recent research paper by National Weather Service meteorologists, is the result of a computer simulation of the storm that helped cause the *Fitzgerald* shipwreck. The highest simulated winds (over 45 knots, from the west and northwest with a long fetch across Lake Superior) and highest simulated waves (7.8 meters, or 25.6 feet) are both located near the X marking the location of the shipwreck. This illustrates the close connection between wind speed, fetch, and high waves. Why the winds were so strong—a central question of Chapter 10—is still under investigation.

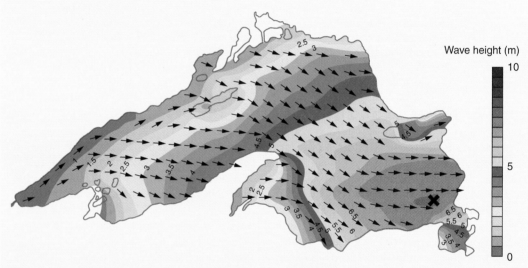

(Reproduced from Hultquist, T.R., et al., *Bull. Amer. Meteor. Soc.*, 87 [2006]: 607–622.)

High waves are hazardous to shipping and are also perilous for coastlines. Waves can cause flooding, crush buildings, scour the soil from under structures, wash away beaches, and breach levees. For example, in 1953, over 1800 people died in the Netherlands because of high water and waves overtopping levees, which was caused by a low-pressure system similar to the one that helped sink the *Edmund Fitzgerald*. This death toll was almost identical to the number of fatalities from Hurricane Katrina in 2005. The rapid rise in sea level associated with tropical cyclones is called *storm surge*, which we study further in Chapter 8. But remember that any time high winds blow over large expanses of water, it can be dangerous, whether or not the storm is a hurricane.

"Balance of Forces."
(this time use friction)

In **FIGURE 6-20**, an upper-level isobaric chart depicts wintertime conditions. Where the isoheights are nearly straight, the wind "flagpoles" point parallel to the lines of equal height. The "flags" on the flagpoles indicate that the strongest winds are observed where the isoheights are closest together. Both of these observations are consequences of geostrophic balance.

Now imagine that you are standing with your back to the wind on this chart. Anywhere on this chart you will find that lower heights, which are analogous to lower pressures, are toward your left. This is Buys Ballot's Law, another offshoot of geostrophic balance.

300 mb Heights (dm) / Isotachs (knots)

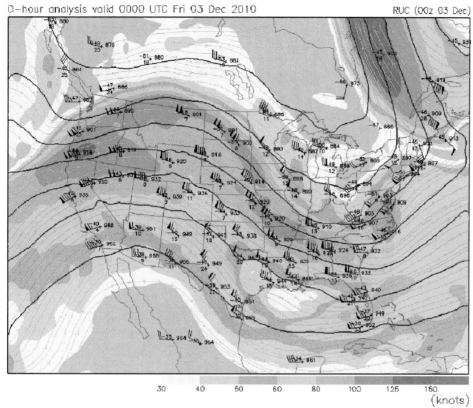

FIGURE 6-20 A typical isobaric chart for North America. The winds are directed parallel to the isoheights, with lower heights to their left. Also, the highest winds (pink and red shadings) occur where the isoheights are closest together and also near the high-pressure areas. All of these features are explained by geostrophic and gradient balances. (2010 Copyright, University Corporation for Atmospheric Research NCAR, Research Applications Laboratory.)

In the curved regions in Figure 6-20, the wind around low pressure is slower than in the high-pressure regions where the isoheights are the same distance apart. As in the straight-wind regions, the wind "flagpoles" point parallel to the isoheights. Gradient balance explains this.

Near the surface, however, friction causes wind to blow across the isobars toward lower pressure via Guldberg-Mohn balance. **FIGURE 6-21** shows this effect very clearly, as does every surface weather map.

Also notice that the midlatitude winds at the surface in Figure 6-21 are much slower and more likely to be from the east than the winds in Figure 6-20, which generally blow from west to east. This is partially explained by the simple fact that friction slows down the wind. Depending on the surface, friction usually slows the wind by 25% to 75% of what the speed would be in the absence of friction. However, this increase of westerly wind speed with increasing altitude is also seen all the way up in the middle and upper troposphere, far away from the friction near the Earth's surface. Why is this true? To explain this, we need to put together not forces, but two force-balances.

PUTTING FORCE-BALANCES TOGETHER: THE THERMAL WIND

We can explain our observation that winds increase throughout the troposphere by using geostrophic and hydrostatic force-balances. Let's start from the very beginning, with energy. An energy imbalance can lead to a force that generates a wind. Energy imbalances are, in fact, the trigger for atmosphere and ocean circulations.

FIGURE 6-21 The isobars (purple lines) at the surface drawn over a satellite image for an October, 2010, cyclone over the upper Midwest. The wind direction, as indicated by the direction the arrows are pointing, cross the isobars. The color of the arrow represents the wind speed in knots. Blue: 5–9 kn; dark green: 10–14 kn; light green: 15–19 kn; orange: 20–24 kn; yellow: 25–29 kn; pink: 30–34 kn; and red: 35–39 kn. (Image created by Prof. Joshua Durkee, Western Kentucky University, using GREarth software.)

Consider a layer of air between two pressure surfaces, say, 850 and 300 mb (**FIGURE 6-22**). Initially the isobars are parallel to one another. If we heat a column of air (make its energy gains greater than its energy losses), the density decreases. To reduce the air density within the column, the vertical height difference between the two pressure levels must increase. By heating just a portion of the layer of air, we generate a horizontal pressure gradient. In Figure 6-22 an outward PGF is generated near 300 mb and an inward PGF near 850 mb. This leads initially to a wind pattern with air flowing toward the region at 850 mb and outward at 300 mb.

Another way of saying this is that hydrostatic balance requires that pressure must decrease more rapidly with increasing altitude in cold air than in hot air—that is, cold air is more compressed than hot air. Therefore, the 300-mb pressure surface is at a higher altitude at 30°N than near the pole, and a PGF acts from the south to the north.

Moving air at middle latitudes feels the Earth's rotation, so we must also discuss the role of geostrophic balance in this situation. The geostrophic wind is proportional to the slope of the pressure surfaces. The greater the slope, the stronger the geostrophic wind. **FIGURE 6-23** shows that the thickness between pressure surfaces in warm, tropical air is greater than in cold, polar air, so the slope of the pressure surfaces keeps increasing all the way up in the troposphere. For this reason, the geostrophic wind increases with altitude when the temperature decreases toward the pole. This explains why winds at 300 mb are usually stronger than winds at 500 mb and also why winds at 500 mb are usually stronger than winds at 850 mb.

Finally, Buys Ballot's Law tells us the direction in which the wind blows. At high altitudes in the middle latitudes, lower pressure is toward the pole. Buys Ballot's law tells us that the winds in this region must blow with low pressure on the left in the Northern Hemisphere—in other words, from west to east. (Upper-level winds in the Southern Hemisphere middle latitudes also

"Jet Streams and Horizontal Temperature Gradient."

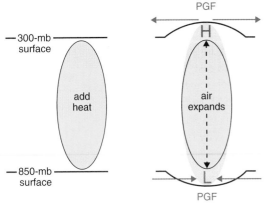

FIGURE 6-22 When the air between two pressure surfaces is warmed, the distance between the two pressure levels, or thickness, increases. Cooling the air would reduce the thickness. In either case, horizontal PGFs are created in the troposphere.

blow west to east. Can you figure out why?) In the tropics, the pressure gradient is reversed and upper-level winds usually blow from east to west.

This relationship between the vertical changes in the geostrophic wind and the horizontal temperature changes is called the **thermal wind** because it relates temperature and winds to each other. Its main result can be summarized for the middle latitudes of both hemispheres as follows: *The winds are more westerly as you go up wherever it's colder toward the poles.* It is also an example of how force-balances can be combined just like forces to explain how and why the wind blows. Remember this when we study large-scale circulations, jet streams, and fronts in the atmosphere in Chapters 7, 9, and 10.

PUTTING HORIZONTAL AND VERTICAL WINDS TOGETHER

We've learned that at the surface, the wind blows across the isobars away from high-pressure areas and into low-pressure areas. But what happens to the air once it reaches the center of the low? The atmosphere, like a balloon that is being inflated, prefers to expand in other directions rather than increasing its density in one spot. However, the air at the low's center cannot go into the solid ground. Neither can it retrace its steps and move away from the low, which would violate the principle of balance. Therefore, its only option is to go *upward*. This partly explains why air in a low-pressure center rises. Similarly, air in a high-pressure center sinks because if it didn't, the diverging air at the surface would create a vacuum at the center. FIGURE 6-24 illustrates this connection between horizontal and vertical wind for both highs and lows.

Figure 6-24 demonstrates that the vertical motion of air in pressure systems can be explained as a consequence of Guldberg-Mohn balance. This balance, along with our study of adiabatic processes and clouds in previous chapters, helps explain a fundamental fact of weather: *Low-pressure areas are usually cloudy and wet, whereas high-pressure areas are usually clear and dry.* This is because rising air cools adiabatically, causing the water vapor in it to condense and form clouds and precipitation. Sinking air warms adiabatically, lowering its relative humidity and evaporating any clouds and precipitation. Voilà—using just three forces combined in one force-balance, plus a few details from earlier chapters, we have explained one of the most basic and universal observations of weather!

SEA BREEZES

If you have spent much time at the beach during the summer you've probably noticed that in the late afternoon there often is a strong, steady breeze blowing in from the water. This steady wind, the **sea breeze**, is a result of the uneven heating during the daytime between the land and the adjacent water. Late at night the wind often reverses direction and blows from the land to the water (a **land breeze**). Let's learn how these winds work, using the concepts that we've developed in this and other chapters.

The idealized life cycle of a sea breeze is shown in FIGURE 6-25. Imagine, to begin with, that in the morning the land and the sea are the same temperature and the wind is calm (Figure 6-25a). Later in the day the land surface, which has a low specific heat and is a poor conductor of heat (see Chapter 2), warms much more quickly than water. As the land warms up, the air next to it also warms by conduction and rises. This warmth is carried upward by convection (see Chapter 2), as shown schematically in Figure 6-25b. Just as in our discussion of the thermal wind, this leads to a bulging upward of pressure surfaces over the land (Figure 6-25c) creating a PGF pushing toward the ocean (Figure 6-25d).

Figure 6-25e completes the picture of the sea-breeze circulation. At the surface over land, low pressure is developing. This is suggested in the figure by the lowering of isobars over the land.

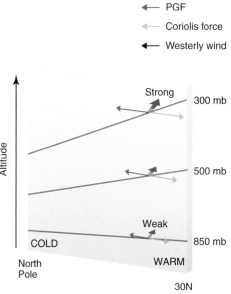

FIGURE 6-23 The thermal wind visualized. On average, the tropospheric temperature decreases toward the pole. This causes sloping pressure surfaces, via hydrostatic balance. The sloping pressure surfaces cause a PGF pushing toward the pole. Winds caused by the PGF are directed eastward by the Coriolis force; by geostrophic balance, the end result is a westerly wind. Because the pressure surfaces slope more steeply at higher altitudes, the westerly wind is stronger with increasing altitude.

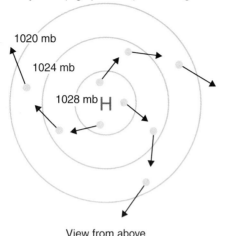

Surface winds blow clockwise around an anticyclone (high pressure) and diverge.

1020 mb
1024 mb
1028 mb H

View from above

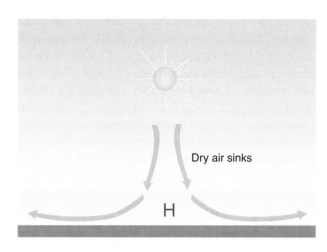

Dry air sinks

H

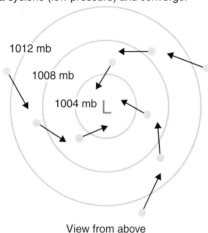

Surface winds blow counterclockwise around a cyclone (low pressure) and converge.

1012 mb
1008 mb
1004 mb L

View from above

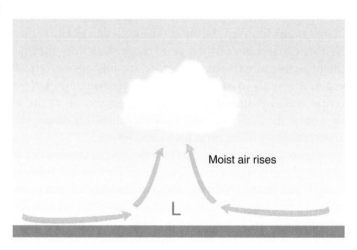

Moist air rises

L

FIGURE 6-24 How surface wind patterns induce vertical wind motions. The divergence of air out of a surface high causes air above it to sink, leading to dry and clear conditions (top). The convergence of air into a low causes rising air above it, leading to moist, cloudy conditions (bottom).

Meanwhile, over the ocean, high pressure is developing. This generates a horizontal PGF at the surface acting from over the ocean toward the land. Air moves toward the land, creating a sea breeze. To replace the surface air that is moving from over the water toward land, air sinks from above, completing the circulation.

Notice the slope of the pressure surfaces in Figure 6-25e. As the temperature difference between the land and water increases throughout the afternoon (see Chapter 3), the pressure difference causes the winds to increase, reaching a maximum in the middle to late afternoon. Over land, the distance between two isobars (e.g., 980 and 960 mb) is greater than over the ocean. This difference is what keeps the circulation developing and is the result of the air over land being warmer than the air over the ocean.

So far we've described the sea breeze in terms of the PGF. Other forces are less important but can affect it. For example, friction slows down the breeze, especially over land. Because the breeze lasts for many hours, the Coriolis force will gradually deflect the wind toward the right in the Northern Hemisphere. Because the Coriolis force is larger nearer the poles, this means that the sea breeze is turned away from the shore more quickly at higher latitudes. At lower latitudes—for example, along the Florida coastline—the sea breeze can last into the evening because the Coriolis force is weaker there. The important concept, however, is that heating (or cooling) of a column of air leads to horizontal differences in pressure, generating a PGF, which in turn causes the air to move and a circulation to develop.

If you are not at the beach to feel the sea breeze, you can observe it by analyzing satellite imagery (FIGURE 6-26). A rising parcel of air expands and cools, and the relative humidity increases—conditions favorable for the formation of clouds (see Chapter 4). For this reason, the upward branch of the sea breeze is often visible in satellite pictures (see Chapter 5) in the form of cumulus clouds. During the day, the upward branch moves inland and is an indication of the strength of the sea breeze. If the atmospheric conditions are favorable for the formation of thunderstorms, the sea breeze may provide just enough lifting to cause thunderstorms to develop.

Whenever large land and water bodies are adjacent to one another, sea breezes may develop and may cause thunderstorms. Florida's abundant summertime rainfall is a result of sea breezes. One sea breeze advances from the Atlantic Ocean on the east and another from the Gulf of Mexico on the west coast of Florida.

At night, the land cools more quickly than the water, and the process is reversed (FIGURE 6-27). The net result is a land breeze: surface winds that blow from the land out to sea. This gives coastal regions an alternating cycle of winds: onshore during the day and offshore at night.

SCALES OF MOTION

Atmospheric motions span an enormous range of space and time. The size of an atmospheric weather system is related to how long it exists. Small swirls of wind may last for a few seconds, whereas a hurricane may last several days. In general, as the size of the phenomenon increases, so does its life span. Meteorologists divide weather systems into four primary size categories:

- **Microscale is generally applied to circulations that are less than 1 kilometer in size, generally smaller than a puffy cumulus cloud.** At these scales, the pressure gradient and centrifugal and frictional forces are usually important, and the Coriolis force is negligible.
- **Mesoscale systems, such as thunderstorms, fronts, and sea breezes, are between about 1 and 1000 kilometers in size.** The Coriolis force becomes increasingly important at the large end of this category (e.g., in hurricanes).

Together, microscale and mesoscale weather systems are referred to as "**subsynoptic-scale,**" meaning that they are smaller than the largest features most often shown on continent-wide weather maps. We look at specific mesoscale systems such as tropical cyclones, fronts, and thunderstorms in Chapters 8, 9, and 11. Then we examine a variety of other subsynoptic-scale winds and weather systems in Chapter 12.

- **Synoptic-scale systems, such as midlatitude low-pressure systems, are at least 1000 kilometers in size.** In this category, the geostrophic balance between the pressure gradient and Coriolis forces is dominant. Chapter 10 examines midlatitude lows and highs in detail.
- **Planetary-scale systems are roughly 10,000 kilometers in size, and again, geostrophic balance helps explain them.** We examine these systems in depth in the next chapter.

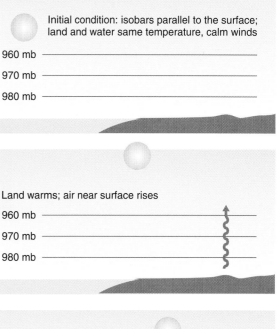

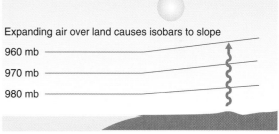

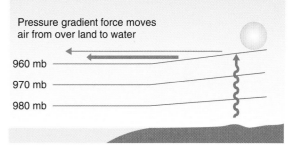

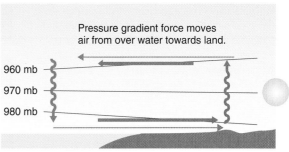

FIGURE 6-25 Sequence of schematic images depicting the idealized development of the sea breeze.

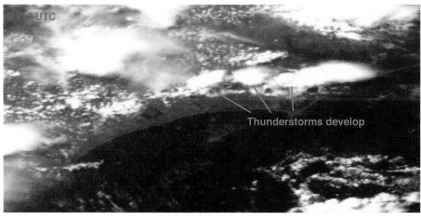

FIGURE 6-26 A sequence of satellite images demonstrating the sea breeze along the east coast of North Carolina. The first image is at 1545 UTC, or 11:45 AM EDT. Along the North Carolina coast, there are few clouds over the Atlantic Ocean, and there are many scattered clouds over the land. There is a distinct cloud-free boundary at the coastline. By 1815 UTC, the clouds have moved inland, marking the boundary of the upward branch of the sea breeze, or the sea-breeze front. As the day progresses and the temperature difference between the land and ocean increases, the circulation gets stronger and the sea-breeze front penetrates farther inland (2015 UTC). On this day, the atmosphere rises easily, and the lifting associated with the sea breeze is enough to trigger thunderstorms by 2145 UTC, or 5:45 PM local time.

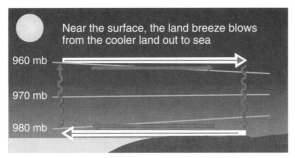

FIGURE 6-27 A land breeze forms during the night as the land cools faster than the sea.

PUTTING IT ALL TOGETHER

Summary

The wind is air in motion. Newton's Second Law relates the acceleration of air to forces. The important forces at work, besides gravity, in the atmosphere are as follows:

PGF: directed from higher pressure toward lower pressure at right angles to lines of constant pressure, or constant height on an isobaric chart. The greater the change in pressure over a given distance, the stronger the PGF.

Coriolis force: an apparent force caused by the rotation of the Earth. The Coriolis force acts to the right of motion in the Northern Hemisphere and to the left of motion in the Southern Hemisphere. The Coriolis force is zero at the equator and increases in magnitude toward the poles. This force is zero if the velocity of the air parcel is zero and increases as the air parcel's velocity increases.

Centrifugal force: directed outward from the center of curving motion. It increases as the speed of the object increases, the radius of curvature decreases, or both.

Frictional force: acts in the direction opposite to movement. The magnitude of this force depends on the type of surface and the wind speed. Friction is weaker over a lake and stronger over a forest or mountain.

Winds arise as a result of different balanced combinations of these forces. These balances are as follows:

Hydrostatic balance: a balance of the gravitational force and the vertical PGF. This balance explains the lack of strong vertical winds in most of the atmosphere.

Geostrophic balance: a balance between the horizontal pressure gradient and Coriolis forces. Geostrophic winds blow parallel to the isoheights. Buys Ballot's Law states that in the Northern Hemisphere, lower pressure is to the left when your back is to the wind. In addition, the wind in geostrophic balance is stronger when the isobars or isoheights are close together, and weaker when they are far apart. Because of this balance, winds blow counterclockwise around lows and clockwise around highs in the Northern Hemisphere. The reverse is true in the Southern Hemisphere.

Gradient balance: a three-way balance of horizontal pressure gradient, Coriolis, and centrifugal forces. This balance explains why the winds in highs and lows are different in terms of their speeds (supergeostrophic in highs, subgeostrophic in lows).

Guldberg-Mohn balance: a three-way balance of horizontal pressure gradient, Coriolis, and frictional forces. This balance explains why winds blow at an angle toward lower pressure, rise upward at the center, and cause cloudy, wet weather in lows. It also explains why highs are sunny regions of sinking, dry air.

The thermal wind, a combination of hydrostatic and geostrophic balances, explains why winds in the middle latitudes generally become more westerly with increasing altitude. This is because the surface temperature usually decreases toward the poles. The sea breeze is a similar, although much smaller scale, circulation that is driven initially by pressure gradients created by unequal heating of land and ocean.

Winds and wind patterns occur at all scales in the atmosphere. In the next chapter we look at wind systems that span the globe.

■ Key Terms

You should understand all of the following terms. Use the glossary and this chapter to improve your understanding of these terms.

Acceleration	Gravitational force	Pressure gradient
Anticyclones	Guldberg-Mohn balance	Pressure gradient force
Buys Ballot's Law	Hydrostatic balance	(PGF)
Centrifugal force	Isobaric chart	Sea breeze
Centripetal acceleration	Knots	Speed
Coriolis force	Land breeze	Subgeostrophic flow
Cyclones	Law of Momentum	Subsynoptic-scale
Force	Mesoscale	Supergeostrophic flow
Frictional force	Microscale	Synoptic-scale systems
Geostrophic balance	Momentum	Thermal wind
Geostrophic wind	Newton's Second Law of	Velocity
Gradient balance	Motion	Wind gust
Gradient wind	Planetary-scale systems	

■ Review Questions

1. You hear on a weather report that the wind is "15 miles per hour" or "15 meters per second." To describe fully the velocity of the wind, what other information must be reported?

2. If air has zero velocity, is the wind blowing? If air has zero acceleration, does this mean that the wind is calm?

3. According to Newton's Second Law, what causes air (or anything else) to accelerate?

4. Which forces that act on the atmosphere are dependent on wind speed? Which forces are not? Which forces act in the vertical direction? Which forces act in the horizontal direction?

5. Wind is triggered in the beginning by which force?

6. If the pressure difference over a fixed distance doubles, how does this change the PGF? If the pressure difference is fixed, but the distance across which the pressure changes doubles, how does this change the PGF?

7. Using a calculator, compute the magnitude of the Coriolis parameter at the latitude of the Madison, Wisconsin, area (latitude approximately 43.3°N). Use 0.0000729 per second for the rotation rate of the Earth. Then multiply this number by 3,600. Your result is the change in eastward wind speed caused by the Coriolis force acting on a 1 meter per second southerly wind for an hour. What would your result be at the equator?

8. If you drive a car around a sharp curve at 40 mph and a friend of yours in another car drives around the same curve at only 20 mph, which one of you will have experienced a stronger centrifugal force? What is the ratio of the force in your situation versus your friend's situation?

9. Relate the discussion of cars and centrifugal forces in Question 8 to the atmosphere. Where is the centrifugal force important? Where is it zero? In what direction does it push air parcels curving around a low-pressure system?

10. If the surface wind blows toward the north, in what direction does the frictional force act on it? If the wind reverses direction and blows toward the south, in what direction does the frictional force act now?

11. The Empire State Building in New York City is approximately 306 meters tall. Assume that the air density from the bottom to the top of the building is a constant all of the way up and is equal to 1 kilogram per cubic meter. Then, using hydrostatic balance, compute what the pressure change is from the bottom to the top of the skyscraper. (Hints: insert 1 for the air density in the hydrostatic balance

equation, and then insert 306 for the distance; solve for the pressure difference, and divide by 100 to get your answer in millibars.) Your answer is an approximation because density actually decreases in the vertical. Even so, it explains why your ears "pop" (see Chapter 1) when riding the high-speed elevators up to the observation deck of the Empire State Building.

12. A hurricane has very strong winds. On a weather map, would the isobars around a hurricane be very close together or far apart? Explain using the concept of geostrophic balance.

13. If an airplane is flying over the United States with a strong "tailwind" pushing it forward, on which side of the plane are you more likely to see bad weather off in the distance: the left-hand side or the right-hand side of the plane?

14. Explain in words how the effect of the frictional force on surface winds causes lows to be regions of bad weather and highs to be regions of good weather.

15. What forces help cause the sea and land breezes?

16. Does Buys Ballot's Law have to be modified to apply to winds in the Southern Hemisphere? Why or why not?

■ Observation Activities

1. On a windy day, estimate the location of the low pressure by applying Buys Ballot's Law. Verify your observations by analyzing a weather map drawn at nearly the same time as your observations.

2. Fill your sink with water. Open the plug and observe which way the water spirals down the sink. Repeat for a total of five times. Did the water spiral in the same direction all five times?

3. Sit in a swivel chair and have a friend spin you. With a heavy book in your hand, extend your hand forward and back. Does this change your rotation?

This rain cloud icon is your clue to go to the *Meteorology* Web site at http://physicalscience.jbpub.com/ackerman/meteorology/. Through animations, quizzes, web exercises, and more, you can explore in further detail many fascinating topics in meteorology.

7 Global-Scale Winds

CHAPTER OUTLINE, CONTINUED

PUTTING IT ALL TOGETHER
- Summary
- Key Terms
- Review Questions
- Observation Activities

AFTER COMPLETING THIS CHAPTER, YOU SHOULD BE ABLE TO
- Use a conceptual model of the atmospheric circulation to explain the existence and location of global-scale high- and low-pressure areas
- Explain why jet streams occur and where they form
- Describe the season-to-season changes in wind patterns such as the Indian monsoon
- Explain how the Hadley cell is affected by the Earth's rotation

INTRODUCTION

What do the Europeans' discovery of the "New World" and Darwin's theory of evolution have in common? An understanding of global-scale winds helped lead to both discoveries.

Mariners have known for centuries that the trade winds off the Atlantic coast of North Africa blow steadily from the northeast direction, out to sea, whereas farther north along the coast of Europe the winds typically blow from west to east. In 1492 Christopher Columbus took advantage of this knowledge at the start of his famous voyage by sailing with the wind from Spain to the Canary Islands and then west across the Atlantic (**FIGURE 7-1**). When he returned to Spain the next year, Columbus sailed north toward the Azores, where westerly winds quickly guided his ships homeward.

Nearly three and a half centuries after Columbus's first voyage, the *HMS Beagle* sailed around the world. On board were Charles Darwin and others on a global scientific expedition. Throughout its 5-year sea trek, the *Beagle* used the prevailing winds of both the Northern and Southern Hemispheres to complete its journey (**FIGURE 7-2**). In particular, the crossings of the Pacific and Indian Oceans were accomplished at around 20°S latitude, where the winds are steady from the southeast.

Both ships steered clear of the subtropical regions around 30°N and 30°S latitudes. Columbus and the captain of the *Beagle*, Robert FitzRoy, knew the winds in these **horse latitudes** were usually light or calm, forcing mariners to throw horses overboard when their ships were stranded and short of drinking water. Columbus and FitzRoy's understanding of winds helped them accomplish their objectives. (Later, FitzRoy would try his luck at weather forecasting, with less success [see Chapter 13].)

Using observations of global-scale wind patterns, this chapter develops a conceptual model of global-scale winds that helps explain the location of the world's largest deserts, tropical rain patterns, and even the existence of the jet stream. We also learn how atmospheric motions help correct the energy imbalances caused by the Earth's tilt, warming the poles and cooling the tropics.

Thanks to Anders Persson of the UK Met Office for his insights that have been incorporated into this chapter.

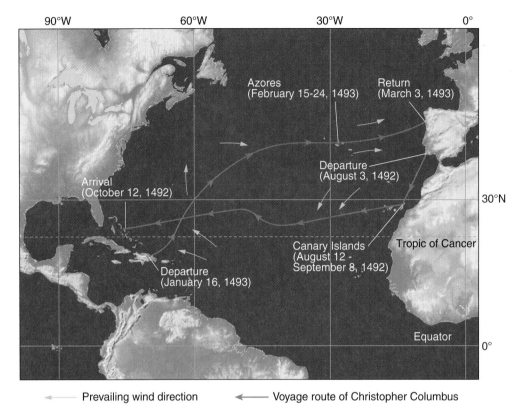

Prevailing wind direction   Voyage route of Christopher Columbus

FIGURE 7-1 Christopher Columbus's route to and from the New World shows an understanding of prevailing wind directions.

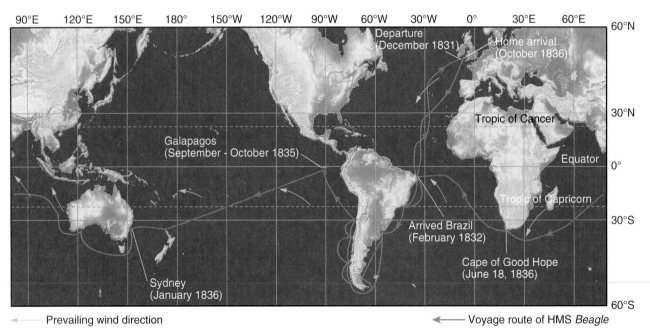

Prevailing wind direction   Voyage route of HMS *Beagle*

FIGURE 7-2 The journey of the *HMS Beagle*. Note the effort made to stay in the region of the trade winds near 20°S latitude while crossing both the Pacific and Indian Oceans.

WHAT ARE CONCEPTUAL MODELS?

A simple drawing of a face is not very sophisticated (FIGURE 7-3), yet it successfully demonstrates the major features and the overall pattern of a human face: two eyes and ears, a nose, a mouth, eyebrows, and a chin. When you describe a friend to someone, you give details about those features. A description such as "he has brown eyes, bushy eyebrows, a big nose, and small ears" may not give a very complete picture of someone, but it might be enough to recognize that person in a crowd. It is the same with describing the atmosphere. A general description of a few current atmospheric features might allow you to recognize the overall weather pattern.

We can draw simple faces based on observations. For example, we all know from experience that the nose lies above the mouth and between the eyes. To make a simple, conceptual model of global wind patterns requires similar observations. We will rely on our own observations, historical records, and data from satellites to identify the major features of a simple model that represents the atmospheric circulation.

Although all faces have the same basic features, few people look identical. It is the same with the atmosphere. You can identify many of the basic features of the atmosphere on most days, but you will not find 2 days with identical patterns. An exercise on the text's Web site is designed to help you identify these general atmospheric patterns on satellite images. Go investigate!

FIGURE 7-3
A drawing like this one represents the major features of a human face in a simple fashion. In this chapter, we create a similar simplified picture of the global wind patterns.

"Discover Features in a Satellite Image."

OBSERVATIONS OUR MODEL SHOULD EXPLAIN

A simple conceptual model of the global wind pattern must explain the steady winds that have long been observed by mariners. A model of global atmospheric circulation must also account for regions that consistently lack winds. Such a model must also be consistent with other observed patterns, such as the position of deserts and regions of high precipitation in January and July (FIGURE 7-4).

Another observation our conceptual model should explain is the global pattern of cloudiness that we first noted in Chapter 1. FIGURE 7-5 is a general cloud pattern as observed from weather satellites. We can see several distinct patterns in these images. Note the lack of clouds near the horse latitudes, consistent with the precipitation maps. A band of thunderstorms extends around the world near the equator. Tracking this cloud band throughout the year, we find that when it is summer in the Northern Hemisphere, this band is located at approximately 8°N latitude. The average position of this cloud band is approximately 4°S latitude when the Southern Hemisphere experiences summer. The maps of average precipitation also depict these patterns in January and July.

The midlatitude cloud patterns summarized in Figure 7-5 have characteristics distinctly different from the cloud systems we observe in the tropics. The midlatitude cloud systems move quickly, change from one day to the next, and are further poleward in the hemisphere that is experiencing summer than in the hemisphere that is experiencing winter. In both summer and winter, the midlatitude cloud systems tend to move from west to east. You can observe these patterns for yourself on weather reports on television or the Internet.

Today's weather reports often discuss the position of the **jet stream**. If you listen to these reports carefully and often, you will realize that there is more than one of these swift "rivers of air" above us. Our conceptual model will predict the existence of two of these jet streams. Both are fast-moving currents of air that flow from west to east and are closely related to the midlatitude cloud patterns discussed above.

The existence of fast winds moving from west to east was long suspected because of the movement of storm and cloud systems. But the suspicions remained unconfirmed until World

January

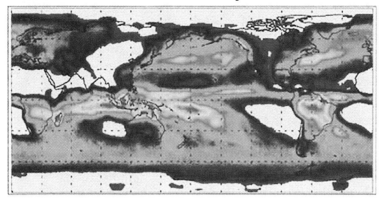

July

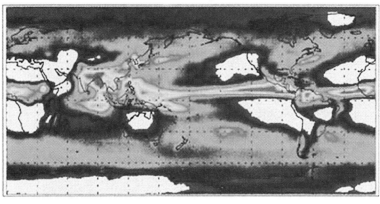

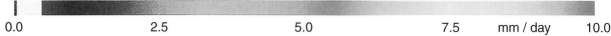

0.0   2.5   5.0   7.5   mm / day   10.0

FIGURE 7-4 Global climatology of precipitation in January (top) and July (bottom) in millimeters per day. (One millimeter is approximately 0.04 inches.) Measuring precipitation in polar regions is difficult, and there are few observations there (white regions). (Reproduced from Adler, R., et al., *J. Hydrometeor*, 4 [2003]: 1147–1167.)

War II, when the U.S. prepared for major air raids against Japan. The B-29 airplanes flew from east to west at altitudes of 10 kilometers, where they encountered a strong stream of westerly winds that slowed or even stopped the planes in mid-air! Our model must describe all of these observations. Let's begin building our model.

A SIMPLE CONCEPTUAL MODEL OF GLOBAL CIRCULATION PATTERNS

A conceptual model makes sense of nature by simplifying wherever possible. Initially we assume that the Earth is entirely covered with water. With no land present, we can assume that what happens in one hemisphere happens in the other, and we can ignore the variations in circulation patterns that land masses cause. Although this water world is unrealistic, it will explain the basic observations we have discussed.

Let's begin our conceptual model by explaining the cloud band that nearly encircles the tropics. These are convective clouds that require upward vertical motions. This cloud band is nearly always present, so there must be steady upward motions near the equator. Accordingly, our conceptual

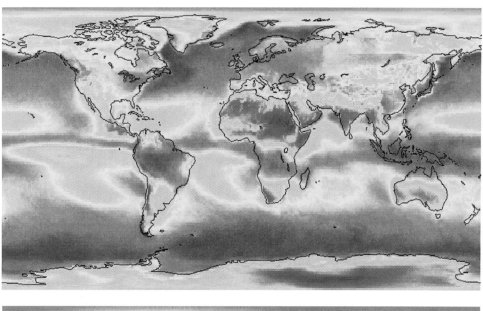

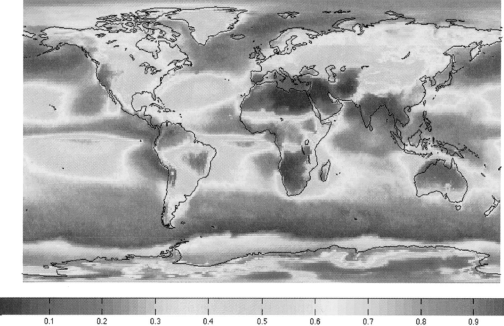

FIGURE 7-5 Satellite climatology of cloud cover from 2000 to 2009 for January (top) and July (bottom). Cloudiness is plotted as a fraction (1 = completely cloudy). Note the cloudy regions near the equator and in the midlatitudes (red regions) and the relatively cloud-free regions in the subtropics (green and blue regions).

model has upward vertical motions near the equator (**FIGURE 7-6**). This rising air produces the cloud and precipitation patterns of the tropics. Eventually, the rising air encounters the stable stratosphere, stops rising, and converges, creating a poleward pressure gradient force. The air then accelerates northward and southward along the tropopause. Let's follow the northward branch of air.

As the air accelerates northward, the Coriolis force turns it to the right (eastward). As discussed in Chapter 6, the magnitude of the Coriolis force increases as the air is accelerated toward the pole. When this upper air stream reaches 30°N latitude, the Coriolis force has turned the flow, causing the air to move from west to east—a westerly wind. As the air accelerates poleward, it gets closer to the Earth's axis of spin (**FIGURE 7-7**). Before going on, we need to understand a concept known as **conservation of angular momentum**.

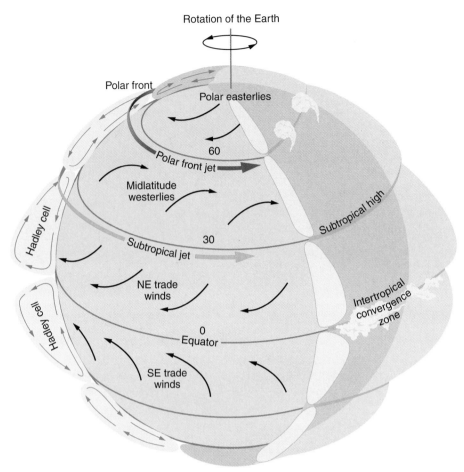

FIGURE 7-6 A conceptual model capable of explaining weather patterns spanning the globe.

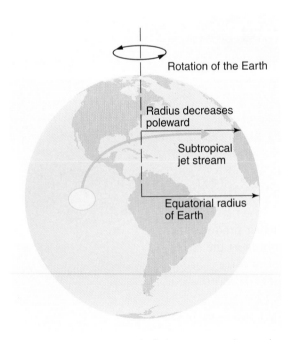

FIGURE 7-7 As a parcel of air moves poleward, its distance from the Earth's spin axis decreases. As it conserves angular momentum, the parcel increases its west-to-east speed, creating a subtropical jet stream.

In Chapter 6 we saw that the momentum of an object moving in a straight line is equal to its mass × velocity. A rotating body has **angular momentum**, defined as follows: mass × rotation velocity × perpendicular distance from the axis of rotation. A rotating object will conserve its angular momentum. In other words, the product of mass, velocity, and distance from the rotation axis is a constant value. If the distance from the axis of rotation decreases, the velocity of rotation increases to conserve angular momentum. For example, figure skaters and divers spin faster as their arms and legs are moved closer to the axis of rotation (**FIGURE 7-8**). Similarly, as a parcel of air moves north or south from the equator, its distance from the Earth's axis of rotation decreases. The parcel's angular velocity increases as a result to conserve angular momentum. This means that the air travels faster in the same direction as the Earth's rotation as it accelerates poleward.

Returning to our model, the ascending air near the equator accelerates poleward after reaching the tropopause. Deflected by the Coriolis force on its poleward trek, it becomes a westerly wind in the upper troposphere between 20°N and 30°N latitude. Because of the combined effect of acceleration and the Coriolis force, the air blows rapidly from west to east. This stream of air is referred to as the **subtropical jet stream** (**FIGURE 7-9**).

The air that accelerates into the subtropical jet stream tends to converge at high altitudes. Why? The pressure gradient force accelerates the air poleward while the Coriolis force turns the wind to the right. The result is a balanced stalemate of forces, while still more air is accelerated into the region. This leads to a rise in air pressure at the surface because surface pressure is the weight of the air above the surface. This implies that there should be surface high-pressure systems between near 30° latitude in both hemispheres.

Air molecules cannot continue to converge into the subtropical jet stream region forever without some "release valve." Stymied in the horizontal direction by the convergence of air into the region, the air can still move vertically. But because the horizontally converging air is near the tropopause, the stable stratosphere above it resists upward motions. The only other option is downward motion. So in our model, the converging air must sink toward the surface. This is consistent with our discussion in Chapter 6 of sinking air in high-pressure systems. Sinking air typically suppresses tall cloud development and precipitation. It is therefore logical that this sinking air is above our current desert locations and the horse latitudes (Figure 7-6), at about 30°N and 30°S latitudes. Stratus clouds often appear over oceans below this sinking air, as discussed in **BOX 7-1**. After the subsiding air reaches the surface, it spreads north and south out of the high pressure. Now let's follow the air near the surface that is flowing toward the equator.

As the air flows south toward the equator, the Coriolis force pulls it to the right (west) (to the left in the Southern Hemisphere). The Coriolis force weakens as the air approaches the equator,

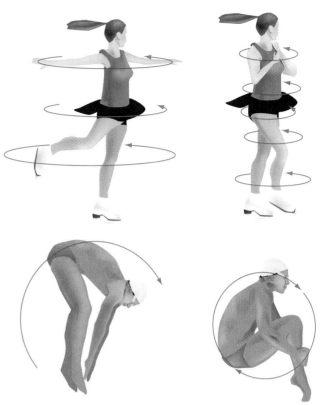

FIGURE 7-8 Keeping his or her limbs close to the axis of rotation (the body) allows a diver or skater to conserve angular momentum and thereby spin faster. The conservation of angular momentum is important in explaining the existence of the subtropical jet stream.

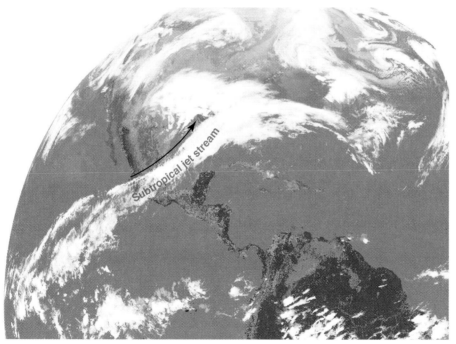

Subtropical jet stream

FIGURE 7-9 The subtropical jet stream can be recognized in infrared satellite images by the cirrus clouds that often accompany jet streams. In this example, a strong subtropical jet extends northeastward from the equatorial Pacific Ocean across Mexico and the Gulf of Mexico and over Florida. Compare this real-life example to the path of the air parcel in our conceptual model in Figure 7-7. (Courtesy of CIMSS, University of Wisconsin-Madison.)

Box 7-1 Marine Stratocumulus Cloud Regions

High-pressure systems are supposed to be cloudless, according to Chapter 6. So why aren't the world's subtropical highs always sunny? It is common to find stratocumulus clouds in the vicinity of the descending branch of the Hadley cell that lies over cold ocean regions. This is evident in the satellite image of the Western Hemisphere below. The marine air near the surface is cool and humid. As the air in the upper troposphere descends, it warms adiabatically. When a deep layer of the atmosphere sinks, a temperature inversion can develop as a result of adiabatic compression. This is illustrated in the chart on the opposite page. The upper region of the layer descends over a greater distance than the lower region and thus warms more. After subsiding, the top of the layer is therefore warmer than the bottom. This is called a *subsidence inversion* because it results from descending, or subsiding, air. In meteorology, subsidence denotes sinking motions. This particular subsidence inversion is also called the *trade-wind inversion* because it occurs in the region of the trade winds. The temperature increases sharply with altitude in the trade-wind inversion. Stratus clouds form when the air approaches saturation as a result of mixing of the dry subsiding air with the moist air near the ocean surface. The stratocumulus clouds observed in the satellite image below lie just below the trade-wind inversion, where there is enough instability for the stratus to become partly convective, forming stratocumulus clouds. Because of the trade-wind inversion, which inhibits vertical motion, these clouds cannot grow taller.

First GOES–15 fulldisk visible image
6 April 2010   1733 UTC

Stratocumulus

Stratus

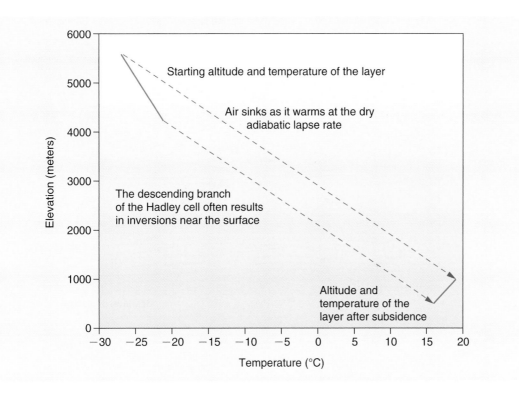

resulting in a northeast wind in the Northern Hemisphere and a southeast wind in the Southern Hemisphere. These are the **trade winds** sought by traders sailing from Europe to the Americas, as well as by FitzRoy and Columbus. Notice how the trade winds from the two hemispheres converge, supplying moist air for cloud development. The region where the two trade winds come together is called the **Intertropical Convergence Zone (ITCZ)** (**FIGURE 7-10**).

In our conceptual model, a circulation cycle extends from the equatorial to subtropical regions. Air flows upward at the ITCZ and accelerates poleward at the tropopause to form a jet stream, in the vicinity of which the air sinks. When the sinking air reaches the surface, some of

FIGURE 7-10 The bright white clouds in the center of this visible satellite image are the ITCZ over the Pacific Ocean. The ITCZ is normally found just north of the equator across the Pacific, as in this image.

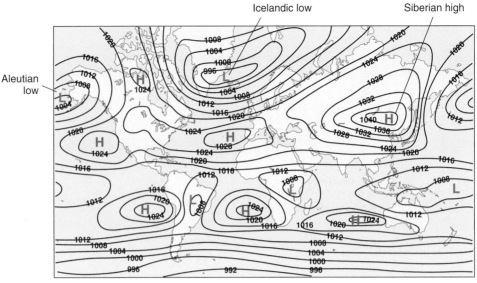

Sea-level pressure—January

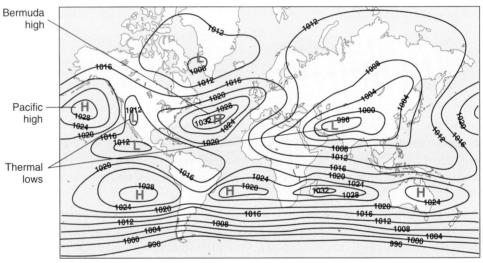

Sea-level pressure—July

FIGURE 7-11 Sea-level pressure and wind maps for a typical January and July. Notice how our conceptual model of Figure 7-6 is consistent with the observed surface winds. In the wind maps on the next page, the thicker the wind arrow, the more constant the wind. (Reproduced from *The Science and Wonders of the Atmosphere*, Gedzelman, S.D. © 1980, John Wiley & Sons. Reproduced with permission of John Wiley & Sons, Inc. Courtesy of Stanley David Gedzelman.)

the air flows toward the equator and converges into the ITCZ (Figure 7-6). This circulation cell is called the **Hadley cell**, after 18th-century scientist George Hadley, who in 1735 was the first person to propose a reasonable explanation of the trade winds.

FIGURE 7-11 shows the average sea-level pressures observed over the globe for January and July. Notice the high pressures at the surface near 30°N and 30°S latitudes, particularly over the oceans, confirming our model. These semipermanent pressure systems are referred to as **subtropical highs**. These subtropical highs, or *anticyclones,* have a major influence on the weather and climate of the subtropics and middle latitudes. The pressure gradients are weak within the subtropical highs, and the surface winds are, therefore, light or calm over large regions of the subtropical oceans—the horse latitudes. This is consistent with the observations of mariners.

Air near the surface flows outward from the subtropical highs under the influence of friction (see Chapter 6). Equatorward of these highs are the trade winds. In equatorial regions where

Wind—January

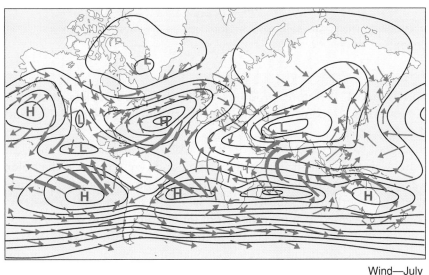

Wind—July

FIGURE 7-11 Continued.

neither trade wind dominates, the wind is calm. These wind regions are known as the **doldrums**; crews of sailing ships dreaded the doldrums because of the light winds, hot temperatures, and high relative humidity.

Let's turn our attention to the polar regions, where (as we learned in Chapter 2) more energy is emitted to space than is gained from solar radiation. As a consequence, the air above the poles is cold and sinks. Sinking air warms adiabatically. Warmer air over the cold polar surface is an inversion and inhibits precipitation. This explains the small amount of precipitation observed near the poles (Figure 7-4).

When the sinking air over the poles reaches the ground, it must flow toward the equator. Because this air is flowing southward from northern polar regions, the Coriolis force acts to turn these winds westward, forming **polar easterlies** at the surface poleward of 60° latitude (Figure 7-6). Eventually, air must rise and flow poleward to replace the sinking air over the poles, completing the circulation of air through the polar cell (Figure 7-6). The sea-level pressure maps show that the polar easterlies are well developed in the winter and over continents. Eventually, as these winds flow equatorward, they are met by subtropical air flowing poleward.

On the poleward sides of the descending branch of the Hadley cell, surface air moves poleward (Figure 7-6). The Coriolis force, which increases in magnitude as the poles are approached, turns the surface wind to the east in the midlatitudes of both hemispheres.

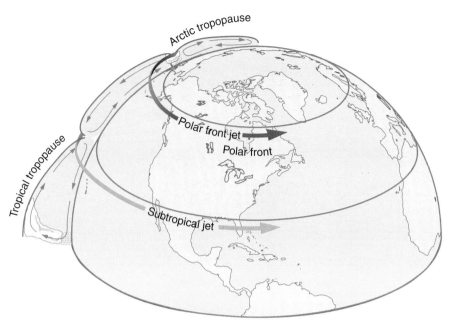

FIGURE 7-12 The subtropical and polar jet streams in relation to our conceptual model of global wind circulations.

Our conceptual model, therefore, produces **midlatitude westerlies** at the surface. The midlatitude westerlies are consistent with the observations in Figure 7-11, validating our conceptual model.

The midlatitude westerlies encounter the polar easterlies around 60° latitude. This clash of winds separates warm tropical air from cold polar air and is referred to as the **polar front**. As we saw in Chapter 6, strong winds should exist above regions of strong temperature gradients (see the applet on the text's Web site). Therefore, we expect to find another westerly jet stream above the polar front, referred to as the **polar front jet stream** (FIGURE 7-12). The polar front jet stream is displaced further poleward than the subtropical jet stream (FIGURE 7-13).

Our conceptual model explains important circulation features of the atmosphere: the ITCZ, the subtropical jet stream, the subtropical deserts, the trade winds, midlatitude westerlies, and the polar front jet. This model is a simplification of actual global circulation. In reality, there are landforms, which tend to disrupt our simple model. In addition, neither the polar front jet nor the subtropical jet flows directly west to east. Both meander like rivers, producing a wavelike pattern of *troughs* and *ridges*. These upper-air jet streams are extremely important for understanding midlatitude weather. These meandering winds are discussed in more detail in the next section.

"Jet Stream and Horizontal Temperature Gradients."

UPPER-AIR MIDLATITUDE WESTERLIES

The winds in the upper troposphere, above approximately 500 millibars (mb), flow in wavelike patterns with troughs and ridges (FIGURE 7-14). The air flow through these upper-level waves results in storms that move warm air poleward and cold air toward the equator. The Northern Hemisphere typically is encircled by several of these waves at any particular time; Figure 7-14 shows three troughs and three ridges in the Northern Hemisphere in January. Each trough–ridge combination is called a **Rossby wave**, named after Carl-Gustaf Rossby, the famous 20th-century meteorologist who discovered them (see Box 6-2). Rossby waves usually drift slowly eastward. As discussed in Chapter 6, rising motions tend to occur near lows (e.g., wave troughs), with sinking motions occurring near highs (e.g., ridges). The movement of these waves is very important in determining the development of surface weather systems, so understanding these waves is central to weather prediction.

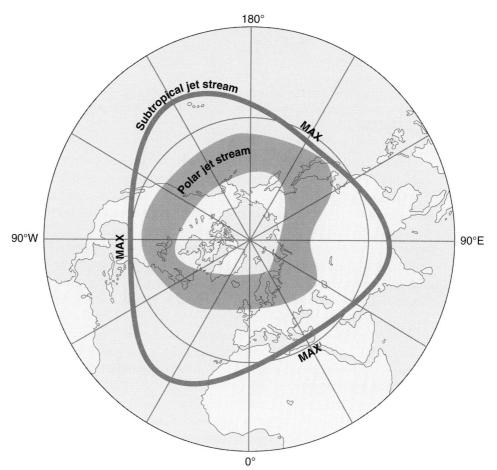

FIGURE 7-13 The approximate positions of the polar front jet stream and the subtropical jet stream over the Northern Hemisphere during winter. Locations of maximum wind speed (MAX) along the subtropical jet are indicated. (Modified from S. Lee and H.-K. Kim, *J. Atmos. Sci* 60 [2003]: 1490–1503.)

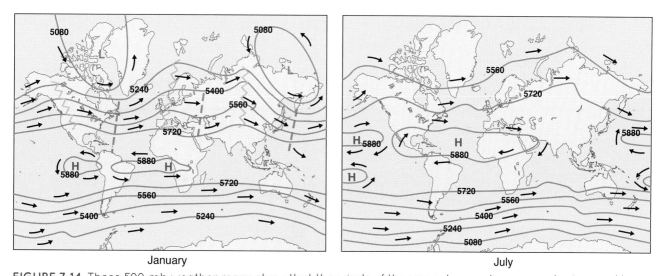

FIGURE 7-14 These 500-mb weather maps show that the winds of the upper troposphere meander in wavelike patterns with troughs and ridges. Troughs (dashed orange lines) occur where the height contours dip equatorward, and ridges (jagged yellow lines) occur where the height contours bulge poleward. Numbers are geopotential heights (in meters).

(a)  (b) 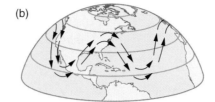 (c)

FIGURE 7-15 (a) In a zonal flow pattern, the air flows nearly parallel to latitudes and moves from west to east. (b) In a meridional flow pattern, the air moves north and south as it flows eastward. (c) In a split flow pattern, zonal and meridional flow patterns are combined.

Waves are described by their wavelength (distance between successive troughs or ridges) and amplitude (north–south extent). Amplitude and wavelength determine the type of weather associated with these waves. In the case of Rossby waves, a small-amplitude pattern results in mostly west-to-east winds known as a **zonal flow pattern**, nearly parallel to the lines of constant latitude (**FIGURE 7-15a**). In zonal flow, cold air masses tend to stay toward the polar regions, and the warm air remains equatorward. A **meridional flow pattern** occurs when the waves have large amplitude with deep troughs and peaked ridges (Figure 7-15b). In a meridional flow pattern, cold air flows equatorward toward the subtropics, and warm air flows poleward. The usual westerly jet may be absent, replaced by strong northerly and southerly winds.

A simple way to determine the type of flow pattern is the **zonal index**. The zonal index, as first defined by Rossby, is calculated as the sea-level pressure at 35° latitude minus the sea-level pressure at 55° latitude. High zonal index situations (with lower pressure toward the pole) indicate zonal flow patterns; low or negative zonal index values (with higher pressure at higher latitudes) indicate meridional flow patterns.

Different configurations are also possible. A **split flow pattern** (Figure 7-15c) can occur where zonal flow exists near the pole with a meridional flow pattern further to the south. Sometimes in these cases the meridional pattern becomes so strong that masses of air separate and become "cut off" from the main westerly air flow. Cut-off lows or cut-off highs represent large pools of cold or warm air that block the eastward progression of weather systems. These **blocking patterns** can persist for extended periods and can result in extreme weather events such as floods and droughts.

For example, in summer 1988, a blocking pattern created a warm high-pressure system over the central United States with troughs over the east and west coasts. This particular blocking pattern contributed to the spring and summer drought in the Midwest, the Northeast, and the Great Plains that summer (**FIGURE 7-16**).

(a) 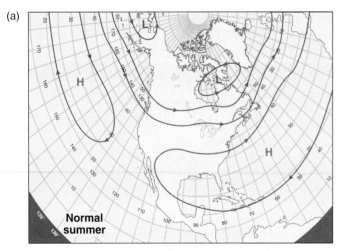 (b)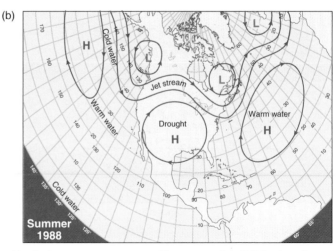

FIGURE 7-16 Normal (a) and "blocking" (b) wind patterns above North America in summer. The upper-level "blocking" high over the U.S. in 1988 diverted the jet stream into Canada and led to drought conditions over the central U.S. (Modified from "Written in the Winds: The Great Drought of '88." J. Namias, *Weatherwise*, Jan. 4 1989, vol. 42, pp. 85–87. Reprinted by permission of the publisher [Taylor & Francis Group, http://www.informaworld.com].)

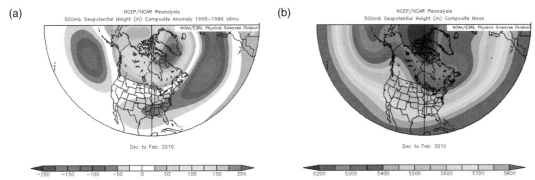

FIGURE 7-17 (a) Anomalies in 500-mb heights in the vicinity of North America during December 2009 to February 2010. Notice the large positive (high pressure, red) anomaly over Greenland and Canada and the negative (low pressure, blue) anomaly over the eastern U.S. and the Atlantic Ocean. (b) Actual 500-mb heights in the vicinity of North America during December 2009 to February 2010. A deep trough extended across the eastern U.S. and Canada, bringing cold air and storms to the East Coast. A ridge over western North America brought warmth to the Pacific Coast. (Courtesy of ESRL Physical Science Division/NOAA.)

More recently, the bitter-cold winter of 2009–2010 in the eastern U.S. was caused by unusual blocking high over Greenland (**FIGURE 7-17A**) that replaced the typical wintertime Icelandic low (Figures 7-11 and 7-14). The blocking high was part of a meridional flow pattern that brought cold air and stormy lows to the eastern U.S. (Figure 7-17B).

If you look at upper-level weather charts every day for several weeks, you will notice that high zonal index periods are punctuated by low-index spells, and sometimes split flow patterns develop. This irregular oscillation between patterns is well known, but poorly understood, and is called the **index cycle**. Current research is examining atmospheric oscillations on a variety of timescales, which are often related to interactions with the oceans (see Chapter 8).

Superimposed on the Rossby long waves are ripples referred to as **short waves** (**FIGURE 7-18**). Short waves travel rapidly through the longer Rossby waves. The short waves travel along with the Rossby waves and are sometimes difficult to observe because of their relatively small scale and fast forward motion. It is, therefore, difficult to predict their position and the associated weather. Both short waves and long waves are needed for the development of storms. Rossby waves, short waves, and long waves are examined in more detail in Chapter 10.

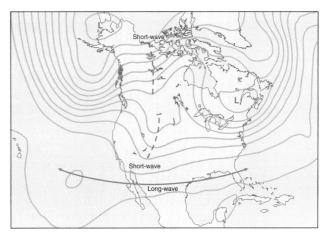

FIGURE 7-18 Short waves are weather disturbances that are smaller than and move through the atmosphere faster than longer Rossby waves.

THE POLEWARD TRANSPORT OF ENERGY

In Chapter 2 we learned that radiative energy gains exceed the radiative losses at the top of the atmosphere in the tropical regions of the globe. In the polar regions, the radiative losses exceed the gains. Circulations in the atmosphere help transfer heat poleward to compensate for these regional differences in energy budgets. The movement of warm air poleward and of cold polar air equatorward is one way that this heat transfer occurs. The air flow of our conceptual model includes these patterns. Thus, in addition to explaining the main weather features of the globe, our model of atmospheric circulation implies a heat transport from the tropical regions to the poles.

If there were no poleward heat transfer, the poles would be much colder and the tropics much warmer. **FIGURE 7-19** shows the annual average poleward energy transport accomplished by the atmosphere and the oceans. The region of maximum energy transport by the atmosphere falls

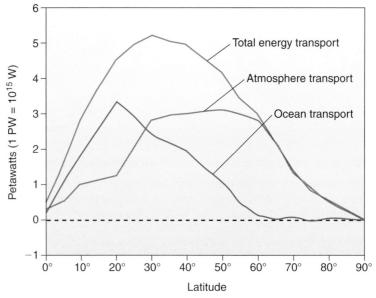

FIGURE 7-19 Atmospheric winds transport heat from the tropics to the North Pole, helping to maintain an energy balance around the globe. The oceans are the primary transport mechanism in the low latitudes.

between 30° and 60° latitude in both hemispheres. This poleward heat transport is accomplished by Rossby waves and midlatitude cyclones (see Chapter 10 for more details).

This figure reveals why the traditional explanation for western Europe's mild winters is wrong (see Chapter 3). At the latitude of London, England (51.5°N), the atmospheric heat transport is more than three times the magnitude of the oceanic heat transport. The oceans move very little heat poleward at such high latitudes. And so the mild winters are actually the result of the mild southwest winds that blow into Europe (as shown in Figure 7-11), which are ultimately caused by the persistent wintertime Rossby wave pattern across the Atlantic (as seen in Figure 7-14).

SEASONAL VARIATIONS

There are seasonal variations in the amount of solar energy received by the Northern and Southern Hemispheres, and these variations account for seasonal changes in weather. How does our simple model handle these seasonal differences? The global circulation pattern—made of the ITCZ, the polar front, the subtropical highs, and the jet streams—shifts with the Sun, moving poleward in spring and equatorward in autumn. This is reflected in the mean sea-level pressure maps for January and July (Figure 7-11). The subtropical highs are shifted further poleward in each of the summer hemispheres.

The position of the ITCZ has a definite seasonal variation. The average position of the ITCZ is approximately 4°N. Seasonal averages show that the ITCZ is shifted toward the summer pole, for example, closer to 10°N in July in the Pacific and Atlantic oceans (**FIGURE 7-20**). North–south variations in the position of the ITCZ are larger over land than over the oceans because continents warm and cool more dramatically than the oceans.

In each hemisphere, the polar jet stream is displaced further poleward in summer than in winter. During summer, the positions of the subtropical highs also shift poleward. As a result, the boundary between the surface air flowing poleward from the subtropical highs and the polar easterlies is displaced poleward. This boundary indicates the position of the polar jet stream. The poles are warmer in summer than in winter. Because the polar air is warmer, the temperature difference between the polar air and the subtropical air is smaller. According to the

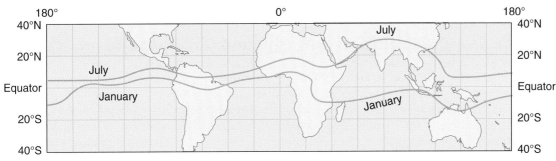

FIGURE 7-20 The approximate position of the ITCZ in January and July.

thermal wind relationship (see Chapter 6), this means that the polar jet stream can be expected to be weaker during the summer. The transport of energy from the middle latitudes to poles should be smaller during the summer than during the winter because the summer poles are illuminated and thus gain solar energy. Observations generally confirm these expectations, again validating our conceptual model.

The seasonal variation of the subtropical jet stream can also be explained with our model. Remember that this jet stream is caused by the Coriolis deflection of air accelerating poleward out of the rising branch in the Hadley cell, which is the sum total of all the rising air in the thunderstorms of the ITCZ. When the ITCZ is near the equator in wintertime, this air travels poleward for about 20° to 30° latitude before the Coriolis force and PGF reach balance. This air is accelerated over a long distance, and the winds in the subtropical jet are fast as a result; However, in summertime, the ITCZ is located farther poleward, and the air is accelerated over a shorter distance before converging at the latitude of the subtropical jet. This implies that the wintertime subtropical jet should be stronger in winter than in summer. This, too, is confirmed by observations.

The semipermanent low-pressure systems also vary seasonally in their position (Figure 7-11). In January there is a string of low-pressure areas around 8°S latitude over South America, Africa, and Indonesia. This corresponds to the average position of the ITCZ. During July there is a low-pressure region west of Central America that is associated with the mean position of the ITCZ in this region. During January there are also low-pressure regions near Iceland and near the Aleutian Islands. These are the subpolar lows and are regions that often spawn storms. During July there is one low near the North Pole, and it is weaker than during winter. The low-pressure region around the South Pole is associated with the cold Antarctic continent and its topography.

There are also low-pressure regions in the vicinity of Iraq and the southwest U.S. in July. These lows are associated with deserts and are referred to as heat lows or **thermal lows**. These heat lows develop because of intense surface heating caused by absorption of solar radiation that occurs in the absence of clouds. It is no accident that these lows form at about the same latitude as the subtropical highs, where sinking air leads to persistently clear skies. The lows form when surface heating warms the air above, causing it to expand both vertically and horizontally throughout a deep layer. This expansion reduces the air density, making the pressure near the surface lower than it would be if the air were cooler.

MONSOONS

A **monsoon** is a weather feature driven by seasonal differences in the heating of land and ocean along with seasonal shifts in global-scale circulations. Our original model of atmospheric circulation assumed that the Earth was covered with water. To examine an important seasonal variation in precipitation, we will have to add land masses to our conceptual model. The distribution of land, especially large mountain ranges, has a strong influence on weather patterns (**BOX 7-2**). One good example of a place where this effect is prominent is the Indian subcontinent. The Indian monsoon that occurs there is a complex interaction of seasonal shifts in the global

Box 7-2 Precipitation Patterns and Topography

As discussed in Chapter 4, orographic lifting (lifting of air as it travels over a mountain) is one mechanism of lifting air to reach saturation, cloud formation, and thus precipitation. Careful examination of global precipitation patterns indicates a correlation between total precipitation and topography. Precipitation, for example, is increased on the windward side of a mountain. However, the correlation is not perfect. Precipitation is also related to the temperature and humidity of the rising air, the speed and the direction of the winds relative to the mountain topography, and the stability of the air.

The coastal mountain ranges of the Pacific Northwest are a good example of the relationship between topography and precipitation. Because of their location in the midlatitude westerlies, the upwind side (the side the wind strikes first) of these mountains receives a lot of rain, whereas the downwind side can resemble a desert. To explain this observation, let's consider a simplified model of the weather situation, which consists of a mountain peak 3 kilometers high in the midlatitude westerlies.

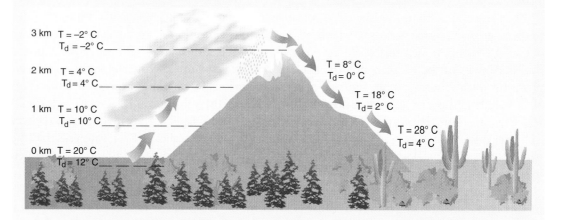

Suppose that the air temperature (T) of the wind approaching the mountain is 20° C with a dew point (T_d) of 12° C. Air whose temperature is higher than its dew point is unsaturated, or "dry." As the air is orographically lifted on the windward side, it expands and cools at the dry adiabatic lapse rate—approximately 10° C for each kilometer it is lifted.

As the parcel cools, the relative humidity increases. As the parcel is forced to rise, the mixing ratio remains the same until a cloud forms. The parcel also expands as it rises, thus decreasing the vapor pressure. The air must therefore be cooled even further to reach the point of saturation. In other words, the dew point temperature decreases. The dew point decreases approximately 2° C for each kilometer it is lifted.

After being lifted 1 kilometer, the temperature and the dew point are the same, so the air is saturated. The mixing ratio then decreases because water vapor molecules are removed from the air to form cloud droplets. Further lifting causes the air to cool at the moist adiabatic lapse rate, approximately 6° C per kilometer. Collision and coalescence result in large droplets and precipitation as the air ascends the mountain. At the summit, assumed here to be 3 kilometers, the air has cooled to approximately −2° C. As long as the air remains cloudy, the dew point and the temperature are equal. As the cool air begins its descent on the downwind side of the mountain, the cloud droplets evaporate as the air warms because of compression heating (which lowers the relative humidity). The cloud quickly dissipates after reaching the summit. The now-heated sinking air is depleted of moisture, resulting in a local minimum in rainfall on the downwind, or leeward, side. This local minimum of precipitation is called a *rain shadow*.

In addition, as the air moves down the mountainside, the temperature increases at the dry adiabatic lapse rate, and the dew point increases at about 2° C per kilometer. When the air returns to sea level, its temperature is 28° C, with a dew point temperature of 4° C. At the end of its trip across the mountain, the air has warmed and is drier than when it began its journey. In this way, topography can dictate the weather and climate of a region.

circulation patterns, thermal heating differences between land and sea, and the interaction of winds with the Himalayan mountains.

In spring and summer, the Sun warms the land and the air above it. With cooler air over the surrounding water, a horizontal pressure gradient, directed from ocean to land, is established near the surface, bringing humid air inland. The solar heating inland triggers convection. This causes hot, humid air to rise, expand, and cool, leading to condensation and rain. Orographic lifting by the Himalayas cools the air when it reaches the mountains, generating additional rain. Air rushing out of the tops of the thunderstorms flows out over the Arabian Sea and the Bay of Bengal, where it sinks, completing the monsoon circulation (**FIGURE 7-21**). The air flowing over the water remains over the ocean a long time, which causes it to gain moisture. (This summertime monsoonal circulation shifts the location of the ITCZ northward across Asia, as seen in Figure 7-20.)

During autumn and winter, the air above the land cools faster than the air over the water, establishing a pressure gradient force from land to sea. The winds are reversed from the summer monsoon flow—from land to sea instead of from sea to land (**FIGURE 7-22**). Sinking air above the land suppresses cloud development and precipitation. As a result, the winter monsoon is a dry season, and the summer monsoon is a wet season.

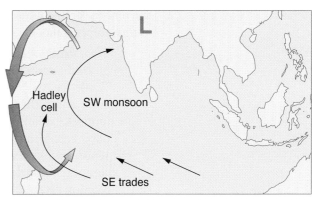

FIGURE 7-21 A depiction of the complete monsoon circulation during summer. Low-level winds travel over the warm Arabian Sea and supply the moisture for the torrential rains.

BEYOND CONCEPTUAL MODELS: CURRENT RESEARCH IN GLOBAL-SCALE WINDS

Meteorological research has progressed beyond the conceptual models of Hadley and his successors. Using advances in theory, the results of computer models of the atmosphere (see Chapter 13) and new data from past climates (see Chapter 14), scientists have expanded our understanding of global-scale winds during the past few decades. At the same time, however, the

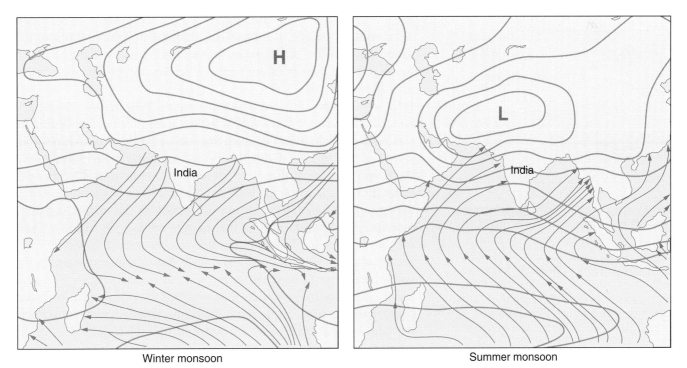

Winter monsoon   Summer monsoon

FIGURE 7-22 The surface wind (arrows) and pressure patterns (contours) associated with the summer and winter Indian monsoons. During summer, the winds blow from sea to land. This pattern reverses in winter.

current research questions are deceptively simple and can often be understood in the context of our conceptual model. Some research questions of today include the following:

- *What determines the location of the ITCZ?* A 2009 study led by scientists at the University of Washington found that the ITCZ was closer to the equator in the Pacific during the "Little Ice Age," when the planet received slightly less solar radiation than it does now. Since about the year 1600, the ITCZ has been moving northward at a rate of 1.4 km (almost 1 mile) per year. Why? No one knows for sure, but it may be related to stronger cross-equatorial temperature gradients since 1600.

- *What controls the poleward extent of the Hadley cell?* During the 1980s, theorists and computer modelers at Princeton University revealed an inverse relationship between the rotation rate and the poleward reach of the Hadley cell, that is,

$$\text{Hadley cell width is proportional to } \frac{1}{\text{rotation rate}}$$

Why does this relationship exist? The faster the rotation, the stronger the Coriolis force is and the lower the latitude at which the PGF and Coriolis come into balance and form the subtropical jet on the poleward edge of the Hadley cell. In effect, a stronger Coriolis force "squeezes" the Hadley cell closer to the equator. For an Earth that spun twice as fast as at present, the Hadley cell would be only half as wide (15 degrees versus about 30 degrees at present). On a very slowly rotating Earth, the Hadley cell could extend from the equator to the pole. This can help explain the global-scale winds on other planets, too, such as slowly rotating Venus and fast-rotating Jupiter. Recent research has also explored the relationships between the Hadley cell, north–south temperature gradients, and the stability of the atmosphere.

"Hadley Cell Dynamics."

- *What causes the summertime subtropical highs in the Northern Hemisphere?* In 1995, meteorologist Brian Hoskins of the University of Reading (UK) gave an invited talk to the American Meteorological Society in which he identified the following dilemma: why is it that the subtropical highs of the Northern Hemisphere are strongest in summertime when the sinking branch of the Hadley cell is at its weakest? Hoskins's observation suggested that our textbook explanation of why the subtropical highs exist—due to convergence of air at the subtropical jet—might need tweaking. Since then, researchers have examined other possible causes for subtropical highs, including the ability of monsoon circulations to trigger Rossby waves that lead to sinking air into the summertime highs. Interestingly, the subtropical highs of the Southern Hemisphere apparently fit our textbook explanation. Research continues on this subject.

- *What controls the locations of the jet streams?* Why are the jet streams typically configured as they appear in Figure 7-13? Penn State University meteorologist Sukyoung Lee asked this question in a provocative research journal article in 2003. Using theory and computer model simulations, she found that the atmosphere "prefers" two situations: (1) for weak convection in the tropics, a weak subtropical jet at low latitudes and a strong polar jet, or (2) for strong convection in the tropics, a strong, more poleward subtropical jet that merges around 30° to 35° latitude with an equatorward-plunging polar jet (**FIGURE 7-23**). Could Lee's findings explain why the polar jet typically dips southward wherever the subtropical jet is strongest and the farthest northward in Figure 7-13, near the U.S. East Coast, the Middle East, and Japan? Possibly. Her work also raises many fascinating questions about how wind circulations at lower latitudes, such as the Hadley cell and subtropical jet, can influence the weather in the midlatitudes.

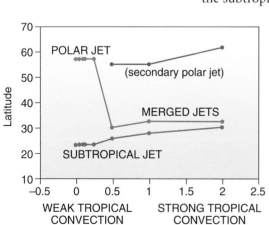

FIGURE 7-23 Computer simulations of the location of the polar and subtropical jet streams for varying amounts of tropical convection. Stronger convection in the tropics causes a stronger, more poleward subtropical jet, and the polar jet merges with it. Compare this result with Figure 7-13, in which the polar jet is closest to the subtropical jet where it is strongest and farthest poleward. (Modified from S. Lee and H.-K. Kim, *J. Atmos. Sci.* 60 [2003]: 1490–1503.)

More than 275 years after Hadley's paper on the trade winds, the study of global-scale winds is still full of unanswered questions!

PUTTING IT ALL TOGETHER

Summary

In this chapter, we developed a conceptual model to explain patterns of global circulation and their associated weather and climate.

Upward vertical motions near the equator explain the ITCZ, which can be identified in global satellite imagery and precipitation maps. When rising air reaches the tropopause and accelerates poleward, it comes closer to the Earth's axis of rotation and increases speed while conserving angular momentum. This produces the subtropical jet stream.

Convergence of air near the subtropical jet stream helps cause the subtropical highs. This descending air of the Hadley cell is compressed and warms, lowering the relative humidity. This sinking air explains the large deserts of Africa, Saudi Arabia, and Australia. The sinking air of the Hadley cell results in nearly calm winds at the surface, producing the "horse latitudes" dreaded by ancient mariners who used wind power to travel. Some descending air that reaches the surface moves toward the equator to supply moisture to the ITCZ. As air moves toward the equator, the Coriolis force acts to produce the steady northeast and southwest trade winds sought by sailors.

Some of the air associated with the descending branch of the Hadley cell moves poleward and clashes with cold polar air masses that are moving toward the equator. Fronts exist where the cold air meets the warm subtropical air, producing midlatitude storms that tend to move from west to the east. These storms are embedded in the midlatitude westerlies, causing them to move eastward. The polar jet stream exists in the vicinity of fronts.

In our simple conceptual model, the globe is covered with water—a good approximation, because water covers 70% of the Earth's surface. However, the differences between land and water are important in weather and climate. The Indian monsoon provides an example of how ocean, land, and topography interact to form clouds and precipitation or to bring about a dry season. A monsoon is a seasonal reversal in wind patterns associated with changing regional energy budgets. The change in wind direction results in dry and wet seasons. Over the Indian subcontinent, the wet season occurs in summer and the dry season occurs in winter.

The general circulation of the atmosphere transports heat from the equator toward the poles, especially in the midlatitudes. The world's oceans also transport heat, especially at low latitudes.

Current research on global-scale winds is trying to understand the physical mechanisms governing features such as the ITCZ, the Hadley cell, the subtropical highs, and the jet streams.

Key Terms

You should understand all of the following terms. Use the glossary and this chapter to improve your understanding of these terms.

Angular momentum	Jet stream	Subtropical highs
Blocking pattern	Meridional flow pattern	Subtropical jet stream
Conservation of angular	Midlatitude westerlies	Thermal lows
momentum	Monsoon	Trade winds
Doldrums	Polar easterlies	Zonal flow pattern
Hadley cell	Polar front	Zonal index
Horse latitudes	Polar front jet stream	
Index cycle	Rossby wave	
Intertropical Convergence	Short wave	
Zone (ITCZ)	Split flow pattern	

Review Questions

1. What might the global circulation of the atmosphere be if the Earth did not rotate and if it were covered with water? (Hint: Start with an ITCZ.)
2. Explain the seasonal shift in the polar front jet over the continental U.S. from January to July.

3. Why are the subtropical highs referred to as semipermanent highs rather than permanent high-pressure systems?

4. Draw a picture of the global air flow near the surface of the Earth.

5. Explain the existence of the ITCZ.

6. Discuss the differences and similarities between a land or sea breeze and a monsoon.

7. Using a current satellite image of the Earth's weather, locate the major features of the atmospheric circulation pattern.

8. Explain why the Coriolis force does not influence how your sink drains but does affect the subtropical jet stream.

9. Based on your knowledge of how the Coriolis force affects the Hadley cell, how far poleward (in degrees of latitude) do you think the Hadley cell would extend if the Earth rotated three times as fast around its axis as it does now (about 30° latitude)? What would your answer be if Earth rotated only half as fast as at present?

10. Just south of the Tropic of Cancer on the island of Kauai lies Mount Waialeale (elevation, 1.6 km), which receives approximately 1170 centimeters (460 inches) of rain a year. Explain why this region receives so much rain.

11. Why are the Southern Hemisphere westerlies slightly faster than the Northern Hemisphere westerlies?

■ Observation Activities

1. What large-scale circulation patterns determine the weather in your region?

2. Sketch the position of the jet streams at 300 mb on a map over a week (see http://squall.sfsu.edu/scripts/nhemjetstream_model.html to create an animation). Describe their variation during this time period. Would you describe the jet stream pattern as "zonal," "meridional," or "split flow"?

This rain cloud icon is your clue to go to the *Meteorology* Web site at http://physicalscience.jbpub.com/ackerman/meteorology/. Through animations, quizzes, web exercises, and more, you can explore in further detail many fascinating topics in meteorology.

Atmosphere–Ocean Interactions: El Niño and Tropical Cyclones

8

AFTER COMPLETING THIS CHAPTER, YOU SHOULD BE ABLE TO

• Explain how the atmosphere affects the ocean and leads to ocean circulations
• Describe how the atmosphere and ocean combine to create the worldwide El Niño–Southern Oscillation phenomenon
• List the stages of the life cycle of a hurricane and recognize these stages from weather satellite images
• Explain the different ways in which tropical cyclones can threaten life and property

INTRODUCTION

In the early 1990s, beachcombers along the Pacific coasts of Alaska, Canada, and the northwest United States made some surprising discoveries. Instead of seashells, they found hundreds of plastic bathtub toys and Nike tennis shoes washed up on the beach! Where did this flotilla of flotsam come from? Two separate accidents in the north Pacific in 1990 and 1992 caused 61,000 shoes and 29,000 toys to spill off of ships into the ocean. Once overboard, the shoes and the toys floated, in some cases for several years. However, they didn't just bob up and down in one place in the Pacific. Driven by strong winds at the surface, steady ocean currents carried the shoes and toys eastward toward the North American coastline.

The atmosphere interacts with an ocean across 70% of the Earth's surface. In this chapter, we learn about the structure of the ocean and how the force of wind on water creates **ocean currents**. Then we examine how cross-talk between the atmosphere and oceans creates oscillations in weather and climate such as the El Niño phenomenon. Finally, we investigate how air and ocean can combine to form nature's deadliest storm system, the tropical cyclone. These ferocious atmospheric storms, often called hurricanes, cannot survive without the moisture supplied by warm ocean water. Killer tropical cyclones are a far cry from plastic toys washing up on an Alaska shore, but both are signs of the strong linkages between the atmosphere and oceans.

OCEANOGRAPHY

Oceanography is the study of the oceans and is essential to an understanding of weather and climate. The oceans play three important roles in determining weather and climate: (1) they are a source of atmospheric water vapor; (2) they exchange energy with the atmosphere; and (3) they transfer heat poleward.

Each year approximately 396,000 cubic kilometers (almost 95,000 cubic miles) of water vapor circulate through the atmosphere through the hydrologic cycle. Most of the water that recycles into the atmosphere comes from the oceans (see Chapter 1). About 334,000 cubic kilometers (almost 80,000 cubic miles) of water evaporate from the oceans every year. At any given time, the atmosphere contains approximately 15,300 cubic kilometers (about 3500 cubic miles) of water. It takes approximately 2 weeks for all of the water in the atmosphere to be recycled. The oceans provide the majority of water needed to form precipitation.

Exchanges of heat and moisture occur at the interface between the atmosphere and the ocean. FIGURE 8-1 shows the net energy gains and losses at this interface for Northern Hemisphere winter and summer. On average, the ocean gains energy during summer and loses energy to the atmosphere in winter. Therefore, the oceans normally warm in the summer and cool during the winter. Energy budget maps like Figure 8-1 are a measure of the average interaction between the atmosphere and ocean. Maximum exchanges of energy occur in the Northern Hemisphere winter to the east of North America and Asia. Warm ocean currents flow northward in these regions (see Chapter 3) and supply energy to winter storms that are leaving these continents heading eastward. As we will learn later in this chapter, warm ocean waters also supply the energy that fuels hurricanes.

The rate of heat and moisture transfer, as discussed in Chapter 2, depends on the temperature difference between the air and the water, as well as the strength of the winds. Warm sea surface temperatures (SSTs) and strong winds are favorable for large heat exchanges between the ocean and atmosphere. It is therefore important to understand where the warm waters of the oceans are and to know why they are there. We learned in Chapter 3 that warm currents of the middle and high latitudes border the eastern coasts of some land masses. This chapter discusses the origins of these cool and warm currents.

Ocean currents are similar to jet streams in the atmosphere, but the currents are much slower than wind speeds in a jet stream. Like the winds, the ocean currents transport heat from the tropics to higher latitudes. FIGURE 8-2 shows the poleward transport of heat in the Northern Hemisphere by the atmosphere and oceans required to balance the radiation budget, as discussed in Figure 2-21 (reproduced as an inset of Figure 8-2). The poleward transport of heat by the atmosphere is discussed in Chapter 7. At about 30° latitude, the atmosphere and ocean each transport about the same amount of heat. Equatorward of 30°, the ocean transfers

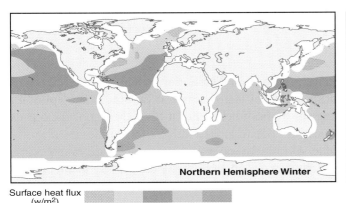
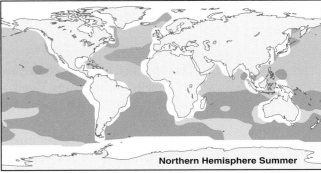

Surface heat flux (w/m²)
−150 −100 −50 50 100

FIGURE 8-1 The energy gains and losses of Earth's oceans. Positive values indicate energy gains by the ocean, while negative values represent energy losses by the ocean.

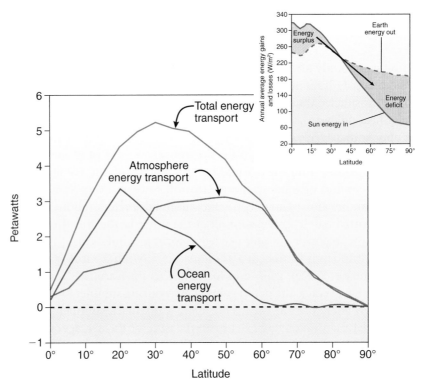

FIGURE 8-2 The yearly mean transport of energy by the atmosphere and ocean as a function of latitude for the Northern Hemisphere. The inset indicates the need for a poleward energy transport in order to maintain energy balance.

the majority of heat required to maintain balance. The atmospheric transport has a broad peak at around 30° to 60° north latitude. The oceans and the atmosphere redistribute the Sun's energy, transporting energy toward the poles. This means that we cannot study the atmosphere without taking the oceans into account. For example, any changes in the ocean circulation patterns may change the heat transport by the oceans, which in turn may change heat transport by the atmosphere and, thus, change the world's climate.

The teamwork between the atmosphere and oceans is a two-way street. Changes in the ocean and its circulation patterns affect the atmosphere and vice versa. The Sun is the primary source of energy that warms the oceans, and atmospheric conditions determine the amount of solar energy reaching the ocean. Cloud-free conditions allow more solar energy to enter the oceans, while cloudy conditions reduce the amount of solar energy reaching the surface. Winds also play an important role in determining ocean temperatures. The solar energy that is absorbed at the surface of the water can be mixed to deeper layers by the winds, which stir the upper layer of the ocean. Next we investigate the temperature patterns of the world's oceans.

Ocean Temperature

Chapter 1 described the atmosphere in terms of vertical distribution of temperature. We can do the same with the oceans, based on observations of temperatures at various ocean depths at a given location. The Sun warms the atmosphere from below, but it heats the top of the oceans; therefore, the vertical temperature profile of the oceans looks a little like an upside-down version of the atmosphere's temperature profile, with warm water at the top and colder water below, as seen in **FIGURE 8-3**. Based on measurements of temperature, we can classify the ocean into three vertical zones.

The characteristic feature of the top 100 meters of the ocean at a given location is a constant temperature. The uniform layer is called the **surface zone** or **mixed layer.** Wind-driven waves and currents mix this layer so that the temperature is relatively uniform. Only about 2% of the world's ocean waters are within the surface zone. The bottom layer, or **deep zone**, lies below about 1000 meters and is composed of cold water at a temperature between 1° and 3° C (33.8° F–37.5° F). The temperature in the deep zone is uniform. The transition zone between the surface and deep layers is the **thermocline** (*therme* means "heat" and *clinare* means "to lean"). In the thermocline, temperature decreases rapidly with depth down to a depth of about 1000 meters.

Figure 8-3 is a typical temperature distribution of the ocean with depth. This temperature variation with depth is a function of latitude. FIGURE 8-4 shows the temperature profile with depth for oceans typical of tropical, midlatitude, and polar regions. The deep zone temperatures are similar for all three regions. The tropical waters have the warmest surface zone temperatures, and the polar regions have the coldest. This is because the polar regions, on average, receive less solar energy. The tropical and midlatitude regions have the steepest thermocline. For our purposes, the important point is that the world's oceans, even those that are warm at the surface, are cold beneath the surface.

SST Distributions

Interactions between the atmosphere and ocean occur at the surface and result in the transfer of heat and moisture. The distribution of SSTs is, therefore, important in meteorology. Indeed, as we shall see later in this chapter, the SST determines where tropical cyclones can develop and grow.

FIGURE 8-5 shows the SST distribution in the world's oceans, with red regions representing warm water and blue regions cold water. We can draw the following general conclusions from this figure:

1. Western coasts in the subtropics and middle latitudes are bordered by cool water.
2. Western coasts in tropical latitudes are bordered by warm water.
3. Eastern coasts in the middle latitudes are bordered by warm water.
4. Eastern coasts in polar regions are bordered by cool water.

To explain these temperature distribution patterns, we have to understand the circulation of the world's oceans.

◾ Ocean Currents

An ocean current is a massive, ordered pattern of water flow. The world's ocean currents are shown in FIGURE 8-6. The global-scale wind patterns such as the trade winds and the midlatitude westerlies (Chapter 7) blow steadily over large regions of the ocean and push the ocean surface in the same general direction as the wind. As a result, the ocean current patterns in Figure 8-6 resemble the surface wind patterns discussed in Chapter 7.

The swirling ocean gyres that lie underneath the subtropical anticyclones are a striking example of how the atmosphere affects the oceans. A **gyre** is an ocean circulation that forms a closed loop and stretches across an entire ocean basin. These immense ocean circulations spin in the same direction as the anticyclones in the atmosphere, causing warm water to flow poleward along the eastern coasts in strong currents such as the Gulf Stream (FIGURE 8-7). In addition, the winds of the subtropical highs cause cold water to flow equatorward near the western coasts, such as the Canary Current near Africa and the Peru Current off the Pacific coast of South America. Thus, the effect of wind on the oceans helps

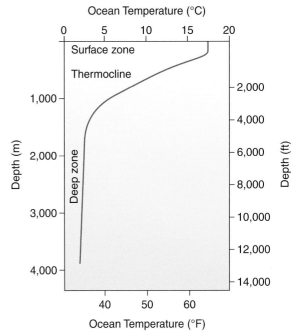

FIGURE 8-3 Vertical profile of temperature in the oceans, indicating the surface zone, the thermocline, and the deep zone.

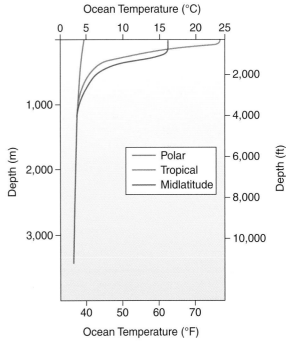

FIGURE 8-4 Although all three zones—the deep zone, the thermocline, and the surface zone—appear at all ocean regions, their depths and the steepness of the thermocline are functions of latitude.

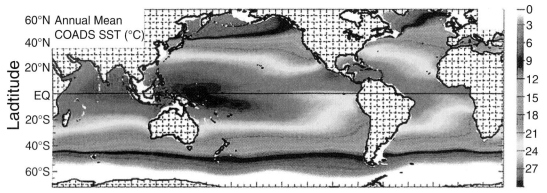

FIGURE 8-5 SST distributions across the globe. Western coasts in the subtropics and middle latitudes are bordered by cool water, and western coasts in the tropical latitudes are bordered by warm water. Eastern coasts in the middle latitudes are bordered by warm water currents, and eastern coasts in polar regions are bordered by cool water. (Reproduced from Kara, Wallcraft, Hurlburt, *J. Atmos. Oceanic Technol.*, 20 [2003]: 1616–1632.)

explain many of the features of the temperature distribution in Figure 8-5. The fact that warm water flows poleward and cold water flows equatorward in these gyres also helps explain how oceans transport heat poleward, as we saw in Figure 8-2.

We can explain even more about ocean circulations by considering the balance of forces on an ocean current, just as we did for winds in Chapter 6. In **FIGURE 8-8**, the ocean currents that make up the North Atlantic gyre are shown with the winds that blow over them. These currents actually flow somewhat to the right of the wind direction. Why? As we learned in

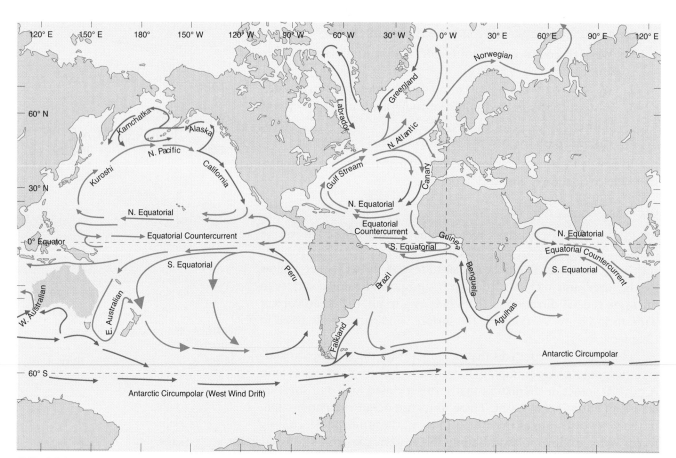

FIGURE 8-6 The major ocean currents. Warm currents are represented by red arrows and cold currents by blue arrows. In general, warm waters tend to flow poleward. Cold currents, except the Antarctic Circumpolar Current, tend to flow toward the equator.

Chapter 6, moving air is subject to the Coriolis force that pushes to the right in the Northern Hemisphere and to the left in the Southern Hemisphere. The same holds true for moving ocean water. The frictional force created by the wind pushes the ocean in the direction of the wind, but in combination with the Coriolis force, it leads to a current that is at an angle to the wind direction. Surface currents therefore flow at an angle of approximately 45° to the wind, to the right of the wind in the Northern Hemisphere and to the left in the Southern Hemisphere.

These surface currents also have a major impact on ocean flow beneath the surface, causing two key atmosphere–ocean interactions known as the **Ekman spiral** and **Ekman transport** in honor of their discoverer, who investigated why ice in the ocean drifted at an angle to the wind.

To explain the Ekman spiral, let's divide the ocean into slabs of water as in FIGURE 8-9, starting with the top layer and working our way down into the ocean. As noted, the combination of friction caused by the wind and the Coriolis force leads to a surface ocean current in the top layer that flows at about 45° to the right of the wind in the Northern Hemisphere. This is not the end of the story, however. The slab of water underneath the top layer feels the surface current as friction. This slab moves initially in the direction of the surface current, but, it too is then affected by the Coriolis force, and its resultant motion is to the right of the surface current. The third slab feels the motion of the second slab as friction, and it too moves and is affected by the Coriolis force, resulting in a motion that is to the right of the slab above it. This clockwise turning of the ocean currents proceeds to a depth of about 100 meters, with the current gradually reducing in strength from the surface to this depth.

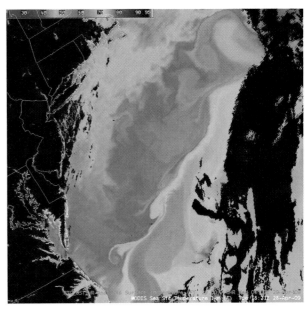

FIGURE 8-7 A high-resolution infrared satellite image of the Gulf Stream (dark orange and red area; see the legend in upper left corner). The undulations of this current, known as the Gulf Stream eddies, are reminiscent of cut off-lows and highs that form in association with the atmosphere jet stream (see Chapter 7).

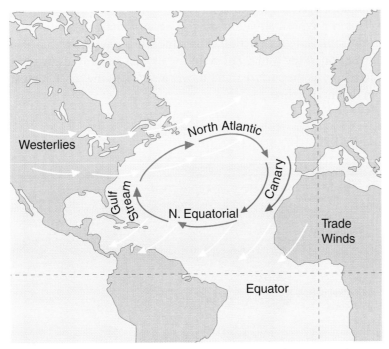

FIGURE 8-8 The North Atlantic gyre comprises four currents: the North Equatorial Current, the Gulf Stream, the North Atlantic Current, and the Canary Current. These surface currents are driven by the trade winds and the westerlies but are deflected to the right of the wind direction by the Coriolis force.

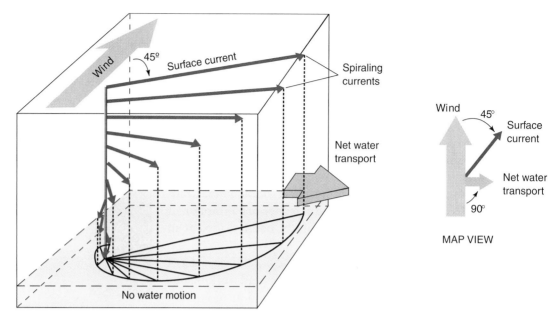

EKMAN SPIRAL IN THE NORTHERN HEMISPHERE

FIGURE 8-9 The Ekman spiral in the water is a change of ocean current direction and speed with depth. The body of water can be thought of as a set of layers. The top layer is driven by surface wind and each layer below is driven by frictional drag of the layer above. The direction that the water in each layer flows rotates to the right in the Northern Hemisphere.

The cumulative effect of the Ekman spiral, from the surface to 100 meters, is to move water at a 90° angle to the direction of the wind, as shown in Figure 8-9. This is called Ekman transport and occurs to the right of the wind in the Northern Hemisphere and to the left of the wind in the Southern Hemisphere. This horizontal movement of surface layer water generates an important vertical motion in the oceans called **upwelling**.

To understand why upwelling occurs, consider a wind blowing parallel to the shoreline (**FIGURE 8-10**). This is a typical situation off the western coast of continents because of the presence of the subtropical highs. Friction causes the surface waters to move and the Coriolis force deflects the motion of the water. The resulting Ekman spiral leads to a transport of water away from the shoreline. As the surface water moves away from the coast, it must be replaced by water rising from the thermocline.

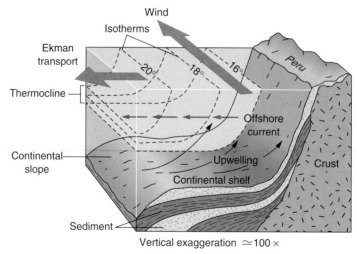

Vertical exaggeration ≃100×

FIGURE 8-10 Surface winds along coastlines cause upwelling when they create ocean currents that move offshore.

Upwelling has a profound effect on ocean temperatures and life along coastlines. As we saw earlier, ocean temperatures decrease with depth. Therefore, upwelling brings cold water to the surface, and this water is rich in nutrients. These nutrients are an important food source for marine organisms living near the surface. These organisms, in turn, are sources of food for fish and birds, and millions of people across the world depend upon the abundance of life in and near regions of upwelling. When upwelling ceases because of a change in wind patterns, the result can be catastrophic for coastal regions. One example of this, which turns out to have implications for weather and climate worldwide, is El Niño.

El Niño

The term **El Niño**, or "the Christ Child," was first used by the fishermen along the coast of Ecuador and Peru, where the economy of the coastal communities depends on the normally cold, nutrient-rich waters. An invasion of a warm southward ocean current into this region would disrupt fishing and would typically occur around Christmas. The warm El Niño waters disrupt the marine food chain in the region and the local economy suffers from the loss of fish. Today, the term El Niño also refers to the weather effects associated with the periodic warming of the equatorial Pacific Ocean between South America and the Date Line. This warming is a natural variation of the ocean and atmosphere and is not caused by human activities. El Niño is an excellent example of the interaction between the atmosphere and the ocean and how these interactions affect climate.

The waters off the western coast of South America typically experience upwelling because the southerly winds associated with the subtropical high cause water to be transported offshore, to the left of the wind. (In the Southern Hemisphere, the Coriolis force and Ekman transport are to the left.) The upwelling water is cold and rich in nutrients, even in the summer. Every few years, the upwelling slows or ceases in this region, and the ocean turns markedly warmer, sometimes 4° C (7° F) above normal. **FIGURE 8-11** reveals that in the past century, El Niño

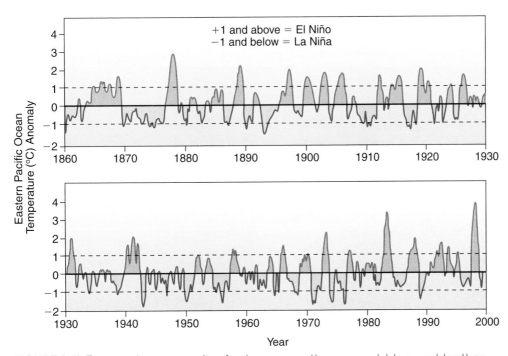

FIGURE 8-11 Temperature anomalies (red = warmer than normal, blue = colder than normal) across the eastern equatorial Pacific Ocean during 1860 to 2009. Red periods above the upper dashed line are El Niño episodes, and blue periods below the lower dashed line are La Niña episodes. (Courtesy of University of Washington, Joint Institute for the Study of the Atmosphere and Ocean [JISAO].)

warmings (in red) have occurred periodically every 2 to 7 years—sometimes less (2–3 years) and sometimes more (8 to 11 years). An El Niño is defined to occur when a warming of at least 0.5° C (0.9° F) averaged over the east-central tropical Pacific Ocean. During a strong El Niño, ocean temperatures can average 2° C to 3.5° C (4° F–6° F) above normal between the date line and the west coast of South America.

Under normal conditions, the SSTs off Peru are cold because of coastal upwelling, and waters in the western equatorial Pacific are warm. SSTs off South America's western coast are 8° C (14° F) cooler than SSTs in the western Pacific. The steady easterly trade winds push the water toward Indonesia, raising the level of the ocean 0.5 meter higher there versus the South American coast. As the water flows westward, it warms as a result of absorption of solar energy and heat exchanges with the atmosphere. Over the western Pacific Ocean, the warm water evaporates and is accompanied by precipitation (FIGURE 8-12A).

An El Niño event is triggered when, for reasons not fully understood, the trade winds weaken or reverse direction and blow from west to east. This allows the large mass of warm water that has piled up near Indonesia to move eastward along the equator until it reaches the coast of South America. During an El Niño event, two distinct changes occur in the equatorial Pacific Ocean: (1) cold coastal waters are replaced by warm waters, and (2) the height of the ocean surface drops over Indonesia and rises in the eastern Pacific, forcing the thermocline lower near South America and preventing upwelling. Satellites can measure both of these features with an active sensor called an altimeter. FIGURE 8-13 shows global deviations in the height of the ocean surface as measured by a NASA satellite. Above-normal heights over the central and eastern equatorial Pacific, as seen in the top portion of this figure, are the smoking gun indicating an El Niño is in progress.

How Does El Niño Affect Global Weather Patterns?

During El Niño, the western Pacific experiences below-normal precipitation as the heavy rains follow the warm water and move eastward toward South America (Figure 8-12B). The shift in the large tropical rain clouds alters the typical pattern of the subtropical jet stream. Air rising in these thunderstorms moves northward and strengthens the subtropical jet over the southern United States (see Figure 7-9 in Chapter 7 for a satellite image showing the connection between El Niño thunderstorms and the subtropical jet). This change in the jet stream allows El Niño to affect the weather and climate of the midlatitudes as well as the tropics. To determine the impact of El Niño on global weather, meteorologists compare average weather conditions with weather conditions experienced during an El Niño year. There have been ten major El Niño events since

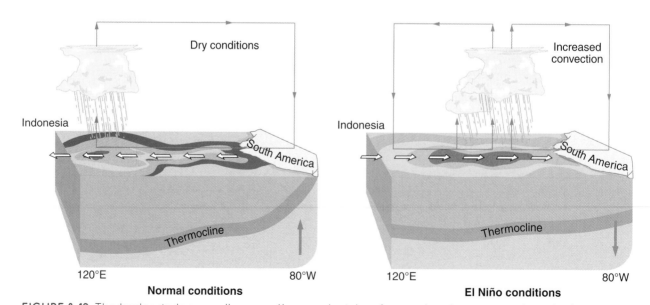

FIGURE 8-12 The trade winds normally cause the equatorial surface waters to move westward, piling up warm water in the western Pacific. During an El Niño event, the winds weaken and the warm water propagates eastward. Thunderstorms follow the movement of warm water into the central Pacific.

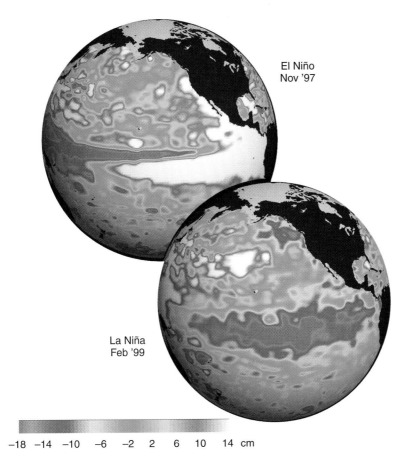

El Niño
Nov '97

La Niña
Feb '99

−18 −14 −10 −6 −2 2 6 10 14 cm

FIGURE 8-13 Satellite observations of changes in sea level height during El Niño (top) and La Niña (bottom) events. During El Niño, tropical waters in the western Pacific Ocean are below normal (blue and purple), while those over the now-warm waters of the eastern tropical Pacific are above normal (red and white). The reverse is true during a La Niña event.

the mid-1950s: 1957–1958, 1965, 1968–1969, 1972–1973, 1976–1977, 1982–1983, 1986–1987, 1991–1992, 1994, and 1997–1998, 2002–2003, and 2006–2007. The 1997 to 1998 El Niño was the strongest recorded in the past century (Figure 8-11). **FIGURE 8-14** compares differences in temperature and precipitation between normal and El Niño years for winter and summer conditions. The impacts of El Niño on climate in temperate latitudes are most evident during winter. A weak polar jet stream forms over eastern Canada, and as a result, a large part of North America is warmer than normal. During winter, the southeast U.S. experiences above-normal precipitation in connection with a strong subtropical jet stream. The increased rainfall across the southern U.S. and in coastal Peru, which is normally a desert, can cause destructive flooding. Drought in the western Pacific has also been associated with devastating fires in Australia and Indonesia. **TABLE 8-1** summarizes the effects of severe El Niños across the globe. Each El Niño event is unique, but its impact on some locations is quite predictable.

Changes in precipitation and temperature patterns caused by El Niño affect snowfall in the U.S. (**FIGURE 8-15**). In the Southwest, there is a slight tendency toward cooler winters and a strong tendency toward wet winters, which makes higher elevation snowpack deeper. In the Pacific Northwest, El Niño winters are warmer and drier than usual so that at a given elevation there is less precipitation and the freezing level is at a higher altitude. The type of precipitation is more likely to be rain, and the accumulation season is shorter. These factors produce a smaller snowpack accumulation by the end of winter in the Pacific Northwest. A significant reduction in total winter snowfall also occurs in the Midwest and New England regions during strong El Niño events. Learn more about this on the text's Web site.

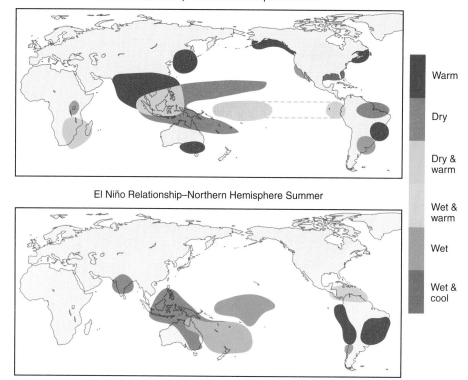

FIGURE 8-14 Changes in global weather patterns associated with an El Niño event. (*Source*: NOAA.)

Legend:
- Warm
- Dry
- Dry & warm
- Wet & warm
- Wet
- Wet & cool

El Niño Relationship–Northern Hemisphere Winter

El Niño Relationship–Northern Hemisphere Summer

TABLE 8-1 Global Impacts of Five Major El Niño Events

Region	1877–1878	1899–1900	1972–1973	1982–1983	1998–1998
India	Century's worst drought	Century's second-worst drought	Century's second-worst drought	Intense drought	[No unusual drought or flooding]
Philippines	Moderate drought	Intense drought	Moderate drought	Century's worst drought	Moderate drought
Australia	Intense drought	Century's worst drought	Intense drought	Intense drought	Intense drought
North China	Century's worst drought	Century's second-worst drought	Intense drought	Intense drought	Moderate drought
Yangzi River, China	Intense flooding	[No unusual drought or flooding]	[No unusual drought or flooding]	Intense flooding	Intense flooding
South Africa	Intense drought	Moderate drought	Intense drought	Intense drought	Intense drought
East Africa	Moderate drought	Moderate drought	Intense drought	Intense drought	Intense drought
Sub–Saharan Africa	Moderate drought	Intense drought	Century's worst drought	Intense drought	[No unusual drought or flooding]
Northeast Brazil	Intense drought	Moderate drought	Intense drought	Intense drought	Intense drought
South Brazil	[No data]	[No data]	[No unusual drought or flooding]	Intense flooding	[No unusual drought or flooding]

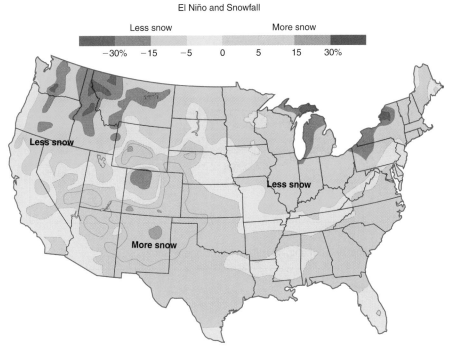

FIGURE 8-15 Percent changes in snowfall in El Niño winters versus the average of all other winters. A decrease in snowfall during an El Niño event can be disastrous to winter recreation areas. (*Impacts of El Nino on Snowfall* by Angel, Jim. Image courtesy of the Midwestern Regional Climate Center, Illinois State Water Survey.)

Atmospheric scientists have discovered that El Niño is associated with a seesaw of atmospheric pressure between the eastern equatorial Pacific and Indo-Australian area. This seesaw in pressure is referred to as the **Southern Oscillation**. Scientists have been studying this oscillation since the 1890s. Generally, when pressure is high over the Pacific Ocean, it tends to be low over the eastern Indian Ocean, and vice versa. The Southern Oscillation is monitored by measuring sea-level pressure at Tahiti in the east and at Darwin, Australia, in the west. The difference in the pressure at these locations is called the **Southern Oscillation Index** (**SOI**). With a high positive SOI, surface pressure is unusually low in the western Pacific and high in the eastern Pacific, the trade winds are strong, and rainfall in the warm western tropical Pacific is plentiful.

The Southern Oscillation is closely linked to El Niño. High negative values of the SOI represent an El Niño, or "warm event" (**FIGURE 8-16**; compare with Figure 8-11). El Niño and the Southern Oscillation often occur together, but they can also happen separately. When an El Niño and Southern Oscillation occur together, the event is referred to as **ENSO**.

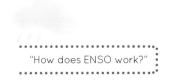

"How does ENSO work?"

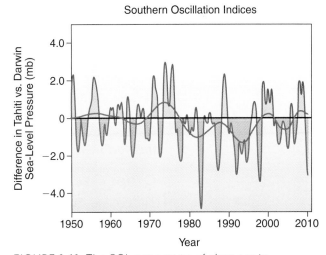

FIGURE 8-16 The SOI, a measure of changes in atmospheric pressure patterns across the equatorial Pacific, for the years 1950 to 2010. The thick, black line is a long-term average. Negative values of the SOI often occur during El Niño events (compare with Figure 8-11). (Copyright, University Corporation for Atmospheric Research, NCAR CGD Fig. 1 [in] Trenberth, K. E., 1997: The definition of El Niño. *Bull. Amer. Met. Soc.*, 78, 2771–2777, updated courtesy of Kevin Trenberth.)

La Niña

La Niña is defined as cooler than normal SSTs in the eastern tropical Pacific Ocean (Figure 8-13). La Niña is the counterpart of El Niño. In a La Niña event, the SSTs in the equatorial eastern Pacific drop well below normal levels. Intense trade winds move warm surface waters of the Pacific westward, while increasing cold water upwelling in the eastern equatorial Pacific. La Niña years often, but certainly not always, follow El Niño years. As

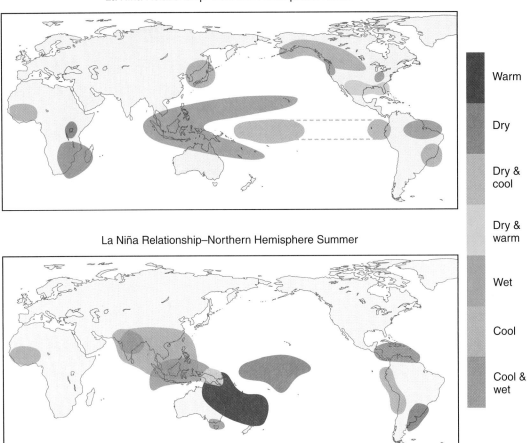

La Niña Relationship–Northern Hemisphere Winter

La Niña Relationship–Northern Hemisphere Summer

Warm

Dry

Dry & cool

Dry & warm

Wet

Cool

Cool & wet

FIGURE 8-17 Changes in global weather patterns associated with a La Niña event. (*Source*: NOAA.)

with El Niño, La Niña conditions typically last for 9 to 12 months but on occasion may persist for as long as 2 years.

Both El Niño and La Niña impact the atmosphere globally. In some locations, especially in the tropics, La Niña produces weather departures opposite of those from El Niño (**FIGURE 8-17**). In the U.S., weather is drier than normal during La Niña in the southwest U.S. in late summer through the subsequent winter. Drier than normal conditions also typically occur in the Central Plains during autumn and in the Southeast during winter. With a well-established La Niña, the Pacific Northwest is wetter than normal in the late fall and early winter. La Niña winters are also on average warmer than normal in the Southeast and colder than normal in the Northwest.

Other Oscillations

The El Niño–Southern Oscillation is a primary way in which the atmosphere and oceans interact with each other. In the past decade, scientists have uncovered other oscillations that can affect weather and climate. One is the **Pacific Decadal Oscillation** (**PDO**), so named because it is a seesaw of atmospheric pressure and SSTs (like ENSO) that occurs over the North Pacific over periods of 20 to 30 years. During a "positive," or "warm," phase, the west Pacific becomes cool and part of the eastern ocean warms; during a "negative" or "cool" phase, the opposite pattern occurs. The North American climate changes associated with the PDO are broadly similar to those connected with El Niño and La Niña, although generally not as extreme. Furthermore, the PDO and El Niño affect each other. When the PDO causes warm equatorial Pacific waters, El Niños can be more severe; this was the case during the 1980s and 1990s. There is no widely accepted theory explaining the cause of the PDO.

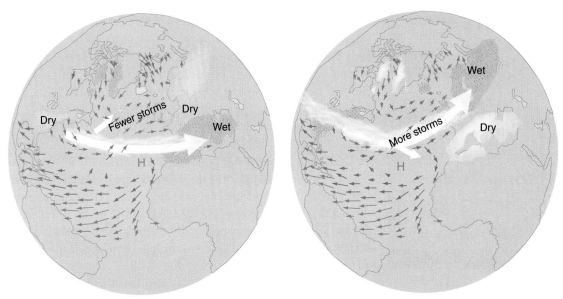

FIGURE 8-18 Surface atmospheric pressure (H and L), SST (shading), and wind (arrows) anomalies associated with the NAO. The positive phase of the NAO is shown to the left, and the negative phase is on the right. (Reproduced from www.ldeo.columbia.edu/NAO, Courtesy of Martin Visbeck.)

The **North Atlantic Oscillation (NAO)** is an oscillation in which the atmospheric pressure seesaws between the polar low near the North Pole and the subtropical high over the Atlantic Ocean, similar to the Northern Hemisphere index cycle that we discussed in Chapter 7. This seesaw affects Arctic Ocean temperatures as well as the strength and location of the jet stream. The positive phase of the NAO (**FIGURE 8-18**) is associated with a stronger than usual, more northerly wintertime jet stream over the Atlantic. In turn, the position and strength of the jet stream cause milder, wetter than normal winter conditions over the eastern U.S. and northern Europe and unusually dry conditions in southern Europe and northern Africa. The negative phase of the NAO, in contrast, leads to a southward-plunging polar jet stream and colder than normal, snowy winters over the eastern U.S. It is not a coincidence that the NAO was in a strongly negative phase during the 2009 to 2010 winter (see Figure 7-17).

For reasons that are still not understood, the NAO was in the positive phase almost continuously during the 1980s and 1990s. This oscillation demonstrates how meteorology cannot be fully understood without studying the interactions between the atmosphere and the oceans. A close relative of the NAO is the **Arctic oscillation (AO)**, which is an index of the sea-level pressure variations north of 20°N latitude.

TROPICAL CYCLONES: HURRICANES AND TYPHOONS

■ What Are They?

The Earth's atmosphere and oceans can interact in all sorts of ways. We have already seen that large-scale oscillations of weather and climate occur when the vast oceans and the air above them "talk" to each other over periods of several years. A different kind of weather feature develops when certain regions of the tropical oceans interact with the atmosphere during the summer and fall of each year. From space, they look like large circular swirls of clouds (**FIGURE 8-19**). They tend to be several hundred kilometers in diameter. These swirls are clearly much bigger than an individual thunderstorm but are also much smaller than the global circulations induced by El Niño.

Because of the location of their birth and the pattern of their clouds, these swirls are given the generic name **tropical cyclones**. In the tropical regions of North and Central America,

FIGURE 8-19 GOES-12 visible satellite image of Hurricane Isabel on 11 September 2003 at 11:45 GMT over the Atlantic Ocean. The horizontal and vertical lines are latitude and longitude, respectively.

the most powerful of them are called **hurricanes.** Residents of the western Pacific call them **typhoons.** In most other parts of the world, such as the Indian Ocean, they are simply called "cyclones." In an average year, about 5 or 6 hurricanes form in the Atlantic and the Gulf of Mexico, 9 form in the eastern Pacific off of Mexico, and 16 typhoons form in the western Pacific Ocean.

Regardless of their name, these storms are the most highly organized and destructive weather patterns on Earth. A hurricane hit Galveston, Texas, in 1900 and killed approximately 8000 people. A cyclone ravaged Bangladesh in 1970 and killed 300,000 people. In 1992, Hurricane Andrew blew through south Florida, causing about $25 billion in damage (**FIGURE 8-20**). And in 2005, Hurricane Katrina killed over 1300 people and inflicted more than $100 billion in damage on the central Gulf Coast. It became the deadliest U.S. hurricane in nearly 80 years and the costliest natural disaster of any type in American history. **TABLE 8-2** lists the most damaging hurricanes to hit the U.S., including many mentioned in this chapter. To save lives and property, we need to learn more about these remarkable storms.

What Do They Look Like?

As we learned in Chapter 6, a cyclone is a center of low pressure. Weather satellite photographs (**FIGURE 8-21**) indicate that the innermost part of a strong tropical cyclone's center is almost entirely clear of clouds. This region is known as the **eye** of the storm. The eye can be as small as 8 kilometers (5 miles) across, but can also be ten times that size or more. Immediately surrounding the eye is a narrow, circular, rotating region of intense thunderstorms called the **eye wall**. The eye wall winds thrust up as much as a million tons of air every second. **FIGURE 8-22** shows you the cumulonimbus clouds of the eye wall from the perspective of an airplane flying inside the eye itself!

Perhaps the most remarkable aspect of the hurricane is the near-perfect circular pattern of the eye and the surrounding cloudiness, which is especially true for the most intense tropical cyclones. The whirling circles within circles of a strong hurricane are evidence of a rare combination of power and coordination. Next, we learn where the power of tropical cyclones originates, which is intimately connected to the areas where they develop.

FIGURE 8-20 Wind damage caused by hurricane Andrew just south of Miami, Florida, in August 1992.

TABLE 8-2 The Most Damaging Tropical Cyclones to Affect the United States, 1900–2009

Rank	Tropical Cyclone (Unnamed Before 1950)	Year	Minimum Pressure at Landfall	Saffir–Simpson Category at Landfall	Deaths (U.S. Only)	U.S. Damage (Million 2003 U.S. Dollars)
1	Katrina (Louisiana, Mississippi, Alabama)	2005	920	3	1,500	100,000
2	"Great Miami" (southeast Florida, Alabama)	1926	935	4	243	98,051
3	Ike (Louisiana, Texas)	2008	940	2	114	29,600
4	Andrew (southeast Florida, Louisiana)	1992	922	5	58	44,878
5	"Galveston" (north Texas)	1900	931	4	8,000	36,096
6	North Texas	1915	945	4	275	30,585
7	"New England" (New York, Rhode Island)	1938	946	3	600	22,549
8	Southwest Florida	1944	962	3	300	22,070
9	"Lake Okeechobee" (southeast Florida)	1928	929	4	1,836	18,708
10	Betsy (southeast Florida, Louisiana)	1965	948	3	75	16,863
11	Donna (Florida, eastern U.S.)	1960	930	4	50	16,339
12	Camille (Mississippi, Louisiana, Virginia)	1969	909	5	256	14,870
13	Agnes (northwest Florida, northeast U.S.)	1972	980	1	122	14,515
14	Ivan (Alabama, northwest Florida)	2004	946	3	57	14,200
15	Wilma (south Florida)	2005	950	3	35	14,000
16	Charley (southwest Florida, South Carolina)	2004	941	4	30	14,000
17	Diane (northeast U.S.)	1955	987	1	184	13,875
18	Hugo (South Carolina)	1989	934	4	26	12,718
19	Carol (northeast U.S.)	1954	960	3	60	12,291
20	Southeast Florida, Louisiana, Alabama	1947	940	4	51	11,266

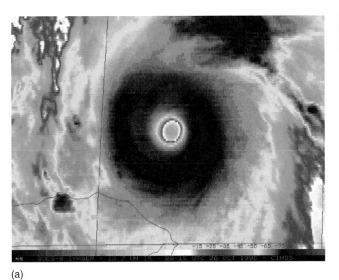

(a)

(b)

FIGURE 8-21 Infrared satellite image of hurricane Mitch in the western Caribbean Sea on October 26, 1998, showing (a) a close-up view of the hurricane's eye and (b) a view of the cyclone (lower left) and other weather patterns at the time, note the clear eyes, the fierce eye-wall ring of thunderstorms, and the difference in the hurricane's appearance compared to other tropical and extratropical weather systems. Mitch later moved inland over Central America; its flooding rains there caused one of the worst disasters in Western Hemisphere history, killing tens of thousands of people.

FIGURE 8-22 Photograph of the eye wall of Typhoon Vera taken from a "hurricane hunter" aircraft.

How and Where Do They Form?

Tropical cyclones are essentially large weather engines, and any engine needs energy to run. In Chapter 7 we learned that the unequal heating of different parts of the Earth by the Sun drives the huge engine of global weather and climate. But the tropics are another matter, as temperature contrasts there are usually small. Warm air, warm water—it sounds more like an advertisement for a vacation cruise than a prelude to stormy weather.

The "secret" energy source of a tropical cyclone is the large latent heat of water, which we first mentioned in Chapter 2. Air over the tropical oceans is drier than you might think. This is because the trade winds are partly made up of air that has descended from above—part of the downward branch of the Hadley Cell we studied in the last chapter. Although both the air and water may be warm and calm, evaporation can take place because the air is not at 100% relative humidity. Silently and invisibly, water changes from liquid to vapor and enters the atmosphere. The energy required to make this change comes from the Sun, and this energy is lying in wait—latent—ready to be released when the vapor is condensed into liquid again. This happens in rising air in a cloud or thunderstorm.

This process alone, however, is not enough to power a tropical cyclone. A tropical cyclone adds fuel to its own fire by drawing surface air toward its low-pressure center (see Chapter 6). The tight pressure gradient nearer the center means that the winds grow stronger as the air approaches the eye. The faster the wind blows, the more evaporation takes place (this is why you blow-dry wet hair or hands instead of merely warming them). Increased evaporation means more water vapor in the air and more energy ready to be liberated in the tropical cyclone's thunderstorms as water vapor condenses. In short, evaporation and condensation of water are the keys to understanding the power of tropical cyclones.

How strong is the engine that powers a tropical cyclone? The energy released by condensation in a single day in an average hurricane is at least 200 times the entire world's electrical energy production capacity! Part of this energy is expended reducing the central pressure of the storm and strengthening the winds. Tropical cyclones can create the lowest sea-level pressures observed on Earth, the record being 870 mb in the Pacific, about the same surface pressure as at Denver, Colorado, the "Mile-High City." You can experiment with the relationship between pressure gradients and high winds using the Hurricane Winds learning applet on the text's Web site.

"Hurricane Winds": use the simple model of a hurricane to change a hurricane's central pressure and see the impact on its winds and temperatures.

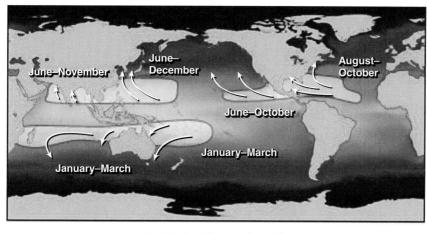

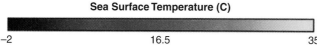

Sea Surface Temperature (C)

-2   16.5   35

FIGURE 8-23 The times and locations of tropical cyclone development across the world, superimposed on global SST distributions. (Courtesy of NASA.)

FIGURE 8-23 illustrates when and where tropical cyclones develop. In regions with SSTs below 26.5° C (80° F), tropical cyclones cannot form. This is the fundamental reason why they are tropical cyclones. But why only in the tropics? As we learned in Chapter 4, evaporation rates increase very quickly as temperatures increase, and evaporated water is the fuel for these storms. A few degrees' difference in SSTs can be the difference between no storm, a run-of-the-mill tropical cyclone, and a record-setting hurricane. As a result of this requirement, tropical cyclone development is not only limited to the tropics, but is also usually limited to the summer and fall seasons, as shown in Figure 8-23. (Remember that January is summertime in the Southern Hemisphere!)

Some low-latitude regions are not conducive to tropical cyclone development. The far eastern boundaries of the tropical oceans are home to cool ocean currents, as we learned earlier in this chapter. The ocean temperatures are below 26.5° C, and therefore, tropical cyclones do not form there.

Tropical cyclones also do not form within about five-degrees latitude of the equator. The SSTs there are definitely warm enough for tropical cyclones. But tropical cyclones never form near the equator. Why is this so? As we learned in Chapter 6, the Coriolis force is negligible near the equator. This means that air tends to flow straight into low-pressure centers there. The low cannot intensify because it "fills up" before the pressure can drop very much.

How Are They Structured?

Now that we know the power source for tropical cyclones, we can explore their remarkably coordinated structure in more detail. The pattern of winds in a tropical cyclone is fascinating, and we can explain it fairly easily. **FIGURE 8-24** is a time series of wind speeds recorded when powerful Hurricane Celia passed directly over a weather station in Texas. The winds of a tropical cyclone are strong and consistently strong for many hours. The bull's-eye of isobars around the eye of the tropical cyclone ensures that it is a big blow; as we learned in Chapter 6, a strong pressure gradient means a strong wind.

The opposite is true inside the eye of the tropical cyclone. As the eye passes over a location, it causes a nearly calm "halftime" between the front and back sides of the storm. Why? The air spiraling into a strong tropical cyclone rises in the eye wall, and it is there that the strong pressure gradient ceases. Instead, air sinks gently from above. We learned in earlier chapters that sinking air warms adiabatically and is very stable. This explains why the eye in a strong storm is cloudless (except for some harmless cumulus and cirrus clouds), calm, and warm.

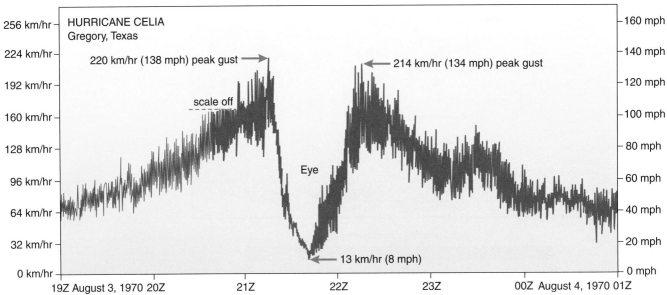

FIGURE 8-24 Wind speeds recorded near Corpus Christi, Texas, during the passage of Hurricane Celia directly over the anemometer on August 3, 1970. (Reproduced from Simpson, Robert H. and Riehl, Herbert. *The Hurricane and Its Impact.* Copyright © 1981, Reprinted by Permission of Louisiana State University Press.)

Sinking air suggests an anticyclone (see Chapter 6). Indeed, one of the hallmarks of a strong tropical cyclone is an intense low at the surface and a high-pressure center or ridge near the tropopause. Generally, the stronger the cyclone, the stronger the upper-level ridge or high. This high is caused in part by the cyclone itself. Without it, the cyclone cannot remove air from the eye quickly enough to keep its pressure low. Therefore, the nearby presence of an upper-level low or trough, or even a high-speed jet stream, is enough to weaken or destroy a tropical cyclone. The weakening in these cases is attributed to **wind shear**, meaning that the winds are changing fast enough in the vertical to disrupt the cyclone's own high-building process at those levels. FIGURE 8-25 is a schematic view of how the winds in a hurricane differ in the lower and upper levels in both strengthening and weakening cases.

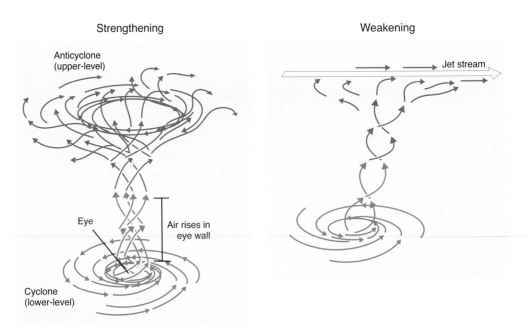

FIGURE 8-25 The lower- and upper-level wind flows in a tropical cyclone in conditions (left) favorable and (right) unfavorable for growth of the storm. (Adapted from Nese, J. and Grenci, L., *A World of Weather: Fundamentals of Meteorology*. Kendall/Hunt, 1998.)

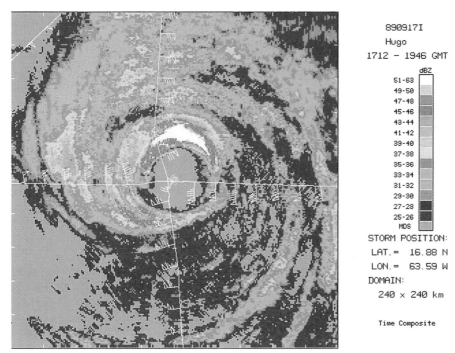

FIGURE 8-26 Airborne radar view of hurricane Hugo at peak intensity over the northeast Caribbean in September 1989. The white lines indicate the flight paths of the "Hurricane Hunter" aircraft (see Box 8-1), which measured total storm winds (shown in white) at various points along the flight paths. The brightest colors indicate the most intense thunderstorms in the eye wall.

The radar image of Hurricane Hugo in **FIGURE 8-26** illustrates two more important features of tropical cyclones. First, the radar-reflecting rain in this hurricane, like most, is not symmetrically distributed as the winds are. Instead, the heaviest rain is contained in spiral **rainbands** that curve into and blend with the eye wall. This characteristic of tropical cyclones, so fundamental that it is incorporated into the weather-map symbol for them, is still not well understood by meteorologists. This is one of many reasons that "hurricane hunters" still fly planes into these storms to gather weather data such as that shown in Figure 8-26; **BOX 8-1** explains the work of hurricane hunters in more detail. **FIGURE 8-27** shows a cross-section through an idealized hurricane based on the discussion above.

The second feature can be seen in the wind measurements from aircraft that are overlaid on the radar image in Figure 8-26 (the white wind barbs). The winds are strongest on the north and east sides of this cyclone. Surprisingly, this is not because the pressure gradient is different on one side of the storm than on the other. It is because the storm's forward motion adds or subtracts from the winds. In the same way, if you throw a baseball out of a moving car, the baseball's velocity is the combination of the velocity of the ball and the velocity of the car. **FIGURE 8-28** shows that the storm's forward speed adds to the wind on the right side of the eye but subtracts from the wind on the left side of the eye. As a result, the difference in wind on the right versus the left side of the eye is approximately twice the cyclone's forward speed.

We now know quite a bit about the organization of a tropical cyclone. The fact that these storms are given names (**BOX 8-2**) suggests that, like people, they are unique and have individual "lives." Next, we examine the typical life cycle of a strong tropical cyclone.

What Are the Different Stages of Their "Lives"?

Tropical cyclones usually begin small, as **tropical disturbances**. These disturbances are not even low-pressure centers, just disorganized clumps of thunderstorms. In the Atlantic Ocean, these disturbances take the form of **easterly waves** in the wind patterns that move along with the easterly jet stream across Africa and into the extreme eastern Atlantic. Easterly waves

Box 8-1 The "Hurricane Hunters"

Why would anyone in their right mind intentionally fly into the middle of a hurricane? During every hurricane season, U.S. Air Force (USAF) and National Oceanographic and Atmospheric Administration (NOAA) pilots and scientists do just that, time and again. Their goal is to obtain up-to-the-minute weather data on storms that cannot be obtained by any other means. They are the "hurricane hunters."

The first exploration of a major hurricane by a "reconnaissance aircraft," as they are called, occurred in 1944 during the Great Atlantic Hurricane. Since 1955, no planes have been lost in the Atlantic while penetrating these deadly storms. (There was one close call during a rough flight into Hurricane Hugo in 1989 when three engines failed and the plane was almost submerged by a high wave!) The pilots carefully choose their altitude (high, away from the roiling ocean) and path (straight through to the eye—no dilly-dallying in the eye wall) to maximize their safety while gathering crucial weather data.

Today, hurricane hunting is a highly coordinated effort between the NHC, USAF, and NOAA. When the NHC needs to know if a tropical disturbance far out in the Atlantic is getting better organized, the USAF pilots fly long missions to search for signs of a closed circulation. They use weather instruments that have been specially installed on their large cargo airplanes. In addition, the planes launch "dropsondes"—the reverse of radiosondes—that drop out of the plane and into the storm from above. These dropsondes provide information about the weather conditions in the very center of the disturbance, below where the planes normally fly. Finally, the pilots and scientists use their eyes to see, literally, if the storm clouds are becoming better organized. This information is relayed to NHC, where meteorologists use it to decide whether a disturbance has achieved tropical depression or tropical storm status.

If a mature hurricane comes closer to the U.S., the NOAA hurricane hunters take over. NOAA's specially instrumented P-3 aircraft (see the photo) include onboard radars. Figure 8-26 is a radar image from one of the NOAA flights into Hugo. The measurements of winds and location of the hurricane's center, as shown on Figure 8-26, are extremely important pieces of information for hurricane forecasters. They are much more accurate than satellite estimates.

For this reason, hurricane hunters will continue to brave the bumpy flights into these storms for the foreseeable future. There is also an added bonus: hurricane hunters often say that a strong hurricane's eye wall, viewed from an airplane circling in its eye (Figure 8-22), is one of the most beautiful and awe-inspiring sights in all of nature.

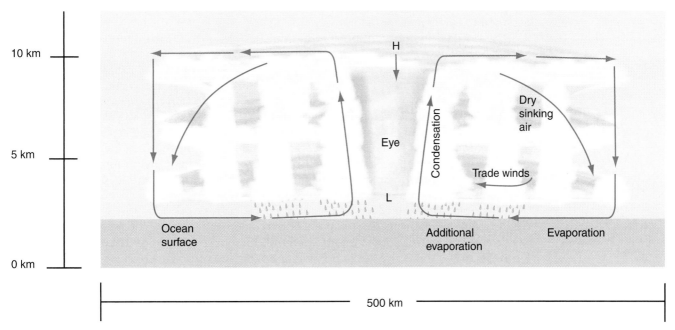

FIGURE 8-27 Cross-sectional view of the processes involved in fueling a hurricane.

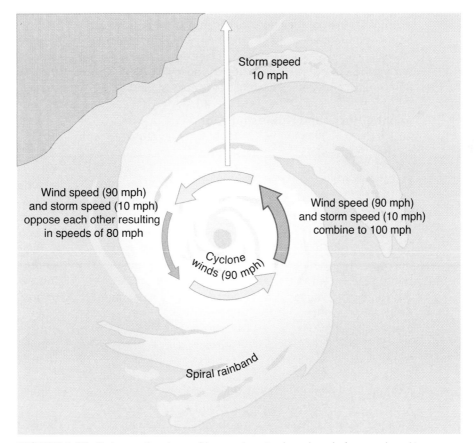

FIGURE 8-28 Schematic view of how a tropical cyclone's forward motion affects the speed of the total wind associated with the storm.

Box 8-2 Naming Hurricanes

Hurricanes come in large and small sizes; some have big eyes, and some have squinty eyes. They are born, grow, and then die. It's not surprising, then, that human names have been assigned to hurricanes for centuries.

The earliest instance of hurricane naming was in the Caribbean islands, where hurricanes were associated with the Roman Catholic saint's holiday when they hit the islands. For example, Hurricane "Santa Ana" struck Puerto Rico on July 26, 1825.

The first known scientific use of hurricane naming arose in the Pacific during World War II. It was an easy and effective way to distinguish one tropical cyclone from another on the weather maps. The system was simple and alphabetical: the name of the first storm of the season would begin with A; the second, B; the third, C; and so on.

Naming hurricanes also gave homesick soldiers a way to recall loved ones. Thus began the practice by predominantly male meteorologists of giving female names to hurricanes. This practice persisted at the NHC in Miami until the late 1970s. One retired NHC meteorologist ironically commented, "It is surely only a coincidence that many names of NHC employees, their wives, and other female relatives appear on the [name] list." This may explain the distinctly Southern flavor of the names of past hurricanes. For example, Hurricane Camille was named after famed NHC forecaster John Hope's daughter Camille.

Beginning in 1979, the list of names for each year's tropical storms and hurricanes was broadened to include male and Spanish and French names. The alphabetical naming system continues, now overseen by the World Meteorological Organization (WMO). The WMO composes different alphabetical lists for different ocean basins (e.g., starting in 2000, northwest Pacific typhoons now have Asian names). The names of severe and/or notable hurricanes are "retired" and taken off the lists. Horror sequels are popular in the movies, but no meteorologist wants to tempt fate by naming a new hurricane Andrew, Ivan, Mitch, or Katrina to cite a few of the recently retired names.

Following is the list of names for Atlantic Ocean, Gulf of Mexico, and Caribbean Sea tropical storms and hurricanes for the years 2010 to 2015. You may be wondering what happens if there are more storms than names. If the English alphabet is exhausted because there are so many named storms, the next names used are the letters of the Greek alphabet. During the record-setting hurricane season of 2005, Greek names were used for the first time ever: Alpha, Beta, Gamma, Delta, Epsilon, and Zeta!

Atlantic Hurricane Names by Year

2011	2012	2013	2014	2015	2016
Arlene	Alberto	Andrea	Arthur	Ana	Alex
Bret	Beryl	Barry	Bertha	Bill	Bonnie
Cindy	Chris	Chantal	Cristobal	Claudette	Colin
Don	Debby	Dorian	Dolly	Danny	Danielle
Emily	Ernesto	Erin	Edouard	Erika	Earl
Franklin	Florence	Fernand	Fay	Fred	Fiona
Gert	Gordon	Gabrielle	Gonzalo	Grace	Gaston
Harvey	Helene	Humberto	Hanna	Henri	Hermine
Irene	Isaac	Ingrid	Isaias	Ida	Igor
Jose	Joyce	Jerry	Josephine	Joaquin	Julia
Katia	Kirk	Karen	Kyle	Kate	Karl
Lee	Leslie	Lorenzo	Laura	Larry	Lisa
Maria	Michael	Melissa	Marco	Mindy	Matthew
Nate	Nadine	Nestor	Nana	Nicholas	Nicole
Ophelia	Oscar	Olga	Omar	Odette	Otto
Philippe	Patty	Pablo	Paulette	Peter	Paula
Rina	Rafael	Rebekah	Rene	Rose	Richard
Sean	Sandy	Sebastien	Sally	Sam	Shary
Tammy	Tony	Tanya	Teddy	Teresa	Tomas
Vince	Valerie	Van	Vicky	Victor	Virginie
Whitney	William	Wendy	Wilfred	Wanda	Walter

(*Source*: NOAA.)

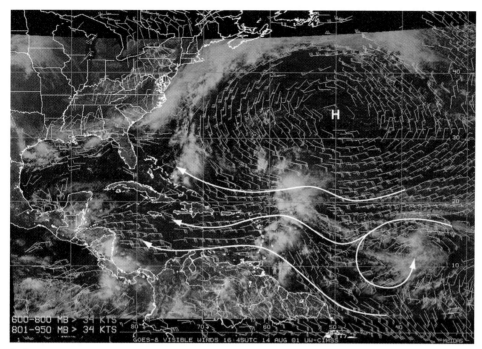

FIGURE 8-29 Visible satellite picture of two tropical disturbances in the Atlantic Ocean on August 14, 2001. Clouds are white, and winds derived from the satellite observations are depicted using the flag/flagpole symbols explained in Chapters 1 and 6. The white arrows indicate the general direction of the wind flow. The disturbance in the lower center of the image is a tropical wave, so named for the hump in the easterly wind pattern. The disturbance at lower right is a tropical depression, which is distinguished from a tropical wave by its closed cyclonic wind circulation. This depression later intensified to become Tropical Storm Chantal and hit the Yucatan Peninsula of Mexico.

are also seen along the Pacific coast of Mexico. **FIGURE 8-29** shows an example of an easterly wave in the Atlantic Ocean east of Puerto Rico. The easterly wave is a region of clouds and rain associated with a wave-like pattern in tropospheric winds that moves from east to west across tropical regions. As the satellite-observed winds and clouds indicate, winds converge into the wave on its east side, leading to clouds and precipitation. Only a few easterly waves develop into tropical cyclones.

The vast majority of tropical disturbances die without growing any stronger. However, about 1 in every 10 develops a weak low-pressure center of roughly 1010 mb. After the low-pressure center is identified and cyclonic rotation is noticed in the winds as in the lower right of Figure 8-29, it is called a **tropical depression.**

Some, but not all, tropical depressions gain more strength. Their central pressures drop below about 1000 mb, the pressure gradient between the center and the edge of the storm increases, and the winds strengthen. When the winds in a depression reach consistent speeds of 35 knots (39 mph or 63 km/hr), a **tropical storm** is "born" and given a name (Box 8-2).

Roughly half of tropical storms intensify further. Their central pressures drop below about 990 mb, and their winds keep strengthening because of the increasing pressure gradient, as shown by the tightening of isobars around the center. When the highest sustained winds in a tropical storm reach 65 knots (74 mph or 119 km/hr), the storm is reclassified as a hurricane or a typhoon depending on its location. As the storm intensifies a little more, an eye may form. In the western Pacific, typhoons with winds above 240 km/hr (150 mph) are called **supertyphoons.**

After 1 or 2 weeks (in rare cases as long as a month), even the strongest hurricanes and typhoons lose their strength, or "dissipate," because they move away from their energy source of warm water. On occasion a hurricane will make the transition from a tropical cyclone to an **extratropical cyclone** (see Chapter 10) with cold air and fronts near its center.

"Hurricane Intensity": view satellite images of hurricanes in different Saffir–Simpson categories.

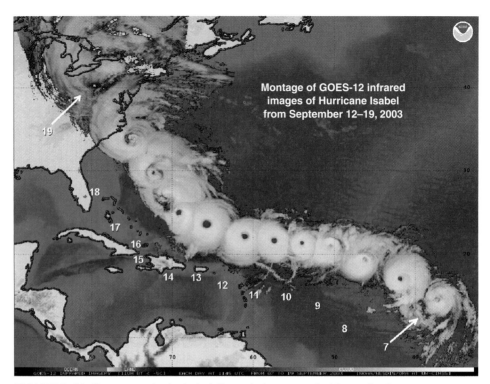

FIGURE 8-30 Hurricane Isabel at different stages in its life cycle during September 2003, as seen in a series of daily infrared satellite images.

What does the mature portion of a hurricane's or a typhoon's life cycle look like? FIGURE 8-30 shows the evolution of Hurricane Isabel over 13 days in September 2003, all superimposed on the same map. Notice how the appearance of the eye changes with time. Meteorologists routinely use the appearance of the eye in satellite images to estimate the strength of a hurricane. Using this figure, we can learn how they do this while studying Isabel's growth and decay.

Figure 8-30 shows an eye forming in the center of Isabel on September 7. Isabel has just attained hurricane status, with the strongest sustained winds in the eye wall around 120 km/hr (75 mph). This matches our expectations; as we learned earlier, eyes generally form in tropical cyclones when they strengthen to hurricane status. By the next day, Isabel strengthens to about 200 km/hr (125 mph), and the eye becomes much better defined.

During the next two days, Isabel maintains its strength but does not intensify. Then on September 11, its eye becomes very distinct and very small. This is often a sign of rapid intensification in hurricanes, and it is the case with Isabel. By 1800 UTC on the 11th, Isabel reaches peak intensity, with winds of nearly 270 km/hr (165 mph) and a very low central pressure of 915 mb!

From September 12 to 15, Isabel's eye widens. A widening eye usually implies some weakening of a hurricane, although it also expands the area of highest winds. In Isabel's case, the winds slowly decrease to about 225 km/hr (140 mph) over this 4-day period.

Judging from the hurricane's appearance after September 15, Isabel rapidly loses its strength as it approaches the U.S. The less-white clouds indicate that Isabel's thunderstorms are weaker and shorter, and only a hint of an eye can be made out. Isabel makes landfall along the North Carolina coast on September 18 with winds of about 170 km/hr (105 mph)—a powerful storm (FIGURE 8-31), but not nearly as dangerous as it was at peak strength a week earlier. In the final image from September 19, the cyclone no longer has an eye or even a clearly circular shape. The remains of Isabel are over the Great Lakes region, where it is turning into an extratropical cyclone.

FIGURE 8-31 Hurricane Isabel ripped this buoy from its mooring and washed it onto Virginia Beach, VA. At least 17 people died and about 4.5 million lost power because of Hurricane Isabel.

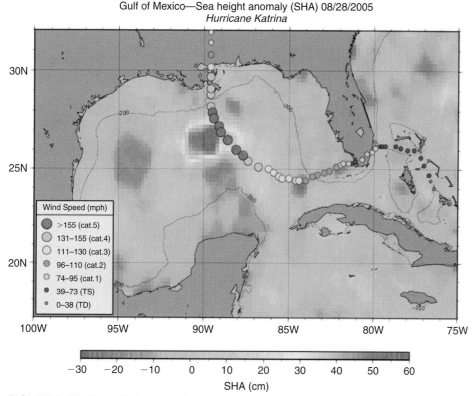

FIGURE 8-32 The relationship between Hurricane Katrina in August 2005 and the SSTs beneath it, as determined by satellite observations of ocean altitude anomalies (largest positive anomalies in red = warmest SSTs). Circles denote Katrina's position and wind speed; shading indicates the ocean altitude anomalies.

Meteorologists have recently discovered that the life cycle of a tropical cyclone over water is strongly affected on a day-to-day basis by the SSTs beneath it. In **FIGURE 8-32**, the evolution of Hurricane Katrina in 2005 is shown as a succession of circles whose color and size indicate the maximum sustained wind speeds at different times. Underlying the dots, satellite observations of ocean altitude anomalies (largest positive anomalies, in red, are where the warmest water is located) are shown for the day before Katrina's fateful landfall along the Gulf Coast. Katrina exploded into a monster hurricane over the central Gulf of Mexico as its center passed nearby unusually warm water. The hurricane weakened somewhat as it left the vicinity of that warm ocean anomaly. But Katrina maintained much of its strength all the way to land-fall because it moved over waters that were significantly warmer than the 26.5° C threshold for cyclone formation and growth. However, Katrina quickly decreased in intensity after landfall because it became cut off from its source of energy, the Gulf of Mexico. The presence of warm water beneath the storm's circulation is a major key to understanding tropical cyclone strength.

"Hurricane Intensity and SST": move hurricane over different SST to see how it impacts hurricane intensification.

What Does a Year's Worth of Tropical Cyclones Look Like? The Record Year of 2005

Now that we have a sense of the life cycle of a single tropical cyclone, let's look at an entire season's worth of cyclones to see how different storms evolve at different places and times in the year. FIGURE 8-33 shows the paths and stages of the tropical storms and hurricanes that affected the Atlantic Ocean, Gulf of Mexico, and Caribbean Sea in 2005. We can use this record-setting year for tropical cyclones to illustrate the typical seasonal cycle of their development.

Hurricane "season" carefully follows the seasonal cycle of SSTs; however, this is not the same as saying "hurricanes form in summer." Because it takes time for the oceans to warm, the Atlantic hurricane season doesn't officially start until June 1 and lasts until November 30. In rare instances, hurricanes can form even in December; the calendar is not the important

Name	Dates	Name	Dates	Name	Dates	Name	Dates
Arlene	Jun 8–13	Harvey	Aug 2–8	Ophelia	Sep 6–17	Alpha	Oct 22–24
Cindy	Jul 3–7	Irene	Aug 4–18	Philippe	Sep 17–24	Beta	Oct 26–31
Dennis	Jul 4–13	Katrina	Aug 23–30	Rita	Sep 18–26	Gamma	Nov 13–20
Emily	Jul 11–21	Lee	Aug 28–Sep 1	Stan	Oct 1–5	Delta	Nov 22–28
Franklin	Jul 21–29	Maria	Sep 1–10	Tammy	Oct 5–6	Epsilon	Nov 29–Dec 8
Gert	Jul 23–25	Nate	Sep 5–10	Wilma	Oct 15–25	Zeta	Dec 30–Jan 6
				Vince	Oct 9–11		

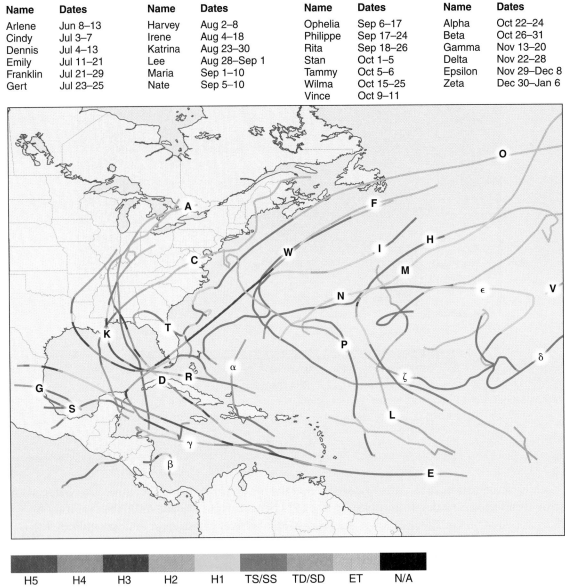

| H5 | H4 | H3 | H2 | H1 | TS/SS | TD/SD | ET | N/A |

FIGURE 8-33 The names, dates, paths, and life-cycle stages of all tropical storms and hurricanes in the Atlantic Ocean basin in 2005, a year that set all-time records for the number of named storms, hurricanes, and major hurricanes. The hurricanes are color coded by Saffir–Simpson category. (Data from NOAA Coastal Services Center: http://csc-s-maps-q.csc.noaa.gov/hurricanes/viewer.html.)

thing—the SSTs and the lack of wind shear are. The peak of hurricane activity in the tropical waters south and southeast of the U.S. is typically in early to mid September.

In Figure 8-33, we see that the 2005 tropical cyclone season began quickly, with Tropical Storm Arlene beginning to form on June 8. A Gulf or western Caribbean birthplace is typical for early-season tropical cyclones. Why? Largely because these relatively confined bodies of water reach the 26.5° C threshold sooner than the open Atlantic.

Hurricane Dennis was the first sign that 2005 would be a record-setting year for major hurricanes. Dennis formed in the southeastern Caribbean Sea on July 5 and quickly became the most intense hurricane ever observed for so early in the season with a minimum central pressure of 930 mb and sustained winds of 241 km/hr (150 mph). It ravaged Haiti and Cuba before landfalling near Pensacola, Florida. Dennis's record was short lived, as Hurricane Emily topped Dennis little more than a week later. Emily packed even higher winds and lower pressure than Dennis along an even longer path from the western Atlantic, through the Caribbean Sea, all the way to coastal Mexico.

Dennis and Emily both resemble early-season versions of the dangerous **Cape Verde hurricanes**, which in a typical year develop during August and September from easterly waves off the western coast of Africa. These storms gather strength as they travel for thousands of miles over open water. Approximately 85% of intense hurricanes in the Atlantic begin as easterly waves.

In 2005, however, only Hurricane Irene was a classic peak-season Cape Verde cyclone, and it did not become one of the strongest hurricanes of the year. Instead, during August and September, two record-breaking hurricanes formed not off Cape Verde, but on the other side of the Atlantic. These storms caused havoc and carnage throughout the central Gulf Coast.

Hurricane Katrina developed just east of Florida on August 23 and, after raking south Florida with minimal-hurricane winds, intensified over the Gulf of Mexico into one of the strongest and largest hurricanes ever observed. Katrina's minimum central pressure of 902 mb and sustained winds of 278 km/hr (175 mph) ranked it as the third most intense hurricane on record in the Atlantic basin, behind only Hurricane Gilbert in 1988 and the Labor Day Hurricane in the Florida Keys in 1935. Even after some weakening, at landfall in southeastern Louisiana Katrina was one of the strongest hurricanes ever to cross the U.S. coastline. Its ferocity, combined with its extremely large size, led to unprecedented devastation in Louisiana, Mississippi, and Alabama. We examine the damage from and the forecasts for Katrina later in the chapter.

Not to be outdone, less than 1 month after Katrina an even stronger hurricane, Rita, formed north of Hispaniola on September 18. Rita followed a similar path to Katrina over the central Gulf of Mexico, and like Katrina, it exploded into a record-setting storm. Rita matched Katrina's winds and its minimum central pressure of 897 mb supplanted Katrina's as the third lowest ever observed in the Atlantic basin.

By October, the Atlantic Ocean is cooling, and wind shear from extratropical cyclones (see Chapter 10) is increasing. As a result, the threat from Atlantic hurricanes usually decreases after September. For this reason, late-season hurricane watchers usually focus on the still-warm Caribbean Sea and Gulf of Mexico. The 2005 season was fairly typical in that regard. Beginning with Hurricane Stan on October 1, most of the year's tropical cyclones formed in or near the Gulf and Caribbean. (It's a little-known fact that Stan caused as many deaths because of rain-fed flooding in Guatemala as Katrina's flooding did in New Orleans.) Hurricane Vince, however, defied all expectations, forming off the coast of Spain and Portugal—the first time a tropical cyclone has ever been observed in that region of the far eastern Atlantic, and in October to boot!

Tropical cyclones that approach the midlatitudes, especially during September and October, often take paths that look like a large letter "C." Meteorologists say that these storms exhibit **recurvature**. Recurving tropical cyclones begin their lives by moving east to west around the edges of the **Bermuda high** that dominates the large-scale wind patterns over the subtropical Atlantic Ocean. (This high is part of the belt of subtropical anticyclones that are associated with sinking air in the Hadley Cell that we studied in the previous chapter.) Then, as the storms turn northward around the western flank of the Bermuda high, they come under the influence of the west-to-east jet-stream wind patterns of the midlatitudes. The jet stream and associated extratropical cyclones help accelerate tropical cyclones ahead of them toward the northeast, disrupting their circulation and weakening them as the cyclones head out over cold Atlantic waters.

The 2005 hurricane season's all-time record-holder, Hurricane Wilma, exemplifies a recurving and accelerating tropical cyclone. Wilma formed on October 15 over the western Caribbean Sea. By the 19th, Wilma's pressure had plummeted to an all-time Atlantic basin record of 882 millibars. Wilma dealt a major blow to resorts along the coast of the Yucatan Peninsula, hovering over the region for several days. Then, as an approaching extratropical weather system and associated upper-level winds affected Wilma's direction of motion, Wilma suddenly turned to the northeast and raced across Florida, causing billions of dollars in damage from Key West to Fort Lauderdale. Wilma continued to accelerate after it passed over south Florida, retaining hurricane status while zooming northeastward at forward speeds over 80 km/hr

(50 mph)! Fortunately, Wilma threatened only shipping interests after its damaging departure from Florida.

Unfortunately, some recurring hurricanes can hit land while moving at high speeds. The Great New England Hurricane of 1938 roared from off the coast of North Carolina to Vermont in only 12 hours, steamrolling across central Long Island at an estimated forward speed of nearly 100 km/hr (60 mph)! Based on our earlier discussion of the effect of forward motion on overall storm winds, this implies that people to the right of the hurricane's eye in Rhode Island experienced winds up to 192 km/hr (120 mph) faster than did luckier residents (including the co-author's father) near New York City who were to the left of the eye.

After Wilma, a Greek alphabet soup of weaker tropical cyclones ensued. The most notable were Epsilon, which attained hurricane status on December 2 and became the longest lasting December hurricane ever, and Tropical Storm Zeta, which formed on December 30 and churned away until January 6, 2006, setting new records for late-season tropical cyclone longevity.

Every hurricane season is different. In 1995, there were 19 tropical cyclones, a modern record until 2005, but most of the storms recurred before striking the U.S. and only one 1995 hurricane, Opal, caused significant damage. In contrast, 2005 is remembered as a terrible year for hurricanes because of the record number of Atlantic tropical cyclones (27), hurricanes (15), and major hurricanes (7), the stunning death toll along the Gulf Coast caused by Katrina, and the extraordinary amount of damage (over $100 billion) in the U.S. caused by a record number of land-falling intense storms. But many Floridians still remember 1992 as a horrible hurricane season, because Hurricane Andrew ravaged Miami in August. However, Andrew was the only major hurricane in the Atlantic that entire year!

How Do They Cause Destruction?

Tropical cyclones can create havoc in a variety of ways. They can damage and kill with wind, seawater, and even with rainwater. To preserve life and property, we have to understand how to defend ourselves against such a variety of threats. Below we examine the destructive capacity of tropical cyclones in more detail.

Seawater

The winds in a tropical cyclone push ocean water in front of them. The stronger the wind, the more water is "piled up" by the winds. As the cyclone nears shore, the winds to the right of the eye, which blow onshore, push this water inland. This dome of water is usually about 80 km (50 miles) wide and causes massive flooding near and to the right of the eye where it makes landfall. This process of wind-induced seawater flooding is called **storm surge**. It is extremely rapid; in Galveston in 1900, two different observers reported that the sea level rose 4 feet in just a few seconds.

Storm surge is by far the deadliest weapon in the tropical cyclone's arsenal, historically causing as much as 90% of all hurricane-related deaths. Simply put, this is because water is heavier than air. The brute force of even a small storm surge dwarfs the power of the strongest wind gust. Anyone dragged underwater during a storm surge is at great risk of drowning. Hundreds of thousands of people in low-lying Bangladesh have died by drowning in storm surges. The 8,000 or more deaths in Galveston in 1900 were primarily due to the 5-meter storm surge there.

The combined power of water and wind in a major hurricane captured the world's attention in 2005 in the aftermath of Hurricane Katrina in New Orleans (FIGURE 8-34). The winds ripped away portions of the roof of the Louisiana Superdome (at top left in the photo), and the wind-whipped storm surge overwhelmed the levees of the city, submerging 80% of New Orleans with up to 6 meters (20 feet) of water, including thousands of homes and most major roads (at right in the photo). Incredibly, New Orleans was spared the worst of Katrina's record storm surge. Just to the east of Katrina's landfall along the Mississippi Gulf Coast, the storm surge reached 8.25 meters (27 feet) high, causing hundreds of deaths and complete devastation well away from the coastline. Severe wind damage from Katrina extended more than 160 km (100 miles) inland, and tropical storm-force winds reached all of the way northward to just west of Nashville, Tennessee!

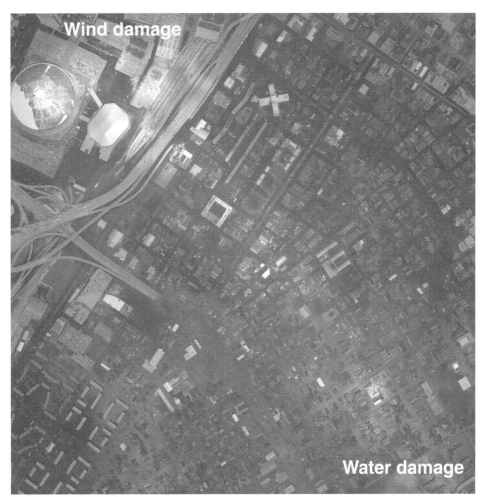

FIGURE 8-34 Overhead satellite photograph of downtown New Orleans in the aftermath of Hurricane Katrina in August 2005. At the top left, the white roof of the Louisiana Superdome has been peeled like an egg by Katrina's winds. At center and right, entire neighborhoods and major roadways are submerged by flood waters (dark in the photo) caused by the failure of levees under the weight of Katrina's storm surge.

Wind

The most obvious threat of a tropical cyclone is the powerful wind associated with its incredibly tight pressure gradient, which as we have seen can blow in a single spot for many hours. Damage is inevitable; a 322-km/hr (200-mph) wind gust in the most severe cyclones can exert a weight of 30 tons against the wall of a house! Wind damage is such a hallmark of hurricanes that hurricanes are classified by meteorologists using the **Saffir–Simpson scale**, which rates hurricanes on a scale of 1 to 5 based on the damage their winds would cause on landfall. TABLE 8-3 explains the Saffir–Simpson scale and relates the observed damage to estimated wind speeds. Major hurricanes are those classified as Category 3 and higher on this scale. FIGURE 8-35 is a unique before-and-after photograph of wind damage caused by Hurricane Andrew—a Category 5 hurricane on the Saffir–Simpson scale. Earlier versions of this scale incorporated central pressure and storm surge as components of the categories. The central pressure was used as a proxy for the winds because accurate wind speed intensity measurements were not available from aircraft. Storm surge was also originally used to determine hurricane category; however, storm surge is very complicated and depends on bathymetry, the size of the storm, and prior history. For example, Hurricane Ike was a Category 2 hurricane that struck Texas with a 20-ft storm surge, while Hurricane Charley, a Category 4 hurricane, struck Florida with one a peak storm surge of 7 ft.

Fortunately, most hurricanes do not produce the extreme winds linked with the highest category on the Saffir–Simpson scale. Since 1900, only three "Cat 5" hurricanes have crossed

TABLE 8-3 The Saffir–Simpson Scale for Hurricanes

Category	One-Minute Sustained Winds in mph (km/h)	Example Storms	Damage
1	74–95 (119–153)	Agnes (1972) Dolly (2008)	*Very dangerous winds will produce some damage*: Damage primarily to trees and unanchored mobile homes; some coastal flooding
2	96–110 (154–177)	Floyd (1999) Frances (2004)	*Extremely dangerous winds will cause extensive damage*: Some damage to roofs, doors, windows, trees and shrubbery; flooding damage to piers
3	111–130 (178–209)	Celia (1970) Ivan (2004)	*Devastating damage will occur*: Some structural damage; large trees blown down; flooding near shoreline and possibly inland; mobile homes destroyed
4	131–155 (210–249)	Hugo (1989) Charley (2004)	*Catastrophic damage will occur*: Extensive damage to doors and windows; major damage to lower floors near shore; terrain may be flooded well inland
5	>155 (>249)	Camille (1969) Andrew (1992)	*Catastrophic damage will occur*: Complete roof failure and some building failures; massive evacuation; flooding causes major damage to lower floors of all shoreline buildings

the U.S. coastline: the Labor Day Storm of 1935 in Florida, Hurricane Camille in Mississippi in 1969, and Hurricane Andrew in 1992. While Camille and Andrew caused major damage, the Labor Day storm did not cause nearly as much damage and does not appear on the list of most damaging hurricanes in Table 8-2.

Why can the damage caused by similarly intense storms differ so much? The amount of damage depends not only on the size and strength of the storm, but also on what is in its way. The Labor Day storm was intense but very small; it caused extensive damage in the Florida Keys, but Miami was left unscathed. In addition, coastlines today are much more densely populated than were the Florida Keys in 1935. Soon, over 50% of the U.S. population will live within 80 kilometers (50 miles) of the ocean. This boom in coastline development means that damage caused by tropical cyclones will continue to increase, even as meteorologists learn more about the storms and how to protect against them.

To add insult to injury, hurricanes also contain smaller whirlwinds inside them that can cause additional damage. Upon landfall, nearly every hurricane produces at least one tornado

FIGURE 8-35 Before and after photograph illustrating Hurricane Andrew's wind damage at Fairchild Tropical Garden, south of Miami. The hand-held postcard shows what the scene in the larger photograph looked like before the hurricane.

(see Chapter 11) that wreaks localized havoc in a narrow path for a few miles at most. In 2004, Frances and Ivan were responsible for over 100 tornadoes apiece across the Southeast U.S. In Hurricane Andrew, meteorologists identified tornado-sized "mini-swirls" that tacked on more speed to Andrew's already powerful winds. These mini-swirls helped cause the worst damage attributed to Andrew, as seen in Figure 8-20. Houses just a few blocks away escaped with much less damage because of the local nature of the mini-swirls.

Rainwater

A tropical cyclone that has moved over land may seem to be dying, but it still has one last knockout punch left: flooding caused by heavy rains. Tropical cyclones, even relatively weak storms that never attain hurricane status, are capable of causing record-setting amounts of rainfall over land. In June of 2001, weak Tropical Storm Allison dropped close to 1 meter (3 feet) of rain on metropolitan Houston, Texas, much of it in just 24 hours (**FIGURE 8-36**). At least 20 people were killed in the massive urban flooding that resulted; over 40,000 homes were damaged, with costs estimated near $5 billion. Allison is the only nonhurricane to make the list of most damaging tropical cyclones in Table 8-2! Hurricane Camille is said to have dumped 76 centimeters (30 inches) of rain in 6 hours on the James River Valley of Virginia 3 days after its 320-km/hr (200-mph) winds raked Mississippi; the resulting floods and mudslides killed 109 people.

From 1970 to 2000, more tropical cyclone-related deaths were caused by rain-fed floods than by wind, storm surge, or other hazards. Hurricane Floyd in 1999 illustrates the damage

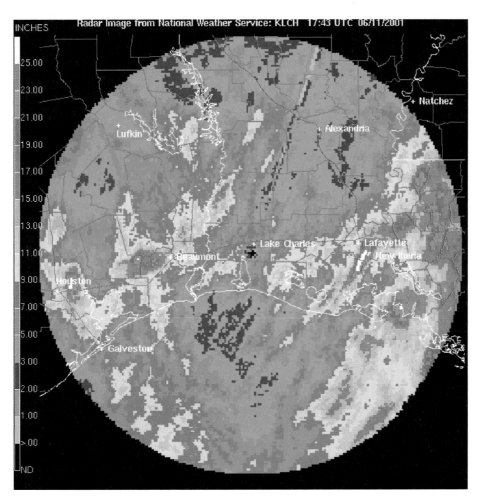

FIGURE 8-36 A Doppler radar estimate of the total rainfall resulting from tropical storm Allison in June 2001, as calculated by the National Weather Service radar at Lake Charles, Louisiana. Purple and white regions indicate areas where more than 51 cm (20 inches) fell, according to radar estimate. In the white area, between the U and S of Houston, onsite gauges measure nearly 92 cm (36 inches) of rain. Allison caused rainfall in every Gulf and Atlantic Coast state, from Texas to Maine.

FIGURE 8-37 A flooded subdivision of Greenville, North Carolina, after the heavy rains of Hurricane Floyd in 1999.

that a tropical cyclone's rainwater can cause. By the time Floyd reached the coastline of North Carolina, its winds had weakened to Category 2 on the Saffir–Simpson scale. Its ability to do wind-related damage had diminished, but its capacity to cause destruction with flooding rains had not. The floods that ensued along the U.S. East Coast (**FIGURE 8-37**) accounted for the majority of Floyd's death toll of 57—almost as many fatalities as were attributed to the stronger Andrew. Over half of the victims died in or while attempting to abandon their vehicles. Floyd was the deadliest hurricane in the U.S. since another flood-producer, Agnes, in Pennsylvania in 1972. Damage caused by Floyd was estimated at up to $6 billion.

Tropical cyclones are part of the overall pageantry of weather, however, and they do good as well as harm. Their winds move energy poleward, blow down forests, and make way for new growth. Storm surges carve new channels and cleanse wetlands. The rainfall from tropical weather systems save crops from summer and autumn droughts. From a human perspective, however, tropical cyclones are generally far too much of a good thing in all respects and must be anticipated and avoided if at all possible.

How Do We Observe and Forecast Tropical Cyclones?

The science of meteorology has made extraordinary advances in both the detection and the prediction of tropical cyclones (**BOX 8-3**). As Katrina demonstrated in 2005, however, thousands can still die from a major hurricane. Let's now examine the details of the forecasts for two killer U.S. hurricanes more than a century apart to understand today's tragedies in light of the past.

Galveston, 1900 and New Orleans, 2005: A Tale of Two Cities and Two Deadly Hurricanes

A major hurricane bears down on a large American coastal city along the Gulf of Mexico. In Galveston in 1900, almost 10,000 people die as a result. In New Orleans in 2005, over 1,000 people perish. Why did so many people die in Galveston? Why was the death toll, while still heartbreaking, much lower in New Orleans? To help answer these questions, we need to examine

Box 8-3 Hurricanes and Dust Storms

Residents of Florida and the Caribbean experience more than just hurricanes. In some years, fine grains of dust turn the sky a pale orange and cover exposed surfaces, but when the dust wafts down, there doesn't seem to be as many hurricanes. Could there be a connection?

Atlantic hurricanes often initially form downwind of the African continent. Also coming off the Saharan desert in Northern Africa are layers of dry air loaded with dust and sand (see photograph). This air mass is called the Saharan air layer, or SAL, and is most active from late spring through early fall. SAL outbreaks can cover areas the size of the lower 48 U.S. states as they traverse the North Atlantic from east to west and can reach as far west as the western Caribbean, Florida, and the Gulf of Mexico. Recent research is finding that the SAL may play a major role in limiting tropical cyclone activity in the Atlantic.

Satellite observations have played a key role in unlocking this puzzle by enabling scientists to track the SAL (see satellite image) and observe how it interacts with a hurricane. The visible satellite imagery can detect the presence of suspended dust in the SAL as it comes off the coast of Africa and moves across the North Atlantic. The infrared channels can also track the dust, in ways similar to how we observe clouds (see Chapter 5). Studying these images, along with flying through the storms, scientists have shown that tropical waves and tropical cyclones that interact with the SAL tend to weaken or have difficulty intensifying into stronger storms.

The SAL appears to suppress Atlantic tropical cyclone activity in three ways:

1. It introduces dry, stable air into the storm, which promotes convectively driven downdrafts in the tropical cyclone.
2. A midlevel easterly jet stream associated with the SAL can dramatically enhance the vertical wind shear. (The thermal wind law from Chapter 6 can explain this jet stream. It is easterly instead of westerly because the air to the north is *warmer* than to the south, instead of the usual situation where air is colder toward the north.)
3. The SAL enhances the pre-existing trade wind inversion (see Chapter 7, Box 7-1) in the Atlantic, which stabilizes the environment and prevents thunderstorm growth.

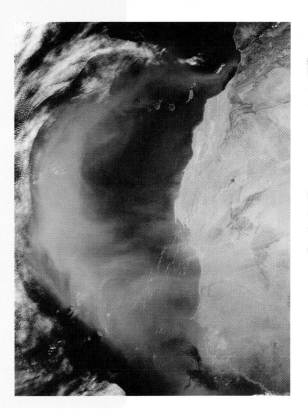

So, hazy, dusty summer skies in Florida might be good for low hurricane activity.

the vast improvements in hurricane detection, forecasting, and warning made during the 20th century and the challenges ahead in the 21st century.

In 1900, weather satellites did not exist, the global network of weather observations was in its infancy, and countries did not routinely share weather data. The U.S. Weather Bureau was even afraid to use the word "hurricane" in its statements to avoid undue concern! In the case of the Galveston hurricane (FIGURE 8-38), the Weather Bureau ignored reports from Cuban meteorologists and expected the storm to recurve as usual to the northeast along the U.S. East Coast. Assumption became "fact" as the official government reports stated, wrongly, that the storm was traveling northeast in the Atlantic. Instead, a high-pressure system shunted the hurricane far to the west through the Gulf of Mexico. The hurricane did not recurve until

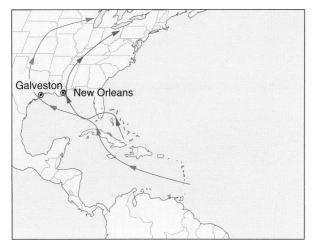

FIGURE 8-38 Tracks of the Galveston Hurricane of 1900 and Hurricane Katrina in 2005. The storms were similar in strength and severity but took different paths because of the different upper-level wind patterns that steered them.

it hit an unprepared Galveston. Even Galveston's chief Weather Bureau meteorologist was forced to ride out the storm in his floating house, so unexpected was the storm's fury. Chicago was surprised by hurricane-force wind gusts by the recurving storm! Failed forecasts of "fair, fresh" weather based on a lack of scientific understanding and observations cost thousands of lives.

In the decades after the Galveston disaster, the Weather Bureau entered the modern era of meteorology. As we have seen, hurricane observations using weather satellites, radar, and "hurricane hunter" aircraft ensure that no major storm escapes detection. The U.S. also supplies nations located in or near the Caribbean Sea with weather balloons and equipment so that wind patterns in those regions can be better observed during tropical cyclone season.

In addition to the improvements in hurricane observation, hurricane forecasting has become a science since the days of the Galveston tragedy. Recently, computer models of hurricanes have improved to the point that they can be used to anticipate, up to several days in advance, what a particular hurricane will do. These models are still not good at capturing the quirky changes in direction and intensity of some hurricanes, such as Charley in 2004, but the models are getting better each year. **FIGURE 8-39** illustrates the sizable reductions in the errors of the official National Hurricane Center

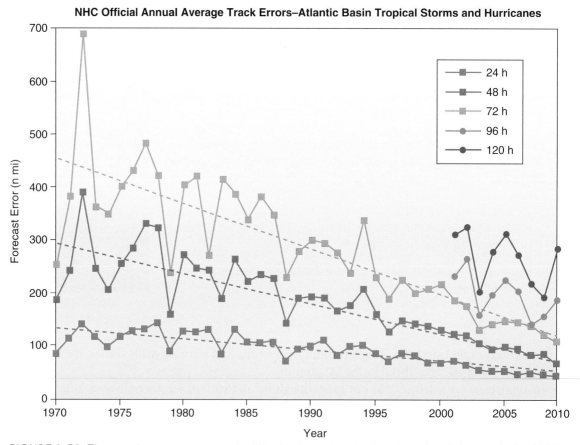

FIGURE 8-39 The yearly average errors in Atlantic basin tropical cyclone track forecasts for NHC from 1970 to 2010. Errors, expressed in nautical miles, are shown for forecasts 1, 2, 3, 4, and 5 days in advance. The straight dashed lines indicate the trends in the forecast errors, all of which are decreasing with time. Today, a 3-day forecast of a tropical cyclone's locations is as accurate as a 2-day forecast made just 10 years ago. (*Source*: NOAA.)

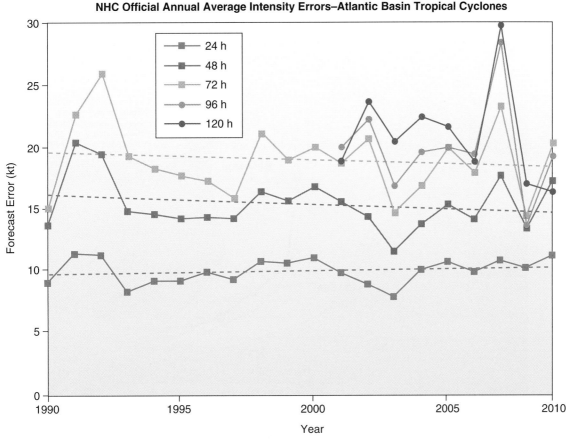

FIGURE 8-40 The yearly average forecast errors in the intensity of Atlantic basin tropical cyclones by the NHC from 1970 to 2010. Errors are shown for forecasts 1, 2, 3, 4, and 5 days in advance expressed in knots. The straight dashed lines indicate the trends in the forecast errors, all of which are decreasing with time. Unlike the forecasts for hurricane locations, there has been little improvement in forecasting the hurricane intensity. (*Source*: NOAA.)

(NHC) forecasts over the past several decades. Unfortunately, improvements in forecasting hurricane intensity have not progressed as rapidly (**FIGURE 8-40**), indicating that we lack a complete understanding of why hurricanes intensify. (In Chapter 13, the science of numerical weather forecasting is covered in much more detail.)

The science and politics of hurricane forecasting came together in a "perfect storm" during Hurricane Katrina. The NHC performed heroically, making a forecast that correctly predicted the Mississippi landfall location of Katrina to within about 16 km (10 miles) as much as 60 hours—2.5 days—before landfall. This nearly perfect forecast allowed plenty of time for evacuations along the central Gulf Coast (**FIGURE 8-41**). Unlike Galveston in 1900, the government forecasters knew exactly where the hurricane was via satellite and "hurricane hunter" observations, and they issued timely watches and warnings for the regions Katrina later ravaged. Thousands of lives were saved as a result.

But the response of local, state, and Federal governments to the NHC forecasts was slow. Partly as a result, hundreds of thousands of people along the Gulf Coast were pinned down and pummeled by Katrina's furious winds and 8.25-meter (27-foot) storm surge. How could this happen in 21st-century America? The answers to this question may take years to uncover and may involve a complex combination of personal and governmental decisions.

For example, the NHC forecast for Katrina shifted landfall from the Florida Panhandle to Mississippi at 11 AM EDT on a Friday in late August, a time when the public's thoughts are often more focused on summer recreation and weekend getaways. In addition, one private-sector forecasting firm did not revise its own Katrina forecasts until up to 6 hours after the NHC made its crucial forecast shift, potentially causing confusion. Katrina did not expand into

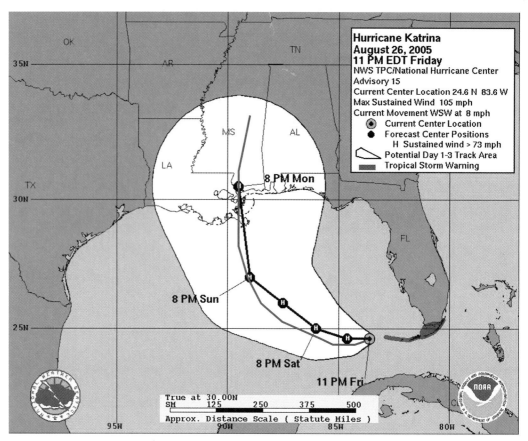

FIGURE 8-41 The NHC forecast for Hurricane Katrina, as of 11 PM EDT on Friday, August 26, 2005. The NCH forecast correctly predicted a Monday landfall along the Southeastern Louisiana Coast and at the Louisiana/Mississippi border more than 2 days in advance when Katrina was just west of Key West, Florida. The white shading indicates the "forecast cone," which is the range of probable locations of Katrina's eye during the forecast period. The black line is the NCH forecast of the most likely track of Katrina's eye. The red line indicates the actual path of Katrina. (Courtesy of National Weather Service/NOAA.)

a Category 5 behemoth nearly 750 km across (460 miles) until about 24 hours before landfall, early on a Sunday morning when most people are not following the news.

Regardless of the causes, the end result was that local, state, and Federal agencies were unprepared for Katrina, even though the official NHC forecasts were pinpoint accurate days in advance. Katrina became the deadliest hurricane in the U.S. since 1928 and the worst natural disaster in U.S. history.

The Galveston tragedy taught the U.S. a lesson: its Weather Bureau was a mess and needed fixing. The nation sorely needed much better hurricane observations and forecasts. Over a century later, the lesson is different: the NHC did nearly everything perfectly, under difficult circumstances, but even the best forecasts cannot save lives unless they are heard and heeded. In 2005, as in 1900, too many lives were lost as a result of lessons not learned.

Long-Range Tropical Cyclone Forecasting

Starting in the 1980s, Colorado State University meteorologist Professor Bill Gray developed and refined a method for forecasting tropical cyclones in a statistical, seasonal sense. His method would not have told you whether Katrina would hit the Gulf Coast on August 29, but it does predict if the coming hurricane season will probably see more than its usual share of hurricanes in the Atlantic, Gulf of Mexico, and the Caribbean Sea.

Professor Gray's forecasts are based primarily on the El Niño cycle, SSTs off of the northwest coast of Europe, sea-level pressures along the west coast of Africa, and 500-mb geopotential heights in the Arctic near Greenland. A novice to tropical cyclones might wonder how these far-flung parameters could help one predict the number of storms in a given year. Based on our discussion of atmosphere–ocean interactions, they make sense. Earlier in this chapter, we saw that El Niño leads to a strong jet stream over the subtropics. This jet destroys the carefully organized circulation of hurricanes. On average then, El Niño years are low-hurricane years, and La Niña years are high-hurricane years. Low pressure along the African coast implies lots of thunderstorm "seedlings" that move east and develop into easterly waves and sometimes into Cape Verde hurricanes. Finally, the pressure pattern in the Arctic is related to the NAO: high pressure means a negative NAO index, a weaker jet stream, and less wind shear over the Atlantic, all of which favor tropical cyclone development.

The accuracy of Professor Gray's forecasts is illustrated in FIGURE 8-42. In most years, his methods can accurately predict months in advance whether a given hurricane season will be more active or less active than normal. This is not the same as predicting a specific hurricane in a specific location in advance—that is a task for computer models (see Chapter 13). However, it is a vast improvement over expectations based on climatological records and is helping coastal residents as well as businesses and insurance companies prepare months in advance for the chance of damaging storms. As we will see in Chapters 13 and 16, these kinds of long-range predictions are the future of weather and climate forecasting.

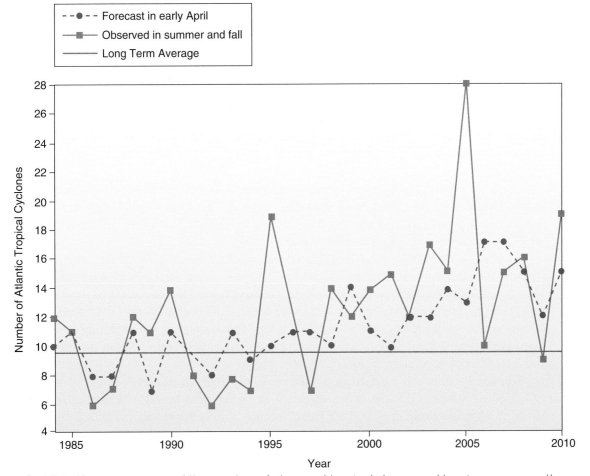

FIGURE 8-42 A comparison of the number of observed tropical storms and hurricanes versus the number predicted by Professor Bill Gray's statistical methods in early April, almost 2 months before the beginning of each hurricane season. The horizontal line represents the number of hurricanes expected based on a long-term average.

PUTTING IT ALL TOGETHER

■ Summary

Based on temperature, the vertical structure of the ocean can be classified into three basic layers: the surface zone, a deep-water zone, and a transition zone. In the bottom layer, or the deep zone, the temperature is uniform with depth. The surface zone is the warmest layer and is warmest in the tropical waters. Ocean temperature decreases toward the poles and with increasing depth.

The atmosphere and ocean interact in a variety of ways, through exchanges of heat and moisture and also by the creation of surface ocean currents. These currents are driven by the winds. Some, like the Gulf Stream, move in well-defined boundaries like a river, while others are broad and diffuse. Surface currents generated by the winds can induce vertical circulations in the oceans. This can cause upwelling of nutrient-rich waters. For example, over the eastern tropical Pacific Ocean, winds typically rotate counterclockwise around a high-pressure system, producing along-coast winds off Peru and easterly winds along the equator in the eastern Pacific. The combined effect of wind and the Coriolis force produces offshore flow off Peru, causing upwelling in which cold, deeper water comes to the surface. This produces the cold SSTs off Peru and along the equator in the eastern Pacific. In the western Pacific on the equator, this upwelling does not occur, and as a result, the SST is higher there than in the east.

El Niño and La Niña are extreme phases of a naturally occurring climate cycle and describe large-scale changes in SST across the tropical Pacific. El Niño and La Niña result from interaction between the surface of the ocean and the atmosphere in the tropical Pacific. During El Niño, warm SSTs replace the cold upwelling waters in the eastern Pacific. This shifts thunderstorm activity eastward across the Pacific and ultimately affects weather and climate across the globe. In many locations, especially in the tropics, La Niña (or a cold episode) produces the opposite climate variations from El Niño. For instance, parts of Australia and Indonesia are prone to drought during El Niño but are typically wetter than normal during La Niña.

The oceanic patterns of El Niño and La Niña are closely related to a Pacific-wide seesaw of atmospheric pressure known as the Southern Oscillation. Recently, similar oscillations have been discovered over the north Pacific and Atlantic oceans, highlighting the importance of atmospheric–ocean interactions.

Tropical cyclones are large, whirling storms that obtain their energy from warm ocean waters. They stand out on satellite photographs because of their circular cloud patterns and, in the stronger storms, a nearly clear eye at the center. The clarity and size of the eye on satellite images helps meteorologists estimate a cyclone's strength.

A tropical cyclone begins as a disorganized tropical disturbance. A few grow to hurricane or typhoon strength with winds up to 360 km/hr (200 mph). Most weaken within a week or two. They typically move west or northwest and then recurve toward the northeast, but each storm's path is unique. Hurricanes affecting the U.S. often form in the Gulf of Mexico and Caribbean Sea in early summer and late fall, with powerful Cape Verde hurricanes often dominating attention in August and September. The location of warm ocean waters largely determines the birthplaces and intensities of tropical cyclones.

Tropical cyclones cause destruction with extremely high winds, storm surges of seawater, and flooding rainfall. Of these, storm surge has been the most deadly and devastating, as in Hurricane Katrina along the U.S. Gulf Coast in 2005, but in many years, flooding rains have killed more U.S. residents than wind or storm surge. Each hurricane packs its own unique combination of these weapons.

Weather satellites have revolutionized the forecasting of tropical cyclones. Today no tropical cyclones go undetected, a vast improvement on the situation a century ago. Thousands of lives are saved as a result. Numerical forecasts are giving increasingly reliable estimates of storm tracks up to several days in advance. Statistical forecasts now provide long-range estimates of the activity of hurricane seasons as much as a year in advance.

Key Terms

You should understand all of the following terms. Use the glossary and this chapter to improve your understanding of these terms.

Arctic oscillation (AO)
Bermuda high
Cape Verde hurricanes
Deep zone
Easterly wave
Ekman spiral
Ekman transport
El Niño
ENSO
Extratropical cyclone
Eye
Eye wall
Gyre

Hurricanes
La Niña
Mixed layer
North Atlantic Oscillation (NAO)
Ocean current
Oceanography
Pacific Decadal Oscillation (PDO)
Rainbands
Recurvature
Saffir–Simpson scale
Southern Oscillation

Southern Oscillation Index (SOI)
Storm surge
Supertyphoon
Surface zone
Thermocline
Tropical cyclones
Tropical depression
Tropical disturbance
Tropical storm
Typhoon
Upwelling
Wind Shear

Review Questions

1. Why is the ocean surface zone thicker over the tropics than over the poles?
2. How can SST distribution affect weather?
3. What is the thermocline?
4. What is the North Atlantic gyre?
5. What is the Ekman spiral, and how is it formed?
6. How can surface winds generate upwelling along coastal regions?
7. What is El Niño, and how does it impact the weather where you live?
8. What is La Niña, and how does it impact the weather where you live?
9. Is a typical tropical cyclone bigger or smaller than the following weather phenomena:
   a. Hadley cell
   b. Air parcel
   c. Thunderstorm
   d. Southern Oscillation
10. What is the NAO? When it is in its positive phase, what kind of winter weather can the eastern U.S. expect?
11. When a tropical cyclone "opens its eye" for the first time, what can you say about its approximate wind speed? Is it strengthening or weakening? When a hurricane narrows its eye and becomes very circular, is it strengthening or weakening?
12. If you could cover the Gulf of Mexico with water-impermeable plastic wrap, what would the effect be on the development and intensity of tropical cyclones over the Gulf?
13. In some El Niño years, the ocean waters off the west coast of Mexico also become abnormally warm, and tropical fish are sighted unusually far north. How might this unusual warming affect the intensity of tropical cyclones that develop in this region?
14. Only one or two tropical cyclones have developed in the South Atlantic Ocean off the coasts of South America and Africa in the last century. What ocean temperature and wind patterns might be responsible for this amazing lack of tropical cyclone activity?
15. Why is it that hurricanes never form off the West Coast of the U.S. but can form off the East Coast?
16. Global warming should lead to warmer ocean temperatures and less wind shear because of a reduced latitudinal temperature gradient, but perhaps more wind shear because of "perpetual El Niño" conditions. Discuss how each of these three changes might affect future tropical cyclone activity and intensity.
17. If a Northern Hemisphere hurricane has maximum sustained winds of 100 mph in its eye wall and it is moving at 20 mph, what will be the maximum winds in mph experienced by locations that are just to the right of the eye?
18. In 1985, Hurricane Gloria recurved northeastward along the U.S. East Coast and traveled quickly just east of the coastline for hundreds of miles. East Coast residents braced for a fearful storm; a Boston

radio station even played a marathon of songs titled "Gloria" in anticipation. But in the end, most inland residents were disappointed, saying that the winds were much less than they had expected. Using a map of the U.S. and your understanding of the relationship between total storm winds and the forward motion of the storm, explain why the inland residents experienced relatively weak winds from this strong hurricane.

19. If you lived or ran a business on the U.S. East Coast and you heard that Professor Gray's forecast was for a very active and damaging hurricane season in the near future, what actions would you take to protect your life and property?

20. True or false: In a hurricane, you want to drive to a safe location in order to stay out of the high winds. Explain your answer.

21. Why don't the damage totals and death tolls from U.S. hurricanes always increase with increasing Saffir–Simpson category number? Discuss the different variables that can make a Category 1 storm more of a killer and destroyer of property than a Category 5 storm.

22. Can tropical cyclones exist on other planets in our solar system? Explain why or why not.

23. Using information in this chapter, explain how El Niño could prevent the formation of hurricanes in the Atlantic by a mechanism other than subtropical jet strengthening. (Hint: See Table 8-1 and refer to Professor Gray's forecast methods.)

▓ Observation Activities

1. If you are taking this course during hurricane season, chart the track of at least one hurricane. Make a table of the intensity of the hurricane and the SST (both are available on the Web, for example, http://www.wunderground.com/tropical/). Do your observations agree with the discussions presented in this chapter?

2. If you live near a lake or the coast, observe the condition of the body of water at different wind speeds. Describe your observations in relation to concepts discussed in this book.

This rain cloud icon is your clue to go to the *Meteorology* Web site at http://physicalscience.jbpub.com/ackerman/meteorology/. Through animations, quizzes, web exercises, and more, you can explore in further detail many fascinating topics in meteorology.

Air Masses and Fronts

9

AFTER COMPLETING THIS CHAPTER, YOU SHOULD BE ABLE TO

- Explain what an air mass is and identify the regions in which different air masses develop
- Explain how and why air masses change once they leave their "birthplaces"
- Characterize the different ways in which clashes of air masses create fronts
- Contrast the varying types of weather that are associated with cold, warm, stationary, and occluded fronts

INTRODUCTION

Lightning bolts around dawn soon are gone. Later in this chapter we explain the science behind this bit of weather folklore, but for the moment, just say it out loud. Then have a friend from another part of the country say it to you. Depending on where you grew up, your pronunciation of these words may differ from your friend's. We call this difference an accent, and it arises because your pronunciation is influenced by the way people around you as a child said them. For example, people who grew up in the rural South might begin by saying "lahtning boats." Residents of Canada and the far northern United States sometimes pronounce "around" as "aroond." Native New Englanders could finish the sentence by saying "dahn soon ah gahn." The words are the same, but how they sound differs from one region to another.

We can draw an analogy between accents and the atmosphere. The composition of the atmosphere is the same worldwide, but where the air "grows up" determines its temperature and humidity. In other words, air develops an "accent" and takes on the characteristics of its surroundings. Just as we can identify a vocal accent and label it with its region of origin, we can do the same with air. This leads to the concept of air masses, the first topic of this chapter.

People sometimes lose or modify their accents when they move away from where they grew up. A New Englander who moves to the South may, over time, find herself saying "dawwn" instead of "dahn." Air masses are modified in a similar way; they move around and in the process become cooler or warmer, drier or moister, depending on where they have traveled. Along the way, these air masses can lead to record cold outbreaks and lake-effect snows in winter and deadly heat waves in summer.

Finally, different air masses, like different accents, don't mix together easily. In the atmosphere, air masses with differing characteristics clash in regions called fronts. Much of the stormy weather observed in the middle latitudes, from clouds and rain to thunderstorms and tornadoes, occurs in frontal regions as a result of air masses interacting with each other. In this chapter, we also explore the different types of fronts and the typical weather associated with them.

WHAT IS AN AIR MASS?

An **air mass** is an extremely large body of air whose properties of temperature and moisture content (humidity) are similar in any horizontal direction. Air masses can cover hundreds of thousands of square miles. Air masses are formed when air stagnates for long periods over a uniform surface. The characteristic weather features (temperature and moisture) of air masses are therefore determined by the surface over which they form. An air mass acquires these attributes through heat and moisture exchanges with the surface. As a result, an air mass can be warm or cold and moist or dry.

▨ Observations

In a typical year, air mass weather kills more people in the U.S. than all other weather phenomena combined. Heat waves in the summer can bring perilous weather to cities not accustomed to handling such hazardous conditions (**BOX 9-1**). These heat waves are caused by very hot, stagnant air masses. Heat waves kill more people than any other weather type. For example, a heat wave in the Midwest in late July 1999 resulted in average daily temperatures of greater than 33° C (90° F). More than 200 deaths in the Midwest were attributed to this 1999 heat wave (**FIGURE 9-1**).

The August 2003 heat wave in Europe killed an estimated 35,000 people. Between August 3rd and 13th, temperatures in France regularly exceeded 40° C (104° F); the typical August temperature of Paris is approximately 23° C (75° F). An estimated 900 people died from the heat in London, England, as the city recorded its first triple-digit Fahrenheit temperature ever on August 10, 2003 (see Chapter 10 for more details).

Cold weather is also a concern of and threat to humans. Cold air outbreaks cost American homeowners at least $500 million per year when frozen water pipes burst and inflict water damage. The Centers for Disease Control's National Center for Health Statistics estimates that on average 689 U.S. residents die each year because of the effects of excessive cold, with the highest death rates in Alaska, New Mexico, North Dakota, and Montana. Two-thirds of the victims of hypothermia are men, and over half are at least 65 years of age.

Extremely hot or cold temperatures are both killers. People from warmer winter climates might be at greater risk from cold weather than people living in colder climates. Similarly, people who live in climate regimes with cooler summers may be at greater risk to hot weather than those who live in hotter climates. This is demonstrated in **FIGURE 9-2**, which plots the risk of death as a function of temperature for 11 cities in the U.S. (6 northern and 5 southern). There is a greater risk from cold temperatures in the more southern cities, whereas a greater risk from warmer temperatures exists in the more northern cities. However, this observed relationship may not be the result of only geographic location of the city. Socioeconomic status, cultural backgrounds, and the size of the elderly population can put the city's residents at greater or lesser risk. For example, households with heaters in the north and air conditioners in the south may reflect socioeconomic status, whereas cultural backgrounds may influence the amount of time people spend outdoors.

▨ Air Mass Types

Air masses are classified according to the temperature and moisture characteristics where they develop. Cold air masses originate in polar regions and are, therefore, called polar air masses.

Box 9-1 Deadly Heat Waves

Each summer in the U.S. approximately 175 to 200 deaths are attributable to heat waves. Most of these deaths occur in cities, particularly northern cities. Regions that suffer under intense hot spells are usually dominated by a surface high-pressure system with a midtropospheric ridge aloft. Dew points are also high, and to compound matters, wind speeds are often low. Clear or partly cloudy skies allow intense solar radiation to further heat the ground and the air mass. High humidity and stagnant air reduce the body's ability to cool down through sweating (see Boxes 3-3 and 4-1). When these conditions persist day and night for several days, lives are endangered.

Overexposure to heat leads to giddiness and nausea. Heat cramps are generated when the body loses too much water and salt through perspiration. Symptoms of heat cramps are muscle spasms, fever, and nausea. Heat exhaustion occurs when the body temperature and blood pressure drop and the skin turns cool and clammy. Perspiration is profuse as the body continues trying to lose energy. More dangerous than heat exhaustion is heat stroke, which poses an immediate threat to life. Most victims of heat stroke are people older than 60 years of age, although infants are also susceptible. The first signs of a heat stroke are that the skin becomes dry and hot as sweating ceases, the face becomes flushed, and the pulse accelerates. Once the body loses its ability to cool down, death can occur within a few hours. A victim of heat stroke needs immediate medical attention. Sponging the body with water will help to cool the body.

During a heat wave, the threat to health is increased in large cities by what is called the "urban heat island" effect, which amplifies the heat by 1° C to 4° C (2° F to 7° F) or more (see graph below). During the heat of the day, poorly ventilated urban houses and apartments that do not have air conditioning can become like brick ovens. To avoid the dangers of heat exhaustion, it is important to drink water, stay out of direct sunlight, and minimize

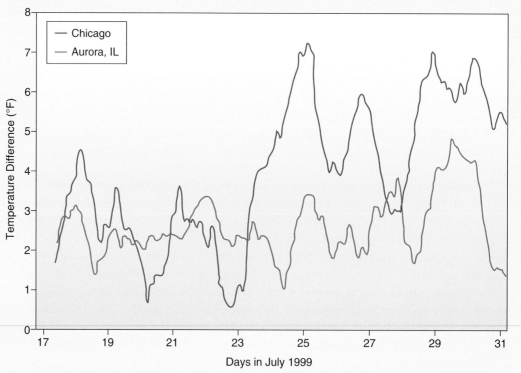

The urban heat island is demonstrated by a comparison of the differences in hourly temperatures between Chicago, IL, and Aurora, IL (a nearby suburban station). During the late afternoon and early evening of July 25 to 31, 1999, the average temperatures were about 3° C to 5° C (5° F to 7° F) higher in Chicago. During this heat wave, most of the people who died on July 29 and 30 lived in large cities. (Urban-Rural Temperature Comparison for July 17–31, 1999; image courtesy of the Midwestern Regional Climate Center, Illinois State Water Survey.)

physical activity. Reduction in the loss of life can be accomplished by monitoring the health of elderly urban residents. The news media helps by not only providing useful information about the weather but also by explaining procedures that reduce heat stress and giving the locations of places where relief is available.

Heat waves also have a strong economic impact even if there is no loss of human life. A prolonged heat wave can cause the widespread use of air conditioning, leading to increased demands for power that stress gas and electric utilities. During the July 1995 heat wave, Chicago residents set a record by using almost 20 megawatts of energy in just 1 day. Excessive demand can leave people without power for hours and sometimes days. Transportation can be stymied when highway surfaces and railways buckle and warp in the heat. All types of outdoor work, such as landscaping and construction, experience reduced productivity. Agriculture is especially vulnerable. Heat waves stunt crops and kill livestock; on July 14, 1995, 850 cows died from heat exposure in Wisconsin.

Intense heat waves are often broken with the invasion of a polar air mass. Violent storms can be associated with cold fronts, resulting in additional economic losses and the potential for further loss of life.

Warm air masses usually form in tropical or subtropical regions and are called tropical air masses. Moist air masses form over oceans and are referred to as maritime air masses. Dry air masses that form over land surfaces are called continental air masses. The five primary air mass regions are summarized as follows:

- Polar (P): formed poleward of 60° north or south—cold or cool
- Tropical (T): formed within about 30° of the equator—hot or warm
- Arctic (A): formed over the Arctic—very cold
- Continental (c): formed over large land masses—dry
- Maritime (m): formed over the oceans—moist

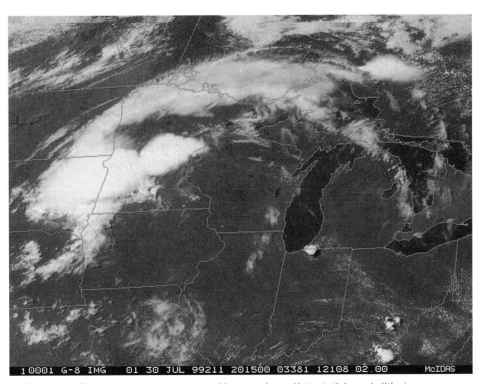

FIGURE 9-1 The thunderstorms over Minnesota in this visible satellite image taken on July 30, 1999, do not pose the major weather hazard on this day. Instead, the clear air in the middle of the image is a hot, cloud-free air mass where an intense heat wave is occurring. This heat wave killed more than 200 people in the Midwest.

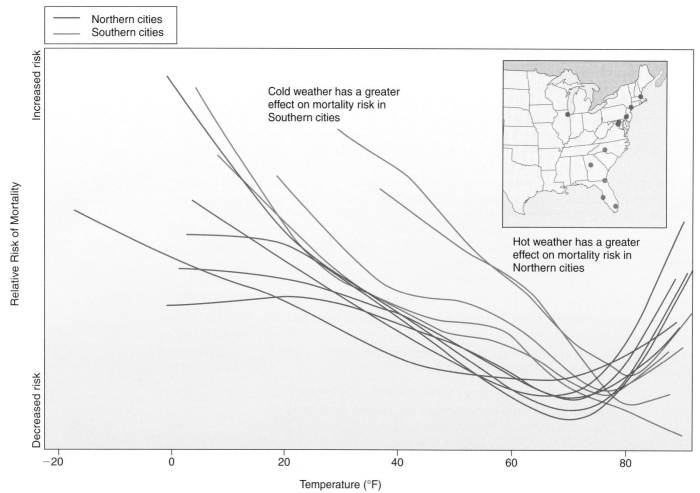

FIGURE 9-2 Risk of death because of hot or cold weather versus temperature for 11 U.S. cities between the years 1973 and 1994. Northern cities: Boston, Massachusetts; Chicago, Illinois; New York, New York; Philadelphia, Pennsylvania; Baltimore, Maryland; and Washington, DC. Southern cities: Charlotte, North Carolina; Atlanta, Georgia; Jacksonville, Florida; Tampa, Florida; and Miami, Florida. (Modified from Curriero, F.C., et al., *Am. J. Epidemiol.*, vol. 155, page 83 by permission of Oxford University Press.)

These regions overlap. For example, a tropical air mass can be warm and dry or warm and moist, depending on whether an ocean or a continent is beneath it. This suggests a two-letter classification scheme for air masses, with the first letter denoting land versus ocean and the second letter indicating the latitude band of origin.

Four main air mass types are dominant: **continental polar (cP)**, **continental tropical (cT)**, **maritime polar (mP)**, and **maritime tropical (mT)**. An air mass's temperature and humidity are a function of the time of year when it forms. Winter air masses will be colder than those that form during the summer. Some classification schemes differentiate cold air masses into polar and arctic air masses. Continental arctic air masses (cA) are much colder than polar air masses and form in winter over snow-covered surfaces in Siberia, the Arctic Basin, North America, and Greenland. The temperature and moisture properties of these air masses as a function of season during which they originate are shown in **TABLE 9-1**.

"Identifying Air Masses"

Air Mass Source Regions

Air mass source regions are the "birthplaces" of air masses. A source region must have light winds or no winds, so the air has time to acquire the temperature and moisture properties of the region's surface. Strong winds would move the air before it could acquire the characteristics related to the surface. Given the requirements of a uniform surface and light or no winds, not all parts of the world can generate air masses. Good source regions for air masses are the subtropical belts, which

TABLE 9-1 Temperature and Moisture Characteristics of Air Masses

Air Mass	Winter Characteristics	Summer Characteristics
Continental polar (cP)	Very cold and dry	Cool and dry
Maritime polar (mP)	Cool and humid	Mild and humid
Continental tropical (cT)	Cool and dry	Very hot and dry
Maritime tropical (mT)	Warm and humid	Warm and humid
Continental arctic (cA)	Bitter-cold and dry	—

have light winds and, thus, favor the development of air masses with uniform temperature and moisture (**FIGURE 9-3**).

A source region must have an extensive and homogeneous surface. Large water bodies or flat uniform lands are potentially good source regions for air masses. Coastlines are not very good source regions of air masses because these locations have both land and water.

Once formed, air masses are not confined to their source regions. They move on to other regions. How they move is determined by upper-air patterns and is discussed in the next chapter. Before we look at the air masses that affect North America, we need to revisit the concept of atmospheric stability (see Chapters 3 and 4).

ATMOSPHERIC STABILITY REVISITED

The **static stability** of the atmosphere is determined by comparing the temperature of a rising parcel with the temperature of the atmosphere at the same altitude as the parcel. This is known as the parcel's "environment." At a given pressure, the colder the air parcel, the greater is its density. If an air parcel is colder than its environment, it is denser than its surroundings and will descend. The atmosphere is stable when an ascending parcel, which cools as a result of expansion, becomes cooler than its environment. Therefore, as discussed in Chapter 3, any region of the atmosphere that has a temperature inversion is stable.

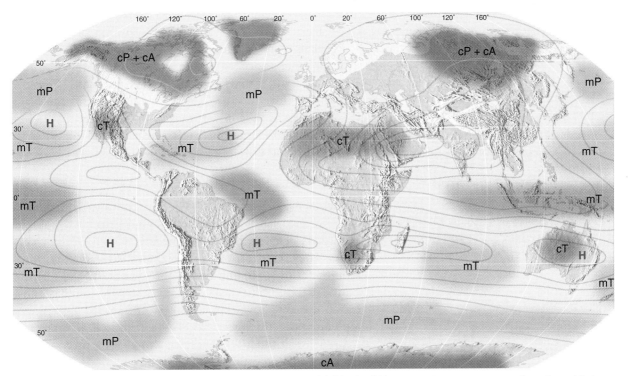

FIGURE 9-3 Major air mass source regions of the world. Most of these regions are areas of surface high pressure. Blue lines are surface isobars.

If an air parcel is warmer than the surrounding air, it will rise. If an ascending parcel within an air mass becomes warmer than its environment, the air mass is said to be unstable. An unstable atmosphere has a temperature and moisture structure favorable for lifting air parcels and thus forming clouds and possibly precipitation. A stable atmosphere inhibits rising motions and is usually less cloudy and drier.

Another important concept is the means by which the stability of the atmosphere changes. Warm air overlying cold air results in a stable atmosphere. So, when the lower regions of the troposphere are cooled, the air mass becomes more stable. This explains why polar air masses are generally stable. However, when cold air moves over a warm surface (e.g., a warm body of water), the air near the surface warms. Warming the air near the ground makes the atmosphere less stable. This is why tropical air masses are, in general, less stable than polar air masses.

To make an air mass unstable, the environment must be more conducive to vertical motions. This happens when the lower layers of the atmosphere warm up through energy exchanges with the surface under the air mass.

AIR MASSES AFFECTING NORTH AMERICA

FIGURE 9-4 shows the major air mass source regions that affect North American weather and climate. The arrows show the typical paths the air masses take once they begin to move. In the following, we examine each type of air mass.

Maritime Polar Air Masses

Maritime polar air is formed over the oceans at high latitudes. The North Pacific is a good source region of mP air masses. These North Pacific air masses often move into the Gulf of Alaska and then to the West Coast of North America. During winter, they can influence the weather as far south as California. The East Coast of North America is also affected by North Atlantic air masses, which approach the coast from the northeast. The stormy weather pattern associated with these advancing air masses is referred to as a *northeaster* (usually pronounced and spelled *nor'easter*). Low-pressure systems that comprise a nor'easter typically move up the East Coast of North America between October and April. The counterclockwise winds around the low draw moist mP air over the Atlantic Ocean inland. At the same time, cold cP air moves eastward toward the East Coast and meets with northward-moving mT air from the Gulf of Mexico. This clash of warm and cold has produced some of the heaviest snowfalls on record along the East Coast (**FIGURE 9-5**).

The weather associated with mP air masses is variable. Winter mP air masses are formed in polar regions over water. Cold mP air that moves over a warm surface is unstable and can bring rain showers. When the mP air moves over a surface that is only slightly warmer, the atmosphere is less unstable and may produce stratus clouds and drizzle.

Continental Polar Air Masses

Continental air masses affecting North America are formed over the interior high-latitude regions of the continent, such as Alaska and Canada. The air masses formed in winter are very cold and dry. They require long, clear nights to form, during which strong radiational cooling causes the temperature of air near the surface to plummet, especially over snow-covered ground. Because the air above the surface is not cooled as strongly, temperatures in the lower troposphere in a cP air mass can increase with altitude. These surface temperature inversions are often observed, indicating the stable nature of these air masses.

In winter, the weather associated with cP air masses can be cloud-free and frigid (**FIGURE 9-6**). From northern Canada, cP

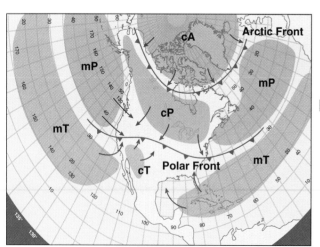

FIGURE 9-4 The major air mass source regions that affect North American weather. The arrows show the typical paths these air masses take once they begin to move. Which air masses affect the weather where you live?

FIGURE 9-5 Visible satellite picture of the mid-Atlantic coast on February 7, 2010, after the first of two blizzards nicknamed "Snowmageddon" by the White House. The white areas in the lower left of the image are snow-covered, all the way from the Virginia mountains and coastline to New York City. This blizzard caused up to 1 meter (40 inches) of snow from Maryland to New Jersey. The two blizzards shut down the federal government in Washington for four consecutive days. Record seasonal snowfall records were shattered from Washington to Philadelphia.

air masses move southward and can result in bitter cold temperatures as far south as Florida. This can lead to crop damage and cause economic hardship. Strong winds and snow in cP air masses can lead to blinding **blizzard** conditions, in which winds of about 55 km per hour (35 mph) or greater and visibilities of 0.4 km (¼ mile) or less exist for at least three consecutive hours. The wintry temperatures and surface temperature inversion can also lead to reduced air quality and poor visibility as pollution levels increase with the burning of heating fuels such as coal, fuel oil, and wood.

FIGURE 9-6 Cold temperatures in a continental polar (cP) air mass make the breath of the red-winged blackbird visible, similar to the formation of a mixing fog discussed in Chapter 4.

When formed in summer, cP air masses have more moderate temperatures. The snow that would cool this air in winter has melted because of the increased warmth from the Sun. Surface temperature inversions are usually absent in cP air masses formed in summer. The air mass has formed over a relatively dry surface, so the characteristic weather is cool, dry, and clear—a pleasant change from a humid summer caused by mT air masses. Daytime heating, however, may lead to puffy cumulus clouds and even showers as the air mass becomes unstable, but clear skies return soon after sunset.

▨ Continental Arctic Air Masses

Arctic air masses are formed over the frozen Arctic and are much colder than cP air masses. This air mass type is confined to a shallow layer near the surface. Because the layer is not very deep (less than 0.6 kilometer [1 mile]), there is little vertical lifting associated with its movement and thus little precipitation. Arctic air can move south and occasionally reach into the U.S. When it does, record-setting cold outbreaks are the consequence. **FIGURE 9-7** shows the hourly temperature at Madison, Wisconsin, during a prolonged Arctic outbreak in 1996.

▨ Continental Tropical Air Masses

Continental tropical air masses are hot and dry and usually form over tropical and subtropical deserts and plateaus. The hot surface temperatures and the lack of vegetation make the cT air mass hot and dry. The southwest U.S. and northern Mexico are summer source regions in North America.

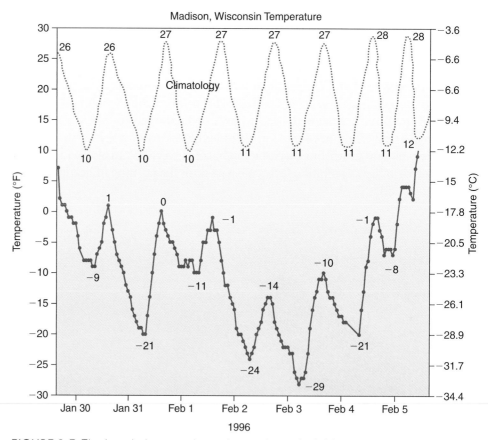

FIGURE 9-7 The hourly temperature observations (solid lines with dots) at Madison, Wisconsin, during the Arctic outbreak of January 30 to February 5, 1996. Numbers indicate daily high and low temperatures (in degrees Fahrenheit), which may be a degree different than the highest or lowest hourly temperatures in some cases. For nearly a week, the daily high and low temperatures at Madison both were 11° C to 22° C (20° F–40° F) below the normal values (dotted line) for a typically cold Wisconsin winter!

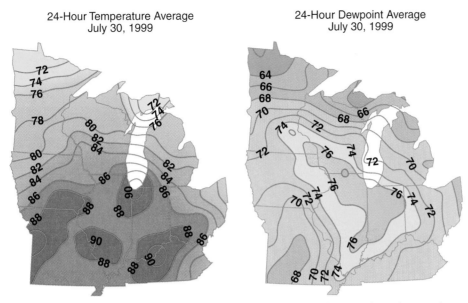

FIGURE 9-8 The high temperatures and dew points of the maritime tropical (mT) air mass during the Midwest heat wave in July 1999 were responsible for 232 deaths.

Continental tropical air influences the southwest and central U.S. in summer as warm, dry air arrives from the Mexican Plateau. The hot temperatures near the ground make the air mass unstable, but the dryness limits cloud formation. The contrast between dry cT air and moist mT air over the south-central U.S. can be so extreme that meteorologists label it as a special kind of front known as a "dryline." At the end of this chapter we discuss an example of the dryline.

When cT air from the Mexican Plateau moves aloft over the southern Great Plains, it can create a thin but strong inversion known as a "capping inversion" or "the cap" to storm chasers. We discuss this type of inversion in Chapter 11.

Maritime Tropical Air Masses

The sultry summer weather of the eastern U.S. is strongly influenced by mT air masses that form over the Gulf of Mexico, the subtropical western Atlantic Ocean, and the Caribbean Sea. The moisture source of precipitation for the midwestern U.S. is the mT air masses, particularly those originating over the Gulf of Mexico. When the air mass is stable, however, an oppressively humid heat wave is the result (FIGURE 9-8). Summertime weather in the southwestern U.S. is influenced by Pacific mT air masses. Moisture supplied by these mT air masses to the Southwest generates rains referred to as the Arizona Monsoon.

AIR MASS MODIFICATION

As air masses move from one place to another, their properties, such as temperature, moisture, and stability, change as the air masses exchange heat and moisture with the underlying surface. This process is called **air mass modification**. There are two primary mechanisms that modify an air mass: heat exchanges with the surface and mechanical lifting.

Heat exchanges with the surface primarily affect the lowest regions of the air mass. The rate at which the air mass temperature and moisture change is determined by the rate of heat and moisture exchanges. The greater the temperature difference between the air mass and the surface over which it is moving, the greater is the heat exchange. Exchanges of moisture are at a maximum when the relative humidity is low and the surface is wet (FIGURE 9-9).

As a cold cP air mass moves over a warm body of water, there is a rapid exchange of heat and moisture. The lowest layer of the air mass warms and moistens, increasing the instability of the air mass. If the temperature difference between the air and water is large, rapid evaporation causes the

FIGURE 9-9 This visible satellite image provides of a view of the East Coast during the Presidents' Day Storm on February 19, 1979. Notice the hurricane-like "eye" in the center of the storm! Officially, the storm dumped 47.5 cm (18.7 inches) of snow at Washington (now Reagan) National Airport. After the passage of the cold front, the cold continental polar (cP) air mass moved over the warm Atlantic waters off the Carolinas. The cP air mass warmed quickly near the surface and became unstable, causing cumulus-type clouds to form over water, as seen in the lower center of the image.

air to saturate and form a fog. These are *steam fogs* (see Chapter 4). Steam fog is common over the Great Lakes and the north Atlantic in fall and winter when the cP air flows over the warm waters (FIGURE 9-10). This fog can be as deep as 1500 meters, with swirling columns of fog called steam devils. The movement of cP air masses over the Great Lakes can increase snowfall downwind. The increased snowfall is referred to as lake-effect snow (BOX 9-2).

Fogs over the Great Lakes also form during summer when mT air moves over the cooler lake waters. This is an example of an advection fog (see Chapter 4). In this weather situation, the mT air near the surface cools, increasing the relative humidity and causing a fog to develop. The cooling of the air occurs rapidly in the lowest 50 meters. In this case, the lower layers of the air mass are cooling, causing the air mass to become more stable. This suppresses vertical mixing, limiting how deep the fog will form.

An extreme example of air mass modification occurs over the Gulf of Mexico during the cool season. A cold cP air mass plunges south across the Great Plains and stalls over the warm Gulf of Mexico. During the next few days, this air mass modifies rapidly while the large-scale weather pattern draws the air mass northward again across the Gulf Coast of Texas. However, by the time the air mass returns to the mainland, it is considerably warmer and moister than before. In fact, it may even be reclassified as an mT air mass. This remarkable transformation is called a "return flow event" and can happen as many as 25 times during the months of November through April. These events supply the southern U.S. with warm, moist air during the winter and help contribute to the secondary maximum in severe weather activity across the Southeast in November (see Chapter 11).

FIGURE 9-10 Steam fog in Wisconsin. A winter continental polar (cP) air mass that moves over a warmer body of water becomes unstable, as energy and moisture exchanges between the air and the water cause the lower levels of the air mass to warm and moisten.

Box 9-2 Lake-Effect Snows and Buffalo's "Aphid" Infestation

Regions with a localized maximum in snowfall on the southerly and easterly sides of the Great Lakes are referred to as *snowbelts* (shaded regions in the figure below). Why does the geographic location of the snowbelts vary for each lake?

Average Annual Snowfall (Inches)—Great Lakes Region

| <20″ | 20-40″ | 40-60″ | 60-80″ | 80-100″ | 100-120″ | 120-130″ | 130-160″ | 160+″ |

The figure above shows the average annual snowfall for the Great Lakes region. Two distinct patterns are discernible. The first is that, in general, snow depth increases northward. This is expected because temperature usually decreases poleward. The other distinct feature is the difference in the amount of snow along the shoreline.

The Great Lakes modify the weather and climate in the region by modifying air masses that move over them. As the cold air moves over the water, the lower layers are warmed and moistened by the underlying lakes (see satellite image on the right). This makes the air mass unstable. Evaporation increases the moisture content of the air mass, which is then precipitated in the form of snow on the land downwind. Maximum heat and moisture exchanges occur when the air is cold and the temperature difference between the air and the water is large. This condition tends to occur during early winter; this is when the most lake-effect snow is produced. A long path across the warm water by the air mass results in heavy precipitation over the land (see illustration on the next page). The longer the path or "fetch," the more the evaporation and greater the potential for large snowfall amounts over the land on the downwind side of the lake. Hills can amplify the snowfall amounts by providing additional lifting. The location of a snowbelt along a particular lake is a function of the temperature difference between the air mass and the water, the fetch, and the terrain on the leeward side of the lake.

A GEO/NASA satellite called SeaWIFS acquired this image of cloud air flowing over the Great Lakes on December 5, 2000. Lakes Nipigon (top left), Superior, and Michigan have cloud streets forming as the cold wind blows from the northwest across the lakes.

(continued)

Box 9-2 Lake-Effect Snows and Buffalo's "Aphid" Infestation, continued

Lake-effect snows are good for the economy of a region, particularly ski resorts. They also provide water for reservoirs and rivers. Too much lake-effect snow can be hazardous, however, as during the winter of 1976 to 1977. During that winter, lake-effect snows helped to produce 40 straight days of snowfall in Buffalo, New York. A blizzard during this time generated 9-meter (30-foot) snowdrifts and resulted in the deaths of 29 people.

More recently, on October 12 to 13, 2006, Buffalo was blitzed with 57.4 centimeters (22.6 inches) of snow in less than 1 day (see figure on the next page). This was almost four times more snow than from any other October snowstorm in Buffalo history! This exceptionally early storm tapped energy from a very warm Lake Erie, creating 9-km-tall (30,000 feet) convective clouds and blinding, wet thundersnow.

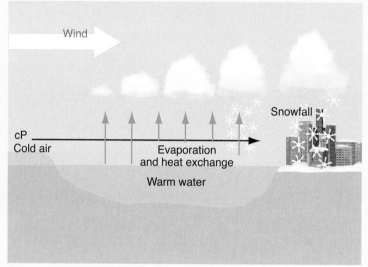

As a cold polar air mass moves over a warmer body of water, it can become unstable. This results in increased snowfall on the downwind side of the lake.

Because trees had not yet shed their autumn leaves, this heavy mid-October snow weighed down and broke tree branches by the thousands. It was said that nearly every tree in Buffalo was damaged in some way from the storm. Nearly 1 million residents lost electrical power at the height of the storm because of falling trees and power lines. Total damage from the storm in the Buffalo area was estimated in the hundreds of millions of dollars.

Ironically, the National Weather Service in Buffalo named lake-effect snowstorms after insects that year, in imitation of hurricane names (see Chapter 8). This early storm was the first of the 2006–2007 season, and so its name began with "A": "Aphid" (see map on the next page).Given the amount of damage to vegetation it caused, "Aphid" was a particularly apt name!

Lake-effect snow can bombard a location as long as all of the ingredients—cold winds, warm water, and a long fetch—are present. This was demonstrated when 141 inches of snow—more than 3.5 meters, or nearly 12 feet—fell on Redfield, NY, downwind of Lake Ontario over a 10-day stretch from February 3 to 12, 2006. The name of the storm that plagued Redfield was, appropriately enough, "Locust."

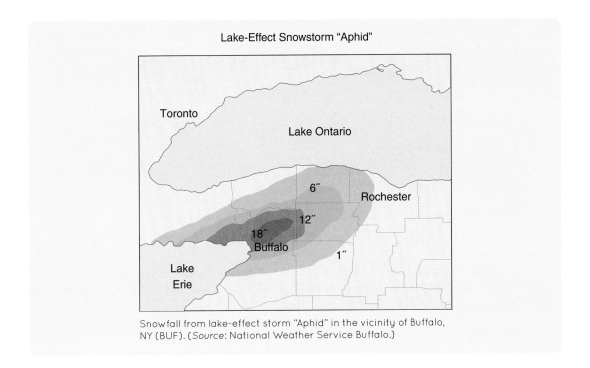

Lake-Effect Snowstorm "Aphid"

Snowfall from lake-effect storm "Aphid" in the vicinity of Buffalo, NY (BUF). (*Source*: National Weather Service Buffalo.)

Air masses can also be modified when lifted or forced to descend by topography. As an mP air mass from the Pacific Ocean moves over the northwest coast of the U.S., the mountains lift it. This causes the air to cool, increasing the relative humidity and leading to cloud formation and snowfall in the mountains. A cP air mass formed over the Greenland Plateau will descend as it moves off the high plateau, warming along the way.

FRONTS

Although an individual air mass can change over time, different air masses do not mix readily. Therefore, when two air masses come in close contact, they retain their separate identities for several days. The transition zone between two different air masses is called a **front**. Fronts can be hundreds of miles long and exist as long as the air masses they separate remain distinct.

Norwegian meteorologists around the time of World War I laid the foundation for our concepts of fronts and their movements. They observed large air masses with different temperature and moisture properties that advanced and retreated versus one another. Clashing air masses often led to disruptive weather conditions. The boundary between air masses was called a "front," analogous to the boundaries used on military maps to separate battling armies.

A frontal zone is a sloping surface that separates two air masses. FIGURE 9-11 is a three-dimensional sketch of a front. The area where the front meets the ground is called the *frontal zone*. The frontal zone is featured on surface weather maps because this is where the air mass contrasts are usually most prominent. Although fronts are represented on weather maps as lines, the front is actually a zone where weather conditions may rapidly change across distances of a few miles as one goes from one air mass to the other.

Fronts are classified by the temperature changes that result after an air mass passes over a given location. A **cold front** indicates that colder air will follow the front's passage; a **warm front** means that warmer air will follow. But there are more weather changes than just temperature during a frontal passage. In this section, we will investigate the characteristic changes in temperature, pressure, wind direction, and cloudiness of a well-developed cold front. Then we will do the same for a warm front. We will also discuss **stationary fronts** and **occluded fronts**.

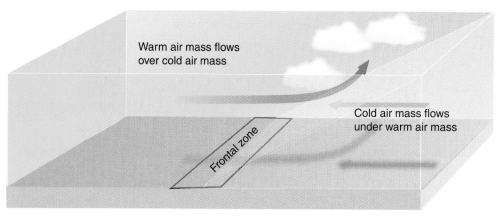

FIGURE 9-11 A schematic three-dimensional representation of a generic frontal zone.

"Identifying Fronts"

Nature is often much more complex than a textbook. Chapter 10 presents examples of real-life fronts associated with a severe storm; look for similarities and differences between the idealized cases below and the observations in Chapter 10.

Cold Fronts

FIGURE 9-12 shows an idealized structure of temperature, pressure, wind direction, and precipitation associated with a cold front. The blue triangles point in the direction of the movement of the front. To imagine how temperature changes as a cold front approaches, observe how the isotherms change along the line in the figure that is perpendicular to the front. As a cold front approaches, the temperature remains steady or rises under the influence of warm southwesterly winds (Figure 9-12a and 9-12d). The temperature then quickly drops in the frontal zone region as the cold air mass moves into the area.

As the cold front approaches, a decrease in pressure is observed (Figure 9-12b). Fronts develop in regions of lower pressure because, as we learned in Chapter 6, air at the surface tends to blow toward lower pressure. This convergence of air helps intensify temperature and moisture contrasts, creating and sustaining fronts. Changing frontal weather is therefore a good bet when the barometer shows a consistent drop in pressure.

After cold frontal passage, the atmospheric pressure begins to increase as a cold, high-pressure air mass moves into the region (Figure 9-12b). The largest pressure changes usually occur in the frontal zone.

Precipitation associated with our idealized cold front is shown in Figure 9-12c. Precipitation is confined to the frontal zone, usually in the form of scattered showers and thunderstorms. When temperature and moisture contrasts are strong, a squall line is sometimes observed in advance of the cold front. The reasons for this are discussed in Chapter 11. After the front passes, the

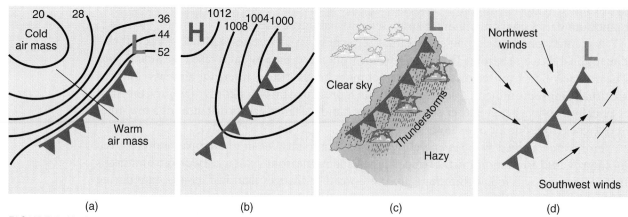

(a)   (b)   (c)   (d)

FIGURE 9-12 Surface weather associated with a cold front. (a) Temperature in degrees Fahrenheit. (b) Pressure in millibars (mb). (c) Clouds and precipitation. (d) Wind direction. The L marks the location of the lowest pressure. The thin line in A is the cross-section used in Figure 9-13.

sky clears. Scattered cumulus or stratocumulus clouds sometimes accompany the cold air mass behind the front. Showers behind the cold front can occur when a cP air mass moves over a warm surface (Figure 9-12c).

Strong wind shifts often accompany fronts. Air flows into the low-pressure area in a counter-clockwise direction in the Northern Hemisphere and because of friction crosses the isobars. Thus, ahead of a cold front the winds tend to be from the south or southwest (Figure 9-12d). Wind direction is variable in a frontal zone as the winds shift to a northerly or northwesterly direction after the cold front passes—that is, the wind turns clockwise in direction because of the frontal passage. These winds advect the cold air into the region. If a cP air mass is behind the front, it has lower humidity, and the visibility is usually good.

Fronts extend above the surface because air masses are three-dimensional phenomena. FIGURE 9-13 is a vertical slice through our idealized cold front along the line shown in Figure 9-12a. The front slopes upward, toward the pool of cold air. The altitude of the cold frontal zone typically increases 1 kilometer for each 50 to 100 kilometers of horizontal distance along the surface. The slope is steepest where the air mass meets the surface, sloping more gently behind the surface front. The cold air often moves fastest near the surface for a variety of reasons, including the fact that the cross-frontal winds that push the cold front along are greatest near the surface. This gives the characteristic curved vertical profile of the cold front (Figure 9-13).

The steep slope near the surface frontal zone forces the warmer air it is replacing to rise rapidly. This strong "frontal lifting" (see Chapter 4) tends to generate thunderstorms. Thus, precipitation associated with the cold front is often intense but brief and restricted to the immediate vicinity of the cold frontal zone near the surface.

The duration of the precipitation depends on the horizontal extent of the vertical lifting and the speed of the front. The precipitation band is often narrow because of the steep cold front. Unlike daytime showers, frontal precipitation can occur 24 hours a day and remain intense even at night. The speed of a cold front is observed to vary from nearly stationary to 50 kilometers per hour (31 mph) or greater, especially over the western Great Plains, where flat ground and the presence of the Rocky Mountains just to the west help funnel cold air rapidly toward Mexico. The fairly rapid speed of cold fronts and their ability to generate stormy weather without the added "juice" of daytime heating together explain the statement at the very beginning of this chapter, "lightning bolts at dawn soon are gone."

FIGURE 9-14 is the meteogram (see Box 5-1) of weather associated with a cold frontal passage experienced in Atlanta, Georgia, during May 2002. Although not an exact match to our conceptual models of Figure 9-12 and 9-13, t he weather associated with this frontal passage shows many of the same features. Initially, there is a warm and moist air mass present at Atlanta. Around 12Z the winds shift to the southwest in agreement with our idealized model (Figure 9-12d). The cold front closes in on Atlanta later in the day as the pressure steadily drops. The thunder-

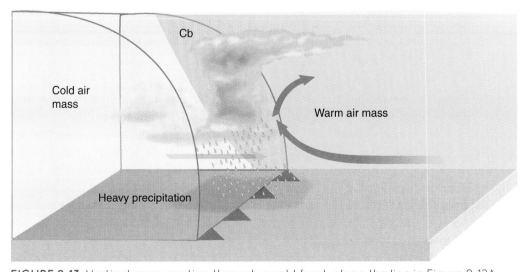

FIGURE 9-13 Vertical cross-section through a cold front, along the line in Figure 9-12A.

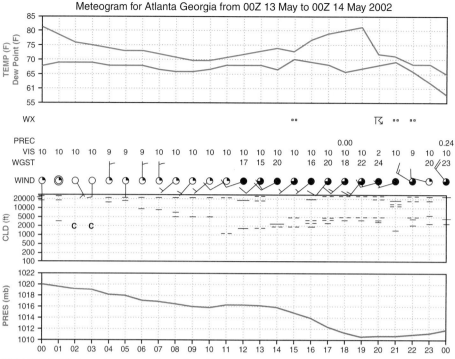

FIGURE 9-14 A meteogram (see Box 5-1) of the weather in Atlanta, Georgia, between 0000 UTC (00Z) May 13 (at left) and 0000 UTC (00Z) May 14, 2002 (far right). A cold front passes through the city at approximately 22Z. (Courtesy of Plymouth State University Weather Center.)

storm at 20Z is just ahead of the cold front; evaporation from the precipitation temporarily cools the air, but the front has not yet reached Atlanta. The cold frontal passage itself is associated with both temperature and dew point decreases, as warm and moist mT air is replaced with cooler and drier cP air. This occurs by 22Z, when the wind direction shifts to northwesterly, as in our idealized model. Shortly after the frontal passage, the pressure begins rising and precipitation ends.

An even more dramatic, "textbook-case" cold frontal passage is depicted in **FIGURE 9-15**. In this case, a very sharply defined cold front passes through Oklahoma City, Oklahoma, immediately dropping temperatures over 6° C (11° F) in the first minute, over 10° C (18° F) in the first 2 minutes, and almost 29° C (52° F) in 10 hours. The dew point drops, and the pressure rises from the moment of frontal passage as well. Meanwhile, the wind direction pivots from southwesterly to northwesterly within just a few minutes of the frontal passage.

Not all real-life frontal passages are this well defined, however, and not all follow this classic pattern. But these examples definitely confirm the usefulness of the conceptual model of a front.

■ Warm Fronts

FIGURE 9-16 depicts the temperature, pressure, wind direction, and cloud patterns typically observed in association with a warm front. The red half-circles point toward the direction of movement of the warm air mass. After the passage of a warm front, the air is warmer and often more humid. As a warm front approaches, the temperature initially remains steady or climbs slowly and then rapidly increases during the frontal passage (Figure 9-16a). As with the cold front, the warm front is located in a region of lower pressure relative to regions ahead and behind the front (Figure 9-16b). Therefore, the pressure drops as the warm front approaches and then increases as it passes by. The changes in pressure associated with the warm front are not as rapid as those associated with the cold front.

The vertical slope of the warm frontal zone is gentler than that of the cold front. Its altitude increases 1 kilometer for every 200 kilometers or so of horizontal distance perpendicular to the surface front. As a result, the warm frontal air slides upward gradually over the cooler air ahead of the front. This upgliding warm air is a form of **overrunning**. Overrunning occurs when an air mass moves over a colder air mass that is denser. Deep layers of stratiform clouds (cirrostratus, altostratus, and nimbostratus) are often the result of overrunning. The gentle rise of the warm frontal slope promotes condensation and clouds but not usually the thunderstorms that typify cold frontal precipitation.

Steady precipitation falls from nimbostratus clouds that result from the overrunning (Figure 9-16c). The precipitation can be far in advance of the frontal zone. The frontal slope also tends to generate stratus-type clouds far in advance of the frontal zone. So, ideally, we could predict the approach of a warm front by watching the changing cloud conditions (Figure 9-16c). Cirrus clouds appear first, followed by cirrostratus, which may result in a halo around the Sun (see Chapter 5). Cirrostratus clouds are followed by altostratus clouds, which make the Sun appear milky or watery. Stratus and nimbostratus clouds follow the middle-level clouds. As discussed in Chapter 4, nimbostratus clouds produce steady precipitation. If the warm air is unstable, cumulus-type clouds may also appear.

The winds ahead of the warm front are usually observed to be easterly or northeasterly (Figure 9-16d). Again, this is associated with the counterclockwise convergence of surface air in toward low pressure. Behind the front, the winds are usually from a southerly direction. As with cold fronts, a warm frontal passage leads to a clockwise turning of the wind direction.

A slice through a warm front revealing its vertical structure is shown in **FIGURE 9-17**. The gentle slope generates stratus-type clouds, starting with cirrus, cirrostratus, altostratus, nimbostratus, and stratus nearest the front. The cirrus that first heralds the approach of the front can be 1000 kilometers (600 miles) ahead of the surface front. Precipitation is steady and moderate and can last for several days, depending on how fast the front is moving. Warm fronts usually move at about half the speed of cold fronts. In mountainous regions such as the Appalachian Mountains, the warm front may stall because the warm air cannot displace cold

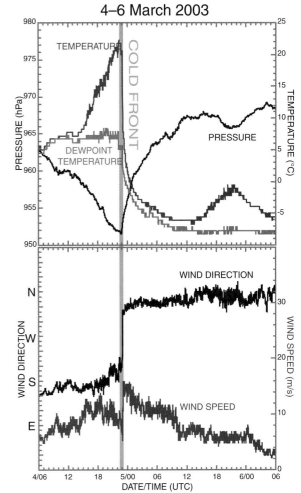

OKLAHOMA CITY, OKLAHOMA (OKC) ASOS
4–6 March 2003

FIGURE 9-15 Graphic depicting a cold frontal passage at Oklahoma City, Oklahoma, using high-resolution 1-minute ASOS observation data. Notice the dramatic changes in temperature, dew point, pressure, wind direction, and wind speed at the moment of frontal passage (vertical blue line) at 2306 UTC on March 4, 2003. (Data from David M. Schultz, University of Helsinki/ASM/NOAA.)

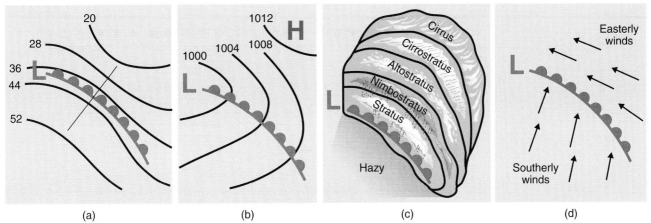

FIGURE 9-16 Surface weather associated with a warm front. (a) Temperature in degrees Fahrenheit. (b) Pressure in millibars (mb). (c) Clouds and precipitation. (d) Wind direction. The L marks the location of the lowest pressure. The thin line in A is the cross-section used in Figure 9-17.

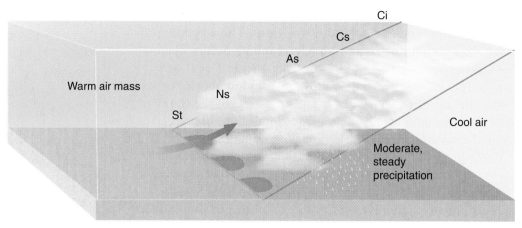

FIGURE 9-17 Vertical cross-section through a warm front, along the line in Figure 9-16A. As the front approaches a location, cirrus (Ci) clouds appear first, followed by cirrostratus (Cs), altostratus (As), nimbostratus (Ns), and stratus (St) clouds.

air entrenched in the valleys. The slow movement, coupled with the large-scale vertical lifting, explains why precipitation can persist for several days.

Hazardous weather can accompany warm fronts. If the air near the ground is below freezing, sleet or freezing rain can occur (see Chapter 4). In addition, as rain falls from the warm air into the cold air near the surface, it evaporates and increases the relative humidity of the air near the surface. If the warm air is significantly warmer than the cold air below, the drops will evaporate quickly and form a fog referred to as a frontal fog.

FIGURE 9-18 is a meteogram demonstrating the weather associated with a warm frontal passage in Athens, Georgia, during a 24-hour period in January 2002. Before the passage of the front the city experiences light winds from the east with fog and near-zero visibility.

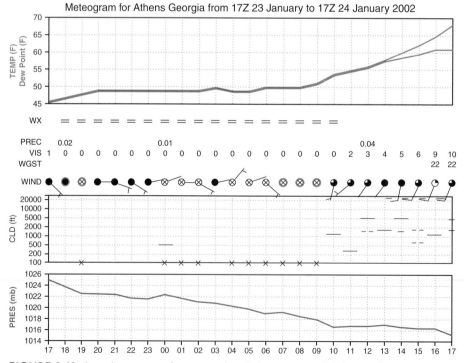

FIGURE 9-18 A meteogram of a warm frontal passage at Athens, Georgia, between 1700 UTC (17Z) January 23 (far left) and 1700 UTC (17Z) January 24, 2002 (far right). The warm front passes Athens at approximately 10 Z. (Courtesy of Plymouth State University Weather Center.)

TABLE 9-2 Typical Frontal Passages

Weather Variable	Cold Front		Warm Front	
	Before Passage	*After Passage*	*Before Passage*	*After Passage*
Temperature	Warm	Cooler	Slowly warming	Warm
Dew point temperature	High	Lower	Slowly rising	Higher
Sea-level pressure	Falling	Rising	Falling	Steady
Wind direction	Southerly	Westerly	Easterly	Southerly
Clouds	Cumulonimbus	Clearing, some stratocumulus	From cirrus to stratus	Cumulus
Precipitation	Heavy near front	Ending	Steady, moderate ahead of front	Ending

Pressure gradually decreases as the front approaches until approximately 10Z, at which time the fog lifts and low clouds remain. Meanwhile, the temperature and dew point begin to rise just ahead of the warm front, as warmer and moister mT air reaches the vicinity of Athens. Shortly after the front passes at 10Z, the winds become steady and from the south-southwest, as in our idealized model (Figure 9-16D). By noon (17Z), the damp foggy morning in Athens is transformed into an unseasonably warm and partly sunny January day because of the warm front's passage. **TABLE 9-2** summarizes the weather conditions prior to and following the passage of typical cold fronts and warm fronts.

"Frontal Passages."

Stationary Fronts

Stationary fronts occur when two air masses collide but move little or not at all at the surface. Although the front appears stationary at the surface, the air above can be moving, causing over-running. This is why the weather conditions along a stationary front are sometimes similar to, although milder than, a warm front with a stable, warm air mass.

The warm air overrunning the colder air mass causes the clouds and precipitation along a stationary front. This overrunning can result in extended periods of cloudiness and light precipitation on the cold side of the stationary front. As with all fronts, the amount of rain depends on the amount of moisture present and the stability of the atmosphere.

A stationary front is depicted on a weather map as a line with blue triangles and red half-circles alternating on opposite sides, suggesting a standoff. Just because a stationary front is not moving, however, do not assume that it is a weakling among fronts. A stationary front may represent a hard-fought stalemate between clashing air masses with sharply contrasting temperatures. By the thermal wind relationship (Chapter 6), this means that midtropospheric winds above a stationary front can be strong. This can be an important factor in the development of thunderstorm-related windstorms known as "derechos," which are covered in Chapter 12.

Occluded Fronts

The Norwegian meteorologists who began the study of fronts hypothesized that fast-moving cold fronts could "catch up" to slower warm fronts and lift the warm air off the ground. This process is called occlusion, and the surface boundary between the two cold air masses is called an occluded front. The occluded front is represented on weather maps by a purple line with alternating triangles and half-circles on the same side of the line. The direction of movement is indicated by the direction the triangles and half-circles point.

There are two basic types of occluded fronts in the Norwegian scheme: the cold-type occlusion and the warm-type occlusion. In the cold-type occlusion, the cold cP air behind the occluded front digs underneath the warm and the cool air masses ahead of it. In the warm-type occlusion, the cool mP air behind the occluded front overruns the colder cP air, as does the warm air aloft

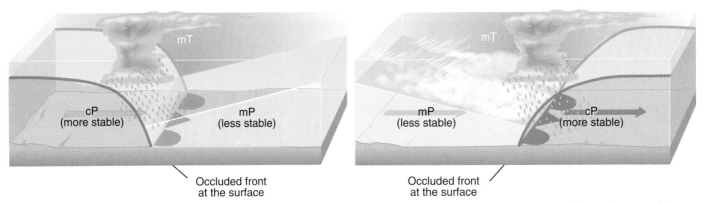

FIGURE 9-19 In the Norwegian model, the occluded front is caused by cold air masses coming together when a cold front moves faster than the warm front. The air mass behind the cold front can be colder or warmer than the air mass ahead of the warm front. As a result, there are two idealized types of occluded fronts: the cold-type occlusion (left) and the warm-type occlusion (right). Recent research has shown that the occluded front actually slopes over the more stable air mass, not necessarily the colder air mass.

(**FIGURE 9-19**). The weather ahead of the occluded front is similar to that ahead of a warm front. The weather behind the occluded front is similar to that of a cold front.

Recent research has modified this traditional view of occluded fronts. Cold fronts do not necessarily "catch up" to warm fronts during the occlusion process, although the weather associated with them may resemble that shown in Figure 9-19. The occluded front slopes over the more stable air mass, not necessarily the colder air mass. Warm-type occlusions are more common than cold-type occlusions because warm-frontal zones are almost always more stable than cold-frontal zones.

Occlusion is also a key part of the maturing process of developing extratropical cyclones, which is examined in detail in the next chapter. Researchers have found that the formation of the occluded front at the surface coincides with a reorganization of the entire cyclone structure aloft.

Drylines

In late springtime, dry cT air from the plateaus of Mexico is drawn northeastward into the south-central U.S. by the counterclockwise circulation of extratropical cyclones in the lee of the Rocky Mountains (see Chapter 10). The difference in temperature can be small between the cT air and the southerly winds bringing moist mT air over eastern Texas. However, the clash in dew points between the two air masses can be literally breathtaking: 5° C (9° F) or more over a distance of a kilometer or two and 28° C (50° F) over 160 kilometers (100 miles) or less. This frontal zone, which is defined by moisture and wind rather than temperature contrasts, is known as the **dryline**. It is depicted on weather maps as a yellow or brown line with clear semicircles in a continuous line that point in the direction of the movement of the dryline.

A damaging dryline event is shown in **FIGURES 9-20** and **9-21**. As seen in Figure 9-20A, on April 9, 2009, a dryline extended from southern Oklahoma throughout most of central Texas, just ahead of a cold front. The exceptionally dry and windy conditions immediately behind the dryline (Figure 9-20B) fueled wildfires in both states that caused 62 injuries in Oklahoma and burned 150,000 acres in Texas, killing three people and destroying several towns. Figure 9-21 is a high-resolution satellite image of the fires and the wind-driven smoke from these fires near the Dallas-Fort Worth, Texas, metropolitan area.

Drylines also help provide a focus for thunderstorm activity over Texas and Oklahoma in late spring, to the delight of storm chasers. As seen in Figure 9-21, converging air at the surface causes clouds ahead of the dryline, sometimes leading to thunderstorms just like a cold front. In a region that averages only about 25 centimeters (10 inches) of rain each year, these thunderstorms are a crucial part of the annual water supply. Therefore, the dry parching wind behind the dryline actually plays a key role in the agriculture of the semiarid regions of central and west Texas.

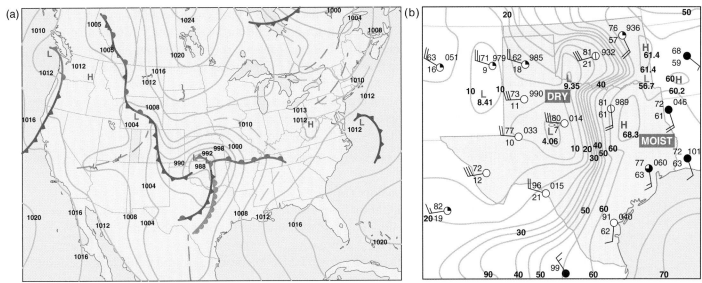

FIGURE 9-20 (a) U.S. surface weather map at 2100 UTC on April 9, 2009, showing a dryline (yellow line) extending from a low-pressure center over Oklahoma to Mexico. (b) Southern Plains weather map at 2000 UTC on April 9, 2009, showing station plots and contours of dew point temperature (green line). Note the extreme gradient in dew point between Dallas-Fort Worth (61° F) and Abilene, Texas (7° F) in north-central Texas. (Part a source: NOAA. Part b courtesy of Plymouth State University Weather Center.)

PUTTING IT ALL TOGETHER

Summary

An air mass is a large body of air with similar temperature and moisture properties. The movement of air masses generates fronts and causes changing weather conditions. These changes are sometimes welcomed, while at other times, they are dreaded because of the weather hazards they bring.

Air masses adopt the characteristics of the source regions in which they form. Cold air masses are referred to as polar air masses (P) because they usually form in the polar regions, where the surface is cold. Warm air masses are of subtropical or tropical origin; both are referred to as tropical air masses (T). Air masses that form over water are referred to as maritime (m), whereas those generated over continents are referred to as continental (c). Maritime air masses are usually cooler and moister than continental air masses formed at the same latitude. Mixing and matching these regional categories leads to the four basic air mass types: cP, cT, mP, and mT. Some classification schemes also use the letter A to denote bitter-cold Arctic air masses.

An air mass eventually moves and exchanges heat and moisture with the ground it migrates over in a process known as air mass modification. When a cold air mass moves over a warm surface, lower layers of the troposphere warm, decreasing the stability of the air mass. This favors rising motion, which increases the possibility of condensation and precipitation. Conversely, when the warmer air mass moves over a cold surface, it is cooled, increasing the stability of the air mass and opposing the formation of clouds and precipitation.

Fronts form when and where air masses collide. The colder air mass pushes or slides under the warmer air mass. When a cold air mass replaces a warm air mass, the boundary between the two air masses is called a cold front. Cold fronts are often associated with a narrow band of clouds and intense precipitation. A

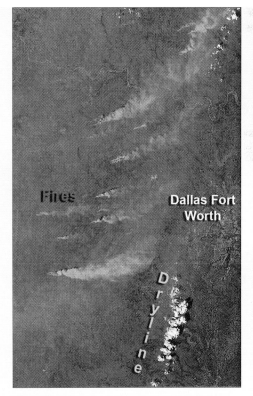

FIGURE 9-21 High-resolution satellite image of fires and windblown smoke just west of the Dallas-Fort Worth area (far right center) at 1928 UTC on April 9, 2009. The puffy white areas at top right and bottom right of the images are cumulus clouds at the leading edge of the dryline.

warm front occurs when the warm air mass replaces a cooler air mass. Warm fronts herald their approach with a large deck of steadily lowering and thickening clouds. Moderate precipitation occurs as a warm front nears. A stationary front occurs when neither air mass is advancing. Occluded fronts are said to form when a cold front catches up with and overtakes a warm front, although this explanation is an oversimplification. Warm-type occlusions, in which a less stable air mass overtakes a more stable air mass, are most common. Drylines are moisture fronts that help trigger thunderstorms over the southwest U.S.

Several general conclusions can be drawn from our study of fronts:

1. Fronts form at the boundaries between air masses of different temperatures and moisture amounts.
2. Warmer air always slopes upward over colder air.
3. Clouds and precipitation form as a warm air mass rises over more dense colder air.
4. The front always slopes upward over the cold (or more stable) air.
5. Pressure drops as a front approaches.
6. In the Northern Hemisphere, wind direction near the ground shifts clockwise as the front passes.

▨ Key Terms

You should understand all of the following terms. Use the glossary and this chapter to improve your understanding of these terms.

Air mass	Continental tropical (cT)	Occluded front
Air mass modification	Dryline	Overrunning
Blizzard	Front	Static stability
Cold front	Maritime polar (mP)	Stationary front
Continental polar (cP)	Maritime tropical (mT)	Warm front

▨ Review Questions

1. What kind of surface pressure pattern is associated with air masses?
2. Why is mountainous terrain a poor source region for air masses?
3. Explain the requirements for a geographic region to serve as a source region for an air mass.
4. Describe the characteristics of each of the four main types of air mass.
5. Use the text's Web site to try your hand at identifying different types of air masses on an actual weather map. What weather variables distinguish one air mass from another?
6. Describe the type of clouds and precipitation associated with an approaching warm front.
7. At the beginning of the chapter, we made an analogy between accents and air masses. Explain why there is no "O'Hare Airport" accent despite the fact that millions of people go through this Chicago airport every year. Then extend your reasoning to explain why there are no midlatitude air mass source regions even though air travels through the middle latitudes.
8. Discuss the differences and similarities between a warm front and a cold front.
9. Explain why a cold cP winter air mass could make it difficult for firefighters to do their job.
10. Would freezing rain be more likely with the passage of a warm front or a cold front? Why?
11. Before the advent of satellites and advanced systems for tracking fronts, weather predictions were often made by observing changes in wind direction, pressure, and cloud types. These predictions drew on knowledge accumulated over years of weather observations. The predictive nature of these observations is sometimes revealed in weather lore. The following weather lore predicts precipitation:

    Ring around the Moon or Sun,
    Rain before the day is done.

    Is there a physical basis for this statement?

▨ Observation Activities

1. When a front is approaching where you live, observe the changing cloud patterns. Try to use these patterns to forecast the time of arrival of the front.

2. Some seasonal recreational activities in your state are profoundly affected by weather. List two activities and describe the air masses that can positively and negatively impact participation.

3. Air mass classification has advanced beyond the days of the Norwegian meteorologists. Visit http://sheridan.geog.kent.edu/ssc.html and read about "Spatial Synoptic Classification," an automated version of air mass classification with air mass types that are slightly different than those described in this chapter. Use this site to look up air mass types for your location on days in the past 50 years or to see a current map of air masses across the entire U.S.

This rain cloud icon is your clue to go to the *Meteorology* Web site at http://physicalscience.jbpub.com/ackerman/meteorology/. Through animations, quizzes, web exercises, and more, you can explore in further detail many fascinating topics in meteorology.

10 Extratropical Cyclones and Anticyclones

AFTER COMPLETING THIS CHAPTER, YOU SHOULD BE ABLE TO
- Describe the different life-cycle stages of the extratropical cyclone, identifying the stages when the cyclone possesses cold, warm, and occluded fronts and life-threatening conditions
- Explain the relationship between a surface cyclone and winds at the jet-stream level and how the two interact to intensify the cyclone
- Differentiate between extratropical cyclones and anticyclones in terms of their birthplaces, life cycles, relationships to air masses and jet-stream winds, threats to life and property, and their appearance on satellite images

INTRODUCTION

What do you see in the diagram to the right: a vase or two faces? This classic psychology experiment exploits our amazing ability to recognize visual patterns. Meteorologists look for patterns in weather observations. Instead of pondering "is that a nose, or the narrow part of a vase?" the meteorologist asks, "Is this a line of rain showers? Could that be a wind shift? Should I be focusing on the areas of warm and cold air or the regions in between?" Finding the "face" in the weather is a lot more confusing than in a textbook diagram!

Early 20th-century weather forecasting reflected this confusion. Some meteorologists focused on the location and shape of low-pressure areas. Others scrutinized temperature patterns; still others looked at cloud types. No one yet saw the overall "face," only these individual weather features.

In 1918, 20-year-old meteorologist Jacob "Jack" Bjerknes (page 302; pronounced BYURK-nizz) discovered the "face in the clouds" of his native Norway: the extratropical cyclone or low-pressure system. His conceptual model of this type of cyclone emphasized the importance of the fronts

that extend like a moustache from the center of midlatitude lows on a weather map. Jack and his father Vilhelm's "Bergen School" of meteorology fleshed out this face using their intuition and observations of the sky. They realized that the extratropical cyclone, like a face, has a distinctive three-dimensional shape. Furthermore, like a person, the cyclone matures from youth to old age in a recognizable and predictable way. At last, meteorology knew what face to look for! Weather maps and forecasts have never been the same since.

In this chapter, we learn how to recognize the face of the extratropical cyclone. To make this face come alive for you, we look at weather data leading up to a tragic day in American history: November 10, 1975. On that day, a cyclone fitting Bjerknes's description ravaged the midwestern United States and contributed to the wreck of the Great Lakes iron ore freighter *Edmund Fitzgerald*. Twenty-nine sailors perished on the *Fitzgerald* without a single "mayday." More than 35 years later, this tragedy lives on through television documentaries, books, and folk songs. We learn here what the face of a classic cyclone looks like from a variety of angles and why it looks that way.

Some faces and people are not so memorable, but the impacts they make are. This is also the case with the cyclone's counterpart, the anticyclone or high-pressure system. Highs look like bland and boring blobs on the weather map, but they can also harbor potentially tragic weather conditions. This chapter ends with an examination of a heat wave in 2003 that quietly killed 35,000 people in western Europe in the space of just a few days.

A TIME AND PLACE OF TRAGEDY[1]

The year is 1975—the Vietnam War ends, and a gallon of gasoline costs 44 cents in the U.S. A small upstart computer software company named Microsoft is founded in Albuquerque, New Mexico. Meanwhile, teenagers are starting to buy platform shoes and boogie to the newest pop music fad, "disco."

On the Great Lakes (FIGURE 10-1) in 1975, large boats deliver iron ore from Minnesota to the steel mills and car factories of Indiana, Michigan, and Ohio. Lake Superior, Earth's broadest lake, serves as a highway for these boats. The pride of the American ore freighters is the 222 meter (729 foot)-long *Edmund Fitzgerald* (FIGURE 10-2). Her respected captain, Ernest McSorley, has weathered 44 years on the lakes. His crew of 29 includes six sailors younger than age 30 years, including 22-year-old Bruce Hudson and 21-year-old Mark Thomas (FIGURE 10-3).

In November 1975, "Big Fitz" is completing just its 17th full shipping season on Lake Superior. It's a young boat by Great Lakes standards. The end of the shipping season comes when the "gales of November" howl across the Lake and winter's icy cold freezes some of the lake surface solid. One trip too many, during a fierce extratropical cyclone, will sink the *Fitzgerald* and its crew in 160 meters (530 feet) of churning Lake Superior water on November 10, 1975.

A LIFE CYCLE OF GROWTH AND DEATH

In the 1920s, Jack Bjerknes helped discover that cyclones such as the one that would sink the *Fitzgerald* often follow a stepwise evolution of development. The left-hand side of TABLE 10-1 shows how Bjerknes himself depicted the **Norwegian cyclone model** life cycle. The cyclone arises as a **frontal wave** along a stationary front separating cold, dry cP air from warm, moist mT air. It is called a "wave" because the **warm sector** region between the cold and warm fronts resembles a gradually steepening ocean wave. In adolescence, the **open wave** develops strong

---

[1]Details of the *Edmund Fitzgerald*'s last voyage used in this chapter are primarily derived from the following books: *The Wreck of the Edmund Fitzgerald* by Frederick Stonehouse (Avery Color Studios, Marquette, Michigan, reprinted 1997), *Gales of November* by Robert J. Hemming (Thunder Bay Press, Holt, Michigan, 1981), and the words of Captain Bernie Cooper in *The Night the Fitz Went Down* by Hugh E. Bishop (Lake Superior, Port Cities, Inc., 2000).

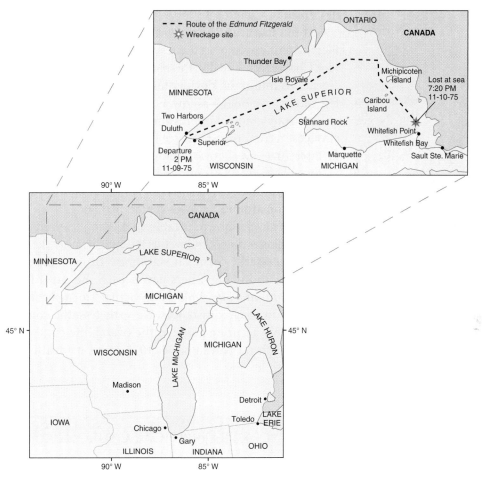

FIGURE 10-1 Map of the Great Lakes region, showing the path of the *Edmund Fitzgerald's* final voyage and locations referred to in this chapter.

FIGURE 10-2 The ore freighter *Edmund Fitzgerald*, in a photograph provided by Ruth Hudson, mother of *Fitzgerald* sailor Bruce Hudson (Figure 10-3).

FIGURE 10-3 *Edmund Fitzgerald* sailors Bruce Hudson (left) and Mark Thomas on the main deck of the *Fitz*.

cold and warm fronts with obvious wind shifts as the entire system moves to the east or northeast. Precipitation (in green) falls in a broad area ahead of the warm front and in a narrow line in the vicinity of the cold front, just as described in Chapter 9.

At full maturity, the **occluded cyclone** sprouts an occluded front, which Bjerknes conceived of as the result of the cold front outrunning the warm front (see Chapter 9). Usually, the barometric pressure at the center of the cyclone reaches its minimum during this stage, sometimes plummeting to 960 mb in a few intense cyclones—as low as in the eye of a Category 3 hurricane! Because of the strong gradient of pressure near its center, the cyclone's winds are usually strongest during this stage (see Chapter 6). The accompanying satellite images in Table 10-1 reveal how the cloudiness associated with the fronts progressively wraps poleward and around the back side of the cyclone. In the final stage, the **cut-off cyclone** (see Chapter 7) slowly dies a frontless death, as clouds and precipitation around the low's center dissipate.

The life cycle shown in Table 10-1 is an idealized conceptual model. Some real-life cyclones do not follow this life cycle in every detail. For example, cyclones do not necessarily form on stationary fronts; some cyclones occlude soon after birth, and many others never reach the cut-off stage. The cyclone that hits the *Edmund Fitzgerald* is an actual storm brought about by a unique combination of circumstances. Nevertheless, the conceptual model provides a remarkably accurate guide to the overall life cycle of the *Fitzgerald* cyclone from its birth on November 8, 1975, until its death several days later, as we see in the following.

DAY 1: BIRTH OF AN EXTRATROPICAL CYCLONE

Saturday, November 8, 1975, is a gorgeous day across much of the U.S. Calm and almost summer-like conditions predominate. High temperatures of above 26° C (80° F) bathe the South and Southwest, with 70s as far north as Indianapolis and Burlington, Vermont. The *Edmund Fitzgerald* glides on smooth Lake Superior waters toward its next load of iron ore at the Duluth (Minnesota)/Superior (Wisconsin) harbor.

Weather maps and satellite pictures (**FIGURE 10-4**) show some cloudiness associated with weak fronts extending from Iowa to Colorado and from Ontario, Canada, to Kentucky. Very little precipitation is falling, however. The fronts will soon stall and will combine into a stationary front lying southwest to northeast from New Mexico to New England. Along this stalling front a very weak (1006 millibar, or mb) low is forming just northwest of Amarillo, Texas, in the northern Texas Panhandle. The low is so weak and the air is so dry (note the low dew point of 30° F [−1° C] at Amarillo) that there is not a cloud in the sky associated with it (Figure 10-4c).

The presence of a low just east, or "downstream," of the Rocky Mountains is a natural consequence of upper-level winds blowing across high mountain ranges. Figure 10-4b shows the observed winds and altitudes (in feet) at the 500-mb level, about midway between the ground and the tropopause. The strong west-to-east winds over the Rocky Mountains are occurring where the solid lines of equal height are closest together, just as we learned in Chapter 6.

During Saturday the 8th, the counterclockwise circulation around this weak low drags colder, drier air southward through the Rockies and draws warmer, moister air northward from the Gulf of Mexico. As a result, temperature and moisture gradients near the low become stronger throughout the day. The occluded front will turn into a "leading" warm front just northeast of the low and a "trailing" cold front that will extend to the southwest of the low.

TABLE 10-1 The Life Cycle of the Extratropical Cyclone, Based on the Bergen School Model

Stage	Weather Map Depiction of Norwegian Cyclone Model	Typical Satellite Image of Life-Cycle Stage	Typical Sea-Level Pressure at Cyclone Center	Corresponding Dates of *Edmund Fitzgerald* Cyclone
Birth (frontal wave)	cP / mT		1000-1010 mb	November 8, 1975
Young adult (open wave)			990-1000 mb	November 9, 1975
Mature (occluded cyclone)			960-990 mb	November 10–11, 1975
Death (cut-off cyclone)			Slowly rising from 960-990 mb up to 1010 mb	November 11–15, 1975

(Modified from Bjerknes, J. and Solberg, H., 1922: "Life cycle of cyclones and the polar front theory of atmospheric circulation." *Geofys. Publ.*, 12, pp 1–61. Courtesy of Norwegian Geophysical Society.)

At the same time, winds above the low intensify for two reasons. First, strong upper-troposphere jet-stream winds are moving into this region. Second, as we learned in Chapter 6, strong winds are usually located above strong surface temperature gradients, and the region around the low is becoming an area of strong temperature gradients. As we'll see shortly, this upper-level wind pattern helps the low grow, which in turn helps the fronts intensify, which intensifies the upper-level winds even more and makes the low grow even faster. . . . It's a cycle of growth!

Meteorologists give this cycle of cyclone birth and growth a special name: **cyclogenesis**. Why cyclogenesis happens at one place and time, but not another, is almost as complicated as explaining how human babies are born. Key ingredients for cyclogenesis include surface temperature gradients, a strong jet stream, and the presence of mountains or other surface boundaries (e.g., a coastline near a warm ocean current). Wherever winds blow across temperature gradients, warm air can glide upward, and cold air can dive and sink, leading to clouds and precipitation; we saw this with both cold and warm fronts in Chapter 9. This tilted pattern of rising and sinking air liberates energy for a cyclone and is called **baroclinic instability**.

The Norwegian cyclone model emphasized the importance of fronts and mountains in cyclogenesis. More recently, meteorologists who, unlike the Bjerkneses, have access to upper-level data from radiosondes have identified short waves (see Chapter 7) in the jet stream as a main triggering mechanism. On November 8, 1975, all three of these ingredients—fronts, a strong jet stream, and mountains—give birth to a memorable extratropical cyclone, a low-pressure system structured completely differently than a hurricane but with the same ability to cause harm.

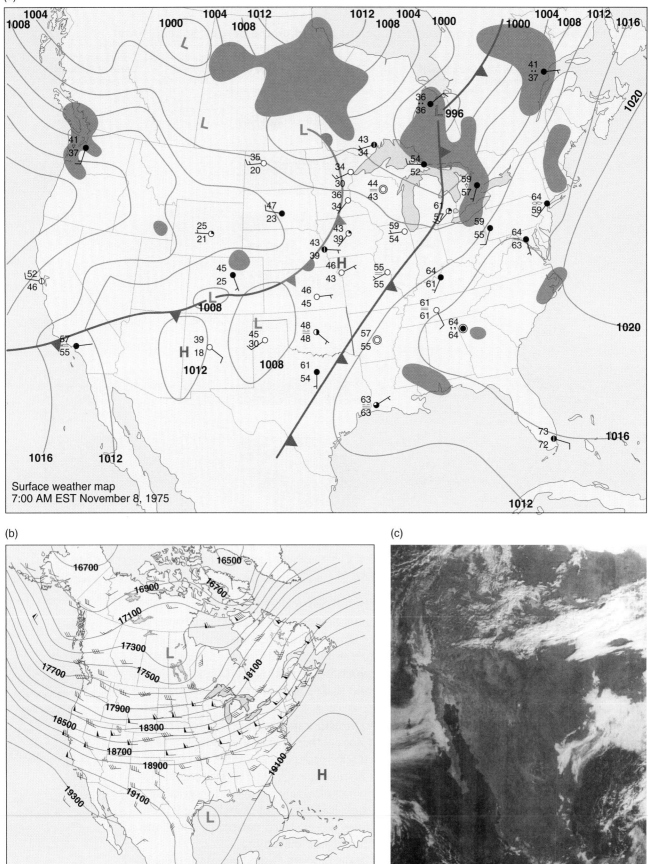

(a)

**Surface weather map
7:00 AM EST November 8, 1975**

(b)

500-mb chart

(c)

FIGURE 10-4 Surface weather (a) and 500-mb conditions (b) for 7:00 AM (EST), Saturday, November 8, 1975. The heights in B are in feet, but the rules relating wind to height patterns are unaffected. The Defense Meteorological Satellite Program visible satellite picture from 10:30 AM (EST) is at the lower right (c). The satellite picture is centered over extreme western Texas; clouds are visible along the west coast of Baja California at the left of the image. (Parts a and b source: NOAA.)

Typical Extratropical Cyclone Paths

For residents of the Great Lakes region, a strengthening cyclone in the Texas Panhandle in November is cause for concern—although the two locations are a thousand miles apart! This is because extratropical cyclones grow and move quickly away from their places of birth, often to the east or northeast.

FIGURE 10-5 shows the typical regions of cyclogenesis (shaded) and typical cyclone paths (arrows) for storms that hit the state of Wisconsin in each season of the year. The figure reveals that the birthplaces of cyclones vary from season to season, moving north in the summer. Why? Because temperature gradients and jet streams, key ingredients for cyclogenesis, move north with the Sun during the summer.

Figure 10-5 indicates that during the fall and winter months the region from southern Colorado to the Oklahoma and Texas Panhandles is a prime breeding ground for Wisconsin-bound cyclones. In fact, this path is so common that these cyclones have a generic name attached to them: **Panhandle Hooks**. The "hook" describes the curved path that these cyclones often take, first bending to the southeast and then curving northeast toward the Great Lakes (**BOX 10-1** explains this bending and curving).

Other parts of the U.S. are hit by extratropical cyclones that form in regions far away from the Rockies (FIGURE 10-6). For example, the Northeast is pummeled by winter cyclones—the **nor'easters** discussed in Chapter 9—that develop along the East Coast over the Gulf Stream near Cape Hatteras, North Carolina. Also, as seen in Figure 10-6a, the Great Lakes region is visited by cyclones born in western Canada that move southeastward. These storms are called **Alberta Clippers** because of their birthplace and their fast forward speed. The Pacific Northwest is often soaked and windblown by cyclones that spin themselves out in the Gulf of Alaska. More rarely (but especially during El Niño and positive Pacific Decadal Oscillation years; see Chapter 8), the West Coast is drenched by extratropical cyclones riding the **Pineapple Express** jet stream blowing northeast from Hawaii. The cyclone paths generally shift northward in summer (Figure 10-6b) along with the regions of strongest temperature gradients.

The Panhandle Hook of November 1975 will also race toward the Great Lakes at a high rate of speed, gathering more energy with every passing hour. The next 2 days will make history, and tragedy, on Lake Superior.

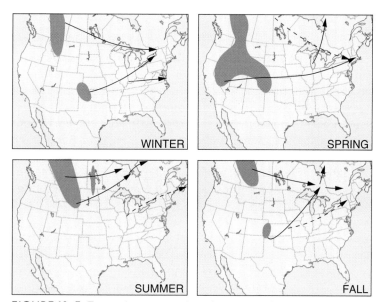

FIGURE 10-5 Typical regions of cyclogenesis (shaded) and paths of cyclones (arrows; likeliest paths in solid lines, other possible paths in dashed lines) that affect the state of Wisconsin on a seasonal basis. (*Source:* Courtesy of Pam Naber Knox, former Wisconsin State Climatologist.)

Box 10-1 Making Cyclones and Waves

Extratropical cyclones, mountain ranges, and upper-level wave patterns in the wind—what's the connection? It's simple if you know a little about the atmosphere and ice skating.

We learned in Chapter 7 that the closer a skater's body is "tucked in" to the axis of rotation, the faster he or she spins. This is why skaters bring their arms and legs in tight against their bodies during fast spins.

What if we could alter the shape, but not the mass, of our skater? Imagine stretching and squashing the skater like a lump of clay. Stretching the skater (see left) has the same effect as pulling the arms and legs in, and so the skater spins faster.

The same principle, with a twist, works for a spinning cyclone. The tropopause acts as a "ceiling" on cyclones and most other weather. The tropopause is, to a first approximation, at about the same altitude throughout the middle latitudes, so a cyclone can be stretched or squashed depending on the altitude of the surface that is beneath it. High mountains squash cyclones; lower oceans and plains help stretch them.

What does this imply about the spin of our cyclones? Just like our skater, a squashed cyclone spins more slowly and a stretched cyclone spins more rapidly. The rate of spin increases as the height of the cyclone increases; spin decreases as cyclone height decreases. Therefore, a meteorological form of the Conservation of Angular Momentum is this: *Spin divided by the cyclone height must be a constant*. (This works for anticyclones too.)

Now, the twist. A skater spins for a minute or so. This is far too short a time for the Earth's spin to affect the skater via the Coriolis force. However, a cyclone spins for days and is strongly affected by the Coriolis force. Because of this, the cyclone, but not the skater, feels the effect of the Earth's spin. So the cyclone has a double dose of spinning. Therefore, the meteorological form of Conservation of Angular Momentum is as follows:

Conservation of angular momentum and skaters.

$$\text{total spin} \div \text{cyclone height} = \text{constant}$$

In meteorological terminology, spin around a vertical axis is called "vertical vorticity," or just "vorticity" for short. The spin of winds in a weather system, such as a cyclone, is called "relative vorticity" because it is measured relative to the ground, but the air and ground are all spinning because of the Earth's rotation too. The Earth's spin is called "planetary vorticity." Vorticity is defined as positive if the direction of rotation is in the same direction as the Earth's spin. Looking down from above the North Pole, the Earth appears to rotate counterclockwise, and so do cyclones; therefore, the relative vorticity of a cyclone in the Northern Hemisphere is positive.

With these definitions, we can put everything together and write a special version of the Conservation of Angular Momentum used in meteorology, known as the Conservation of Potential Vorticity:

$$(\text{planetary vorticity} + \text{relative vorticity}) \div \text{cyclone height} = \text{constant}$$

in which the left-hand side of the equation is a special quantity known as "potential vorticity." Because in this equation potential vorticity is always constant, it is being conserved.

Conservation of potential vorticity lets us link up cyclones, mountains, and upper-level waves in the wind (see the applet "Mountains and Lee-Side Low Formation" on the text's Web site). Imagine a cyclone moving more or less eastward from the Pacific Coast to the Great Plains as depicted in the illustration.

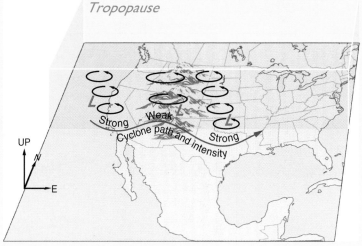

Cyclone strength and path changes as a result of conservation of potential vorticity over the Rocky Mountains.

Over the Pacific, the cyclone is strong. Lows are associated with upper-level troughs. The trough's winds steer the cyclone southeast and then northeast in a cyclonically curved path.

As the cyclone moves over the high Rockies, however, it gets squashed. Conservation of potential vorticity says that in this circumstance the total spin of the cyclone must decrease, too. The once-strong Pacific low appears to decrease in intensity as it moves over the Rockies, and its path curves anticyclonically over the mountains, heading southeast as the weather system crosses the Rockies.

Then suddenly the bottom drops out of the cyclone as it crosses the high Rockies and moves over the lower Great Plains. Over the Plains, the cyclone is stretched. By conservation of potential vorticity, this means the sum of the vorticities needs to increase, too. The cyclone does this in two ways: by "spinning up" (intensifying rapidly), which increases its relative vorticity, and by heading toward the northeast, which increases its planetary vorticity. The northeast path is consistent with the steering winds in the upper-level trough that develops in tandem with the strengthening cyclone.

Because the cyclone seemed to disappear over the Rockies, its reappearance and strengthening just east of the Rockies seem to come out of nowhere. However, they are natural results of potential vorticity conservation, as is the characteristic hook-shaped path of the cyclone (southeast over the Front Range of the Rockies, northeast over the Plains).

Obviously, the cyclone is closely related to the upper-level winds. The wavy path of the cyclone is mirrored in the waviness of the jet-stream winds that guide the storm. These large-scale waves of wind are the Rossby waves that we studied in Chapter 7. Rossby waves move horizontally, like ocean waves. We noted this in Chapter 7, but now we can be more precise about how quickly they move. Using the concept of potential vorticity conservation, famed 20th-century meteorologist C.-G. Rossby found that these waves' west-to-east (W–E) speed is dependent on how fast the jet stream is and how big the wave is (in other words, size matters). Rossby's formula for his waves' speed can be simplified to

Rossby wave W–E speed = 500-mb W–E wind speed
$$- \text{(Rossby wave W–E wavelength)}^2$$
$$\times \text{a constant}$$

Because the two terms on the right-hand side of the equation are subtracted from each other, this means that the speed of a Rossby wave is a push–pull of wind speed versus size. Short waves (the size of the Great Lakes region) move eastward almost as fast as the 500-mb wind because the wind speed factor dominates. However, long waves (the size of a continent) travel very slowly eastward, stop, or even move backwards toward the west. This is because the wavelength term is almost as big, as big, or bigger than the wind speed term.

As you read this chapter, compare the forward speed of the *Fitzgerald* cyclone with Rossby's simple formula: Does the storm move fastest when it is young and small? Does it slow as it matures and grows bigger? Try it and see.

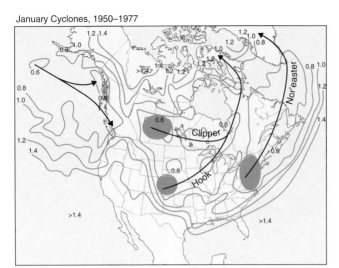

January Cyclones, 1950–1977

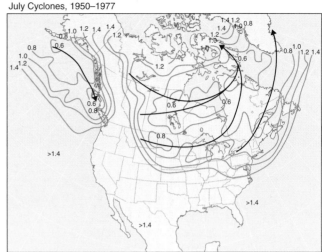

July Cyclones, 1950–1977

FIGURE 10-6 Typical regions of cyclogenesis (shaded) and paths of cyclones (arrows; likeliest paths in solid lines, other possible paths in dashed lines) that affect North America in January and July. (Adapted from Zishka, K. M., and P. J. Smith, *Monthly Weather Review*, April 1980: 391–392.)

DAY 2: WITH THE *FITZ*

Early on Sunday morning November 9th, 1975, workers at the Duluth/Superior docks begin the task of loading the *Edmund Fitzgerald*. On this trip, the *Fitz* will carry enough iron ore to make 7500 automobiles. The iron ore pellets slide down huge chutes like marbles into the 21 hatches in the middle of the boat. The crew then anchors each of the 6350-kilogram (7-ton) hatch covers using 68 special clamps. Shortly before 2:00 PM, the *Fitzgerald* departs into the open waters of Lake Superior.

Less than 2 hours later, the *Arthur M. Anderson,* sister ship to the *Fitzgerald,* leaves Two Harbors, Minnesota, with a load of iron ore for the Gary, Indiana, steel mills near Chicago. The *Anderson*'s captain, "Bernie" Cooper, recalled years later that this Sunday "was one of the special days on Lake Superior—just ripples on the water, sunny and warm for November. As we departed we could see the *Edmund Fitzgerald. . . .*" The two ships will sail together for the rest of the *Fitzgerald*'s last journey.

Portrait of the Cyclone as a Young Adult

The baby cyclone of November 8 has matured overnight—literally (**FIGURE 10-7**). The surface weather map for Sunday morning on the 9th (Figure 10-7a) shows a 999-mb low over Wichita, Kansas, a pronounced cold front digging southward into Texas, and a warm front pushing north toward Iowa. Temperature contrasts across the cold front are now 11° C (20° F), and 5.5° C (10° F) across the warm front. The 500-mb map (Figure 10-7b) shows a deepening short wave trough to the west of the surface low, part of the cycle of growth of the storm. The baby is now a young adult, teeming with energy.

A high-resolution satellite picture taken just 3 hours later (Figure 10-7c, and enlarged in **FIGURE 10-8**) reveals that this cyclone has taken on a classic form at an early age. Seen from space, it has the shape of a **comma cloud**, which is characteristic of mature extratropical cyclones and is quite different from the circular tropical cyclone. The tail of the comma is cloudiness along the trailing cold front, which a tropical cyclone lacks. The comma head consists of clouds and light precipitation circling counterclockwise around the low's center. The cloudless region between the comma head and tail is the **dry slot**, a feature more often seen in more mature extratropical cyclones and not in hurricanes—we discuss it in more detail shortly. The lumpy areas in and just to the east of the comma head are taller convective clouds casting shadows on the lower clouds around them (Omaha, Nebraska, just north of the warm front, reports a thunderstorm at 6:00 AM).

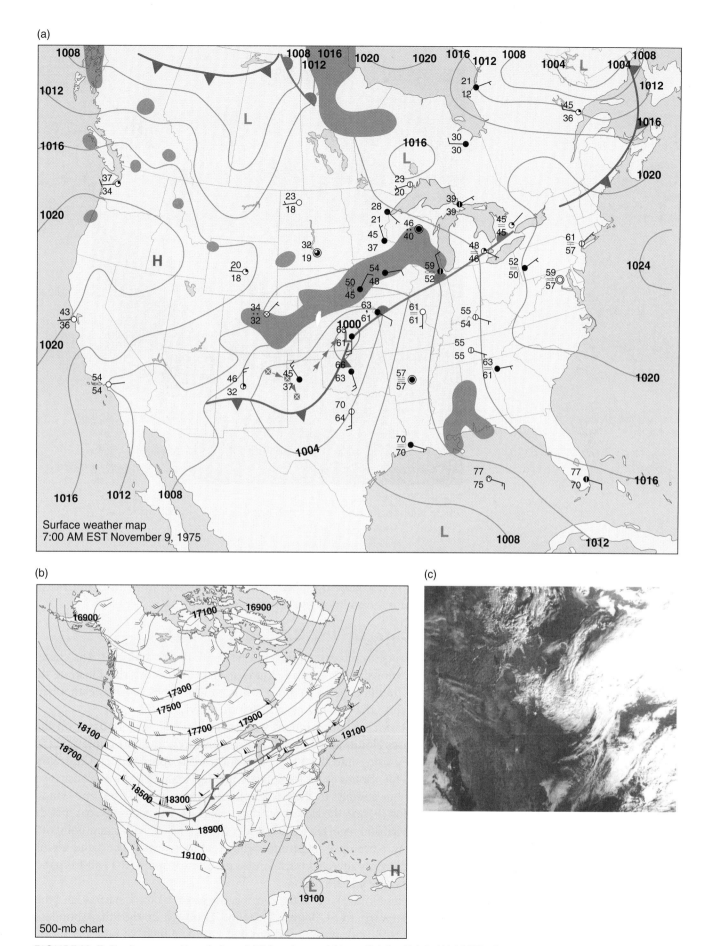

FIGURE 10-7 Surface weather (a) and 500-mb conditions (b) for 7:00 AM (EST), Sunday, November 9, 1975. The small boxes in A denote the location of the surface low 6, 12, and 18 hours earlier. The surface low and fronts are superimposed in B for reference. The Defense Meteorological Satellite Program visible satellite picture at the lower right (c) is from 10:12 AM (EST) that same morning. (Parts a and b source: NOAA.)

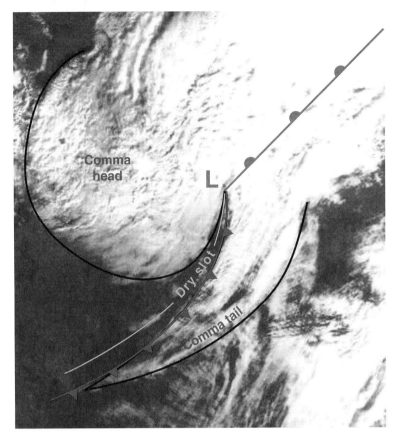

FIGURE 10-8 A close-up of Figure 10-8C. The center of the cyclone, which is underneath the "comma head" clouds, is just northwest of Wichita, Kansas, at this time.

All in all, this picture tells a simple story: This cyclone is mature beyond its tender age. The weather map provides an even more ominous fact: The cyclone is now racing northeastward toward the Great Lakes at up to 64 kilometers per hour (40 mph).

■ Cyclones and Fronts: On the Ground

Madison, Wisconsin, lies just to the east of the cyclone's path on November 9. The changes in temperature, dew point, winds, pressure, and clouds at Madison on this day illustrate that an extratropical cyclone is a *frontal* cyclone. In other words, much of the exciting weather happens along the fronts connected to the low, just as Jack Bjerknes envisioned it.

FIGURE 10-9 tells the story as a timeline of weather at Madison. Before dawn on the 9th, Madison reports altostratus clouds and east winds. Temperatures hover around 10° C (50° F), and dew points are slowly rising into the 40s. The warm front is approaching. Altostratus (As) clouds give way to stratus (St) clouds as the front nears, and light rain falls on Wisconsin's capital. The pressure falls slowly but steadily. In the afternoon, the warm front passes, and winds shift to southeasterly. Despite fog (F) and the lack of sunshine, Madison temperatures soar into the 60s by nightfall. At 9:00 PM, Madison reports a temperature of 18° C (65° F) with a dew point of 14° C (58° F)—unusually balmy November nighttime weather in Wisconsin. The wind circulation around the strengthening low has advected warm, moist maritime tropical (mT) air from the Gulf of Mexico all the way to the Great Lakes region.

Because Madison is close to the path of the cyclone, the warm sector is narrow, and the cold front arrives soon after the warm front. Between 9:00 and 11:00 PM on the 9th, a thunderstorm ahead of the cold front hits the city, dropping about 0.85 cm (0.33 inch) of rain in a short time. The temperature drops 5.5° C (10° F) as a result of evaporational cooling, but the warm Gulf air pushes the thermometer back up to 14° C (58° F) by midnight. Then the cold front rushes through. The winds veer to a more westerly direction and strengthen, peaking at 46 knots (53 mph). The

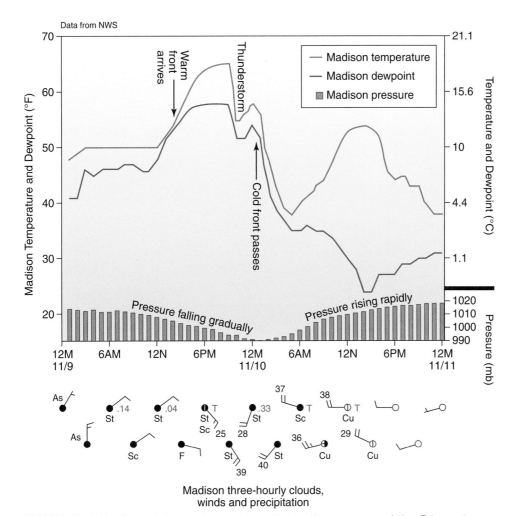

FIGURE 10-9 Madison, Wisconsin, weather during the passage of the *Edmund Fitzgerald* cyclone. Cloud types, wind gusts (in knots), and precipitation (in inches) are shown at the bottom, along with cloud cover and wind speed and direction. (Data from NWS.)

temperature and dew point drop into the 30s by sunrise on the 10th. Stratus clouds break up and are replaced by a few puffy cumulus clouds in the drier air, and the pressure rises rapidly. By nightfall on the 10th, the pressure is high. Winds are slackening. The clouds have dissipated, and the cyclone is past—at Madison.

Cyclones and Fronts: In the Sky

Cyclones and fronts are three dimensional. Jack Bjerknes and the Bergen School used observations from the surface, including cloud types, to determine indirectly the upper-level patterns. We can do the same with the *Fitzgerald* cyclone.

One technique for visualizing a cyclone is shown in **FIGURE 10-10**: take a vertical slice through it, as if the cyclone were a cherry pie. Let's make a cut across the center of the U.S. so that we can slice through both the fronts and the center of the *Fitzgerald* cyclone itself. Figure 10-10a depicts this cut, and in Figure 10-10b, the clouds and weather along this line from Amarillo, Texas, to the Great Lakes is shown graphically for 6:00 AM Central time on the 9th. Notice how well the real-life data compare with our idealized discussion of fronts in Chapter 9. The transitions in clouds, precipitation, temperatures, moisture, and winds across the fronts are almost exactly what you would expect from reading Chapter 9. The cold front is less textbook case, but part of the reason is that clouds ahead of a cold front often need a boost from the Sun to grow. At 6:00 AM in November, the eastern sky is just beginning to glow over the Great Plains, so only stratocumulus clouds hover over Tulsa, Oklahoma.

(a)

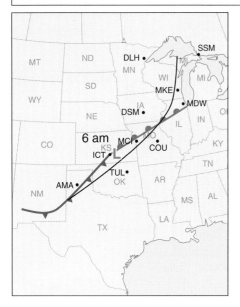

Data from NWS

(b)

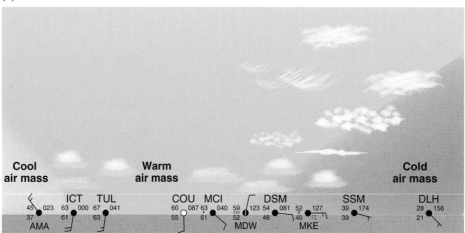

November 9, 1975 6AM CST

(c)

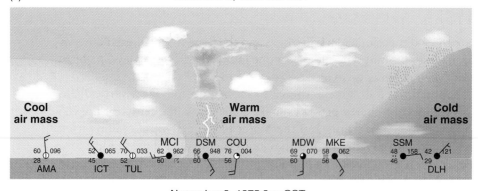

November 9, 1975 3PM CST

FIGURE 10-10 (a) The approximate positions of the *Fitzgerald* cyclone and its fronts at 6:00 AM and 3:00 PM Central Standard Time, November 9, 1975. The line from Texas to the Great Lakes is used in constructing b and c, which are cross-sections through the cyclone of surface weather observations at 6:00 AM (b) and 3:00 PM (c), respectively. (Part a data from NWS.)

By 3:00 PM on the 9th (Figure 10-10c) the strengthening low and the Sun have combined to create severe weather. The Tulsa weather observer can see towering cumulonimbus clouds as the cold front pushes past—so can the observer at Des Moines, Iowa, in the immediate vicinity of the center of the cyclone. Severe thunderstorms are breaking out all over Iowa; winds shred the inflatable stadium dome at the University of Northern Iowa. Meanwhile, the lowering clouds and rising winds on Lake Superior concern the captain of the *Edmund Fitzgerald*.

Back with the *Fitz*: A Fateful Course Correction

Captains McSorley and Cooper pay careful attention to the National Weather Service (NWS) forecasts on the 9th. Cooper recalls, "When the meteorologists become nervous we then start our own weather plots. It did show a low pressure to the south . . . normal November low pressure."

Unfortunately, the rapid growth of the cyclone tests the ability of the NWS and the boat captains to keep up with it. At first the severe weather experts in Kansas City call it a "typical November storm" headed south of Lake Superior. Because of the counterclockwise circulation around a low, this means strong east and northeast winds and waves of 1 to 2.5 meters (3 to 8 feet) on Lake Superior on Monday. By mid afternoon on Sunday the forecast is revised. Gale warnings are issued for winds up to 38 knots (44 mph)—not too unusual, but no longer typical.

The deteriorating weather forecasts prompt Cooper and McSorley to discuss over the radio the first of several weather-related course changes. The "fall north route" the two captains eventually choose to take (Figure 10-1) will add many miles to the usual shipping path. However, by staying close to land and minimizing fetch (see Box 6-2), this route protects boats from high seas caused by north and northeast winds.

By 10:39 PM, the cyclone compels the NWS to revise its forecast again. Now, the winds are supposed to howl at more than 40 knots (46 mph) across Lake Superior on Monday. More significantly, these winds will be coming from the west and southwest, generating waves of 2.5 to 5 meters (8 to 15 feet). The forecast of a storm near the Great Lakes has been consistent, but in just 12 hours the wave heights forecast for Monday have doubled and the forecast wind direction has turned 180°.

At midnight, the *Fitzgerald* is 37 kilometers (23 miles) south of Isle Royale and reports 52-knot (60 mph) north–northeast winds, heavy rain, and 3.5-meter (10-foot) waves. Shortly the NWS will upgrade the gale warning to a storm warning. The cyclone is strengthening rapidly and aiming northward across Lake Superior. The strongest winds over the lake will soon be coming out of the west, not the northeast. This and later course corrections, based on the best available weather information at the time, will leave the freighter exposed to hurricane-force west winds and high seas on the 10th.

Cyclones and Jet Streams

Why is the cyclone strengthening? Why is it moving so quickly and changing direction? Why is the forecast changing every few hours? To understand this better, we need to look closely at what is happening above the surface low.

During the 9th, the upper-troposphere wind pattern has been affected by two factors: the jet-stream winds moving east across the Rockies and the increasing temperature gradients along the surface fronts. Cold air moving southward near the surface helps the upper-level trough "dig" or extend southward. Similarly, warm air moving northward leads to a "building" upper-level ridge (i.e., extending north and northwestward). These upper-level changes, in turn, affect three characteristics of the surface low: its speed, direction, and intensity.

Surface lows tend to move at about half the speed of the 500-mb winds above them. During the 9th, 85-knot (98 mph) winds over New Mexico increased and moved east toward the Great Lakes. This helps explain the rapid movement of this storm.

Surface lows and fronts also generally move in the same direction as the "steering winds" at the 500-mb level. During the 9th (Figure 10-7), the combination of the digging trough and the building ridge bends the wind pattern, causing the direction of these steering winds to shift from northeastward to a more northerly direction, across Lake Superior. In other words, the storm's steering wheel keeps turning on the 9th, and this is why the storm's future path is hard to forecast precisely.

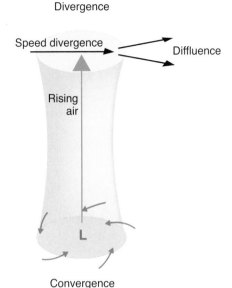

FIGURE 10-11 The relationship between the two types of divergence (speed divergence and diffluence) at upper levels and the development of a surface cyclone. The length of arrows at top is proportional to wind speed.

A better understanding of cyclone–jet-stream relationship requires us to delve more deeply into concepts from Chapter 6 and the Norwegian cyclone model. Most important is the relationship between the surface low's strength and the jet-stream winds. A simple law of meteorology explains the connection: *Surface pressure drops when there is **divergence** of the wind in the column of air above the low.* Ordinarily, winds converge into a low near the surface, as explained in Chapter 6. Therefore, strong upper-level divergence must exist so that the net effect is divergence above the low. If this happens, then the low will intensify because there is less air over it to exert pressure on the surface (**FIGURE 10-11**).

Upper-level divergence can occur in two different ways. Either the winds can accelerate in a straight line, or they can spread out rapidly in a variety of directions. The straight-line acceleration is called **speed divergence**, and the spreading out is called **diffluence**, as illustrated in Figure 10-11.

The intertwined connections between the surface cyclone and the jet-stream pattern are summarized for the first three stages of the idealized Norwegian model in **FIGURE 10-12**. Upper-level divergence exists above and ahead of the low. (Remember from Chapter 6 that for equal spacing of height lines, the winds in a trough are slower than winds in a ridge. Therefore, divergence commonly occurs east of a trough. Figure 10-12 also shows some speed divergence and diffluence, especially for the mature cyclone.) This divergence helps the cyclone deepen and propagate northeastward because the surface pressure drops along and ahead of the low. Upward motion brought about by this divergence helps in the formation of clouds and precipitation. The upper-level convergence behind the cyclone causes sinking air, which suppresses clouds. This sinking air also drags down warm air from the stratosphere above and behind the cyclone (as discussed in **BOX 10-2**). These interrelationships between surface and jet-stream patterns make it clear that the

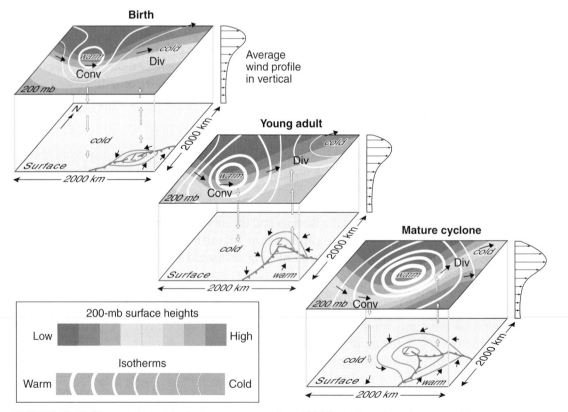

FIGURE 10-12 The relationship between upper-level (200 mb) and surface conditions during the first three stages of the Norwegian cyclone model. (Adapted from Carlson, T. N. *Mid-Latitude Weather Systems.* American Meteorological Society, 1998.)

Box 10-2 Cyclone Winds in 3D: Belts and Slots

The Bergen School saw the cyclone as a face in the clouds and used military metaphors (e.g., fronts) to describe it. Today, meteorologists visualize the cyclone the way a cartoonist would, using just a few deft pen strokes to capture the essence of that face. Along with this new way of seeing the cyclone comes a new metaphor borrowed from industry, not war: "conveyor belts" of air that carry air parcels through an ever-moving, ever-evolving cyclone.

The conveyor belts of an extratropical cyclone, overlaid on top of the weather satellite image for 5:00 PM Eastern time, November 10, 1975. The red arrow indicates the WCB, the blue arrow the CCB, and the yellow arrow the DCB. The red X underneath the blue arrow indicates the position of the *Edmund Fitzgerald* at the time the satellite picture was taken.

The figure above illustrates the approximate locations of these conveyor belts, superimposed on a rare weather satellite picture of the *Fitzgerald* cyclone at 5:00 PM Eastern time on November 10, 1975. The "warm conveyor belt" (WCB) is essentially the warm air that rises up and over the warm front. It then peels off toward the east as it is pushed by the strong jet stream above the surface cyclone. The WCB is the main cloud- and precipitation-making air flow in the cyclone.

The "cold conveyor belt" (CCB) is a flow of air ahead of the warm front that is wrapped into the center of the cyclone, causing clouds. These clouds are the distinctive "comma head" of the extratropical cyclone.

The "dry conveyor belt" (DCB) or "dry slot" is the key to understanding at least some cyclone-related windstorms. It is a tongue of air dragged down by the jet stream from high aloft, even as high as the lowest part of the stratosphere. This air is very dry; as a result, it appears as a cloudless region in a visible satellite picture—helping to form the "dry slot," as seen in Figure 10-8. The strong westerly and southwesterly winds above the cyclone descend in the dry slot and make the surface beneath it a blustery place.

To positively identify a dry slot, meteorologists can try at least two different approaches. One is to look for narrow bands of dark (i.e., dry) air leading into a cyclone on water vapor satellite images. Another way is to look for signs of stratospheric air being dragged down into the troposphere.

Water vapor imagery didn't exist in 1975, but surface ozone concentrations were being measured in large metropolitan regions, such as Chicago, in order to keep track of daytime

(continued)

Box 10-2 Cyclone Winds in 3D: Belts and Slots, continued

ozone pollution. These observations can also be used to look for signs of stratospheric air descending in the dry slot of the cyclone.

On the night before the *Fitzgerald* sank, surface ozone values were rising all over the Midwest, particularly in a band stretching through Chicago. The figure below, using data gleaned from several Chicago high schools by student researcher Stino Iacopelli, tells the story. The ozone values peaked around midnight, right around the time that thunderstorms hit the area (compare with the Madison, Wisconsin, observations in Figure 10-9). Meanwhile, ozone values were also above normal back in Iowa, behind the low. At this same time, winds gusted to 61 knots (70 mph) in the Chicago area, and a man was killed in northeast Iowa when winds behind the cyclone overturned his airplane. It is possible the *Fitzgerald* cyclone had a well-developed dry slot, and high-speed, high-ozone air plunged toward the surface.

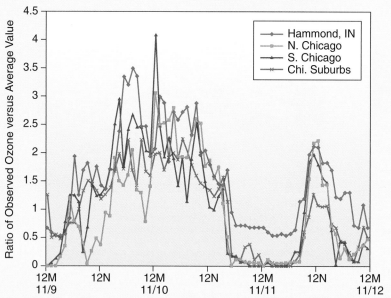

Surface ozone levels in and near Chicago on November 9–11, 1975. The values are ratios of the observed ozone concentration versus the November 1975 average for that time and site. Values greater than 1 indicate greater-than-usual ozone concentrations. Notice the unusually high ozone values around midnight on the 10th as the extratropical cyclone passed by Chicago.

Is this why the *Fitzgerald* and the *Anderson* were ravaged by high winds on Lake Superior the next day? No surface ozone data exist in the vicinity of the wreck, but we do know that the *Fitzgerald*'s last hours were spent behind the low in heavy lake-effect snow squalls. These squalls, like the Chicago thunderstorms the night before, would help drag any high-speed air above them down to the surface. This question won't be settled until the evidence is presented as a formal journal article and the article's results are discussed and debated by today's leading extratropical cyclone experts.

extratropical cyclone is a complex three-dimensional phenomenon, as illustrated in Box 10-2. This explains why observations of upper-level winds are so crucial to weather analysis and forecasting in the middle latitudes.

Figure 10-12 is an idealized model, however. FIGURE 10-13 shows the actual winds measured by radiosondes at 300 mb at 6:00 PM Central time on November 9, 1975. Following the lines of constant height from Nebraska to Minnesota, the winds double in speed going from the trough to the ridge. As a result, on the 9th speed divergence exists over Iowa and Minnesota.

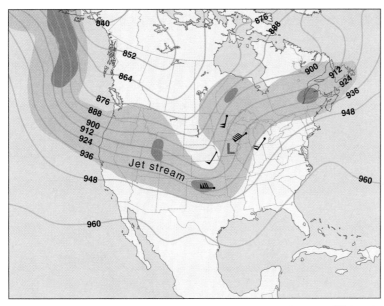

FIGURE 10-13 Winds and heights (in tens of meters) at the 300-mb level for 6:00 PM (CST) on November 9, 1975. The pattern of winds shown here helps the surface cyclone (the L over Iowa) strengthen rapidly. (Data from NWS.)

The winds over Minnesota are southerly, whereas the winds over Michigan are southwesterly. This spreading out of the wind is an example of diffluence.

The combination of these two types of divergence leads to sudden pressure drops over the upper Midwest. Therefore, this cyclone accelerates to the northeast and deepens rather rapidly, from a central minimum pressure of 999 mb at 6:00 AM (CST) to 993 mb 12 hours later. The lowest pressure of the cyclone is dropping 0.5 mb per hour. This cyclone is on its way to becoming the "Storm of the Year" on the Great Lakes.

DAY 3: THE MATURE CYCLONE

Early on November 10, the *Fitzgerald* sails into the Eastern time zone and heads for an unexpected rendezvous with a dangerous, mature, extratropical cyclone. The 7:00 AM (EST) surface weather map (**FIGURE 10-14**) reveals that this cyclone (and its associated fronts) now dominates the eastern half of the U.S.! Just 48 hours ago, it was a cloudless swirl over Texas.

The cyclone's power and scope of influence indicate that it has reached full maturity. The low itself is centered directly over Marquette, Michigan, in that state's Upper Peninsula. Its lowest pressure has plummeted overnight to 982 mb, a drop of nearly 1 mb per hour. To the west of the low, cold air spiraling in toward the storm across tightly packed isobars is causing blizzard conditions—heavy wet snow blown by 60-knot (69 mph) winds is tearing down power lines and piling up as much as 36 centimeters (14 inches) of snow in parts of northern Wisconsin. Meanwhile, the cyclone's warm front has crossed over into Canada, and its cold front has barreled south to the Gulf of Mexico. From Iowa to Tennessee, residents are combing through debris and tending to casualties after 15 tornadoes spawned ahead of the cold front injured 25 people. In the life cycle of this extratropical cyclone, Saturday's baby and Sunday's adolescent is Monday's ferocious adult.

Bittersweet Badge of Adulthood: The Occlusion Process

If you look closely at Monday's weather map in Figure 10-14, you will notice that the cold and warm fronts no longer join like halves of a moustache at the center of the cyclone. Instead, an occluded front joins the low and the cold and warm fronts.

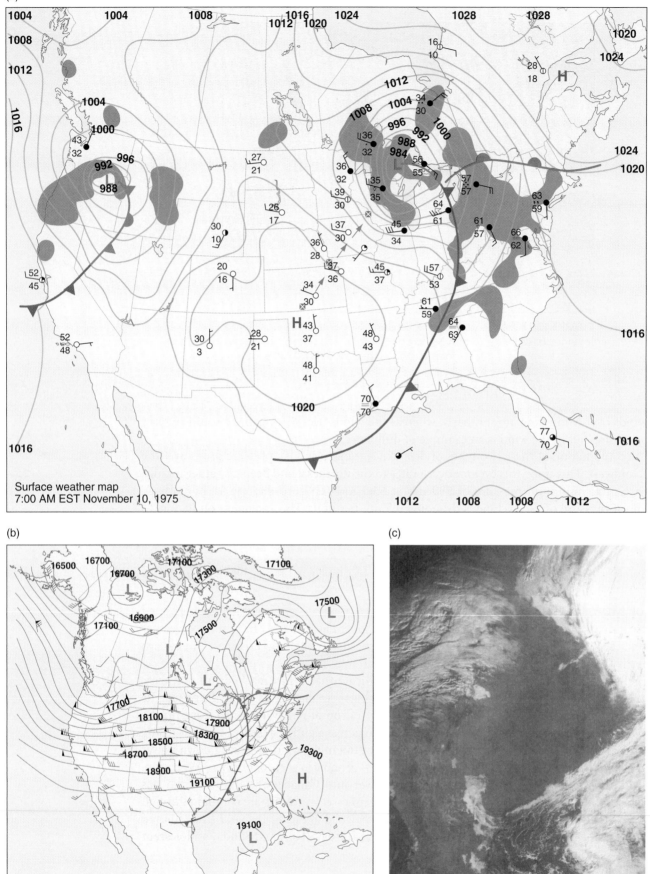

FIGURE 10-14 Surface weather (a) and 500-mb conditions (b) for 7:00 AM (EST) on Monday, November 10, 1975. The Defense Meteorological Satellite Program visible satellite image in the lower right (c) is from 9:55 AM that morning and is centered over Kansas. (Part a source: NOAA.)

What is an occluded front? The Bergen School envisioned it as the cold front "catching up" with the warm front to form an occluded front. As noted in Chapter 9, the modern understanding of occluded fronts is more complicated. Cyclone researchers David Schultz and Geraint Vaughan describe the process of **occlusion** in this way: If the cyclone becomes strong enough, its structure may begin to change, both at the surface and aloft. As the warm and cold fronts wrap around the cyclone, they lengthen, just as ribbons of milk lengthen when stirred into coffee. Cold air ahead of the warm front travels around the low center, and the warm air ascends over the warm front (Box 10-2), which removes warm air from the surface. Eventually, the narrowing region of warm air is completely lifted from the surface near the low center, leaving a boundary between two cold air masses called the occluded front.

During occlusion, the surface low gradually retreats from the zone of sharp temperature contrasts and isolates itself in the cold air to the north and west of the intersection of the cold and warm fronts. The surface low ultimately ends up directly underneath the upper-level low— far removed from the tilted "baroclinic" situation of its youth. In an extratropical cyclone over a large body of water, the occlusion process can lend it some of the characteristics of its tropical cousin, the hurricane (BOX 10-3).

The irony is that the occlusion process eventually isolates the mature extratropical cyclone from its fuel source, the fronts with their strong temperature gradients, and (once the surface cyclone is underneath the upper-level low) the deepening effects of the jet stream. By analogy, it's like a pop music star who makes it big, surrounds himself with bodyguards and barricades himself in a big mansion, and then loses touch with his fans. In both cases, success is bittersweet. The pop star is cut off from his inspiration, and the hits quit coming. Similarly, the mature cyclone's growth eventually cuts it off from its ability to grow further. (Cyclones can keep on growing for a time, primarily because of latent heat energy from the condensation of water vapor.) The cyclone gradually turns into a frontless cut-off low and slowly dies. The occlusion process is therefore not only a badge of adulthood—it is also a sign that an extratropical cyclone is approaching the end of its life cycle.

An occluding cyclone is still at the peak of its powers initially and may even initially intensify further, as the crew of the *Fitzgerald* learns on the 10th. As the cyclone races toward Lake Superior before dawn on the 10th, captains Cooper and McSorley adjust course again, "hauling up" near the eastern Canadian shore of the lake. There the boats will be less vulnerable to northeast winds because the winds will be blowing over only a short distance or "fetch" of water.

Unfortunately, the cyclone outmaneuvers the ore freighters. The cyclone is now moving northeast at 30 mph; even the powerful "Big *Fitz*" cannot match this speed. The low and the boats cross paths in Canadian waters in northern Lake Superior.

"The low had reached Lake Superior and was intensifying dramatically," Cooper recalls. "Shortly after noon, we were in the eye of the storm. The Sun was out . . . light winds, no sea." The *Anderson*'s barometer, adjusted for sea level, bottoms out at 28.84 inches of mercury or 976.6 mb, as low a pressure as will be recorded for this cyclone.

The good weather cannot last because of the raging contrasts near the heart of an extratropical cyclone. As the low and its occluded front race past, strong winds from the west will soon lash at the *Anderson* and *Fitzgerald*. At 1:40 PM, Cooper and McSorley agree to yet another weather-related course correction, cutting between woody Michipicoten Island and Caribou Island on a southeastward heading toward safe haven in Whitefish Bay and Sault Ste. Marie, Michigan. But it's too late.

By 2:45 PM, the *Anderson* reports snow with northwest winds at a steady 42 knots (48 mph), a 180° reversal of direction and a doubling of speed in less than 2 hours. Instead of being sheltered from northeast winds, they are now fleeing rising winds and waves that are rushing at them from behind across the vast expanse of Lake Superior.

Hurricane West Wind

Meanwhile, a severe, localized windstorm is developing directly in front of the Fitzgerald. FIGURE 10-15 shows the hourly barometric pressure readings for four weather stations in or

Box 10-3 Cyclones and Water: Bomb and Bust

Is an extratropical cyclone the same thing as a hurricane? No. Extratropical cyclones thrive on ingredients such as fronts and strong jet streams, which we learned in Chapter 8 would kill off a hurricane by cutting off its fuel sources and disrupting its circulation.

Even so, it's true that the same fuel sources that feed the engines of hurricanes—warm surface waters and lots of water vapor—can also stoke the fires of their extratropical cousins. As a result, extratropical cyclones that are over oceans or that have access to lots of warm, moist air often become more intense than "drier" cyclones. This rapid intensification even has a meteorological nickname—"bomb"—that is reserved for midlatitude cyclones whose central pressure drops approximately 1 mb per hour every hour for 24 hours. It's not typical to see the pressure drop a whole millibar in just 1 hour, let alone keep up that pace for an entire day! Such explosive growth sounds more like a hurricane than an extratropical cyclone, and there are some similarities.

An extreme example is the maritime "polar low" that develops over the seas north of Europe. The water there is cold, but the air is even colder. The water transfers warmth and energy to the air at the surface; this is called sensible heat. Latent heat is also released into the air when the seawater evaporates and then later condenses into cloud droplets. These two water-based energy sources give the polar low its energy, just as in a hurricane. In fact, the polar low is called an "arctic hurricane" by some meteorologists because of its hurricane-like appearance in satellite pictures.

In the middle latitudes, cyclones live and die with fronts and jets. But even here, water plays a key role and makes forecasting very difficult. Midlatitude cyclones born over water are a little like hurricanes and are especially responsive to water-borne energy. One study of a February 1979 nor'easter that became a "bomb" off the U.S. East Coast revealed that the presence of water helped drop the cyclone's pressure by 32 mb (nearly 1 full inch of mercury). That's the difference between a weak cyclone and a record setter! The forecasters were taken completely by surprise, which is called a "bust," short for a "busted" or broken forecast. (See Figure 9-9 for a satellite image of this storm, which had an eye at its center. Chapter 13 has more details on this "bust.")

The Great Lakes are also big enough to affect the strength of extratropical cyclones that pass over them. Below is a satellite picture of the September 14, 1996, "Hurricane Huron." Notice how it seems to have an "eye" and spiral rainbands, a little like a hurricane. It's not a hurricane, but the presence of warm water underneath the cyclone fueled its fires. This extratropical cyclone stalled over Lakes Michigan and Huron for 4 days and deepened 19 mb over that time—not a "bomb," but impressive nonetheless.

Did the Great Lakes play a role in the cyclone that helped sink the *Edmund Fitzgerald*? Yes, most likely on the back side of the storm where cold air passed over warmer water. There, just as in a polar low, sensible and latent heat would be released into the air. This energy release helped cause the lake-effect snow squalls that enshrouded the *Fitzgerald* in its final minutes.

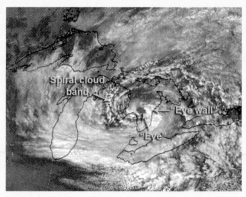

"Hurricane Huron," a mid-latitude cyclone influenced by the waters of the Great Lakes.

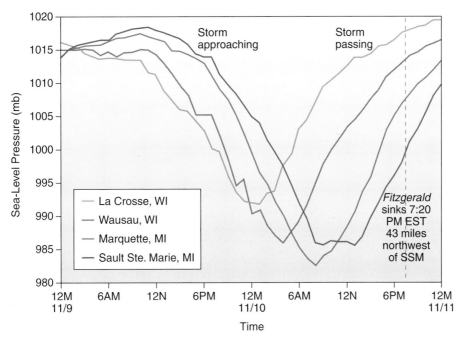

FIGURE 10-15 Hourly sea-level pressures at four stations in the general path of the *Fitzgerald* cyclone. Notice the pattern of falling and then rapidly rising pressure at each city as the storm first approaches and then passes. The *Edmund Fitzgerald* sank during rising pressure northwest of Sault Ste. Marie, Michigan. (Data from NWS.)

near the path of the cyclone. At most of these locations, the pressure rises rapidly after the storm passes. Mariners recognize this as an ominous sign. According to a saying at least as old as 19th-century British meteorologist Admiral FitzRoy (the captain of the *HMS Beagle;* see Chapter 7):

"Fast rise after low
Foretells stronger blow."

This folklore forecast is based partly on the force–balance ideas in Chapter 6. Rapid pressure changes behind a low imply that there is a strong horizontal pressure gradient in the area. As we saw in Chapter 6, geostrophic and gradient wind balances both require strong horizontal winds in such cases.

FIGURE 10-16 proves the wisdom of FitzRoy's rhyme. In the figure, hourly wind observations at Marquette and Sault Ste. Marie, Michigan, tell a tale of increasing, and increasingly gusty, winds after the passage of the cyclone. Marquette is hit first, shortly after noon on the 10th; FIGURE 10-17 is a rare photograph showing the tempest at Marquette that very afternoon. The pressure rises move east, and Sault Ste. Marie is hammered with gusts of 62 knots (71 mph) after nightfall, knocking out electricity over much of the region.

By the time high winds blow through Marquette and Sault Ste. Marie, the cyclone is well into Canada. These damaging west and northwest winds are far removed from any thunderstorms or tornadoes, nor are they concentrated in an eye wall around the lowest pressure as in a hurricane. These winds stand in the path of the *Fitzgerald,* which is now in a race to safe harbor before the winds and waves sink it.

■ One of the Worst . . .

As daylight dims on the 10th, the *Anderson* and *Fitzgerald* are laboring to make headway in worsening weather. All around the *Anderson,* the winds and seas rise in fury: 43 knots (49 mph) and 3.7- to 4.9-meter (12- to 16-foot) waves at 3:20 PM, 58 knots (67 mph) and 3.7- to 5.5-meter (12- to 18-foot) waves at 4:52 PM. It is profoundly bad timing. At this point

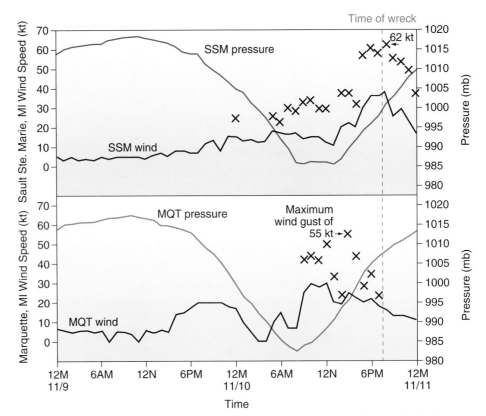

FIGURE 10-16 Hourly wind speeds (average and gusts, in knots) at Sault Ste. Marie and Marquette, Michigan, during the approach and passage of the *Fitzgerald* cyclone. Hourly barometric pressure readings are also depicted. (Data from NWS.)

in the voyage, treacherous shoals near Caribou Island demand precise navigation as the boats thread the needle between Michipicoten and Caribou (Figure 10-1).

In the middle of the howling winds, snow, sea, and spray, something goes wrong. Around 3:20 PM, McSorley calls Cooper on the radio:

"Anderson, this is the *Fitzgerald*. I have sustained some topside damage. I have a fence rail laid down, two vents lost or damaged, and a list. I'm checking [slowing] down . . . Will you stay by me 'til I get to Whitefish [Bay]?"
"Charlie on that *Fitzgerald*. Do you have your pumps going?"
"Yes, both of them."

FIGURE 10-17 *The Wreck of the Edmund Fitzgerald*
Author Frederick Stonehouse's photograph of heavy surf at Marquette, Michigan on November 10, 1975.

Something has damaged the *Fitzgerald*, inflicting wounds worse than anything that has happened to the boat in 17 full years on the Great Lakes. It is tilting to one side, or "listing," presumably as a result of a leak that is letting lake water pour into the boat. The leak is bad enough that even the *Fitzgerald*'s large pumps, capable of removing thousands of gallons of water every minute, apparently do not correct the list for the remainder of its journey. The captain is concerned enough to ask his companion boat to stay nearby. But what happened, exactly, is a mystery forever.

The windstorm intensifies. Around 4:10 PM, gusts blow away the *Fitzgerald*'s radar antenna. Then winds knock out power to the remote navigation station at the entrance to Whitefish Bay, making it that much more difficult for storm-tossed ships to gain their bearings. At 4:39 PM, the National Weather Service in Chicago finetunes its forecast for eastern Lake Superior, calling for northwest winds 38 to 52 knots (44 to 60 mph) with gusts to 60 knots (69 mph) early Monday night. Waves are still expected to be 2.4 to 4.9 meters (8 to 16 feet).

Once again the storm exceeds expectations, however (see Box 10-2 for reasons why this might be so). At 5:00 PM, the lighthouse at Stannard Rock north of Marquette, the closest observing station to the *Fitzgerald* at that moment, records a gust of 66 knots (77 mph). Cooper, on board the *Anderson,* estimated wind gusts of more than 100 mph. Before 6:00 PM, McSorley tells another ship captain via radio the following:

> "I have a bad list, lost both radars. And am taking heavy seas over the deck. One of the worst seas I've ever been in."

Blind and ruptured, the *Fitzgerald* steams into the teeth of raging wind and snow, the *Anderson* 10 miles behind her.

"Nosedive"

> "Sometime before 7 PM . . . we took two of the largest seas of the trip. . . . The second large sea put water on our bridge deck! This is about 35 feet above the waterline!"
>
> —Bernie Cooper, captain of the Anderson

At 7:00 PM, 50-knot (58-mph) winds and 4.9-meter (16-foot) waves are still buffeting the *Anderson*'s crew. They are in touch with the *Fitzgerald* by radio and by following the *Fitz*'s blip on the radar. The *Fitzgerald* has struggled to within 24 kilometers (15 miles) of shore and about 32 kilometers (20 miles) of Whitefish Bay.

But even the bay is not safe harbor. Six Native American commercial fishermen are tending their nets in the western part of Whitefish Bay when, in the words of one of the fishermen, "all of a sudden, the lake began to boil and churn enormously. It was like nothing I'd seen before." Another fisherman barely survives his boat's capsizing from "a giant wall of water coming at us." There is little shelter from this storm.

At 7:10 PM, the *Anderson* gives navigation instructions to the radarless *Fitzgerald* up ahead of it. As an afterthought, the *Anderson*'s first mate asks, "Oh, and by the way, how are you making out with your problems?" The *Fitzgerald* replies, "We are holding our own."

Immediately after this conversation, another severe snow squall enshrouds the two boats. The *Fitzgerald* is hidden from the radar beam of the *Anderson,* lost in a chaos of snow and sea. Just as suddenly, around 7:30 PM, the snow ends. For the first time in many hours, visibility is excellent. Lights from ships coming north from Sault Ste. Marie shine through clearly on the horizon. A TV/radio tower blinks across the water in Ontario. But the *Fitzgerald* is nowhere to be found: no lights, no radar blip, no radio contact.

Captain Cooper searches for the *Fitzgerald.* Did the Fitz duck into a harbor during the snow squall? How can you lose a 222-meter (729-foot) ore freighter in the middle of a lake? Cooper calls the U.S. Coast Guard in Sault Ste. Marie with his worst fears:

> "This is the Anderson. I am very concerned with the welfare of the steamer *Edmund Fitzgerald.* . . . I can see no lights as before, and I don't have him on radar. I just hope he didn't take a nosedive."

The *Fitzgerald* is gone, without a mayday, an eyewitness, or a survivor.

DAY 4 (AND BEYOND): DEATH

The Cyclone

The surface weather map for 7:00 AM Eastern time on Tuesday, November 11, 1975 (FIGURE 10-18a), betrays little hint of the maelstrom of the past 24 hours. On Monday, cyclone winds blew Lake Michigan waters over a pier and drowned two people in Grand Haven, Michigan. That same day the cyclone's southwest winds blew across the entire length of Lake Erie from Toledo, Ohio, to Buffalo, New York, piling up Erie's waters a full 9 feet above normal in Buffalo and killing a Buffalo woman by blowing her off a second-story porch. Showers and thunderstorms spawned

(a)

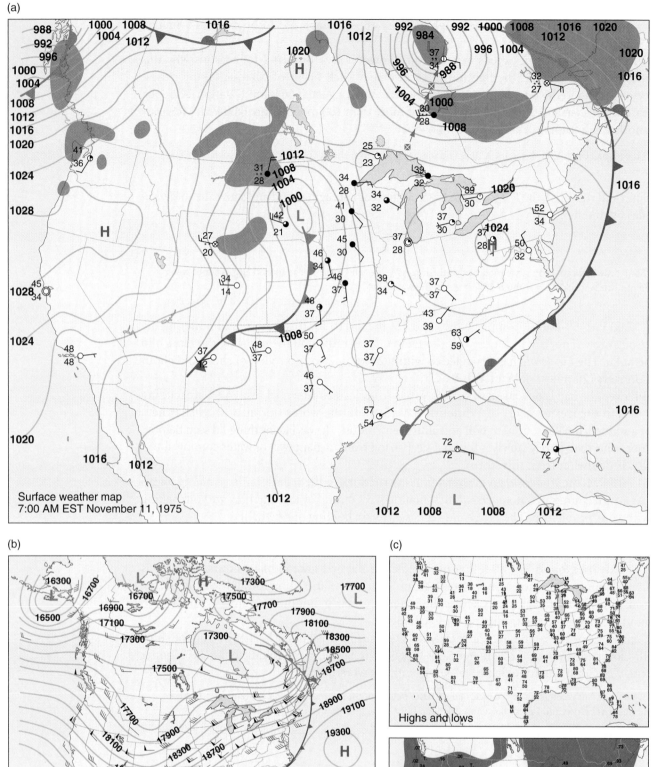

Surface weather map
7:00 AM EST November 11, 1975

(b)

500-mb chart

(c)

Highs and lows

Precipitation

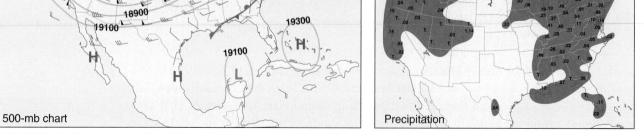

FIGURE 10-18 Surface weather (a) and 500-mb conditions (b) at 7:00 AM (EST) on Tuesday, November 11, 1975. The past day's high and low temperatures are shown in the lower right (c). (*Source*: NOAA.)

by the storm brought rain to every state east of the Mississippi (Figure 10-18c). But on Tuesday, light and variable winds are now firmly in place over the entire Great Lakes.

Bergen School member Sverre Petterssen once said, "Extratropical cyclones are born in a variety of ways, but their appearance at death is remarkably similar." So it is for the storm that helped wreck the *Edmund Fitzgerald*. Figure 10-18a shows an occluded cyclone skirting the eastern shore of Hudson Bay—it is the *Fitzgerald* storm. The surface low is underneath the 500-mb low, and the nearest strong temperature gradients are now hundreds of miles away over the Atlantic. The storm will die a slow death in the coming days as it crosses northern Canada, its fronts dissolving, the face becoming unrecognizable.

Another low, a youngster, is speeding across the Great Plains toward the Great Lakes. Like its predecessor, it has a potent upper-level trough associated with it (Figure 10-18b), but unlike the *Fitzgerald* storm, it has no warm, moist air to feed on; dew points ahead of it hover around the freezing mark. It has also reached the occluded stage early in its lifetime (see also Figure 10-14a). Lacking the energy sources of temperature gradients and moisture, it will not grow into a killer.

Yet in death the *Fitzgerald* storm, like all extratropical cyclones, accomplishes tasks needed to keep weather and climate in balance. For example, on this day at 7:00 AM, Great Whale, Quebec, reports drizzle with a temperature of 2.7° C (37° F), up from −8.8° C (16° F) the previous morning. Meanwhile, cooler and drier air has invaded the Gulf Coast of the U.S. This poleward transport of heat and moisture is vital to Earth's overall energy balance, as we discussed in Chapters 3 and 7. Viewed from this perspective, the cyclone is not a meaningless killer—instead, it is part of Nature's broader design for a stable, habitable world.

The tale of this storm and other similar storms (**BOX 10-4**) should illustrate how useful the Norwegian model of the frontal cyclone is for explaining the wide variety of weather associated with it. Vilhelm Bjerknes summed up the Bergen School's contribution in this way:

> During 50 years meteorologists all over the world looked at weather maps without discovering their most important feature. I only gave the right kind of maps to the right young men, and soon they discovered the wrinkles in the face of the Weather.

The *Fitzgerald*

Early on the 11th, the crew of the *Anderson* sights the first debris from the wreck of the *Edmund Fitzgerald:* part of an unused life jacket. Other "flotsam" washes up near shore: more life jackets, a severely damaged lifeboat (**FIGURE 10-19**), a stool, a plastic spray bottle. The *Fitzgerald* itself is not officially located at the bottom of Lake Superior until the next spring, with the help of underwater camera equipment. The end of the *Fitzgerald* was swift and violent, probably a nosedive straight to the bottom. The freighter was torn into two main pieces as it plummeted, the front half landing right side up, the back half upside down. It lies in 162 meters (530 feet) of water just 27 kilometers (17 miles) from the entrance to Whitefish Bay (**FIGURE 10-20**).

Why did the *Fitzgerald* sink? There is no one simple answer. U.S. government investigators blame non–water-tight hatch closures where the iron ore was loaded into the belly of the boat. A dissenting opinion by one of the government's own investigation board members suggests the possibility that the *Fitzgerald* ran aground on the shoals near Caribou Island. Others implicate the unusually high waves that pounded the *Anderson* and others early on the evening of the 10th. Meteorologists are not to blame; the NWS forecasts are deemed "excellent." The storm was anticipated more than a day in advance, even if its exact intensity and path were not forecast precisely. Theories and controversy abound; probably no one factor is solely to blame. Sadly, the wreck of the *Edmund Fitzgerald* has become the maritime equivalent of the assassination of President John F. Kennedy, complete with disbelieved government explanations and endless intrigue among its latter-day investigators.

What is rarely, if ever, pointed out is that regardless of the proposed cause—leaky hatch covers, shallow shoals, high waves—the *Edmund Fitzgerald* never would have sunk on a calm, sunny day. An overriding reason for this shipwreck is actually quite simple: a powerful extratropical cyclone and a vulnerable boat crossing paths at the worst of all possible times. Over 30 years after the storm, NWS meteorologists proved this point by simulating this storm using high-resolution computer

Box 10-4 Weather History Repeats Itself (Almost)

Does history repeat itself? Sometimes it seems to be the case, even in weather history.

Nature provided a near-perfect twin of the *Fitzgerald* cyclone exactly 23 years later, on November 10, 1998. Following a slightly more westerly track than the 1975 storm (see figure below), this 1998 "Witch of November" storm was even more intense, setting low pressure records of 963 to 966 mb in Iowa and Minnesota (see the text's Web site). Ten deaths, 34 injuries, and at least $40 million in property and crop damage were caused by this storm.

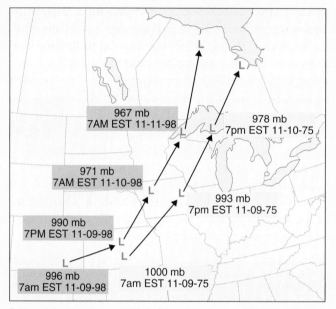

Paths of the November 1998 (left) and November 1975 (right) extratropical cyclones that hit the upper Midwest. (*Source:* Don Rolfson, National Weather Service Marquette/NOAA.)

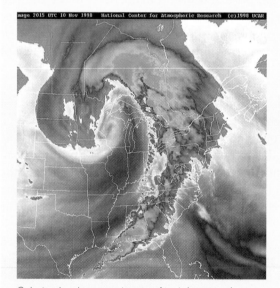

Colorized water vapor image of an intense cyclone over western Wisconsin at 3:15 PM Eastern time, on November 10, 1998 (the 23rd anniversary of the sinking of the Edmund Fitzgerald). The colors indicate the temperature of the clouds (dark green = cold thunderstorm clouds). The dry conveyor belt is denoted by the dark region across Illinois and Michigan and the hook-shaped area over extreme southeastern Minnesota and western Wisconsin.

From a scientific perspective, one big difference in the two storms was the advance of remote sensing techniques from 1975 to 1998. In the water vapor image (left) from the 1998 cyclone, the hook-shaped dark region near the center of the cyclone is a spiral arm of the dry slot that has wrapped around the center of the low. Underneath the dry slot, southwesterly winds gusted to 81 knots (93 mph) at La Crosse, Wisconsin, with 8 gusts over 61 knots (70 mph). The dry-slot winds also caused stunning lighthouse-topping waves at South Haven, Michigan, on Lake Michigan (photo on next page).

Despite many similarities, however, weather history did not repeat itself on November 10, 1998. No large boats sank on the Great Lakes during the 1998 storm. More generally, weather does not repeat itself, which has big implications for how to forecast the weather—as we'll learn in Chapter 13.

Nevertheless, by examining many different cases of high winds due to extratropical cyclones, we can learn more about their common characteristics. According to climatological research by University of Georgia students and faculty, these non-thunderstorm winds in the Great Lakes region are almost always from the south-through-west direction (see figure below)—confirming Gordon Lightfoot's "hurricane west wind" description. The predictable direction of the high winds may be due to the orientation of the dry slot. And so while weather history does not quite repeat itself, sometimes it's close enough for scientists to be able to "connect the dots" and advance our understanding.

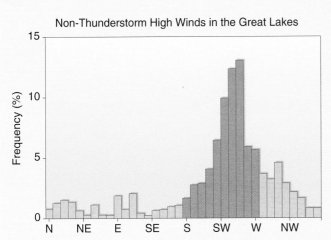

Distribution of nonconvective high (sustained) wind observations as a function of direction, as observed in the U.S. Great Lakes region during the period 1967–1995. Winds from the south-through-west direction are highlighted in green. (Modified from Lacke et al., *Journal of Climate*, 20 (2007): 6012–6022. © American Meteorological Society. Reprinted with permission.)

models. The highest winds and waves converged on the location of the *Fitzgerald* in the model simulations at the very hour of the shipwreck (see Box 6-3).

The Sailors

The bodies of the 29 sailors on board the *Edmund Fitzgerald* (BOX 10-5) have never been recovered or buried. Bruce Hudson's grieving parents preserve his room—books, photos, musical instruments—exactly as Bruce left them in November 1975. He never met his daughter, Heather, born 7 months after the *Fitzgerald* was lost.

The next June, singer/songwriter Gordon Lightfoot's haunting and meteorologically accurate account of the shipwreck (BOX 10-6) was released, spreading the story of the *Fitzgerald* and its sailors to a worldwide audience.

There will always be another extratropical cyclone. The Sun and the tilt of the Earth see to this by the endless re-creation of temperature gradients, and as long as there is cargo to be hauled, there will be Great Lakes freighters. But, people are unique and irreplaceable, and their loss is felt forever. Honor the

FIGURE 10-19 One of the *Fitzgerald's* two mangled lifeboats, now on display in the Museum Ship Valley Camp in Sault Ste. Marie, Michigan.

FIGURE 10-20 *The Wreck Site II*, a painting by David Conklin of the *Edmund Fitzgerald* in its final resting place at the bottom of Lake Superior.

Box 10-5 *Fitzgerald* Sailors Lost at Sea

Ernest M. McSorley, 63, Captain, Toledo, Ohio
John H. McCarthy, 62, first mate, Bay Village, Ohio
James A. Pratt, 44, second mate, Lakewood, Ohio
Michael E. Armagost, 37, third mate, Iron River, Wisconsin
George J. Holl, 60, chief engineer, Cabot, Pennsylvania
Edward F. Bindon, 47, first assistant engineer, Fairport Harbor, Ohio
Thomas E. Edwards, 50, second assistant engineer, Oregon, Ohio
Russell G. Haskell, 40, second assistant engineer, Millbury, Ohio
Oliver J. Champeau, 41, third assistant engineer, Milwaukee, Wisconsin
Frederick J. Beetcher, 56, porter, Superior, Wisconsin
Thomas Bentsen, 23, oiler, St. Joseph, Michigan
Thomas D. Borgeson, 41, able-bodied maintenance man, Duluth, Minnesota
Nolan F. Church, 55, porter, Silver Bay, Minnesota
Ransom E. Cundy, 53, watchman, Superior, Wisconsin
Bruce L. Hudson, 22, deckhand, North Olmsted, Ohio
Allen G. Kalmon, 43, second cook, Washburn, Wisconsin
Gordon F. MacLellan, 30, wiper, Clearwater, Florida
Joseph W. Mazes, 59, special maintenance man, Ashland, Wisconsin
Eugene W. O'Brien, 50, wheelsman, Perrysburg Township, Ohio
Karl A. Peckol, 20, watchman, Ashtabula, Ohio
John J. Poviach, 59, wheelsman, Bradenton, Florida
Robert C. Rafferty, 62, temporary steward (first cook), Toledo, Ohio
Paul M. Riippa, 22, deckhand, Ashtabula, Ohio
John D. Simmons, 60, wheelsman, Ashland, Wisconsin
William J. Spengler, 59, watchman, Toledo, Ohio
Mark A. Thomas, 21, deckhand, Richmond Heights, Ohio
Ralph G. Walton, 58, oiler, Fremont, Ohio
David E. Weiss, 22, cadet (deck), Agoura, California
Blaine H. Wilhelm, 52, oiler, Moquah, Wisconsin

Box 10-6 Gordon Lightfoot, Songwriter— and Amateur Meteorologist

Gordon Lightfoot is a Canadian songwriter whose many hits in the 1960s and 1970s are part of the rich musical legacy of that era. He is also a Great Lakes sailor. Shortly after the *Edmund Fitzgerald* shipwreck, Lightfoot spent two months researching the story behind the disaster. In three days he wrote the eight verses and 430 words, plus music, of his song "The Wreck of the *Edmund Fitzgerald*." The song became one of the top hits of 1976 in North America and has never left the airwaves; it is part of our culture and history. Author Joseph MacInnis writes, "as the years passed, the song and the shipwreck and the men and the lake became one."

Lightfoot's song tells many stories, not the least of which is the meteorology behind the *Fitzgerald* shipwreck. Below are the lyrics to the first five verses of "The Wreck of the *Edmund Fitzgerald*." At the sides are commentaries on the factual accuracy of the lyrics, especially with regard to the weather. May you become as good an amateur meteorologist as Gordon Lightfoot!

The Wreck of the Edmund Fitzgerald
Words and music by Gordon Lightfoot
© 1976 (renewed) Moose Music, Ltd.

The first European ship-wreck on the Great Lakes occurred in 1680.

The legend lives on from the Chippewa on down
of the big lake they call Gitche Gumee.
The lake, it is said, never gives up her dead
when the skies of November turn gloomy.

235 sailors died and 10 boats sank in a November 9, 1913 cyclone on the Lakes.

Were the gales early? Several famous Great Lakes storms hit on Nov. 9-10.

With a load of iron ore twenty-six thousand tons
more than the *Edmund Fitzgerald* weighed empty,
The good ship and true was a bone to be chewed
when the "Gales of November" came early

The boat was named for the president of North-western Mutual Life Insurance Company, whose grandfather and all five of his grand-father's brothers were Great Lakes captains.

The Fitz was the flagship of the shipping division of Oglebay Norton Company, headquartered in Cleveland.

The ship was the pride of the American side
coming back from some mill in Wisconsin.
As the big freighters go it was bigger than most
with a crew and good captain well seasoned,

Its destination is actually Detroit. Oglebay Norton's location caused the confusion.

Concluding some terms with a couple of steel firms
when they left fully loaded for Cleveland
And later that night when the ship's bell rang,
could it be the north wind they'd been feelin'?

Correct: Fitz reports 60 mph NNE winds at 1 AM EST with 10-foot seas; individual waves higher, enough to break over the railing.

The wind in the wire made a tattletale sound
and a wave broke over the railing.
And every man knew as the captain did too
'twas the witch of November come stealin'.

No proof of freezing rain, but not inconsistent with observations: air temperatures drop to near freezing in the afternoon with lake-effect squalls.

Correct on two counts: cyclone clouds obscure the sunrise; Fitz records a gale at 7 am EST.

The dawn came late and the breakfast had to wait
when the gale of November came slashin'.
When afternoon came it was freezing rain
in the face of a hurricane west wind.

Correct: Strongest winds are westerly, behind the cyclone. Even experts miss this key meteorological fact.

When supper-time came the old cook came on
deck sayin' "Fellas, it's too rough t' feed ya"
At seven P.M. a main hatch-way caved in; he said,
"Fellas, it's bin good t'know ya!"

No proof of this damage from radio reports. It is one of the possible causes of the wreck, however.

Captain's reports were from the mid-afternoon.

The captain wired in he had water comin' in
and the good ship and crew was in peril.
And later that night when 'is lights went out of sight
came the wreck of the *Edmund Fitzgerald* . . .

Correct: Fitz vanishes in zero-visibility snow squalls at night.

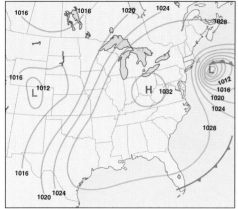

FIGURE 10-21 Surface weather map for 7:00 PM Eastern Standard Time on February 20, 1979 (see Figure 9-9 for a satellite image of the same storm on the previous day). Notice the stark difference in appearance of anticyclones versus cyclones. The bull's-eye–like low off the U.S. East Coast has fronts and a tight pressure gradient near its center. The blobby high over Ohio covers the entire eastern half of the U.S., yet pressures over this vast region differ by only 4 millibars. (Read Box 10-3 and Chapter 13 for more information on this famous storm.) (Adapted from Kocin, P. and Uccellini, L. *Snowstorms Along the Northeastern Coast of the United States: 1955 to 1985.* American Meteorological Society, 1990.)

memory of the *Fitzgerald* and her crew[2] by never losing your life to the many weapons—wind, rain, snow, and wave—of an extratropical cyclone.

THE EXTRATROPICAL ANTICYCLONE

Did you notice the large weather system that moved in immediately behind the *Fitzgerald* cyclone? Probably not. It was an **anticyclone**, or high-pressure system. To complete our understanding of large-scale extratropical weather, we need to learn how to recognize the seemingly boring "face" of the anticyclone.

The anticyclone is the natural complement to the cyclone, the yang to the cyclone's yin:

- Lows form and grow near fronts, the boundaries of air masses where contrasts of temperature and moisture are strongest. Highs are air masses, with temperature and moisture varying little across hundreds of miles.
- Lows are usually cloudy, wet, and stormy. Highs are often clear, relatively dry, and calm.
- The pressure pattern of a low looks like a bull's-eye on a weather map, with tightly packed, concentric isobars. A high looks like a large blob with weak pressure gradients near its center (**FIGURE 10-21**).
- Lows live for a few tumultuous days. Highs can loiter, sometimes for several languorous weeks in summer.

We can explain these features of an anticyclone the same way we did for the cyclone. For example, in Chapter 6, we learned that air in a high diverges at the surface. Diverging air weakens temperature and humidity gradients by spreading out the lines of constant temperature and moisture. Weak gradients mean no fronts in a high; fronts exist instead at the periphery of highs.

Diverging air at the surface in a high requires air to sink toward the surface from above. Sinking air warms adiabatically (see Chapter 2) and dries out in a relative sense (see Chapter 7). A temperature inversion forms above the ground as a result of the compressional warming of this sinking air. This makes highs stable (see Chapter 9) and often cloudless.

Lows can develop strong pressure gradients (bull's-eyes) near their centers because, as we saw in Chapter 6, the pressure gradient force in a low can be counterbalanced by both the Coriolis and centrifugal forces. In a high, the pressure gradient force teams up with the centrifugal force, and the Coriolis force must counterbalance both of them. The stronger the pressure gradient, the harder it is for the Coriolis force to do this. The end result is that nature prefers highs with weak pressure gradients, which implies a spreading out of isobars over large distances. Therefore, highs are large and blobby compared with lows; by the geostrophic wind approximation (see Chapter 6), this makes the winds almost calm near a high's center.

Highs, being air masses, form where air can sit around for a period of time before being steered elsewhere by jet-stream winds. **FIGURE 10-22** shows these regions of **anticyclogenesis**. Many wintertime highs in the U.S. originate in northern Canada—they are the continental polar (cP) air masses discussed in Chapter 9. Cold air outbreaks such as the one depicted in Figure 9-7 occur when strong anticyclogenesis over Canada is combined with jet-stream winds from the northwest that guide the high into the U.S. If a strong, cold high surrounds even a typical low in wintertime, the resulting tight pressure gradient and high winds can lead to blizzards (**BOX 10-7**).

Summertime highs deserve special attention. They share something in common with wintertime lows—both can become cut off from the main jet-stream winds and the surface temperature gradients beneath the jet. Because an extratropical cyclone thrives on temperature contrasts, a cut-off

---

[2] To honor the *Fitzgerald*'s crew and thank their families for the use of the story of their loved ones, a portion of the proceeds from this textbook project has been donated to the Great Lakes Mariners Memorial Project of the Great Lakes Shipwreck Historical Society, located at Whitefish Point, Michigan.

January Anticyclones, 1950–1977

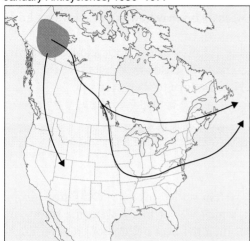

July Anticyclones, 1950–1977

FIGURE 10-22 Typical regions of anticyclogenesis (shaded) and anticyclone paths (arrows) that affect North America in the months of January and July. Compare this figure with the corresponding data for cyclones in Figure 10-7. (Adapted from Zishka, K. M., and P. J. Smith, *Monthly Weather Review*, April 1980: 394–395.)

Box 10-7 The Groundhog Day Blizzard of 2011

As this book was in production in early February 2011, the central and eastern U.S. experienced one of the epic blizzards of the past several decades. Dubbed the "Groundhog Day Blizzard," it was caused by a fairly average (996 mb) but sopping-wet extratropical cyclone moving northeast from Texas. However, this low was surrounded by an intense (up to 1054 mb) and bitter-cold anticyclone to its north and west. The resulting very tight pressure gradient caused high winds gusting over 60 knots (69 mph) across parts of the upper Midwest.

The exceptional amount of Gulf of Mexico moisture in the system rode up and over the cold air, leading to heavy snow accompanied at times by thunder and lightning. Record or near-record snow amounts occurred from Tulsa, Oklahoma (35.5 cm, or 14″, its biggest snowstorm ever) to the Chicago-Milwaukee area (for example, 51.3 cm or 20.2″ at Chicago's O'Hare International Airport, the third-largest snowstorm there).

The snow, combined with the high winds, created blizzard conditions with near-zero visibility and snow drifts up to 1.5 meters (5 feet). Interstates, airports, and businesses were shut down from New Mexico and Texas to the Northeast, and the blizzard turned Chicago's famed Lake Shore Drive into a parking lot of abandoned cars (see photo).

Computer forecast models (see Chapter 13) anticipated this major snowstorm about a week in advance. Meteorologists knew 3 to 4 days ahead of time that there would be a wide swath of heavy snow, ice, and wind across the southern Plains extending to the Great Lakes region and even into New England. Nevertheless, at least 20 people died in the Groundhog Day Blizzard, and damages were estimated initially in the billions of dollars. However, the storm had an impact on over 100 million Americans in over 30 states; in this context, the losses were far less than they would have been without the ample warning provided by meteorologists. No groundhog could have foreseen this monster storm!

cyclone weakens and dies. A summer anticyclone, however, which is an air mass without significant horizontal temperature gradients, can thrive and intensify in this situation. A cut-off high is a form of "blocking," which we discussed in Chapter 7. A blocking high over land in summer can trap and recirculate hot air around and around its center for weeks. This may lead to a heat wave, a drought, or an episode of air pollution (see Chapter 15).

We've recognized and explained the identifying features of the anticyclone. Underneath its calm exterior, however, lurks a potential killer. Did you sense any danger? Neither did the residents of western Europe in the summer 2003. Now we turn to their story.

HIGH PRESSURE, HIGH HEAT: THE DEADLY EUROPEAN HEAT WAVE OF 2003

It began innocuously enough, as a high-pressure system settled over western Europe in June. Residents of London and Paris noticed the heat, which was at first a welcome respite from the cooler temperatures of a European spring. But then the heat became oppressive; England and Wales experienced their hottest June since 1976. The peak of summer had not yet arrived, however, so even record heat for June was not necessarily deadly heat.

Then, in July, the heat abated a little near London as the focus of the high pressure and heat shifted southeast toward Italy and the Mediterranean. In Switzerland, ice melting on the Matterhorn in the sweltering heat led to rockslides, requiring rescues of dozens of trapped mountain climbers. In Croatia and Serbia, the dry heat drained rivers to their lowest levels in a century or more. So much French wheat wilted in the hot sun that flour prices in Britain rose dramatically. A crisis was in the making.

Then on August 3rd, the high pressure and high heat returned with a vengeance to western Europe. An intense blocking high formed above the English Channel (**FIGURE 10-23**), baking much of western Europe in temperatures as much as 10° C (18° F) warmer than typical clear-sky late-summer conditions (**FIGURE 10-24**). Daytime temperatures exceeded 38° C (100° F) for the first time on record in the United Kingdom. In parts of France, Belgium, and Switzerland, temperatures also reached historic peaks. Nighttime temperatures remained unusually warm, preventing people and animals from cooling down adequately overnight before the next day's

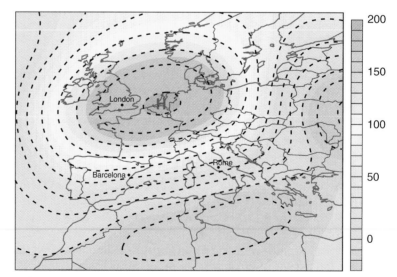

FIGURE 10-23 Anomalies in 500-mb geopotential height over western Europe during August 1 to 13, 2003, versus an average August. Red regions over Great Britain and France indicate much higher than normal heights, where high pressure was concentrated. (From G. A. Meehl et al., *Science* 305, 994–997 [2004]. Reprinted with permission from AAAS.)

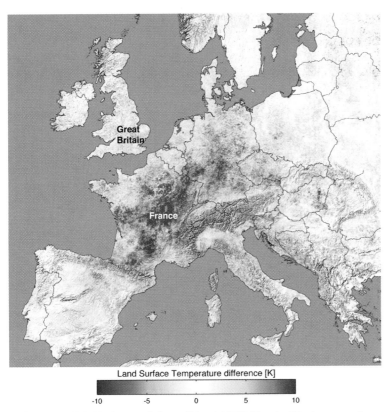

FIGURE 10-24 High-resolution (1 km) MODIS satellite image of western European land surface temperature differences during July 20 to August 20, 2003, versus clear-sky conditions in July and August of 2001, 2002, and 2004. The dark red regions over France were 10° C (18° F) warmer than in comparable clear-sky conditions in other years.

heat. Pollution trapped in the stagnant high made it difficult for residents with asthma or other respiratory conditions to breathe. Forest fires raged in Spain and Portugal, causing $1 billion in damage and adding still more smoke and pollution to the toxic air.

A cold front finally ended the heat wave on August 17th, but by then the damage had already been done. When it was over, the summer of 2003 ranked as by far the hottest in at least a century across much of Europe (FIGURE 10-25), causing over $12 billion in crop losses. Statistically speaking, the summer's heat was as far off the chart as the IQ of an Einstein or da Vinci.

The death toll slowly mounted across Europe; heat waves, unlike tornadoes and hurricanes, are silent killers. Ultimately, at least 35,000 and perhaps over 50,000 residents of Italy, France, Germany, Spain, Great Britain, Portugal, the Netherlands, Belgium, and Switzerland had died due to the heat—an astonishing death toll compared even with the deadliest hurricanes. In towns and cities all across western Europe, the elderly and those with heart and lung conditions died in large numbers during the peak of the heat wave during August (FIGURE 10-26).

The 2003 European heat wave was the deadliest weather-related disaster of the first decade of the 2000s, probably 20 to 30 times deadlier than Hurricane Katrina, and all because of a very persistent high-pressure area.

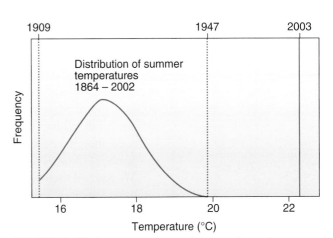

FIGURE 10-25 Average summer temperatures in Switzerland for the years 1864 to 2003. Each vertical line represents the average temperature of one summer. The summer of 2003 was more than 2° C warmer than any previous summer. (Adapted from C. Schär et al., *Nature* 427, 332–336.)

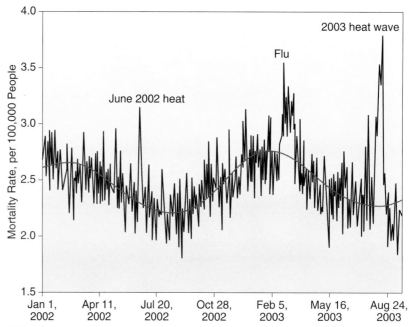

FIGURE 10-26 Daily mortality rate in the state of Baden-Württemberg, Germany, from January 2002 through August 2003. Notice the huge spike in deaths in August 2003, associated with the deadly European heat wave. (Lesser spikes in June 2002 and February 2003 were due to a heat wave and a flu outbreak, respectively. On average, mortality is greater in winter.) The August 2003 heat wave caused about 1000 more deaths than usual in Baden-Württemberg, which with a population of 10.7 million is slightly larger than the Chicago metropolitan area. (Adapted from Koppe, C. and Jendritzky, G. *Gesundheitliche Auswirkungen der Hitzewelle.* Socialministerium Baden-Wurttemberg, Stüttgart, 2004.)

PUTTING IT ALL TOGETHER

■ Summary

Extratropical cyclones are low-pressure systems that cause wet and often windy weather, but they are very different than tropical cyclones. Norwegian meteorologists discovered that extratropical cyclones are associated with fronts and that they have a definite life cycle, growing from birth as a frontal wave to maturity as an occluded cyclone to death as a cut-off cyclone over the course of several days.

Extratropical cyclones are born on the downwind side of tall mountain ranges, near warm ocean currents, and beneath strong jet-stream winds. Further growth of these storms may occur if the upper-tropospheric air is diverging above the cyclone. The age and strength of extratropical cyclones can be estimated by looking at satellite pictures and weather maps. Strong cyclones often have comma-cloud shapes with dry slots and well-defined cold and warm fronts; the presence of an occluded front indicates a mature, but soon-dying, cyclone.

We learned these facts in the context of the storm that helped wreck the *Edmund Fitzgerald* on Lake Superior in 1975. This storm demonstrates that the Norwegian cyclone model, although simplified, can help explain most of the characteristics of a real-life, deadly extratropical cyclone.

The extratropical anticyclone is in many ways the complement to the cyclone. An extratropical cyclone lives at the clashes of different air masses. In contrast, an anticyclone is one big, often slow, fairly calm and stable air mass. Even so, a high can become a killer, for example during the European heat wave in summer 2003.

Key Terms

You should understand all of the following terms. Use the glossary and this chapter to improve your understanding of these terms.

Alberta Clippers	Diffluence	Open wave
Anticyclogenesis	Dry slot	Panhandle Hooks
Anticyclone	Frontal wave	Pineapple Express
Baroclinic instability	Nor'easters	Speed divergence
Comma cloud	Norwegian cyclone model	Warm sector
Cut-off cyclone	Occluded cyclone	
Cyclogenesis	Occlusion	

Review Questions

1. What are the four stages in an extratropical cyclone's life cycle, according to the Bergen School conceptual model used in this chapter? During which stage(s) would you expect to see an occluded front? Strong warm and cold fronts? A cut-off low? Make a sketch of the low and the fronts at each stage of the life cycle.

2. Why do the birthplaces of cyclones vary in location throughout the year? Would this be true if the Earth was not tilted on its axis?

3. Can a cyclone develop out of a clear blue sky? If so, how?

4. Intense cyclones are more common east of the Appalachian Mountains than east of the much higher Rocky Mountains. Why (see Boxes 10-1 and 10-3)?

5. Which one of the cyclone types discussed in this chapter is not named for the region in which it is "born"? It didn't get its name from the region it affects either. So how did it get its name? (Hint: Think about how wind flows around a cyclone.)

6. In Figure 10-9, you can see that Madison, Wisconsin's, temperature rose but its dew point fell on the day after the cold front passed. Using air mass and radiation concepts from this textbook, explain why this happened.

7. Using the real-life cross-sections in Figure 10-10 as a guide, explain how temperature, dew point, pressure, cloud altitude and type, and wind direction change as a warm front approaches and passes a location. Do the same for the cold front. Compare your answers with the idealized discussion in Chapter 9.

8. It is 8:00 AM on Monday, your first day of work at the National Weather Service office in Chicago, Illinois. You notice a developing low-pressure system centered over Wichita, Kansas, about 1000 kilometers (600 miles) to the southwest. The winds at 500 mb above the cyclone are blowing toward the northeast at 100 kilometers per hour (60 mph). If the storm maintains its current intensity, when do you forecast that the center of the cyclone will pass near the Chicago area? If the cyclone and the upper-level trough strengthen, do you think the low will pass to the north or the south of Chicago?

9. What are two causes of upper-level divergence? Why does upper-level divergence help intensify a cyclone?

10. According to Box 10-3, was the Fitzgerald cyclone a "bomb"? What key fuel source did this cyclone have in limited supply compared to a cyclone just east of Cape Hatteras, North Carolina?

11. Name five different ways that weather associated with an extratropical cyclone can injure or kill people.

12. Assume that you have the power to remove extratropical cyclones from the Earth's weather. How might this affect the wintertime weather in eastern Canada? Could this cause a long-term trend in temperatures in the tropics?

13. Why is it that wintertime Canadian high-pressure systems are colder when they are in Canada than when they reach the Gulf of Mexico several days later?

14. Based on the wind circulation around a high, describe the changes in temperature, dew point, and wind direction as an anticyclone moves across the central U.S. from west to east.

■ Observation Activities

1. Keep track of the development and movements of fronts across the continents. Make a cross-section through a front (see Figure 10-10 for an example) and discuss how these observations match and don't match our simple conceptual model of frontal weather.

2. Memorable weather events can shape the lives of communities, and people recall these events via stories and song. What weather stories do people in your town tell? Why do you think these stories endure? Can you explain the cause of this weather to your neighbors?

This rain cloud icon is your clue to go to the *Meteorology* Web site at http://physicalscience.jbpub.com/ackerman/meteorology/. Through animations, quizzes, web exercises, and more, you can explore in further detail many fascinating topics in meteorology.

Thunderstorms and Tornadoes

11

AFTER COMPLETING THIS CHAPTER, YOU SHOULD BE ABLE TO
- Describe the different types of thunderstorms and the weather conditions that lead to their formation
- Explain why scientists think tornadoes occur and how the winds in a tornado are estimated using the Enhanced Fujita scale
- Characterize the different stages of the life cycles of thunderstorms and tornadoes
- Identify likely areas for the occurrence of various types of severe thunderstorm weather

INTRODUCTION

It was supposed to be a relaxing spring-break weekend for the Knoxes. Your co-author John, his climatologist wife Pam, son Evan and sports journalist brother David converged on the tall Four Seasons Hotel in midtown Atlanta, Georgia on March 14, 2008 to unwind and attend the Southeastern Conference college basketball tournament later that weekend.

Shortly after arriving that evening, however, thunder interrupted their unpacking. A quick check of TV weather revealed that the National Weather Station (NWS) had issued a severe thunderstorm warning for the Atlanta area. Just a minute or two later, the warning was upgraded to a tornado warning!

John and Pam, being weather nerds whose love of meteorology was inspired by close calls with tornadoes in their youth, immediately headed for the windows. Their 14th-floor room just happened to face south toward the skyscrapers of downtown Atlanta about 2 miles away. John and Pam both knew from the TV weather radar images that the tornado was moving away from their hotel, toward the southeast. Their knowledge of thunderstorms told them that any wind, rain, and hail would be directed away from their windows. It was dark, however, and there was almost no lightning to illuminate the storm.

Then John saw what seemed at first to be lightning, except that it was only near the ground. He recalled severe weather videos of broken power lines arcing in greenish glows of sparks and made the connection. He told Pam, "Look over there, over that building . . . watch for the glow of a power line break." Seconds later, there was another greenish arcing glow, just to the southeast of the previous one—then another, still closer to downtown. And another. And another. The power line breaks were tracing out the path of the tornado as it created a path of destruction aimed directly at the heart of Atlanta.

John and Pam watched in amazement, with a front-row seat at a tornado, until the rain and hail of the thunderstorm enveloped their hotel. Then Evan's cries of "get away from the window!" were finally heeded. The Knoxes watched TV as the tornado came within just a few hundred yards of the SEC basketball tournament game between Alabama and Mississippi State, tearing the roof of the Georgia Dome and swaying the scoreboard back and forth over the court.

Major downtown buildings in Atlanta sustained damage, and one person was killed as the result of the tornado (see more about this storm later in this chapter). The next day, the basketball tournament was moved to a smaller arena and closed to most of the public, including the Knoxes. But from a selfishly meteorological perspective, the weekend was memorable: "We don't chase tornadoes, we just let them come to us," observed Pam.

In this chapter, we examine tornadoes and other severe weather in the context of the weather feature that creates them: the thunderstorm. Like the extratropical cyclones we studied in the last chapter, thunderstorms and tornadoes are associated with recognizable large-scale weather patterns that allow meteorologists to predict when and where they may occur. Both thunderstorms and tornadoes also exhibit identifiable structures and life cycles that help us understand how they work, just like cyclones, and allow storm chasers to know when and where to view tornadoes safely. Thunderstorm phenomena—not just tornadoes, but also lightning, hail, and floods—are among the most awe-inspiring sights in all of nature, but they can also kill as they did in Atlanta in March 2008 and in Greensburg, Kansas the preceding May. Because of this, we discuss the precautions you need to take to avoid harm from thunderstorms.

WHAT IS A THUNDERSTORM?

A **thunderstorm** is, as the name implies, a cloud or cluster of clouds that produces thunder, lightning, heavy rain, and sometimes hail and tornadoes. This dangerous mix of weather requires great amounts of energy from the atmosphere. This energy is released when saturated air rises rapidly and high into the atmosphere. For this reason, thunderstorms are associated with tall cumulonimbus clouds that form when air rises or is lifted from the surface. The air in thunderstorms rises so rapidly that their tops may briefly overshoot the tropopause and penetrate the stratosphere (**FIGURE 11-1**). These **overshooting tops** can be seen on satellite images either by the visible shadow that they cast on the lower "anvil" cloud top (**FIGURE 11-2a**) or by their especially cold temperatures, which are "seen" in infrared satellite pictures (Figure 11-2b). The air in the anvil may also sink downward to form mammatus clouds (see Chapter 4), as seen in the photo on the cover of this textbook.

Not all thunderstorms are the same. Some thunderstorms are brief rainmakers, whereas others last for hours and form windy lines or clusters. Still others are determined by forecasters to be especially likely to create severe weather such as extremely high winds, hail, and tornadoes. In the following section, we learn where thunderstorms form and the factors that can lead to different types of thunderstorms.

FIGURE 11-1 A photograph of a thunderstorm from the ground, clearly showing the overshooting tops (brightest white clouds at the top) where the thunderstorm's updraft penetrates the tropopause.

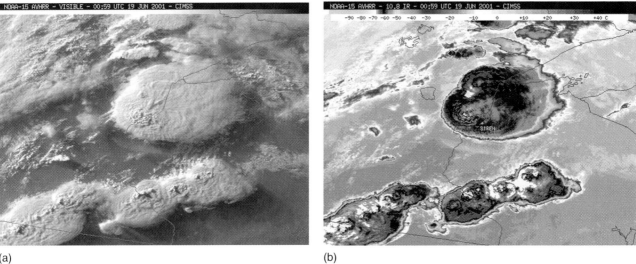

(a)   (b)

FIGURE 11-2 Two low-Earth-orbiting satellite views of thunderstorms, at visible (a) and infrared (b) wavelengths. Overshooting tops are the bubbly features protruding from the broad, white anvils in (a) and that show up as the very coldest cloud-top temperatures in (b). These thunderstorms occurred during the afternoon and evening of June 19, 2001, over the upper Midwest U.S. The large supercell in the top center of the image is crossing the border from Minnesota into northwest Wisconsin. A tornado spawned by this supercell 7 minutes after these satellite images were made was responsible for 3 deaths and 16 injuries along a 0.8-kilometer (0.5-mile) wide path through the small town of Siren, Wisconsin. The line of thunderstorms to the south of the Siren storm produced hail 11.5 cm (4.5 inches) in diameter.

THUNDERSTORM DISTRIBUTION

Thunderstorms require warm, moist air that rises. For this reason, thunderstorms are most prevalent in regions with maritime tropical (mT) air masses and in regions where mountains and frontal cyclones help lift air vertically. **FIGURE 11-3** depicts the global distribution of thunderstorm days. Tropical land regions associated with the average position of the Intertropical Convergence Zone (ITCZ) have the most days with thunderstorms. Africa has the most thunderstorm days—more than 100 per year.

FIGURE 11-4 shows the long-term annual climatology of thunderstorms in the lower 48 United States. Florida leads the nation in thunderstorm frequency because warm, moist air is present year-round and the sea breeze (see Chapter 6) helps create local regions of convergence and lifting even in the absence of large-scale fronts. The far West, in contrast, is too dry, and near the west coast it is too cool for thunderstorms to be common there. Eastern Colorado and New Mexico, although usually dry, rival Florida in thunderstorm frequency. This is because warm, moist air rises as it climbs westward along the sloping Great Plains and because the extratropical cyclone track (see Chapter 10) causes rising motion in this region as cyclones "spin up" in the lee of the Rockies. These factors also determine the location of the 2000 thunderstorms that are seen across the globe at any one time, with most thunderstorms occurring in the tropics and in the midlatitude storm-track regions.

FACTORS AFFECTING THUNDERSTORM GROWTH AND DEVELOPMENT

The condition of the environment around warm, moist air makes all the difference between a clear sky, a garden-variety thunderstorm, and a severe thunderstorm with tornadoes. The crucial factors for thunderstorm development are the environment's temperature, moisture, and wind speed and direction from the ground all the way up to the tropopause. Of course,

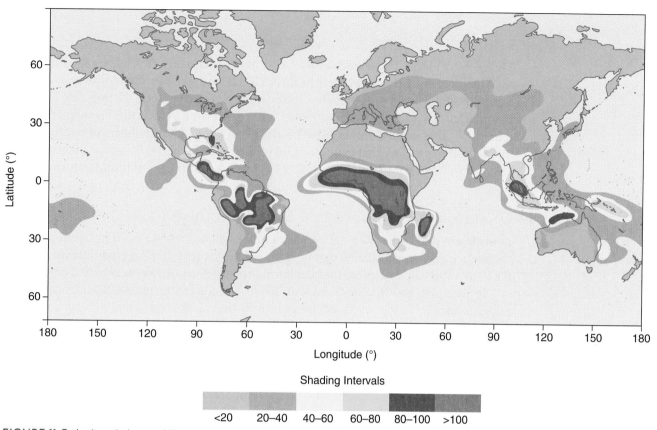

FIGURE 11-3 A climatology of the average number of thunderstorm days in a year. (Adapted from WMO [World Meteorological Organization], 1956: *World Distribution of Thunderstorm Days*. WMO Publ. No. 21, TP. 21.)

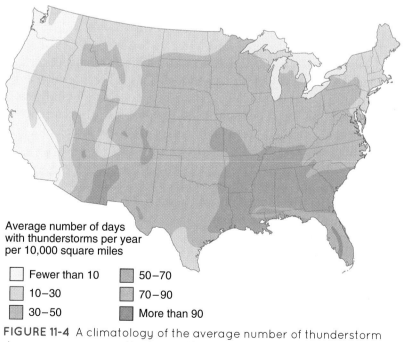

FIGURE 11-4 A climatology of the average number of thunderstorm days per year across the lower 48 United States. The Southeast United States leads the nation with more than 50 thunderstorm days each year—with a maximum in Florida. Note the secondary maximum in eastern New Mexico and Colorado. (Courtesy of Oklahoma Climatological Survey.)

thunderstorms also need an initial lifting mechanism, without which there is no thunderstorm to begin with!

Lifting Mechanisms

Thunderstorms are triggered by various lifting mechanisms. In Chapter 4 we discussed the various ways of lifting air parcels to form clouds. Surface heating by the sun is a common lifting mechanism in summertime thunderstorm development. Lifting also occurs along boundaries. Chapter 9 discussed fronts and the dryline, which are boundaries separating different air masses. Cold fronts provide particularly good lift for thunderstorms, driving upward the warm moist mT air that is often ahead of them. Similarly, the dryline, a moisture boundary between mT and continental tropical (cT) air masses, can lead to thunderstorms and severe weather.

Unstable Atmosphere and Stability Indices

As we learned in earlier chapters, a parcel of air will not rise unless it is forced upward from the surface and/or is unstable, that is, warmer than its surrounding environment. You can use the book's Web site to explore thunderstorm growth in various environmental conditions.

To interpret how the environment affects thunderstorm potential and severity, meteorologists have invented several stability indices that characterize the atmosphere in a single number. Perhaps the most common stability index is the **lifted index** (**LI**). This index follows the same approach we used in Chapters 3 and 4 to discuss how temperature changes in a rising air parcel. Simplified, the index starts with an air parcel from the surface, lifting and cooling it dry adiabatically to saturation and then lifting and cooling it moist adiabatically to 500 millibars (mb). To compute the LI, the temperature of the parcel at 500 mb is subtracted from the environment's temperature at 500 mb, as measured by a radiosonde or satellite. If the observed 500-mb temperature is colder than the lifted air parcel, then the parcel is unstable and will be able to keep on rising and form a tall cumulonimbus cloud. The LI is negative in these cases.

For this reason, negative values of the LI can be related to the potential for thunderstorm severity. A LI of between 0 and −3 (degrees Celsius) indicates that the air is marginally unstable and unlikely to lead to severe thunderstorms (**TABLE 11-1**). Values between −3 and −6 indicate moderately unstable conditions. Values between −6 and −9 are found in very unstable regions. LI values less than −9 reflect extreme instability. The chances of a severe thunderstorm are best when the LI is less than or equal to −6. This is because air rising in these situations is much warmer than its surroundings and can accelerate rapidly and create tall, violent thunderstorms. (However, this index does not tell the probability of occurrence of a thunderstorm, only the chance that a thunderstorm that does form will be severe; other indices predict thunderstorm probabilities. Learn more about them on the text's Web site.)

FIGURE 11-5 shows very negative values of the LI (yellow- and red-shaded areas) derived for regions around severe thunderstorms in the Great Plains region. These regions were placed under severe weather watches (see Chapter 1) because of the extreme instability of the atmosphere. One of these storms produced a record hailstone with the circumference of a soccer ball!

Stability indices.

TABLE 11-1 LI, Stability, and Possible Weather

LI	Stability	Interpretation (if Thunderstorms Form)
> 0	Stable	Thunderstorms unlikely without strong lifting mechanism
0 to −3	Marginally unstable	Severe thunderstorms unlikely
−3 to −6	Moderately unstable	Severe thunderstorms possible
−6 to −9	Very unstable	Severe thunderstorms probable
< −9	Extremely unstable	Severe thunderstorms very likely

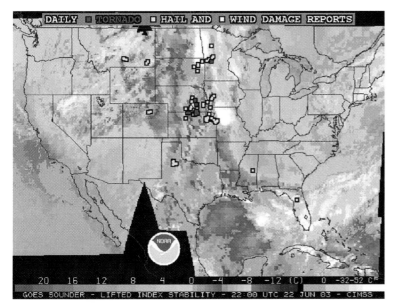

FIGURE 11-5 Satellite-derived values of the LI across the United States on June 22, 2003. The gray and white regions are cloud systems in which the satellite cannot estimate the index. The yellow and red regions in the Midwest and Great Plains regions represent areas with LI values less than −4, indicating the possibility of severe weather. Severe weather reports are indicated with symbols. The storm in Nebraska, which is surrounded by LI values less than −8, produced a soccer-ball-sized hailstone. Most of the rest of the U.S. is shaded in beige, indicating regions with positive LI values where severe weather is unlikely. Black regions are gaps in the satellite data coverage.

Vertical Wind Shear

Another factor affecting thunderstorm type is the change of environmental wind speed and direction from the ground up, known as **vertical wind shear**. Small amounts of vertical wind shear lead to upright and majestic but (as we soon discuss) shorter-lived thunderstorms. Moderate amounts of vertical wind shear, such as are found in and near extratropical cyclones (**FIGURE 11-6**), cause thunderstorm clouds to tilt. If the wind changes direction and increases in speed to a large extent, the thunderstorm itself rotates. Rotating thunderstorms generally cause the worst severe weather, including large hail and violent tornadoes. Although indices exist to estimate the impact of vertical wind shear on thunderstorms, they are beyond the scope of this text. The simple rule of thumb is this: "The more vertical wind shear, the more severe the thunderstorm."

Low-Level Jet Stream and Inversions

Two other factors affecting thunderstorm growth and development are shown in Figure 11-6. A low-level jet stream of air often develops over the Great Plains in the spring and summer. This jet occurs often at night and in that case is called a **nocturnal low-level jet**. This low-level jet is an important ingredient in the formation of severe weather in the United States because it supplies moisture and energy to Great Plains thunderstorms after the Sun goes down.

In addition, a **capping inversion** can develop in which hot dry air at about 700 mb overlies warm, moist air near the surface. As we saw in Chapter 3, inversions usually prevent surface air from rising. During some hot summer days, however, the hot, dry air acts like a lid on a boiling pot, concentrating energy below it until the near-surface air is warmed enough by the Sun to "blow the cap" and rush upward violently. This can lead to rapidly developing severe weather.

You should review Figure 11-6 to make sure you see how and where all of the different ingredients—lift, instability, vertical wind shear, the low-level jet, and capping inversion—come together to make

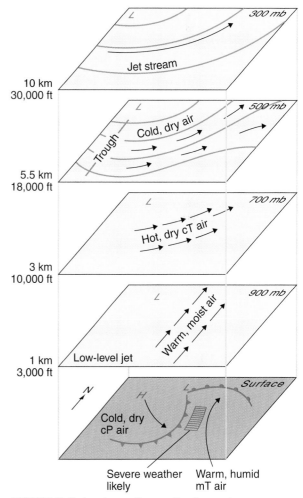

FIGURE 11-6 A schematic profile through a severe weather pattern, illustrating the vertical changes in temperature, moisture, and winds that contribute to severe thunderstorms and tornadoes. (Adapted from Athrens, C. D. *Meteorology Today, Ninth edition*. Brooks Cole, 2009.)

severe weather likely. Now that we have connected thunderstorms with our understanding of the atmosphere from previous chapters, we can categorize the different types of thunderstorms, linking their characteristics to the factors described earlier.

TYPES OF THUNDERSTORMS

Thunderstorms can be classified by either their severity or their structure. For example, the NWS defines a **severe thunderstorm** as a thunderstorm that produces one or more of the following: wind gusts of 50 knots (58 mph or 93 kilometers per hour) or greater, hail 1 inch (2.5 centimeters) in diameter or larger, or a tornado. The damage from these storms can be so devastating that it can be seen from space. FIGURE 11-7 is a satellite view of the region in the path of the tornadic thunderstorm shown in Figure 11-2. The path of damage appears as a whitish blur across northwest Wisconsin, as if a pencil eraser had been dragged across the image.

To understand how thunderstorms, severe and nonsevere alike, are created, however, we must look instead at the structure of these storms. Thunderstorms are composed of basic building blocks referred to as cells. A **cell** is a compact region of a cloud that has a strong vertical updraft.

There are two basic categories of thunderstorm cells: **ordinary cells** and **supercells**. Ordinary cells are a few kilometers or miles in diameter and exist for less than an hour, whereas supercells are larger and can last for several hours. **Multicell** storms are composed of lines or clusters of thunderstorm cells, either ordinary cells, supercells, or a mixture of both. Ordinary single-cell thunderstorms are the most common, but multicell and supercell storms are responsible for the vast majority of severe weather reports associated with thunderstorms. TABLE 11-2 summarizes the characteristic features of the ordinary single-cell, the multicell, and the supercell thunderstorms. Next we examine the structure of ordinary single-cell, multicell, and supercell thunderstorms.

◼ Ordinary Single-Cell Thunderstorms

Ordinary single-cell thunderstorms are short-lived and localized single-cell thunderstorms. An individual thunderstorm cell has a life cycle with three distinct stages: the cumulus, the mature, and the dissipating stages (FIGURE 11-8).

The **cumulus stage** is the initial stage of a thunderstorm. Warm air near the ground rises and cools initially at the dry adiabatic lapse rate. The rising air parcel approaches saturation as the relative humidity increases, until condensation occurs. The altitude to which a parcel of air needs

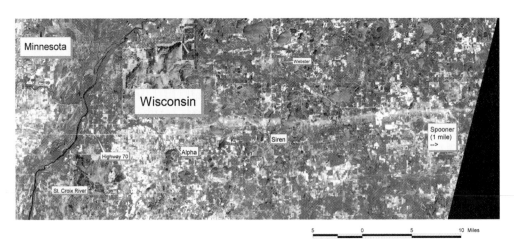

FIGURE 11-7 A LANDSAT photograph of the damage path of the tornado spawned by the supercell thunderstorm in Figure 11-2. This image was taken the morning after the tornado hit Siren, Wisconsin. Vegetation shows up as red in this satellite image. Lakes and rivers are blue, and roads and towns are generally aqua or white. The fuzzy whitish blur extending left to right from the small town of Alpha through Siren and toward Spooner is the damage path of the tornado.

TABLE 11-2 The Three Thunderstorm Types*

Type	Appearance	Vertical Wind Shear	Chance of Severe Weather
Air mass/ ordinary single-cell	"Popcorn" in visible satellite image	Small	Unlikely
Multicell	MCC: state-sized circular cloud in infrared satellite image	Small	Likely (nontornadic high winds)
	Squall line: line of thunderstorms in radar or satellite images	Moderate	Likely (20% of tornadoes from nonsupercell storms)
Supercell	Hook echo in radar reflectivity image	Large	Very likely (80% of tornadoes from supercells)

*Severe thunderstorms have winds of at least 58 mph, hail with 1 inch or greater diameters, or a tornado.

to be lifted before saturation is reached is called the *lifting condensation level,* or LCL. The LCL marks the bottom of the cloud, its base. The cloud grows as moist air continues to be lifted and the growing cell expands both vertically and horizontally. On the text's Web site, you can "grow a thunderstorm" and experiment with how the temperature and dew point change the LCL.

Grow a thunderstorm.

As the cloud is growing, dry air from its surroundings mixes into the cloud at the cloud edges. This is called **entrainment.** This mixing of drier air temporarily lowers the relative humidity and can cause droplets to reduce in size because of evaporation. The evaporation quickly saturates and cools the air within the cloud. Entrainment helps to produce different sizes of drops within the cloud; this is necessary for cloud particle growth by collision and coalescence (see Chapter 4).

The moist air that flows upward through the cloud forms the updraft. As the cloud grows beyond the freezing level (the altitude where the air temperature is below freezing), it is composed of both water drops and ice particles, which supports further growth of the cloud particles (see Chapter 4). Many clouds never develop beyond the cumulus stage. However, when conditions are right, the cumulus cloud builds into a cumulonimbus as precipitation forms and the mature stage in the life cycle of the thunderstorm cell begins.

The **mature stage** of the ordinary-cell thunderstorm begins when precipitation starts to fall from the cloud. During the mature stage, the thunderstorm produces the most lightning, rain, and even small hail. The updrafts in the cumulonimbus become organized and strong, providing the vertical motion needed for cloud-droplet growth as discussed in Chapter 4. Eventually the particles get too large for the updraft to support them in the air, and the cloud particles fall and form the downdraft. As the particles fall out of the cloud base, where the air is unsaturated, they begin to evaporate. This evaporation causes a cooling of the surrounding air, making the air denser and

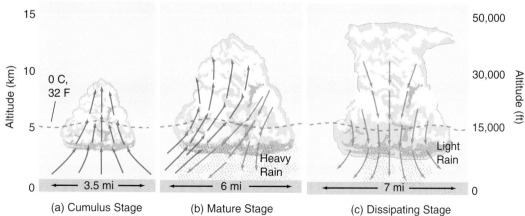

FIGURE 11-8 The life cycle of an ordinary thunderstorm cell contains three stages (a, b, and c respectively): cumulus, mature, and dissipating. Arrows at bottom indicate approximate widths of the base of the storm at each stage.

FIGURE 11-9 A photograph of an ordinary thunderstorm.

thereby enhancing the downdraft. The particles also drag air down with them, enhancing the downdraft. When the downdraft reaches the ground it spreads out and interacts with the updraft.

The **dissipating stage** of a thunderstorm occurs when the updraft, which provides the required moisture for cloud development, begins to weaken and collapse. During this stage of the thunderstorm life cycle, the downdraft dominates the updraft, and the cumulonimbus begins to disappear. Without an updraft, the precipitation ends, and the cloud begins to evaporate as dry environmental air is entrained into the cloud. In this way, the ordinary-cell thunderstorm snuffs itself out, eliminating the upward supply of high-humidity air needed to maintain the thunderstorm.

The **air-mass thunderstorm** is an example of a single-cell thunderstorm that does not produce severe weather. Air-mass thunderstorms develop in warm humid mT air masses and are very common in the southeastern United States, particularly in Florida. They are most prevalent in the summer and complete their life cycle in about 1 hour (**FIGURE 11-9**). Air-mass thunderstorms cannot produce severe weather because the precipitation falls into the updraft. The cooling caused by evaporation and the drag produced by the falling raindrops eventually extinguish the updraft before severe weather can occur.

▇ Multicell Thunderstorms

Many thunderstorms are multicell storms. Multicell storms are composed of several individual single-cell storms, each one at a different stage of development: cumulus, mature, and dissipating (**FIGURE 11-10**). With some cells in the dissipating stage and others in the cumulus stage, a multicell thunderstorm can last for several hours. The key difference between an air-mass thunderstorm and a multicell storm is the presence of moderate amounts of vertical wind shear. This shear tilts the thunderstorm and prevents the precipitation from falling into the updraft and quenching it, as happens in the single-cell thunderstorm. This allows the updraft and downdraft to co-exist. The dense, cold air of the downdraft forms the **gust front**, which helps to lift the warm, moist air flowing toward the storm and then form new cells, as shown in Figure 11-10.

Groups of these thunderstorms often join into larger systems and are generically referred to as **mesoscale convective systems** (**MCS**s). Two classic MCSs are squall lines and **mesoscale convective complexes** (**MCC**s), each of which we discuss individually.

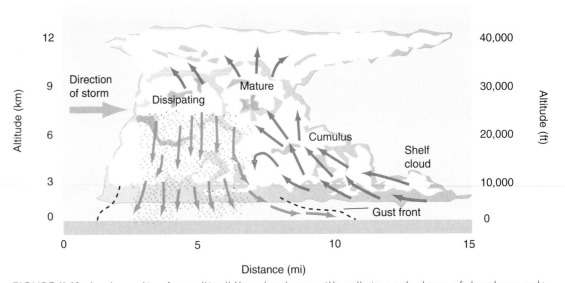

FIGURE 11-10 A schematic of a multicell thunderstorm with cells in each stage of development: cumulus, mature, and dissipating. The red arrows represent the warm updraft, and the blue arrows are the cool downdraft. The storm is moving from left to right.

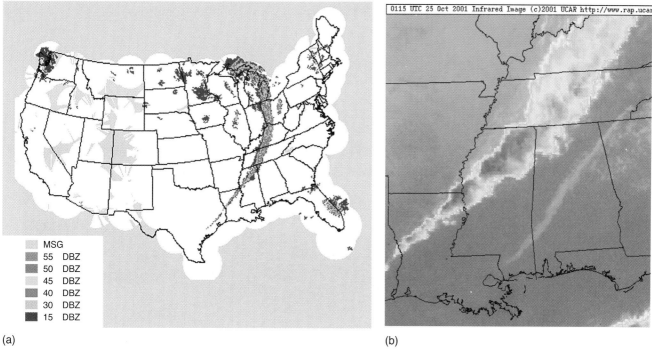

(a)   (b)

FIGURE 11-11 (a) A composite national radar image of a squall line at 8:15 PM Central Daylight Time on October 24, 2001. The most intense thunderstorms in the squall line are in yellow and red and extend from Detroit, Michigan, to just northwest of Houston, Texas. Gray regions are those without radar coverage, usually because of reflection by high mountains. (b) A close-up colorized infrared satellite image of the same squall line at the same time as in (a). Here, the squall line is bearing down on the Southeast United States. The coldest cloud tops associated with the highest thunderstorms are in blue, and the warmer surface is in red.

Squall Line

A **squall line** is composed of individual intense thunderstorm cells arranged in a line, or band, as seen in radar and satellite imagery (**FIGURE 11-11a and b**). They occur along a boundary of unstable air, which gives them a linear appearance. Squall lines often have life spans of approximately 6 to 12 hours or more and extend across several states simultaneously.

Clouds are sometimes observed above the gust front. One such cloud, called a **shelf cloud**, is formed as the gust front forces air near the surface to rise and is caused by the cool downdraft air lifting warm surface air to its condensation level. A shelf cloud is very ominous (**FIGURE 11-12**), but it does not produce damaging weather by itself, although it can precede severe weather by a few minutes.

Weather conditions favorable for the formation of a squall line also include upper-level divergence and a broad, low-level inflow of moist air. Squall lines are often observed ahead of a cold front, as was the case in Figures 11-6 and 11-11. The low-level winds ahead of the front supply the moist air required to develop and maintain thunderstorms. Divergence aloft induces lifting near the surface, causing the storms to develop and grow.

Mesoscale Convective Complex

The MCC is another severe storm composed of multiple single-cell storms. An MCC is a complex of individual storms that covers a large area (100,000 square kilometers or about 40,000 square miles) in an infrared satellite image, roughly the size of the entire state of Iowa. MCCs, like squall lines, are long lived and last for more than 6 hours. MCCs often begin forming in the late afternoon and evening and reach mature stages during the night and toward dawn. In satellite images, MCCs appear as a cluster of thunderstorms that gives the appearance of a large circular storm with cold cloud-top temperatures below −40° C. **FIGURE 11-13** is an infrared satellite image of a MCC that formed late in the day on July 7, 1997, over Nebraska. Lifted indices for the region were less than −8, indicating that any thunderstorms that formed would be severe.

MCCs often form underneath a ridge of high pressure, as indicated in Figure 11-13. This is because, as we learned in Chapter 6, upper-level divergence can occur in a ridge. This promotes

FIGURE 11-12 The shelf cloud extending out and beneath the main thunderstorm cloud signifies the arrival of the gust front.

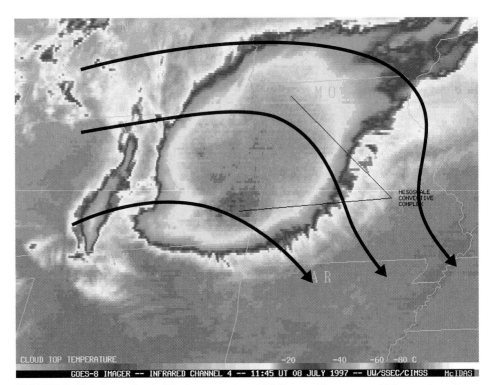

FIGURE 11-13 Infrared satellite image of an MCC over Kansas and Missouri early in the morning of July 8, 1997. The image is color enhanced to highlight the cold cloud-top temperatures. Note the circular shape of the coldest areas, which is the classic signature and definition of the MCC. Overlaid are the wind directions at the 250-mb level, indicating that the MCC is occurring in the vicinity of an upper-level ridge.

rising motion underneath the ridge, leading to thunderstorm growth if conditions nearer the surface are unstable. The MCC on July 7, 1997, produced several reports of heavy rainfall amounts of 10 to 15 centimeters (4 to 6 inches) across parts of Kansas, hail up to 4.5 centimeters (1.75 inches) in diameter, and damaging wind gusts of up to 95 kilometers per hour (60 mph).

MCCs do not require as large amounts of environmental vertical wind shear to survive as do squall lines. The MCC is a multicell storm comprised of convective cells in different stages of their life cycles. For MCCs to exist the individual thunderstorms that comprise the system must support the formation of other convective cells. The downdraft of individual cells of the MCC forms and enhances the updraft of neighboring cells, as shown in FIGURE 11-14.

MCCs require a good supply of moisture from low levels of the atmosphere in order to keep going. As we noted earlier, this is accomplished over the Great Plains by the nocturnal low-level jet. MCCs are unique in that they are maintained by this low-level jet. The MCCs often move eastward as the southerly low-level jet turns eastward overnight under the influence of the Coriolis force. The low-level jet is lifted over the downdrafts of mature cells, and this maintains thunderstorm development. As the low-level jet weakens at sunrise, so does the thunderstorm complex. MCCs usually reach peak intensity in the early morning hours (midnight to 3:00 AM), causing an otherwise inexplicable *nighttime* peak in thunderstorm frequency and intensity over the Great Plains and upper Midwest.

Despite their severity, MCCs also do good. The heavy rains of MCCs are an important source of water for the corn and wheat belts of the United States.

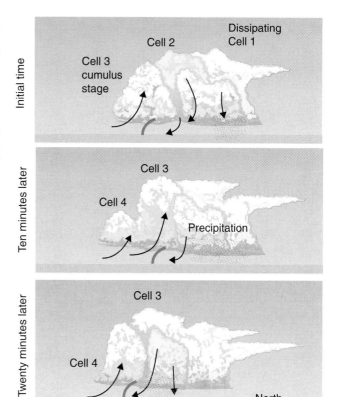

FIGURE 11-14 A time-lapse schematic showing how the downdraft of thunderstorm cells in an MCC assist in the development of new updrafts, leading to new thunderstorms.

Supercell Thunderstorms

The supercell thunderstorm is a large single-cell storm, sometimes 32 kilometers (20 miles) or more across that almost always produces dangerous weather. Supercell thunderstorms produce one or more of the following weather conditions: strong wind gusts, large hail, dangerous lightning, and tornadoes. The severity of these storms is primarily a result of the structure of the environment in which the storms form.

The development of a supercell requires a very unstable atmosphere and strong vertical wind shear (in both speed *and* direction). Often in supercell environments, wind direction at the surface is southerly, whereas the winds aloft are much stronger and from the west. Wind speed may be 25 kilometers per hour (15 mph) at the surface and more than 160 kilometers per hour (100 mph) at the 500-mb level.

Research has shown that this vertical wind shear causes supercell thunderstorms to rotate around a vertical axis. FIGURE 11-15 illustrates how this can occur. First, to picture how this process works, put a pencil between the palms of your hands, with the pencil lying across the lower palm from thumb to pinky and both hands parallel to the ground. Slide your top hand back and forth over the other and notice that the pencil rotates around a horizontal axis (Figure 11-15a). Now, keep sliding your hands while you turn them so that your thumbs are on top. Notice that the pencil is now rotating around a vertical axis (Figure 11-15b).

A similar process is at work inside a supercell thunderstorm. The changes in wind speed in the vicinity of the storm are like your hands in Figure 11-15a: They cause spin around a horizontal axis (Figure 11-15c). In the case of the thunderstorm, it is the air, not a pencil, that is spinning. The vertical updraft inside the thunderstorm then tilts this spinning air so that it spins in the vertical, just as you did with the pencil.

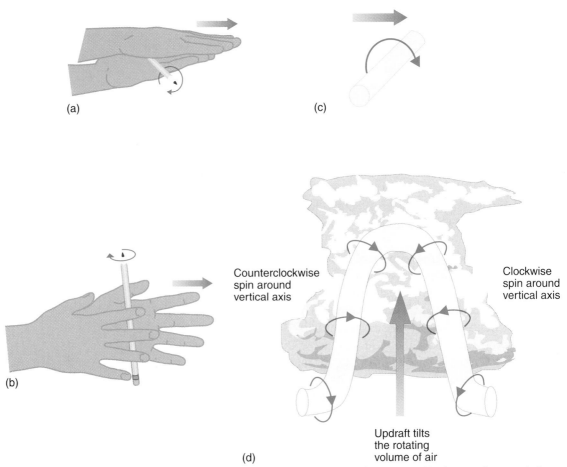

FIGURE 11-15 By moving your hands back and forth, you can make a pencil between them rotate around a horizontal axis (a) or, by turning your wrists, around a vertical axis (b). Similarly, vertical wind shear in the vicinity of a thunderstorm can cause a volume of air to spin horizontally (c). When pushed upward by the thunderstorm's updraft, it can become two vertically rotating columns of air (d). In the presence of winds that change from southerly at the surface to westerly aloft, the cyclonically rotating column on the left becomes the dominant circulation and is called the mesocyclone.

Air is not rigid like a pencil, however, and the result is *two* columns of vertically spinning and rising air, one rotating clockwise and the other rotating counterclockwise. In some cases, the thunderstorm will then split into two parts, with the clockwise portion moving to the left of the original storm path and the counterclockwise cyclonic part moving to the right. The cyclonic part is called a **right-mover**. In the usual case where the wind turns from south to west from the surface to higher altitudes, the cyclonic right-mover is usually the more intense of the two parts. (This is only indirectly due to the Coriolis force, which is too weak on the short time scales of a supercell to have a direct effect.)

This spinning air in the right-moving supercell is like a miniature extratropical cyclone: it has a lower pressure than its surroundings because of air flung outward from it by the centrifugal force, plus the divergence aloft and the latent heating caused by condensation that warms the air in it. It is therefore called a **mesocyclone**, "meso" meaning "middle" (i.e., mid-sized) in Greek. This mesocyclone is 5 to 20 kilometers (3 to 12.5 miles) wide and can extend well up toward the top of the thunderstorm. As it stretches vertically, the mesocyclone becomes narrower and rotates more quickly, a consequence of the Conservation of Angular Momentum we discussed in Chapters 7 and 10. However, the mesocyclone is too large and doesn't spin fast enough to be a tornado, and many tornadoes form without mesocyclones high above them.

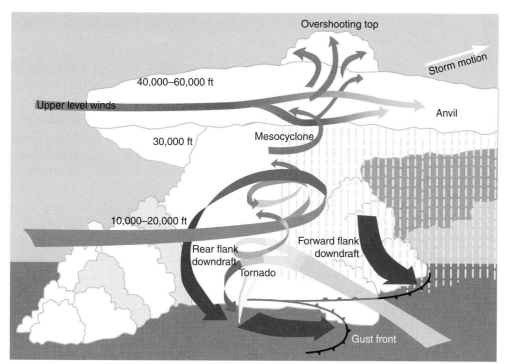

FIGURE 11-16 A schematic of a supercell thunderstorm, as explained in the text. (Courtesy of NOAA.)

The structure of a rotating supercell is illustrated in **FIGURE 11-16**. The turning of the wind direction with height causes the updrafts and downdrafts to wrap around one another. The updraft enters the supercell and slants upward. Overshooting tops indicate where the intense updraft penetrates the stratosphere at the top of the storm. Strong winds aloft carry air from the updraft downwind where it exits through the anvil, helping to maintain one strong and local updraft. Downdrafts both ahead ("forward flank") and behind ("rear flank") the center of the storm are caused by precipitation; their leading edges are denoted by gust fronts.

The surface weather conditions associated with a supercell are depicted in **FIGURE 11-17**. The region of heaviest rain and hail forms a hook around the updraft, which is near the junction of the gust fronts. At this junction, a tornado may form as vertical wind shear increases near the surface and is tilted upward, a possibility we explore shortly. Once initiated, supercell storms sustain themselves for several hours because the updrafts and downdrafts do not interfere with each other.

The favored region for the formation of supercells is the southern Great Plains of the United States in spring. This is because of the presence of lifting mechanisms such as fronts, plus the extreme instability and the combination of low-level and upper-level wind shear conditions that can exist during this time in these locations. Supercells often generate severe weather, but they are not the only thunderstorms that create life-threatening weather. As we see in the following section, all thunderstorms can be hazardous.

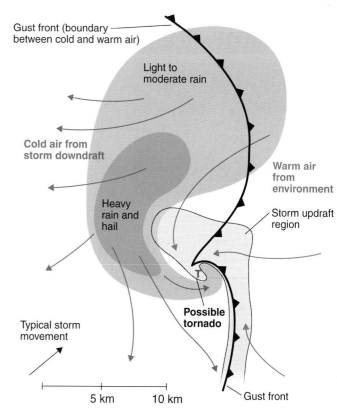

FIGURE 11-17 A schematic diagram of the surface conditions associated with a typical supercell thunderstorm. A tornado is most likely to be found at the intersection of the downdrafts whose leading edges are depicted as cold (gust) fronts. Heavy rain and hail form a hook-shaped region on radar, usually wrapping around and to the west of the intersection of the downdrafts. (Courtesy of Oklahoma Climatological Survey.)

FIGURE 11-18 A downward-plunging microburst with rain (at right) "splashes" against the ground near Wichita, Kansas, on July 1, 1978. This sequence of photographs, taken at intervals of 10 to 60 seconds, helped confirm that microbursts develop vortex-like spins on their leading edges after hitting the ground (upward curl of rain near center of last two photographs).

MICROBURSTS

One particularly dangerous thunderstorm situation develops when rain falling from a thunderstorm evaporates underneath the cloud, cooling the air beneath it. This cold, heavy air plunges to the surface and "splashes" against the ground like a bucket of cold water. The air then rushes sideways and swirls upward as a result of the pressure gradient between the cold air and the warm surroundings. This wind is a **microburst**. It is also sometimes called a "downburst" or a "macroburst" if its path of destruction exceeds 4 kilometers (2.5 miles). FIGURE 11-18 illustrates the microburst life cycle in a series of famous photographs.

The winds from a microburst can cause as much damage as a small tornado, flattening trees and power lines. Microbursts that occur near airports are particularly dangerous. Strong winds from above, below, and sideways buffet aircraft in just a few seconds. Planes that are landing or taking off are pushed into the ground, causing deadly crashes (FIGURE 11-19). Microbursts have led to major air disasters in New York City, Charlotte, New Orleans, and Dallas, killing many hundreds of people. (Microbursts also occur frequently near Denver. The higher frequency of thunderstorms in the East and South, combined with heavy airline travel in these regions, focuses attention on microbursts in those regions.)

The scariest microburst-related aviation event was a disaster that *didn't* happen. On August 1, 1983, Air Force One was ferrying President Ronald Reagan back to Andrews Air Force Base near Washington, DC. The President and his plane landed on the dry runway, uneventfully, at 2:04 PM Eastern time. An approaching thunderstorm then generated a massive microburst that, less than 7 minutes later, caused winds of more than 240 kilometers per hour (150 mph) to blow across that same runway (FIGURE 11-20)! The microburst even had a nearly calm "eye" and a second round of high winds, a little like a minihurricane. If President Reagan's plane had approached the airport just a few minutes later, a crash would have been unavoidable.

Spurred by disasters and close calls such as President Reagan's, the U.S. government has spent millions of dollars on microburst detection equipment at airports. Fewer microburst-related disasters occur today, thanks to this technology and extensive pilot training. Crashes related to microbursts today are rare and are usually attributed to poor decisions by pilots.

THE TORNADO

Tornadoes are rapidly rotating columns or "funnels" of high wind that spiral around very narrow regions of low pressure beneath a thunderstorm (FIGURE 11-21). The funnel itself is visible because of moisture that condenses in the rapidly rising and cooling air in the funnel and also because of dust and debris that are sucked into the vortex. Tornadoes nearly always rotate cyclonically and often move toward the northeast along with the parent thunderstorm. The name "tornado" may stem from the Latin word *tornare,* meaning "to turn," and from the Spanish word for thunderstorm, *tronada*. In the United States, they are also called *twisters* and, as in *The Wizard of Oz, cyclones*. A tornado whose circulation does not extend to the ground is called a **funnel cloud**.

■ Tornado Formation and Life Cycle

Tornadoes are usually less than 1.6 kilometers (1 mile) wide and are short lived, rarely lasting more than an hour. Because of their small size, brief existence, and violent nature, tornadoes are particularly difficult to understand. Therefore,

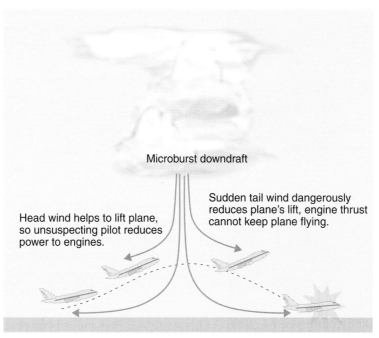

FIGURE 11-19 Schematic illustrating the flight path of an airplane flying into a microburst during takeoff or landing.

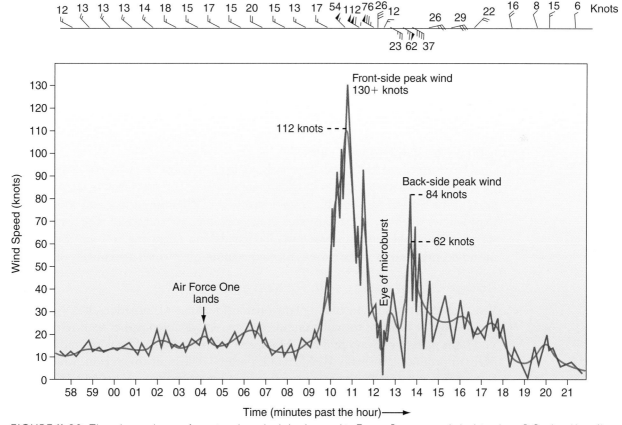

FIGURE 11-20 The chronology of a microburst at Andrews Air Force Base near Washington, DC, shortly after President Reagan landed there on Air Force One on August 1, 1983. Time goes from right to left in this figure, and observed winds are shown in the heavy solid line. (Adapted from Fujita, T. *The Downburst*. University of Chicago Press, 1985.)

FIGURE 11-21 Why chase tornadoes when they can come to you? Audra Thomas poses in front of a majestic tornado passing across the Thomas farm near Beaver City, Nebraska, on April 23, 1989.

emphasis has been placed on up-close observations of tornadoes, including those made by "storm chasers" (BOX 11-1).

No one yet knows exactly how tornadoes form, but close observations of them have revealed many clues about their genesis. Most tornadoes, and the vast majority of damaging tornadoes, develop underneath supercell thunderstorms. Supercells are, as we have seen, rotating thunderstorms. However, the mesocyclone in the supercell is too large and is rotating too slowly to explain the fast spinning of tornadoes. Furthermore, only about 20% to 30% of supercell thunderstorms cause tornadoes. So, while most tornadoes do form with supercells, most supercells do not cause tornadoes.

One explanation for tornado genesis involves the stretching and spinning-up of the mesocyclone in the vertical, based on Conservation of Angular Momentum ideas. In this explanation, the tornado extends downward from the mesocyclone. Another explanation that has gained favor recently is a "bottom-up" approach, in which low-level vertical wind shear near the gust front is tilted up into the updraft and stretched into a tornado funnel. The bottom-up explanation might explain why so many supercells do not produce tornadoes (if there is not enough low-level shear to create a tornado) and also how some nonsupercell thunderstorms do produce tornadoes (if there is enough low-level shear to create a tornado in spite of weak thunderstorm rotation). Research continues on this subject.

Whatever the explanation, a tornado often forms in the following way: The cloud base underneath the updraft on the rear side of the thunderstorm may lower, forming an ominous-looking, rotating **wall cloud** (FIGURE 11-22). As air continues to flow into the wall cloud, a rapidly rotating column of air much smaller than the mesocyclone may protrude below the wall cloud. As water vapor condenses in the air rushing up into this column, a funnel cloud may form and reach the ground, becoming a tornado.

Once formed, tornadoes exhibit a fairly regular four-stage life cycle, as shown in FIGURE 11-23 for a carefully observed tornado in Oklahoma in 1973. The first stage is the organizing stage, during which a funnel cloud picks up debris as it reaches the surface and widens. The mature stage follows when the tornado is often at its peak intensity and width. The tornado reaches the shrinking stage when its funnel narrows, and it ends with a decaying or "rope" stage. At this point, the funnel thins out to a very narrow, ropelike column, after which it eventually dissipates.

■ Radar Observations of Tornadoes

Visual observations help explain many outward features of tornadoes, but weather radar lets us see into the heart of tornadic thunderstorms. As we noted earlier with reference to Figure 11-17, the pattern of heavy rain inside a supercell forms a kind of hook around the region most likely to produce a tornado. Weather radar beams reflect off this region, which appears as a **hook echo** and usually prompts a tornado warning when detected by the NWS.

In addition, modern Doppler radar is able to detect the rapid change in direction of wind around a spinning vortex. This appears as a couplet of red and green colors on Doppler radar images of the velocity of particles in the funnel itself. This couplet signifies winds next to each other that are moving away from and toward the radar beam, respectively. This couplet is called a **tornado vortex signature**, or **TVS**, and is a reliable indicator that a tornado is forming. FIGURE 11-24 shows both the hook echo and the TVS from a deadly tornadic supercell that hit Enterprise, Alabama in March 2007.

Box 11-1 Storm Chasers

Severe thunderstorms produce damaging, and sometimes deadly, weather events. Because of the danger these storms pose, they are studied at a distance, and unlike hurricanes, they are not probed by aircraft. Because they are still not fully understood, eyewitness observations by "storm chasers" have been instrumental in allowing us to learn more about severe thunderstorms and tornadoes.

Storm chasers are trained individuals who seek to gather data, such as video or radar, of severe thunderstorms and tornadoes to further understand their formation and movement. In the photo below, storm chaser Don Lloyd captured a F5 tornado raking eastern Wisconsin on July 18, 1996.

Storm chasing became well-known following the Verification of the Origins of Rotation in Tornadoes Experiment (VORTEX) in 1994–1995 and its sequel VORTEX 2 in 2009–2010. These scientific storm chasing campaigns in the Great Plains, broadly fictionalized in the movie *Twister*, have attempted to answer questions about the formation of tornadoes that cannot be addressed in other ways. However, the campaigns just happened to take place during relative lulls in tornado activity in the Great Plains. In 2010, VORTEX 2 also encountered problems with too many amateur chasers clogging the roads.

Storm chasers are exposed to the hazards that often accompany a tornado, such as large hail, lightning, and heavy rains that make driving difficult and dangerous. Storm chasers must have excellent knowledge of thunderstorms and geography. The most professional storm chasers know that it is better to approach a storm from the southeast to west quadrants so that they can see the tornado without being in its path. They avoid approaching a severe thunderstorm from the north or east because it exposes them to the region of heaviest rain and hail, a region known to storm chasers as the *bear's cage*. They also know it is better to chase in the flat Great Plains, where tornadoes are visible far away, than in the hills or in the moist Gulf Coast region where cloud bases are low, obscuring the tornado funnel. They chase during the day but not at night, when tornadoes are hard to see.

Although it may often seem like thrillseeking, the best chasers practice sound science: They make their own forecasts, use advanced onboard equipment when in the vicinity of severe weather, and share their observations and the data they gather with forecasters and researchers.

The relatively low odds of seeing a tornado, the long hours of driving and sitting around waiting for severe weather to happen, and the cost of travel all make storm chasing a hobby for severe-weather fanatics. Even so, observations from storm chasers have enabled us to understand better the conditions needed for tornado development.

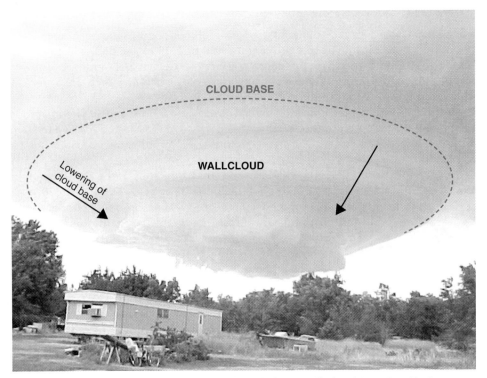

FIGURE 11-22 The ominous approach of a rotating wall cloud is a sign that a tornado may develop at any moment. Mobile homes are dangerous places to be in such situations.

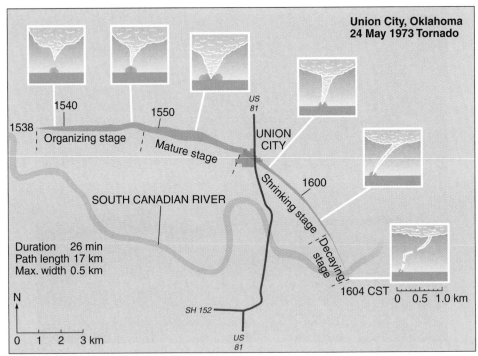

FIGURE 11-23 The life cycle of the Union City, Oklahoma, tornado on May 24, 1973, one of the best-studied tornadoes in history. The boxed line diagrams show what the tornado looked like at various times (shown in 24-hour military time on the graph) during its lifetime, including any dust or debris witnessed near the ground (stippled areas in boxes). The path and width of the tornado are denoted by the purple line stretching across and down the right-hand side of the figure; the tornado is widest just to the west (left) of Union City. The four stages of the life cycle are indicated beneath the tornado path. (Modified from Golden, J. H., and D. Purcell, *Mon. Wea. Rev.*, 106 [1978]: 3–11.)

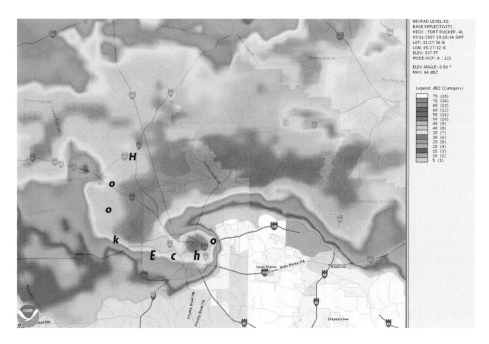

(a)

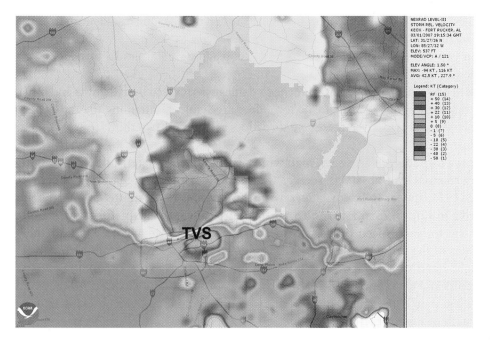

(b)

FIGURE 11-24 Doppler radar reflectivity (a) and velocity (b) images for a tornadic supercell as it hit Enterprise, Alabama, on March 1, 2007. The hook echo is evident in the lower center of the image in (a); the bright red blob over Enterprise is due to debris reflecting the radar beam. In (b), the red and blue couplet near the center is the TVS over Enterprise. Nine people died in this EF4 tornado, including eight students at Enterprise High School, which was destroyed. Fifty people were injured, and damage exceeded $300 million in this town of 20,000 in southeast Alabama.

Even in the absence of hook echoes and tornado vortex signatures, it is possible to identify severe thunderstorms using radar. As noted earlier, rotating supercells often are right-movers. On a time-lapse radar loop they stand out because their paths are to the right of the paths of other nonsevere thunderstorms and rain areas. Watch a radar loop on a television weather program during severe weather, look for the right-movers, and make your own amateur severe weather forecasts.

■ Tornado Winds

Tornado winds are the stuff of legend; some early estimates (incorrectly) gauged them to be faster than the speed of sound! The modern era of wind estimation in tornadoes is based on observations of tornado damage.

On May 11, 1970, a destructive tornado in Lubbock, Texas, hit downtown and even the local NWS office. How fast were the tornado's winds? To answer this question, meteorologist Ted Fujita created the *Fujita scale* to estimate the winds of a tornado after the fact based on the damage caused by them, on a scale from F0 (light damage) to F5 (incredible damage).

Over time, researchers determined that the original Fujita scale wind estimates were too high, and the scale needed to account for the differences in the structural integrity of different kinds of buildings as well as vegetation (especially for tornadoes in rural areas far away from buildings). The new-and-improved scale, which was adopted by the NWS in February 2007, is called the **Enhanced Fujita (EF) scale** (TABLE 11-3). The EF scale uses a set of 28 damage indicators, such as barns, schools, and trees; the degree of damage to each one is used to determine the EF scale of a particular tornado. A tornado rated according to the old Fujita scale has the same number in the EF scale, for example, F3 translates to EF3, although the wind speed ranges are different in the two scales. As in the old Fujita scale, the higher the "EF number" on a scale from 0 to 5, the more severe are the tornado's winds and damage.

You can use the text's Web site to practice assigning EF scales to tornadoes; see if your tornado can fling a cow!

Based on painstaking surveys of tornado damage performed by Ted Fujita and others, it was found that only about 1% of all U.S. tornadoes are "violent," attaining the top two categories on the Fujita scale. However, these rare tornadoes account for two thirds (67%) of all deaths by tornadoes (FIGURE 11-25).

How a tornado achieves such high winds is still not completely understood. Some of the worst tornadoes ever observed, such as the F5 that destroyed large parts of Xenia, Ohio, on April 3, 1974, appear to be tornadoes-within-tornadoes (FIGURE 11-26). This phenomenon is called a **multiple-vortex tornado**. As in a hurricane (see Chapter 8), winds are strongest where the forward speed of the storm adds to the tornado's winds, just to the right of the center of the funnel. Where the winds of a suction vortex are combined with this effect (see right-hand side of Figure 11-26), the total winds can exceed 320 kilometers per hour (200 mph) and cause incredible damage. A few houses or a block away, however, the winds may be less than half as fast. This explains why tornado damage can be so hit-and-miss even in the path of a giant twister.

The intense horizontal and vertical winds in a tornado can cause unheard-of damage. Objects weighing a kilogram or two can be sucked up into the parent thunderstorm and travel more than 160 km (100 miles) downwind from an EF5 tornado. In this respect, the images of airborne cows and houses in the movies *Twister* and *The Wizard of Oz* are not quite as unrealistic as they might initially seem.

■ Tornado Distribution

Tornadoes form in regions of the atmosphere that have extremely unstable moist air (e.g., large negative values of the LI), large amounts

Tornadoes and the EF Scale.

TABLE 11-3 The EF Scale for Tornadoes

EF Scale	Three-Second Wind Gust	
	MPH	km/hr
EF0 (Weak)	65–85	105–137
EF1	86–110	138–177
EF2 (Strong)	111–135	178–217
EF3	136–165	218–266
EF4 (Violent)	166–200	267–322
EF5	> 200	> 322

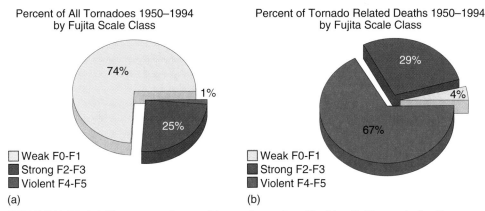

Percent of All Tornadoes 1950–1994 by Fujita Scale Class

74%
1%
25%

☐ Weak F0-F1
■ Strong F2-F3
■ Violent F4-F5

(a)

Percent of Tornado Related Deaths 1950–1994 by Fujita Scale Class

29%
4%
67%

☐ Weak F0-F1
■ Strong F2-F3
■ Violent F4-F5

(b)

FIGURE 11-25 (a) The percentage of tornadoes classified by Fujita scale for the years 1950 through 1994. (b) The percentage of tornado-related deaths during the same period, classified according to F-scale. Nearly 75% of all tornados are classified as weak, whereas 67% of the deaths result from the violent tornadoes that rarely occur. (The results for the new EF scale should be similar.)

of vertical wind shear, and weather systems such as fronts that force air upward. The United States provides these three ingredients in abundance, so it is not surprising that the majority of the world's reported tornadoes are "born in the USA."

Within the United States, tornadoes can occur in nearly every state and in every month of the year. **FIGURE 11-27** shows the average number of tornadoes per year and reveals a **tornado alley** of highest frequency in the Great Plains centered on Texas, Oklahoma, and Kansas. However, the greatest risk due to killer tornadoes is a bit different (**FIGURE 11-28**), with an additional "alley" extending eastward from Oklahoma to west Georgia.

Tornado season is based on when the ingredients for severe weather come together in a particular place. Because vertical wind shear is closely related to the presence of a jet stream,

03 May 1999—
Oklahoma Tornado
Outbreak

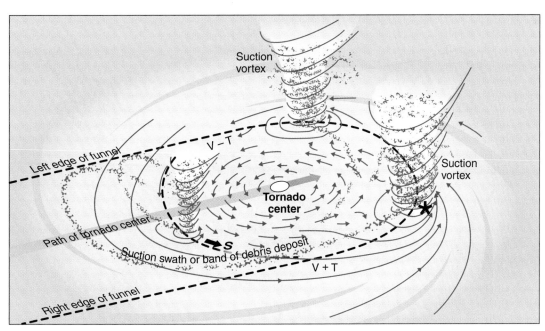

FIGURE 11-26 Dr. Ted Fujita's schematic of the multiple-vortex tornado, showing how the whirls-within-whirls of the suction vortices can lead to intensified winds in certain locations of the main tornado vortex. The effects of the main vortex, the suction vortex, and the forward speed of the tornado itself all combine for maximum winds at the location of the asterisk at far right. (Modified from Tom Grazulis, *The Tornado: Nature's Ultimate Windstorm*, Oklahoma University Press, 2000, p. 111.)

Average Number of Tornadoes Per Year

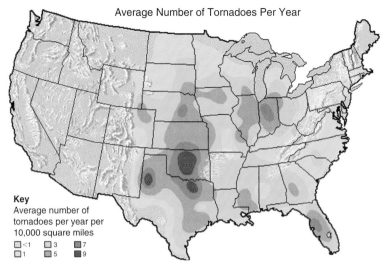

Key
Average number of
tornadoes per year per
10,000 square miles

☐ <1 ☐ 3 ■ 7
☐ 1   ■ 5 ■ 9

FIGURE 11-27 The climatology of the average yearly number of tornadoes per 10,000 square miles across the lower 48 United States. The traditional "tornado alley" from Texas to Kansas stands out as the region of highest tornado occurrence.

tornado season moves north and south during the year with the polar jet. **FIGURE 11-29** shows this seasonal pattern. Notice in this figure that tornado season peaks in March and April in the Southeast but not until July in the upper Midwest and Northeast. A secondary peak occurs in the deep South in November, when the jet stream and strong low-pressure systems occur along the Gulf Coast in late autumn.

Tornadoes can also happen at any time of day or night. However, they benefit from solar heating and in some cases the ability of warm, moist air at the surface to penetrate the capping inversion. Therefore, the most likely times for tornadoes are late afternoon or early

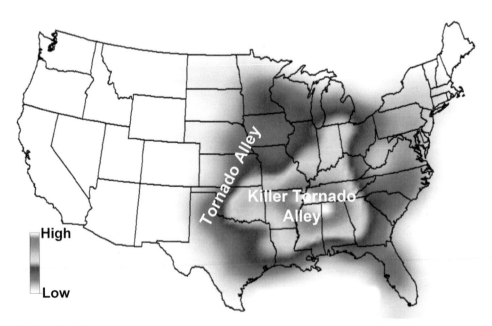

FIGURE 11-28 A climatology of the relative frequency of killer tornado events from 1950 to 2004. The traditional tornado alley is labeled in this figure, but the region of highest risk due to killer tornadoes stretches from Oklahoma eastward into Georgia. (Courtesy of Dr. Walker Ashley, Meteorology Program, Department of Geography, Northern Illinois University.)

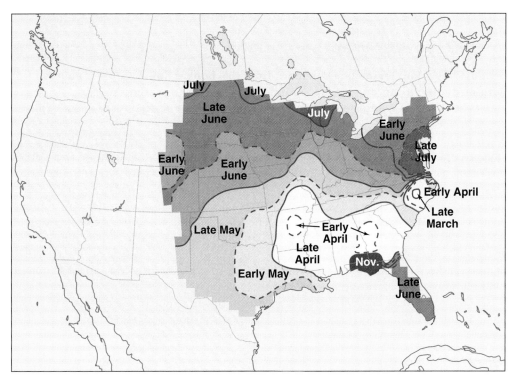

FIGURE 11-29 The geographic distribution of the month of maximum tornado threat for the continental United States. (Adapted from H. E. Brooks et al., *Wea. Forecasting*, 4 [2003].)

evening. More than half of all U.S. tornadoes occur during the hours of 3:00 PM to 7:00 PM local time.

It is sometimes said that tornadoes must be attracted to mobile homes because killer tornadoes often hit mobile home parks. However, the reverse is actually true: for economic reasons, the percentage of people living in mobile homes in the killer tornado alley from Oklahoma to Georgia is the highest of any region east of the Rocky Mountains. Mobile homes are especially vulnerable in high winds; close to 50% of all tornado-related fatalities occur in mobile homes (**BOX 11-2**). The fact that the Southeast has more mobile homes may therefore partly explain why killer tornadoes are concentrated in the South, although overall tornado occurrence is highest in the Great Plains.

Advances in tornado watches and warnings have saved countless lives, however. From 1875 to 2000, over 18,000 Americans died in tornadoes. But the last single tornado to kill more than 100 people occurred in Flint, Michigan in 1953, just a year after the U.S. government began issuing tornado watches. **FIGURE 11-30** reveals that the risk of dying in a tornado in the U.S. has dropped by more than half during the era of NWS watches and warnings, despite some record-setting tornado outbreaks during that period (see below).

Tornado Outbreaks

In May 2003, a cluster of severe weather events occurred in the United States between May 3 and May 11. A persistent warm air mass over the central United States, coupled with a strong jet stream and an intense low-pressure system, led to 361 tornadoes over 26 states. Twelve "killer" tornadoes in 5 states resulted in 41 deaths, 642 injuries, and $829 million in damage. At least 7 of the tornadoes were category F4 or higher, and 50 were category F2 to F3. More severe weather broke out during the week of May 4 to 10, 2003, than during any other week in U.S. history. In addition to the tornadoes, there were 1587 reports of large hail and 740 reports of wind damage.

As amazing as the "tornado week" in May 2003 was, it was not the most violent tornado outbreak in American history. That title belongs to the April 1 to 4, 1974, "Superoutbreak"

Box 11-2 Severe Weather Safety

The most important lesson you can learn from this book is how to protect yourself from injury or death during dangerous weather. Here we discuss some safety rules for the three biggest severe weather killers in the United States today: tornadoes, lightning, and flash floods.

First, you need access to reliable weather information. Commercial television and radio are often good sources of current weather updates. Keep in mind that the watches and warnings they report are issued by government meteorologists. The national system of severe weather watches and warnings has saved untold numbers of lives in the United States since the 1950s. National Oceanic and Atmospheric Administration's (NOAA) improved Weather Radio network puts nearly everyone in the nation within range of government weather broadcasts. These radios can sound an alarm whenever severe weather warnings are issued for your area. You should respond quickly when these warnings are issued. For example, today's tornado warnings from the NWS provide, on average, about 13 minutes of lead time before the tornado strikes.

Once you know severe weather is approaching, what should you do? Here are some safety guidelines. More information is also available from your nearest NWS forecast office.

Tornado

In general, try to get as low as you can. This is because tornado winds decrease close to the ground because of friction. Go into a tornado shelter or the basement or into a small interior room on the lowest floor of a building, such as a bathroom or closet. Protect yourself from flying debris and stay away from windows. Don't bother opening or closing them because you put yourself at risk of being hit by flying debris. Also, opening windows may increase tornado damage. In particular, protect your head. Wear a bicycle or motorcycle helmet if one is available.

If you are away from home during a tornado, you are at greater risk. As always, you need access to current weather information. In 1994, 20 people died in an Alabama church when a tornado blew in a wall of the church during Palm Sunday services. The death toll was heavy because news of severe weather watches and warnings for the area never reached the church.

If you are away from a sturdy home, you must seek adequate shelter immediately in the event of a tornado. At school or in a dorm, follow the severe weather safety plan in place for that building. Avoid auditoriums, gymnasiums, and eating areas; their large, high roofs can blow off, and the walls can collapse. If you are in a mobile home or car, leave it and go to a strong building. Many people are killed when cars and mobile homes are overturned in high winds. If there are no shelters nearby, get into the nearest ditch or depression and protect yourself from flying debris.

Television videos have wrongly popularized the notion that it is safe to drive at high speeds away from a tornado and, if necessary, to hide under a highway overpass as the tornado passes overhead. *Don't do it.* It is far safer to take shelter in a sturdy building instead. A highway overpass creates a "wind tunnel" effect underneath it and can actually increase the amount of damage from a tornado. Three people died and many others were injured during the May 3, 1999, Moore, Oklahoma, tornado because they intentionally chose to take shelter under an overpass. It was a fatal mistake.

Lightning

Because of the vast differences in the speed of light and the speed of sound, the flash of lightning precedes the rumble of thunder. It takes sound waves 5 seconds to travel 1 mile (or 3 seconds per kilometer), whereas the flash of lightning travels the same distance in less than 1/100,000th of a second. For example, if 15 seconds elapse between a lightning bolt and the arrival of its thunder, the bolt was 5 kilometers (3 miles) away. Practice this with an applet on the text's Web site, which simulates the relationships between lightning and thunder.

However, just because a lightning bolt is a few miles away doesn't mean you are safe. The NWS advocates the simple rule: "When thunder roars, go indoors!" Follow these safety rules in the event of lightning. If you are outside, get into an enclosed building. Avoid being in or near high places or in open fields. If outside, avoid all metal objects, such as flagpoles, metal fences, golf carts, baseball dugouts, and farm equipment. If you are in a forest, seek shelter in a low area under a thick growth of bushes or small trees. In open areas, go to a low place and crouch on the balls of your feet. Don't lie down; if lightning strikes nearby, you minimize your risk of burns by crouching instead of lying. Stay away from open water. On the water, you are usually taller than your surroundings and vulnerable to a direct lightning strike. In the water you can be shocked because water can conduct electricity from a lightning strike over long distances. Fully enclosed, all-metal vehicles with the windows rolled up usually provide good shelter from lightning—but avoid direct contact with any metal.

Flash Floods

Stay away from streambeds, drainage ditches, and culverts during periods of heavy rain. Move to high ground when threatened by flooding. Stay out of flooded areas. Never drive your car across a flooded road, even if you think the water is shallow. Most flash-flood-related deaths occur when people drive into floodwaters. Never underestimate the power of moving water! Just 18 inches (less than a half-meter) of water on a roadway can be enough to carry your car or SUV away. As the NWS says, "turn around don't drown."

Keep safety in mind even after the storm passes. Stay clear of downed power lines. Do not touch them. If you smell gas or suspect a gas leak, turn off the main valve; get everyone out of the structure quickly, and open the windows. If there is a power outage, use flashlights instead of candles, which have open flames that might start a fire. After high winds, use caution when walking around trees because trees and tree limbs may be weakened and could fall unexpectedly. Deal with immediate problems, such as helping injured people, until professional help arrives.

Following these safety rules can save lives. See **BOX 11-3** for a tornado safety success story.

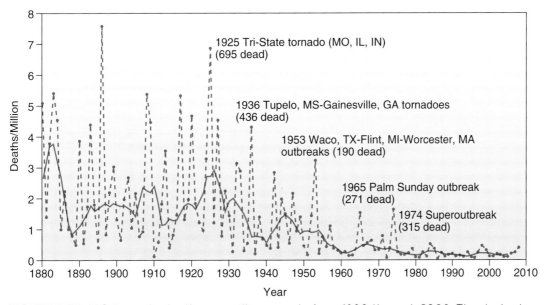

FIGURE 11-30 U.S. tornado deaths per million people from 1880 through 2008. The dashed line is yearly; the solid line is a smoothed multiyear average. The risk of dying in a tornado has been drastically reduced over the past century. (Courtesy of Dr. Charles A. Doswell III, www.flame.org/~cdoswell/Tornado_essay.html.)

Box 11-3 Tornado Safety: A Success Story

July 13, 2004: An F4 tornado completely destroys the Parsons Company manufacturing plant in north-central Illinois with 150 people inside.

Now, the good news. The death toll: zero. The injury count: zero!

How can over one hundred people walk away without a scratch after a direct hit from a violent tornado 400 meters (a quarter-mile) wide?

This tornado safety success story begins with the company owner, Bob Parsons. A severe weather survivor himself, Parsons is a perfectionist when it comes to weather safety. His safety plan takes all severe weather warnings seriously, not just tornado warnings. When a warning is received, trained spotters at his plant look for approaching tornadoes. The company conducts carefully evaluated tornado drills twice a year. The drills are designed to get everyone to safety within just 4 minutes. And "safety" is more than a hallway: Parsons, at his own expense, built steel-reinforced concrete bunkers inside his plant to protect his workers.

All of this preparation paid off on July 13, 2004. The weather on this day was clearly going to be dangerous: the LI in central Illinois was –11 at 7:00 AM and an incredible –14 by 1:00 PM! A tornado watch was issued for much of north and central Illinois before 11:00 AM. By 2:29 PM, a thunderstorm developed rapidly due to lifting by a windy "outflow boundary" from another thunderstorm. At this time, the NWS issued a severe thunderstorm warning for the county including the Parsons plant. The Parsons Company severe weather plan kicked into high gear. The NWS tornado warning went out five minutes later, at 2:34 PM. By 2:37 PM, the last of the Parsons employees and visitors were safe inside the concrete bunkers; the evacuation to shelter had taken only 3 minutes. At 2:41 PM, the Parsons plant was turned into rubble (see the photograph)—everything, that is, except for the concrete tornado shelters! One shelter can be seen a bit below center in the photograph.

Lives were saved because of many factors. Bob Parsons valued his employees' safety more than short-term profits. He invested in shelters and preparation plans that put most other organizations' efforts to shame. The NWS issued timely watches and warnings during a rapidly developing severe weather situation. The employees at the Parsons plant heard the warnings and then executed the severe weather safety plan without hesitation and to perfection.

The Parsons plant was rebuilt within a year of the 2004 tornado, and every employee was kept on the payroll during the rebuilding effort. The plant now has five additional storm shelters that are built to withstand an EF5 tornado.

Bob Parsons and the Parsons Company turned tragedy to triumph because they are "weather-wise." What would you do if a tornado were approaching your home, school, or place of employment?

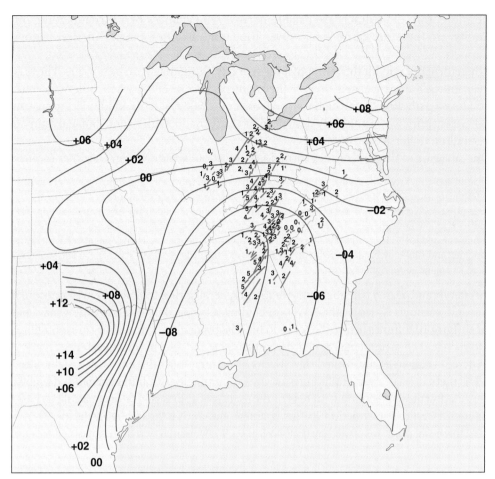

FIGURE 11-31 The Tornado Superoutbreak of April 3–4, 1974, depicted in terms of tornado paths (lines, labeled with the Fujita number of each tornado) that are overlaid on top of the LI values calculated at 1200 UTC on the morning of the tornado outbreak. Notice that most of the tornadoes occurred where the atmosphere was very unstable (negative LI values) even before daytime solar heating destabilized the atmosphere further. (*Source*: NOAA.)

(**FIGURE 11-31**). As the figure shows, the atmosphere over the southern and central United States on the morning of April 3 was very unstable with large negative values of the LI. An approaching low-pressure system and cold front, plus daytime heating, provided the triggers for an unprecedented explosion of severe weather. During a terrible 24 hours on April 3 and 4, 1974, 148 tornadoes—an incredible 30 of them F4 or F5 in intensity—ravaged North America from Alabama to Canada (their paths are the colored lines in Figure 11-31). More than 300 people were killed, including 33 at Xenia, Ohio. Nearly 5500 people were injured.

Recently, research meteorologists calculated that more violent tornadoes developed on the *one day* of April 3, 1974 than during any *4-week* period during the past 130 years. The "Superoutbreak" was probably a once-in-a-millennium event and is likely the single most extreme severe weather event we discuss in this textbook.

When Tornadoes Hit Towns and Cities: Greensburg and Atlanta

It's a popular misconception that tornadoes do not hit cities. In reality, tornadoes have struck the downtowns of numerous U.S. cities, including St. Louis, Missouri (4 times, in 1871, 1896, 1927, and 1959), Louisville, Kentucky (twice, in 1890 and 1974), and Dallas/Fort Worth, Texas (twice, in 1894 and 2000). The reason that it doesn't happen more often is simple and obvious to anyone who takes a cross-country flight: city centers are a tiny fraction of the overall landscape.

Furthermore, tornado funnels are usually narrow. The chances of a tornado going directly into a city are therefore quite small. Over time, and as cities expand in size, it does and will happen.

When tornadoes do hit the center of a town or city, the damage can be horrific. The residents of Greensburg, Kansas learned this on the night of May 4, 2007, when mile-wide-plus EF5 tornado tore a path directly through the town of about 1500 residents, killing 11. Approximately 95% of Greensburg was completely destroyed by the tornado, with a total damage estimate of a quarter-billion dollars (**FIGURE 11-32**).

One survivor of the Greensburg tornado, high-school senior Megan Gardiner recalled:

> Around 9:15 pm, I took the trash out [at work] and that was the most scary lightning I have ever seen. . . . Later the tornado sirens started to go off. . . . When I busted through the front door my dad . . . said, "Grab your purse and let's go downstairs [to the basement]. . . ." All of a sudden my ears started to pop really bad [due to low pressure near the tornado]. . . . I was like, Oh My Gosh, this can't be good. . . . [then] the windows exploded! . . . I heard the walls tearing and ripping off into pieces. Then something fell on my left shoulder, and I had my head covered with my hands (like the drills we do in school). . . . The sound was like a jet engine going right over us. . . .

Megan's family and nearby neighbors survived, largely because they took shelter in response to timely watches and warnings issued by the NWS—including very specific "Tornado Emergency" message for Greensburg 10 minutes before the tornado hit. The residents of Greensburg are rebuilding their town to stringent environmental standards, emphasizing the "green" in Greensburg.

Different challenges confronted the citizens of Atlanta on a different Friday night, March 14, 2008. The metro area of over 5 million residents was pummeled by an EF2 tornado, at night, in the middle of a downtown teeming with tens of thousands of sports spectators at a college basketball tournament (see the Introduction to this chapter) and an NBA game (**FIGURE 11-33**). A severe thunderstorm warning was issued by the NWS at 9:09 PM, which was then upgraded to a tornado

FIGURE 11-32 Near-total destruction due to the March 1, 2007, Greensburg, Kansas, tornado, 12 days later.

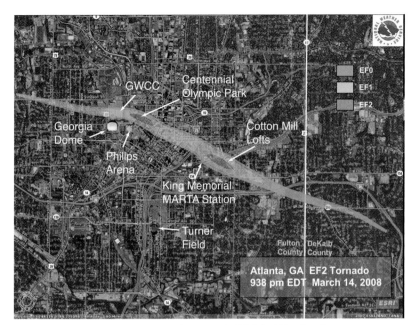

FIGURE 11-33 The path of the March 14, 2008, Atlanta, Georgia, tornado. The width of the tornado is indicated by the size of the path, and the amount of damage (and therefore the strength of the winds) is indicated by the EF scale ratings (color shadings).

warning at 9:30 PM. The tornado first touched down in west Atlanta 8 minutes later, moving east-southeast. The tornado narrowly missed the Georgia Dome and Philips Arena, where the basketball games were being played, causing exterior damage to both. The CNN Center, home to the Cable News Network, suffered extensive damage while shaken reporters inside wondered what to do. The towering 220-meter (723-foot) Westin Peachtree Plaza Hotel swayed more than 2 feet as the tornado passed by, and lost hundreds of windows that were not replaced for over two years (FIGURE 11-34).

In Atlanta, unlike in Greensburg, tens of thousands of people were at risk without knowing it, far away from the safety of basements. No tornado sirens exist in downtown Atlanta. Fortunately, the tornado just missed the two basketball arenas, and the basketball tournament game had gone into overtime because of a last-second shot—without which thousands of fans would have been milling about outside instead of inside the Georgia Dome. Also, at EF2 strength, the worst damage in Atlanta did not begin to compare with what was witnessed in Greensburg. Even so, because of the expensive real estate hit by the Atlanta tornado, the damage was estimated at a quarter-billion dollars—the same as in Greensburg. One man died during the Atlanta tornado when a brick wall fell on him; thirty others were injured by flying glass and debris.

What if the Greensburg tornado had hit downtown Atlanta or a similarly large city? Researchers who have investigated this type of "worst-case scenario" estimate that the death toll could be in the tens of thousands, with perhaps $40 billion in damage. Although this work remains controversial, this much is true: a very slight change in the path of the Atlanta tornado in 2008, and an upgrading of its intensity to EF5 would have brought the carnage of Greensburg on top of 20,000 or more unprotected spectators. It is a reminder to be weather-conscious at all times and in all places.

FIGURE 11-34 The 73-story Westin Peachtree Plaza Hotel in downtown Atlanta, after the March 2008 tornado. Notice all of the blown-out windows on this glass skyscraper, a result of the tornado.

FIGURE 11-35 A waterspout in the Florida Keys, as photographed by meteorologist Joe Golden.

THE WATERSPOUT

The night-and-morning showers caused by the sea/land breezes along the coast (see Chapter 6) appear benign. However, it is not uncommon along the Gulf and Atlantic coasts to hear of "special marine warnings" because of **waterspouts** sighted just offshore in connection with sea-breeze showers.

Like tornadoes, waterspouts are narrow spinning funnels of rising air that form underneath clouds. Although tornadoes generally develop in association with immense rotating cumulonimbus clouds, waterspouts usually form underneath shorter cumulus clouds that are not rotating. As in a tornado, low pressure at the center of a waterspout sets up a pressure gradient that drives air inward. This air rises and cools, causing condensation and making the funnel visible.

Even at their most intense, waterspouts are only as strong as weak tornadoes, with winds generally less than 160 kilometers per hour (100 mph). FIGURE 11-35 shows an airborne view of a waterspout near the Florida Keys, where waterspouts are quite common because of the combination of warm ocean water; warm, moist air; and sea/land breezes helping to create cumulus clouds.

OTHER THUNDERSTORM-PRODUCED SEVERE WEATHER

Other types of severe weather produced by thunderstorms include lightning, flooding, hail, and high winds. These weather phenomena are some of the most spectacular sights in nature. A single thunderstorm can generate all of these perils, or only one. We explore some of these in the following sections.

Lightning

Lightning is a huge electrical discharge that results from the rising and sinking air motions that occur in mature thunderstorms (FIGURE 11-36). Each year in the United States lightning kills, on average, up to 100 people and injures hundreds more, especially college-aged men who are outdoors. Lightning also causes several hundred million dollars of property damage each year. Although your chances of being struck by lightning in a given year are small (about 1 in 500,000), it is important to understand how nature's fireworks operate.

Lightning can travel from cloud to cloud, within the same cloud, or from cloud to ground. In-cloud lightning discharges are far more common than cloud-to-ground discharges and are not as hazardous. The processes that lead up to this electric discharge, or lightning flash, are the same for these three types.

Although lightning appears to be a continuous, almost instantaneous, flash of light to human eyes, high-speed photography shows that a "lightning bolt" is actually a series of flashes. Let's break down cloud-to-ground lightning into a sequence of split-second events. To explain the sequence, let's consider lightning striking a tall building (FIGURE 11-37).

- *Charge separation in the cloud.* Current research indicates that electric charges get distributed throughout the cloud by the collision of ice particles with ice-covered snowflakes or "graupel" (see Chapter 4) at different temperatures. When the collisions occur at temperatures below −15° C (5° F), the ice crystals become positively charged and the graupel acquires a negative charge. At temperatures

FIGURE 11-36 Lightning brilliantly illuminates the night sky.

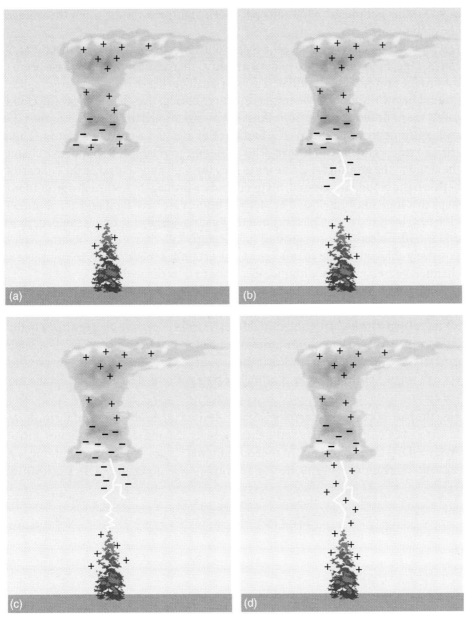

FIGURE 11-37 A sequence of events that leads to a cloud-to-ground lightning strike. (a) Charges collect in the base of the cloud. (b) As negative charges build up near the base of the cloud, the ground repels negative charges and changes from its usual negative to a positive charge. Lightning formation has begun with the pilot leader. (c) The stepped leader connects the cloud to the ground. (d) The bright return stroke surges upward.

warmer than −15° C, the collision induces a positive charge on the graupel. The updrafts that maintain the cloud storm carry the particles to different regions of the cloud. The ice crystals are moved upward to the top of the storm and the graupel collects lower in the cloud. In-cloud lightning is the surge of electric current that passes between the negatively and positively charged regions of the cloud.

- *Ground becomes positively charged.* Opposite charges attract, and like charges repel. As the cloud base becomes negatively charged, the objects on the ground become positively charged. The atmosphere is resistant to the flow of electricity, which allows the development of a very large difference between the charges on the cloud and ground and establishes conditions for lightning strike. The voltage begins to build as the negative charges continue to collect near the base of the cloud. Air is a good insulator; it can separate voltages as great as 9000 volts per meter (3000 volts per foot). Lightning results when the voltages climb above this value.

- *Lightning formation begins.* Once the charge difference becomes so large that the atmosphere can no longer insulate the two regions, negative charges near the cloud base begin to move toward the ground. This is initiated when a small pocket of positive charges collects at the ground below the cloud. The initial discharge of negative charges near the cloud base is called the **pilot leader**. Electrons flow downward toward the ground into the pilot leader and continue to surge down in a sequence of events toward the ground. This flow of charge creates **stepped leaders**, which attempt to establish a conductive channel from the cloud to the ground for electrons to flow through and neutralize the charge difference between the cloud and ground. Stepped leaders propagate toward the ground in distinct steps that look like branches. The stepped leaders are very faint and are about 50 meters (160 feet) long and about 2.5 centimeters (1 inch) wide. By the time a stepped leader surges downward, objects near the ground (particularly tall, sharp, metal objects) have become positively charged. As the channel nears the ground, a spark occurs to complete a narrow channel that serves as a conduit for electrons to flow through.

- *Brilliant flash is observed.* The instant a channel is established from the cloud to the ground, the current flows upward through this charged channel, generating a brilliant flash known as the **return stroke**. After the initial return stroke, negative charges from higher in the cloud move toward the ground; these are called **dart leaders**. The dart leaders may generate additional return strokes if they reach the ground. What appears to the human eye as a single lightning stroke is actually a series of return strokes that occurs too fast for the eye to distinguish.

Thunder and Lightning.

During the past 2 decades, meteorologists have devised methods for recording the location and other characteristics of cloud-to-ground lightning bolts striking the United States. FIGURE 11-38 shows a decade-long climatology of 216 million lightning strikes. Notice that lightning is generally most prevalent where thunderstorms are most common (refer to Figure 11-4). Floridians are at the

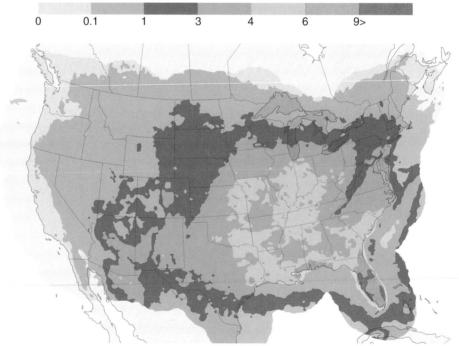

FIGURE 11-38 The average number of cloud-to-ground lightning flashes per kilometer across the lower 48 United States, as measured electronically by the National Lightning Detection Network for the years 1989 through 1998. More than 216 million flashes were recorded during this period. (Adapted from Orville, R., and Huffines, G., *Monthly Weather Review*, May 2001.)

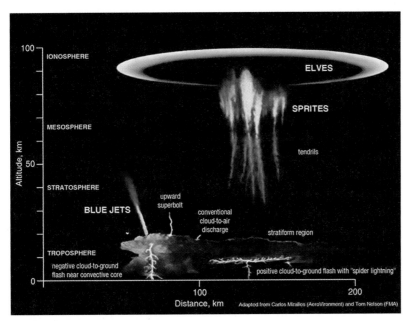

FIGURE 11-39 The curious and largely unexplained upper-atmospheric electrical phenomena of elves, red sprites, and blue jets.

most risk from lightning, not only because of the frequency of lightning but also because of the outdoor lifestyle of Floridians that often makes them the highest objects on the water or on the golf course—natural targets for lightning. Box 11-2 explains how to protect yourself from lightning.

If lightning were not already enigmatic enough, during the past 2 decades scientists have discovered and confirmed the existence of lightning that shoots upward into the upper atmosphere from thunderstorms (**FIGURE 11-39**). **Red sprites** and **elves** occur over cloud-to-ground lightning bolts and can extend to near the top of the atmosphere. They are too quick and weak to be seen by the naked eye. **Blue jets**, in contrast, are limited to the stratosphere and last long enough to be seen by pilots. As their whimsical names suggest, much is still not understood about these electrical phenomena.

Flash Floods and Flooding

A **flood** is a substantial rise in water that covers areas not usually submerged. A flood occurs when water flows into a region faster than it can be absorbed (i.e., soaked into the soil), stored (in a lake, river, or reservoir), or removed (in runoff or a waterway) into a drainage basin. Common causes for floods are high-intensity rainfall, prolonged rainfall, or both.

Floods pose a great weather-related threat to human life, killing more than 130 people each year in the United States. Not all floods are associated with thunderstorms, although most flood-related deaths in the United States result from floods caused by slow-moving thunderstorms or series of thunderstorms that move over the same region. As we have seen, hurricanes (see Chapter 8) and extratropical cyclones (see Chapter 10) can also produce flooding. Thunderstorms are of particular concern because they can produce a very dangerous condition known as a flash flood.

A **flash flood** is a sudden, local flood that has a great volume of water and a short duration. Flash floods occur within minutes or hours of heavy rainfall or because of a sudden release of water from the break-up of an ice dam or constructed dam. The worst flash flood in the United States occurred in Johnstown, Pennsylvania, on May 31, 1889. A dam broke as a result of heavy rain and structural problems, resulting in a 12-meter (36- to 40-foot) wall of water that swept through the town and killed 2200 people.

Rainfall intensity and duration are two key elements of a flash flood. Topography, soil conditions, and ground cover also play important roles. Steep terrain can cause rain water to flow toward and collect in low-lying areas, causing water levels to rise rapidly. If the soil is saturated with water, it cannot absorb more and so the excess water runs off the land quickly. However, extremely dry soil conditions can also be favorable for flooding. Dry soil often can develop a hard crust over which water will initially flow as if the ground were concrete. The impervious asphalt and concrete of cities cause worse and more intense flash floods. When extreme rains due to repeated thunderstorms occur in urban areas, as in Atlanta, Georgia, in September 2009 and Nashville, Tennessee, in May 2010, the result is unprecedented flooding (FIGURE 11-40).

Flash floods occur within about 6 hours of heavy rain. A flood is a longer term event and can last weeks or months. Record flooding along the upper Mississippi River in the summer of 1993 resulted from prolonged rains in the upper Midwest of the United States and caused up to $10 billion in damage. These kinds of floods occur during blocking events (see Chapter 7) in which

FIGURE 11-40 It's a combination rollercoaster and water slide! No, it's a flooded-out Six Flags Over Georgia amusement park on the west side of Atlanta, Georgia, after the record-setting September 2009 floods. Thunderstorms dropped up to 15–20 inches (38–51 cm) of rain in this area in one week, causing extreme flooding on the Chattahoochee River near Six Flags that would not be expected to happen more often than once in at least 500 years.

thunderstorms develop and move over the same region repeatedly. The worst such flood in U.S. history occurred in 1927, when nearly all the tributaries of the Mississippi breached their banks because of thunderstorm rains. Seventy thousand square kilometers (27,000 square miles) of land along the lower Mississippi was flooded to depths of 30 feet; 700,000 people were left homeless, and the $1 billion in damage was nearly one third of the U.S. government's annual budget at the time!

Floods are natural phenomena and do have benefits. Large seasonal floods have resulted in productive farmland, such as in central North America and along the Nile River, by bringing nutrient-rich, fine soil to the flooded region. Floodwaters also refill wetlands and replenish groundwater. Although flooding is a natural event, humans increase the likelihood of flooding by changing the character of the land through such actions as paving with asphalt and removing vegetation on hillsides. This promotes rapid runoff and flooding.

Hail

Hail is precipitation in the form of large balls or lumps of ice (**FIGURE 11-41**). Hailstones begin as small ice particles that grow primarily by accretion and therefore require abundant supercooled water droplets. Hailstones can be as large as baseballs and softballs and can fall as fast as a major-league fastball. How do they get so large?

Remember from Chapter 5 that precipitation particle size depends on how long it stays in the cloud and the amount of water available for growth. When a hailstone is cut in half, rings of ice are often observed. Some rings are milky white; others are clear. To explain this ringed structure, let's consider how hail grows in a thunderstorm.

As a hailstone moves through a storm its temperature changes. As liquid water drops freeze on the hailstone, they release latent heat (see Chapter 2), warming the surface of the hailstone. If the temperature of the hailstone remains below freezing, then the liquid drops they collide with freeze on contact. This growing method is called **dry growth** because the stone has little liquid water on the surface. If the surface temperature of the stone temperature is near 0° C or warmer, then the colliding water droplets do not freeze quickly and have time to spread across the surface of the stone. This mode of growth is referred to as **wet growth** because the stone has a film of liquid water on the surface.

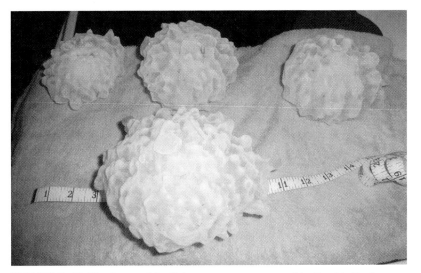

FIGURE 11-41 The largest hailstone ever recovered in the United States fell in Vivian, South Dakota, on July 23, 2010, with a record 8-inch diameter (20.3 centimeters) and a record weight of 1 pound, 15 ounces (0.9 kilograms) even after some hours of melting. Hailstones fall out of the clouds when the updraft no longer can support the stone's weight. Large hailstones can fall at speeds faster than 100 mph (160 km/hr)!

The processes of dry and wet growth explain the rings in hail. In dry growth air bubbles get trapped, which causes the ice to appear *white* because of multiple scattering (see Chapter 5). In wet growth, the water drops freeze slowly, spreading over the hailstone and freezing as a layer of *clear* ice. The ringed structure can result from different accretion rates in the storm. With lots of accretion, the surface of the hailstone warms due to the release of latent heating, leading to wet growth. Small amounts of accretion result in dry growth and white layers of ice.

The production of large hail requires a strong updraft that is tilted and an abundant supply of supercooled water. Because strong updrafts are required to generate large hailstones, it is not surprising to observe that hail is not randomly distributed in a thunderstorm; instead, it occurs in regions near the strong updraft. Supercell thunderstorms, in which the strongest updrafts are created with help from the mesocyclone, often produce the largest hail.

Eventually, though, the weight of the hailstone overcomes the strength of the updraft, and it falls to earth. The curtain of hailstones that falls below the cloud base is called the **hailshaft**. These regions are often said to appear green to observers on the ground, although recent research suggests that heavy rain as well as hail can create this optical phenomenon. As the storm moves, it generates a **hailswath**, a section of ground covered with hail.

Hailstorms can severely damage crops, automobiles, and roofs. As seen in **FIGURE 11-42**, in the United States hail is most common in the Pacific Northwest and southeastern Wyoming and eastern Colorado, not in the Southeast, where thunderstorms are most common. The hail in Colorado and Wyoming is much larger than in other regions of the U.S. This is because the dry air of the high Plains allows falling hailstones to preserve themselves by self-cooling through melting and evaporation. In contrast, the moister air of the Gulf states does not allow as much evaporation, and the rapidly melting hailstones simply turn into large raindrops. (Although the March 2008 supercells in Atlanta dropped hail *inside* the CNN Center, due to roof damage from the tornado!)

For this reason, hail is of more concern to residents of Denver, Colorado, and Cheyenne, Wyoming, than to those in stormier Tampa, Florida. A hailstorm in and near Denver on July 11, 1990, caused $625 million in property damage. The storm caused a power outage, leaving people stranded on a Ferris wheel and exposed to the storm's fury—47 of whom where injured by the storm's softball-sized hail. However, most hail damage is to crops. Hailstorms, which occur

Grow a Hailstone.

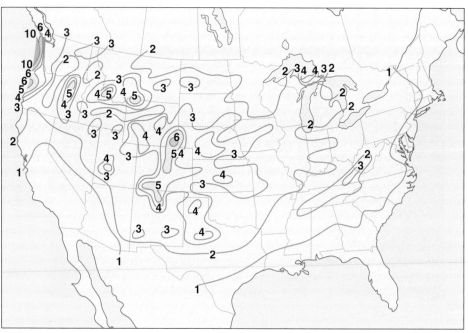

FIGURE 11-42 A climatology of the number of days each year with hail. How does the frequency of hail compare and contrast with the distribution of thunderstorms and lightning in Figures 11-4 and 11-38? (*Hailstorms Across the Nation* by S. Channgon, D. Channgon, and S. Hilbert. Image courtesy of the Midwestern Regional Climate Center, Illinois State Water Survey.)

worldwide and frequently during the growing season, destroy approximately 1% of the world's annual agricultural production. Surprisingly, hailstorms are the single most costly natural hazard in Australia. The great loss of property attributable to hailstorms has generated efforts to suppress or prevent hail. Unfortunately, such efforts have not been fruitful, and many farmers, particularly in the midwestern United States, purchase crop-hail insurance for economic protection. Dual-polarization radar (see Chapter 5) may be able to provide better detection of hail within thunderstorms and at least some additional ability to prevent damage from them.

PUTTING IT ALL TOGETHER

Summary

Thunderstorms produce lightning, thunder, tornadoes, floods, hail, and other severe weather. All thunderstorms form in unstable air masses. Indices such as the LI allow meteorologists to assess quickly the stability of the atmosphere by using observations of the atmosphere above a particular location. Severe thunderstorms grow in unstable environments with lift that also possess vertical wind shear and a change in wind speed or direction with increasing altitude. Environmental vertical wind shear is important for severe thunderstorms because it helps to separate the updraft from the downdraft. This enables the storm to last longer and grow more severe.

The basic building block of a thunderstorm is the cell. A thunderstorm can be composed of a single cell or multiple cells. Air-mass thunderstorms are single, ordinary cells that are not associated with severe weather. Single-cell thunderstorms follow a predictable life cycle from cumulus and mature stages to a dissipating stage in an hour or less. Squall lines and circular MCCs are examples of multicell thunderstorms, both of which can last for hours and produce severe weather. Supercells are single-cell thunderstorms, but they are larger, last longer than ordinary thunderstorms, and often produce severe weather.

Supercell thunderstorms develop when the environmental wind changes in both speed and direction with increasing altitude, causing the supercell to rotate. This rotation promotes the development of tornadoes. Tornadoes and large hail are produced by thunderstorms that grow in unstable atmospheres with vertical wind shear.

Tornadoes are violently rotating vertical columns of air that stretch from the cloud base to the ground in the updraft region of a storm. A sign of tornadic conditions is a rotating wall cloud or a funnel cloud extending downward from the base of the thunderstorm.

Tornadoes appear on radar as hook echo reflections or mesocyclones with particles rushing toward and away from the radar beam, forming a tornado vortex signature. Once formed, the tornado usually grows in size and strength and then narrows to a thin "rope" before dissipating. Tornadoes are usually less than 1.6 kilometers (1 mile) wide and only on rare occasions last for an hour.

Tornado winds can be as high as 320 kilometers per hour (200 mph) and are estimated using the EF scale. The EF scale is damage based and runs from 0 (light damage) to 5 (incredible damage). Only about 1% of U.S. tornadoes are EF4 or EF5 tornadoes, but they cause about two thirds of tornado deaths. Many U.S. tornadoes occur in the Great Plains in spring, but they can occur over nearly the entire country in every month of the year.

Lightning is a huge electrical discharge that results from rapid rising and sinking air motions within the thunderstorm that cause collisions of ice particles with graupel. A typical lightning flash is a composite flash composed of several lightning strokes. The stepped leader with subsequent dart leaders leads each stroke. The return stroke is the part of the lightning flash that we usually observe.

The most life-threatening thunderstorm weather is flooding, particularly flash flooding. A flash flood occurs suddenly and floods a region with a great volume of water in a short time span. Rainfall intensity and duration are two key elements of a flash flood. Topography, soil conditions, and ground cover also play important roles. Severe long-term flooding can occur when thunderstorms repeatedly drench a region.

Hail occurs when the strong updrafts and downdrafts in a thunderstorm cause repeated freezing of supercooled water on small ice particles. The curtain of hail that falls near the updraft, the hailshaft, can cause severe damage to cars and agriculture as the thunderstorm passes over urban and rural areas. Although thunderstorms and lightning are most common in the southeast United States, the Front Range of the Rocky Mountains is a preferred area for hailstorms.

Thunderstorms and their attendant severe weather can threaten your life. Although there are many different dangers associated with severe thunderstorms, we can summarize who is most at risk from severe weather. People who are outdoors in the open, under trees, or on the water are most at risk from lightning. Those in mobile homes and automobiles are at higher risk when tornadoes are nearby. People in automobiles are also at risk in flash-flooding conditions, as are people in low-lying areas and canyons.

■ Key Terms

You should understand all of the following terms. Use the glossary and this chapter to improve your understanding of these terms.

Air-mass thunderstorm	Hailswath	Red sprites
Blue jets	Hook echo	Return stroke
Capping inversion	Lifted index	Right-mover
Cell	Lightning	Severe thunderstorm
Cumulus stage	Mature stage	Shelf cloud
Dart leader	Mesocyclone	Squall line
Dissipating stage	Mesoscale convective	Stepped leader
Dry growth	complex (MCC)	Supercell
Elves	Mesoscale convective	Thunderstorm
Enhanced Fujita (EF) scale	system (MCS)	Tornado
Entrainment	Microburst	Tornado alley
Flash flood	Multicell	Tornado vortex signature
Flood	Multiple-vortex tornado	(TVS)
Funnel cloud	Nocturnal low-level jet	Vertical wind shear
Gust front	Ordinary cell	Wall cloud
Hail	Overshooting top	Waterspout
Hailshaft	Pilot leader	Wet growth

■ Review Questions

1. Explain the differences between a severe thunderstorm and an air-mass thunderstorm.
2. What causes lightning? What causes thunder?
3. What should you do when a tornado warning is issued for your location?
4. Describe how a hailstone can grow to the size of a grapefruit.
5. Are thunderstorms likely if air is converging in the upper troposphere?
6. If the surface temperature is 20° C, the surface dew point is 20° C, the moist adiabatic lapse rate is 5° C per kilometer, and the 500 mb (5.5 kilometer altitude) temperature is −1.5° C, what is the LI equal to? If thunderstorms occur, are they likely to be severe?
7. Draw a vertical profile of temperature, dew point temperature, and wind speed that would be favorable for the formation of a severe thunderstorm.
8. What are the three stages of an ordinary thunderstorm life cycle? When would you expect lightning to occur and why?
9. What should you do if you were caught by surprise by a thunderstorm while you are in a large, open area?
10. Discuss the differences and similarities between a cold front and a gust front.
11. Explain why a tilted updraft is necessary for long-lived thunderstorms and the formation of hail.
12. Why are thunderstorms most common in Florida? Why are tornadoes more common in Oklahoma than in Florida?

13. You are standing outside and you see a shelf cloud. What type of weather phenomenon is nearby? Later you see a wall cloud. What type of weather phenomenon may soon occur?

14. A television meteorologist can tell viewers that a "Cat 5" hurricane is approaching the coast, but an "EF5" tornado can only be identified after the fact. With reference to the Saffir-Simpson and EF scale definitions in Chapter 8 and this chapter, explain why there is this difference between the two scales.

15. What is the "rope" stage of a tornado?

16. "Heat lightning" is the term popularly applied to lightning that is seen on summer nights and is not accompanied by thunder. Do you think that lightning can occur without creating thunder? See the text's Web site for an explanation of this phenomenon.

17. Weather lore states that in thunderstorms you should

Beware the oak,
It draws the stroke;
Avoid the ash,
It draws the flash;
But under the thorn,
You'll come to no harm.

Relate this folklore to the safety precautions discussed in Box 11-2.

■ Observation Activities

1. Many weather Web sites include radar observations to track precipitation (e.g., http://www.nws.noaa.gov/radar_tab.php). Use observations on one of these sites to track the movement of a thunderstorm approaching a town. Use these radar observations to forecast the movement of a thunderstorm. Explain how you did this forecast and the result. Describe additional observations that might help you improve your forecast.

2. Many people have had a frightful experience with severe thunderstorms. If you are one such person, describe your experience, and explain your observations in terms of the concepts presented in this chapter.

3. The next time you are at a sporting event or concert, think about what you would do if severe weather threatened you during the event. Where would the safest place be if high winds developed?

This rain cloud icon is your clue to go to the *Meteorology* Web site at http://physicalscience.jbpub.com/ackerman/meteorology/. Through animations, quizzes, web exercises, and more, you can explore in further detail many fascinating topics in meteorology.

12 Small-Scale Winds

AFTER COMPLETING THIS CHAPTER, YOU SHOULD BE ABLE TO

- Relate the concept of turbulence to "friction" in the atmosphere
- Explain what the dominant force(s) are in most small-scale winds and why
- Name and locate on a map likely locations for the occurrence of various small-scale winds

INTRODUCTION

The phone rings in the office of a government meteorologist. The caller wants to know the wind conditions at a spot along a road at a certain time on a day many months ago. The reason: The caller was moving his new dishwasher in the back of his pickup truck when it blew off the truck and was destroyed. He wants specific weather information that proves to the insurance company that a gust of wind did in his dishwasher.

The meteorologist is stumped. She knows that even in today's high-tech world we simply do not have weather information on such small scales of time and space. Furthermore, we also don't have as good an understanding of small-scale winds as we do of large-scale winds. Ironically, meteorologists comprehend more about the winds in an extratropical cyclone or a hurricane than they do about the local winds that blow in our faces every day! The meteorologist tells the caller, "I'm sorry. The best we can do is an hourly observation of winds 30 miles away from the scene of the disaster." It's more likely that a gust of wind from a passing car, not a weather system, caused the dishwasher's demise, but she can't quite know for sure.

In this chapter, we tackle the topic of subsynoptic-scale weather, especially small-scale winds. In this chapter, "small-scale" usually means weather phenomena that develop and change across distances you can see (a few tens of miles or less). This is the great frontier of meteorology because so little is known about weather on these dimensions. It's not for lack of trying; as we'll see; it is simply a fact of the atmosphere that when meteorologists have to "sweat the small details," they end up perspiring a whole lot. Such is the maddening difficulty of small-scale meteorology, the gentle breezes and sudden windstorms that defy explanation.

Two unifying principles guide our study of these bedeviling winds. These principles are as follows: (1) the balance of forces we learned about in Chapter 6 and (2) the geographic features of the local landscape. The Coriolis force, which is so crucial for explaining the large-scale features of the atmosphere addressed in Chapter 7, is usually negligible for small-scale wind patterns. In the absence of the Coriolis force, the pressure gradient and frictional forces dominate. Knowing this helps us to understand small-scale winds. For all of their complexity, small-scale winds generally boil down to the interaction between a relatively strong pressure gradient and whatever is in its way—a mountain, a valley, a lake, a dusty plain, or even an airplane.

Because geography is so crucial in the way that small-scale winds develop, we will study them by taking a tour of local winds across the United States. The diversity of America's landscapes and meteorology creates a wide assortment of winds, spanning most of the types observed worldwide. After a short introduction to the messiness of turbulence, we will follow the Sun and take an east-to-west journey through America, its small-scale winds, and the small-scale winds of the world.

FRICTION IN THE AIR: TURBULENT "EDDIES"

Friction is a familiar concept to us: Driving a car, sanding furniture, and striking a match all involve one rough surface coming into contact with another. Air, however, doesn't have any rough surfaces. How, then, is there any friction? In Chapter 6 we ascribed friction to the contact between the air and the Earth's surface, but now we will explore this concept in more detail.

The friction in a fluid, such as air, is called **viscosity**. You hear this word on TV in relation to motor oil in cars. In the atmosphere, viscosity means the same as for motor oil: the higher the viscosity, the more friction there is and the slower it flows.

Viscosity comes in two scale-dependent varieties. There is friction at the smallest scales when molecules bump into each other. This happens in particular near boundaries, such as the ground (which is a rough surface). This is called molecular viscosity. If molecular viscosity were the only kind of friction, however, then the atmosphere from just above the ground on up would never feel the effects of friction.

The real "friction" in the atmosphere arises from the jostling of the wind with human-sized swirls of air, not tiny molecules. These swirls are called **eddies**, the same name given to swirls of water in a stream or in the ocean. They arise in the atmosphere when the wind blows over or around obstacles such as trees or buildings. Daytime heating by the Sun also leads to eddies; in addition, the atmosphere naturally develops eddy motions, especially near the Earth's surface. At the smallest scales, the eddies themselves lose their energy to molecular viscosity.

These invisible eddies impede the smooth flow of wind by causing slower moving air to mix with higher speed air. It is similar to traffic merging onto a crowded highway: The right-lane traffic jams up as slower cars from the on-ramp mix into the main flow of traffic. In the same way, eddies mix air from the surface, where winds are slow, with faster-moving air higher up. As a result, the overall wind slows down.

For this reason, the analogy is made between friction and the effect of eddies on the wind. Meteorologists call this slowing down of wind the eddy viscosity. The jostling of air with the swirls, as well as the ever-changing motions within the swirls themselves, leads to very irregular fluctuations in the wind. This irregular, almost random, pattern of wind is called **turbulence**, and the eddies are called *turbulent eddies*. The fluctuations we call gusts. FIGURE 12-1 schematically illustrates the relationship between eddies, turbulence, and wind gusts.

What is turbulence, really? If you know, please tell the world's greatest scientists right away because they don't know yet. The atmosphere is enormously complicated at small scales. A famous physicist (Sir Horace Lamb) once told the British Association for the Advancement of Science:

> I am an old man now, and when I die and go to heaven there are two matters on which I hope for enlightenment. One is quantum [physics], and the other is the turbulent motion of fluids. And about the former I am rather optimistic.

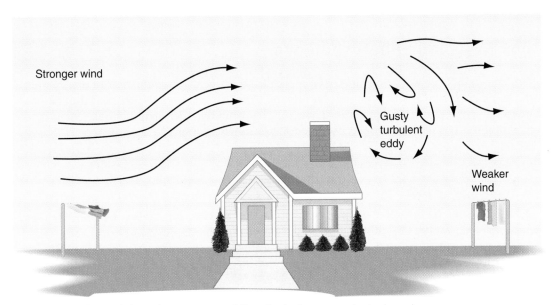

FIGURE 12-1 The relationship among eddies, turbulence, and wind gusts.

Therefore, you can be content with a definition of turbulence that comes from experience with aircraft flights: "bumpiness due to small-scale changes in the wind." BOX 12-1 explores the fascinating and unsolved problem of clear-air turbulence. To summarize what we've covered so far in the context of our understanding of the atmosphere: The atmosphere contains wind patterns at all different scales. At the smaller scales, winds are slowed and made irregular—turbulent—by the effect of eddies. This friction-like process is a "brake" on the natural tendency of the pressure gradient force (PGF) to push air from high to low pressure at all scales. At the tiniest scales, true friction—the rubbing together of molecules—does take place and robs the eddies of the energy they steal from the larger-scale wind. Meteorologist L. F. Richardson, the hero of our next chapter on weather forecasting, expressed this complicated chain of events in a memorable little rhyme:

> Big whirls have little whirls that feed on their velocity
> And little whirls have lesser whirls and so on to viscosity—
> in the molecular sense.

> —Richardson, L. F., *Weather Prediction by Numerical Process*

A TOUR OF SMALL-SCALE WINDS

In Chapter 7 we studied a few circulation systems that together spanned the globe. In this chapter, we will discover a wide variety of small-scale winds that occur locally in parts of the U.S. and across the globe. Time and again we will find that a small-scale wind can be explained by the interaction between a PGF and the topography of the region in which the wind occurs. We will also find, not surprisingly, that the change of seasons plays a governing role in the exact nature and role of the wind.

To explore these winds, let's take a tour of small-scale winds in the lower 48 U.S. (FIGURE 12-2), relating them as we go to these winds' overseas cousins. Our tour follows the Sun from east to west and examines the winds in each region.

THE EAST AND SOUTH

■ Coastal Fronts and Cold-Air Damming

We begin at the northeastern extreme of the U.S., in blustery New England. Cold air formed at these higher latitudes is often trapped between the warmer coast and the high Appalachian Mountains. At the small end of the synoptic scale, this may lead to a *back-door cold front* slipping

Box 12-1 Clear-Air Turbulence

Airplanes and turbulence don't mix. Commercial and especially private aircraft are roughly the same size as large turbulent eddies high up in the atmosphere. This means that planes travel from one bump to another very quickly. The result can be in-flight chaos. One airline pilot in the 1970s recalled from a particularly severe encounter with turbulence:

> I felt as . . . one might expect to encounter sitting on the end of a huge tuning fork that had been struck violently. Not an instrument on any panel was readable to their full scale but appeared as white blurs. . . . Briefcases, manuals, ashtrays, suitcases, pencils, cigarettes, flashlights flying about like unguided missiles.*

For this reason, pilots avoid regions of turbulence. They know to avoid the parallel lines of clouds near mountains, lenticular clouds, and the rainy or dusty swirls of microbursts.

However, nature doesn't always provide a visual indicator for turbulence. Sometimes, in nearly clear skies high up at cruising altitude, planes will suddenly encounter the same sort of jarring bumpiness. This is called clear-air turbulence, abbreviated "CAT." It is one of a pilot's worst nightmares.

What is clear-air turbulence? One of the main theories is that vertical wind shear—the change of wind as you go up—self-develops its own gravity waves. These waves then rapidly break, like ocean waves on a beach. Waves breaking on a beach generate a lot of foam; the "foam" of a breaking atmospheric gravity wave is turbulence, and planes flying through it will encounter bumps and jolts.

What does CAT look like? By definition, you can't see it. Under the right conditions, however, clouds can form and reveal the process outlined earlier. A beautiful example is shown in the photograph.

In the photograph, you can see the different crests of the waves in different stages of breaking. On the left, the wave crest is just beginning to turn over. At right, it's a foamy mess. This process is happening all the time in the atmosphere; only rarely do clouds or a high mountain warn of its presence.

An airplane's encounter with CAT doesn't just make for a good story; CAT is a destroyer and even a killer on occasion. On December 28, 1997, United Airlines Flight 826 carrying 393 people to Honolulu from near Tokyo hit heavy turbulence over the Pacific Ocean. Passengers who happened to be wearing their seat belts at the time described floating "like we were in an elevator falling down," according to the Associated Press. Those not wearing seat belts left dents where their heads crashed into the cabin ceiling. One woman was killed as a result of severe head trauma, and at least 102 people were injured, some of them seriously. Fortunately, CATs this severe doesn't happen very often. Even when it does happen, a CAT persists for only a few minutes in most cases.

A phenomenon as silent, invisible, small, and fleeting as CAT is a major challenge for weather forecasters. Tried-and-true rules-of-thumb exist to steer airplanes around likely areas of wind shear, such as jet streams. This isn't enough. Even today airplanes fly into turbulence on a daily basis, leading to tens of millions of dollars in damage, not to mention passenger and crew injuries. Progress toward understanding this special brand of turbulence has been slow—just as slow as for every other type of turbulence.

Now you know why the airlines tell you to keep your seat belt fastened tightly at all times!

*Lester, P., *Turbulence*. Englewood, CO: Jeppesen, 1994, vi–vii.

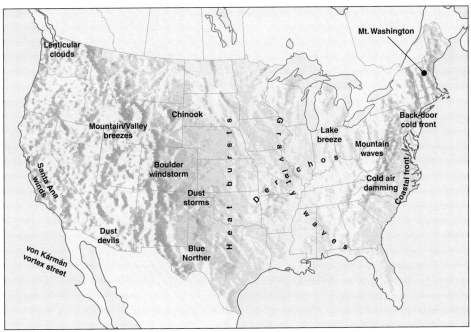

FIGURE 12-2 Geographic summary of small-scale winds across the contiguous (lower 48) U.S.

southward down the Atlantic coast. The interaction between this air and the mountains also has consequences at smaller scales. For example, wintertime extratropical cyclones along the East Coast can draw warmer air above the Gulf Stream onshore, where it clashes with the colder air inland. The boundary between these two air masses is a miniature version of a stationary front and is called a **coastal front**. Coastal fronts add one more layer of complication to the chore of forecasting the impact of extratropical cyclones along the East Coast because a coastal front often separates cold air and heavy snow from warmer air and rain across a distance of only a few miles.

The stubborn entrenchment of cold air that is pinned against high mountains is called **cold-air damming.** This meteorological condition often occurs when a cold air mass is trapped on the east side of mountainous terrain, such as along the Appalachian and Rocky Mountains, by a high pressure system. Warmer air for the west or southwest is then lifted above the cold air instead of reaching the surface, a condition known as overrunning. Cold air damming can result in not only cold temperatures but also freezing precipitation and extensive cloud cover.

Cold-air damming can cause transportation nightmares in winter. As we learned in Chapter 4, freezing rain is likely when warm air overruns a shallow layer of below-freezing air (**FIGURE 12-3**). Cold-air damming is the classic case of shallow cold air. A clockwise circulation around a cold Arctic air mass over the Northeastern U.S. leads to cold air damming along the eastern side of the Appalachian Mountains. When warm, moist air from the Gulf of Mexico or the Atlantic overruns a case of cold-air damming, a damaging ice storm in the Carolinas is a definite possibility. Cold-air damming can also be an ingredient in major East Coast snow storms, such as the blizzards in February 14 to 19, 2003 (also known as the Presidents' Day Storm II), and February 11 to 13, 2006. The cold air helps to keep the precipitation as snow, rather than melting and falling as rain.

In other parts of the world, the **harmattan** of western Africa and the **southerly buster** in southeastern Australia resemble the winds we have just discussed. The harmattan develops when cool air from the Sahara Desert in winter moves south and west and displaces warmer, more humid coastal air. The southerly buster is a mesoscale cold front that causes quick temperature drops (10° C to 15° C) and high winds (up to 100 kilometers per hour) during spring and summer. The cold ocean region south of Australia supplies the cold air, and the mountains of eastern Australia help funnel and intensify the winds of the "buster."

700 mb air flow

Warm air

Cold air

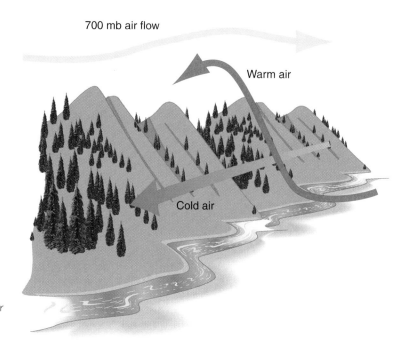

FIGURE 12-3 A clockwise circulation around a cold Arctic air mass over the Northeastern U.S. leads to cold air damming along the eastern side of the Appalachian Mountains. When warm, moist oceanic air overruns the cold air, damaging freezing rain can result (*Source*: SSEC, University of Wisconsin–Madison.)

■ Gravity Waves

"Gravity Waves" to observe a must-see satellite animation of atmospheric waves rippling out of Baja California.

Straight lines are few and far between in nature. However, sometimes long straight lines of clouds will appear in the sky, only to disappear a few minutes later. These clouds occur when the air is jostled. This jostling can be caused by wind blowing over a mountain, by a growing thunderstorm that blocks the wind's path like a mountain, or by complicated changes in winds at the jet-stream level. No matter the cause, the result of this jostling is very similar to throwing a rock into a pond: waves develop.

These atmospheric waves are known as **gravity waves** because their alternating pattern of high and low pressure is maintained with the help of gravity. When made visible by clouds, gravity waves in the atmosphere look a lot like ocean waves, and they are very similar to them in most ways (see the satellite animations on the text's Web site). Air goes up in the crests of the waves, cools, becomes saturated, and forms clouds. Air in the troughs of the waves sinks and dries out. This is why the clouds caused by gravity waves form parallel straight lines.

Mountain-generated gravity waves are visible, even by satellite, many times each year over the central Appalachian Mountains when near-surface winds blow perpendicular to the mountain ridges. The mountainous regions west of Washington, DC, often produce these waves, which are called "lee waves" because they are downwind of the mountains. **FIGURE 12-4** shows a classic example of the clouds produced by these waves; the wind in the figure is blowing toward the southeast. Pilots know to avoid these "wave trains" of parallel lines of clouds because they are likely to harbor clear-air turbulence (Box 12-1).

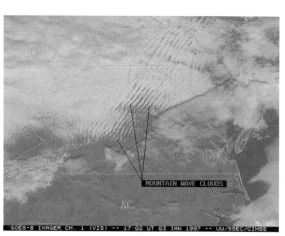

MOUNTAIN WAVE CLOUDS

NC

GOES-8 IMAGER CH. 1 (VIS) -- 17:02 UT 03 JAN 1997 -- UW/SSEC/CIMSS

FIGURE 12-4 Lines of clouds caused by gravity waves in the lee of the Appalachian Mountains, as seen by a weather satellite on January 3, 1997. See the text's Web site for an animation of this and other satellite images of gravity waves.

Other gravity waves form because of wind changes in the jet stream that send out "ripples" of waves. One particularly impressive case occurred in Alabama on the morning of February 22, 1998. A powerful upper-level cyclone (**FIGURE 12-5**) unexpectedly triggered gravity-wave ripples that moved northward across the entire state of Alabama, a distance of more than 400 kilometers (250 miles), in only 3 hours. In downtown Birmingham, the surface pressure dropped 10 millibars (mb) in just 17 minutes (**FIGURE 12-6**). The corresponding tight pressure gradient

FIGURE 12-5 Colorized water vapor image of the upper-level cyclone that helped trigger the Birmingham, Alabama, gravity-wave windstorm at 11:15 AM on February 22, 1998. The reddish hook in the image is a region of dry air from the stratosphere that is wrapping around the cyclone.

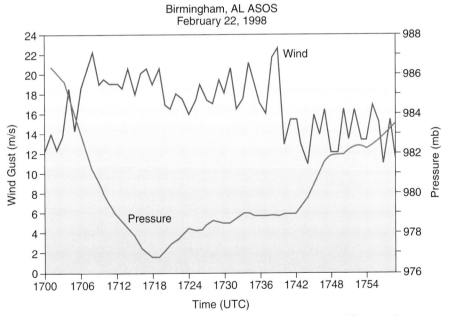

FIGURE 12-6 Automated observations of wind and pressure at Birmingham, Alabama, during the February 22, 1998 gravity-wave–induced windstorm. Notice the rapid changes in both variables. (*Source*: Bradshaw, John T., et al., *The Alabama gravity wave event of February 22, 1998*. NOAA, 1988. Retrieved February 28, 2011, from http://www.srh.noaa.gov/bmx/?n=research_02221998.)

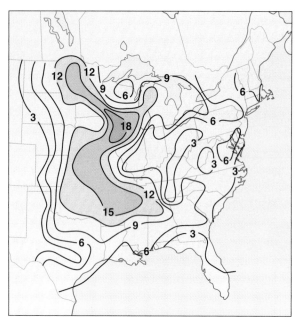

FIGURE 12-7 A 25-year climatology of gravity waves across the lower 48 U.S., based on surface observations of hourly pressure drops of at least 4.25 mb. (Adapted from Koppel, L., et al., *Monthly Weather Review*, January 2000: 58.)

caused winds of more than 22 meters per second (51 mph). Houses and trees exposed on the sides of small mountains in the Birmingham area, unsheltered by the effects of friction, experienced even higher winds; roof and tree damage was extensive. However, no thunderstorm was involved; the winds were all part of the waves generated by the sloshing of the jet stream high above.

Because of the wide variety of ways in which they are triggered, gravity waves can be observed in many parts of the U.S. and the world. For example, the **morning glory** of northern Australia is a spectacular linear cloud up to 1000 kilometers (621 miles) in length that forms on the leading edge of a gravity wave. Closer to home, FIGURE 12-7 shows the observed distribution of non–mountain-related gravity waves across the U.S. They are most common in the southern U.S. and also in our next stop: the Midwest.

THE MIDWEST

West of the Appalachians, we encounter the Midwest. In Chapter 9, we explored lake-effect snow, and in Chapter 10, we learned about the localized windstorm that helped sink the *Edmund Fitzgerald*. Now we investigate other small-scale Midwestern winds, some of which are also related to the presence of the immense Great Lakes.

■ Lake Breezes

"Mountain Lee Waves" to view an animated satellite loop of Figure 12-4, which helps answer the question, "Do mountain-wave clouds always stay in one location with respect to the ground?"

During warm summertime days, the Great Lakes are usually colder than their surrounding coastlines. These lakes are so large that local wind circulations develop because of the resulting pressure gradient, just as they do along the world's ocean coastlines during the daytime. We called the daytime circulations along ocean coastlines the *sea breeze* (see Chapter 6); by analogy, winds that blow onshore during the day around the Great Lakes are called **lake breezes**.

A classic example of a lake breeze is shown in FIGURE 12-8. The cloudless region ringing Lake Michigan is cool lake air; the region encircling it that is dotted with cumulus clouds is the warmer land air. The lake air at this time is 10° F to 15° F (6° C to 8° C) cooler than the air over inland areas.

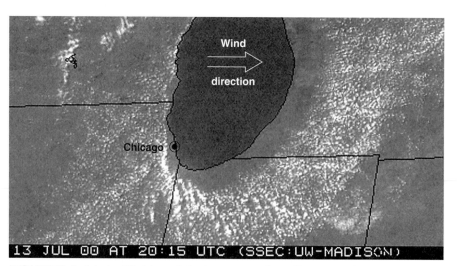

FIGURE 12-8 A Lake Michigan lake breeze on the afternoon of July 13, 2000, as viewed from satellite. Notice the absence of white spots (cumulus clouds) near the lake, especially on the east side over western Michigan. This is the region of the lake breeze.

Because the Great Lakes are right in the middle of extratropical cyclone storm paths in summertime, it is common to see lake breezes in combination with larger scale wind patterns. Notice that the ring of cloudless air in Figure 12-8 is not centered over Lake Michigan; it is shifted to the east. This is because the prevailing winds were from the west at the time of the satellite picture. You can also see that south of Chicago (lowest center part of the satellite image) the cumulus clouds are growing larger. The boundary between the lake breeze and land air can sometimes be a focal point for thunderstorm development, just like a small-scale front.

"Great Lake Breezes" to watch Lake Michigan's lake breezes move cumulus clouds onshore across the upper Midwest.

Derechos

At the opposite extreme from the gentle lake breeze is the **derecho**. A derecho (pronounced deh-RAY-cho, a Spanish word meaning "straight ahead") is an hours-long windstorm associated with a line of severe thunderstorms. It is a result of straight-line winds, not the rotary winds of a tornado—hence its name.

The extreme winds of a derecho—up to 240 kilometers per hour (150 mph) in the worst cases—come about in the following way. Derechos are often associated with a quasi-stationary front (see Chapter 9) in mid-summer. If the atmosphere just north of the front is very unstable, with lifted indices (see Chapter 11) of less than −6, the front may trigger rapidly developing thunderstorms. A line of thunderstorms that forms in the vicinity of the stationary front can, via its cold downdrafts, drag down high-speed air from above. This can cause the high winds of a derecho.

At the same time, the high winds push the line of thunderstorms outward, causing it to bend or "bow." This is called a **bow echo** when it is seen on weather radar. Once they get going, derechos can cover lots of territory—up to 1600 kilometers (1000 miles).

Derechos leave significant property damage, and even entire forests flattened, in their wake. In some cases, derechos wreak as much havoc as a hurricane or tornado, yet because they are less well known, derechos are even more deadly: About 40% of all thunderstorm-related injuries and deaths occur because of them.

FIGURE 12-9 shows a radar image of a severe derecho that caused six deaths, more than 200 injuries, and $300 million in damage from Minnesota to New York on May 31, 1998. Notice the characteristic bowed-out pattern of the squall line on radar. This line of storms moved

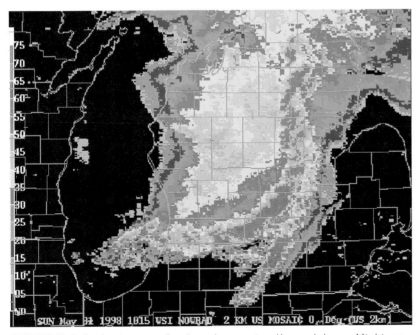

FIGURE 12-9 Radar image of a derecho moving through lower Michigan on May 31, 1998. Notice the curved, bowed-out nature of the red area of strongest storms.

Total number of events

0.1   6   12   18   24   30   36   48

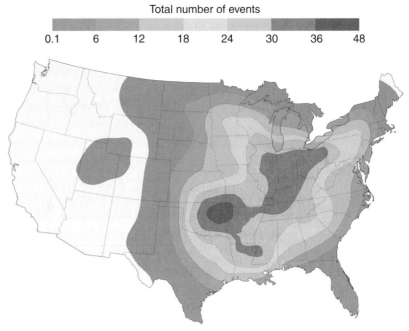

FIGURE 12-10 A climatology of derechos in the U.S. between 1986 and 2003. (Modified from Coniglio, M. C., and D. J. Strensrud, *Wea. Forecasting* 19 [2004]: 595–605.)

at a forward speed of 105 kilometers per hour (65 mph) and generated wind gusts as high as 206 kilometers per hour (128 mph)! Between 1986 and 2003, derechos were responsible for 153 fatalities and over 2600 injuries.

Derechos in the U.S. are most common in the late spring and summer (May through August). The climatology of derechos (FIGURE 12-10) shows that they are generally confined to the eastern two thirds of the U.S. Derechos are common west of the Appalachians in the Ohio Valley and also in our next stop: the southern Great Plains states.

THE GREAT PLAINS

The Great Plains are nearly flat; they tilt upward gradually from east to west, and at their western edge they are close to high mountains, especially the Front Range of the Rocky Mountains. These facts help explain the strong small-scale winds of this region, which farmers try to slow down through the use of windbreaks (BOX 12-2).

Blue Northers

We learned in Chapter 11 that the topography of the Great Plains creates a warm, moist, low-level jet that can blow across the Plains from south to north. In winter, the Plains also provide a clear path for cold air pushing south out of Canada. The Rocky Mountains help channel this continental polar (cP) air rapidly southward. As a result, cold fronts can sometimes zoom across the 2400 kilometers (1500 miles) of the western Plains, from the Dakotas to Texas, in only a couple of days.

As the cold front enters the southern Great Plains, the temperature clash at the fronts can be extraordinary. Temperatures can drop tens of degrees Fahrenheit in only a few hours. However, the air on both sides of the front is extremely dry, having originated in the dry high hills of the southwest to the south and bone-dry Canada to the north. Therefore, the front is accompanied by clear blue skies. The only sign of changing weather is that the mercury in the thermometer drops so fast you can see it move! The classic blue norther occurred on November 11, 1911, when, in just a matter of hours, the temperatures in Oklahoma City dropped from a high of 83° F (28° C) to a low of 17° F (−8° C)—a temperature difference of 66° F (36° C). More recently,

Box 12-2 Using Turbulence to Advantage: Snow Fences and Windbreaks

On the Great Plains and in the nation's snow belts, wind can be an enemy. Snow or dust picked up by small-scale winds can reduce visibility, cover roads, and shut down normal life for days on end. Unfortunately, there is no "off" switch on the winds. However, there is a "slow motion button" that can be used: the effect of turbulent eddies. This is the concept behind two human creations, the snow fence, and the windbreak.

The presence of obstacles such as fences and trees slows the wind by causing turbulent eddies to develop. The wind is broken up into a swirl of eddies, and its overall speed is reduced. This is why a wooded, fenced-in suburb is generally much less windy than a shoreline.

Snow fences use this concept to keep snow from blowing across land and roadways. Snow flying on high winds past a snow fence will, instead of blowing straight downwind, get caught up in the turbulent eddies the fence creates. Some of the snow will slow down just past the fence and drop to the ground (see photograph). As more and more snow collects behind the snow fence, it presents an even larger obstacle to the wind. By the time the wind subsides, a large pile of snow accumulates behind the snow fence. If the fence is located upwind of a road or a farm, this is snow that didn't drift over the asphalt or bury the barn. In addition, in cold, arid regions such as the Dakotas, the snow acts as a blanket in winter and a moisture source in spring, keeping the soil warmer and wetter than it would be without snow cover.

The same idea holds for windbreaks on the dusty Great Plains. Blowing dust in spring can be a great hazard to transportation; it is also precious topsoil leaving the area. To prevent this, lines or belts of trees called "windbreaks" or "shelterbelts" have been planted from place to place across the Plains. The wind blowing past these windbreaks is chopped up periodically into slower turbulent eddies. Windbreaks thus function like the "speed bumps" used in shopping-center parking lots to keep cars from going too fast. They keep the winds from roaring across the hundreds of miles of flat land and lifting freshly plowed soil into the air.

For an example of what can happen when there are no trees, all the soil is plowed, and the winds blow hard, read Box 12-3 on the "Dust Bowl."

on January 3, 2009, Tulsa, OK, reached a high temperature of 78° F (26° C) followed by drop in temperature of 39° F (to 39° F or 4° C) within 5 hours after the norther's passage. This is why the fierce north wind behind a fast-racing cold front in west Texas is known as a **blue norther**. In other parts of the world, blue-norther type winds go by the names of **norte** (Mexico) and **buran** (Russia).

Blue northers can be killers. Farmers or hunters unprepared for their arrival can freeze to death, particularly if the initial clear-sky front is followed by snow. These high winds can, in times of drought, lead to **dust storms**. BOX 12-3 examines dust storms and the infamous "**Dust Bowl**" of the 1930s in more detail.

Box 12-3 Dust Storms and the "Dust Bowl"

Dust storms are to dust devils what a mesoscale convective complex (see Chapter 11) is to a thunderstorm: a larger, longer-lived version that has its own unique look. Downdrafts from thunderstorms over a desert can generate dust and sandstorm called **haboobs**. Dust storms lasting several days are not uncommon in the far western Great Plains of the Texas and Oklahoma Panhandles. As we learned in Chapter 10, extratropical cyclones are often born in the Panhandles, causing windy conditions. Dryline thunderstorms can kick off dust storms with microbursts, and blue northers can usher in dust on their leading edges as they race southward from Canada. A pressure gradient, plus dry ground, is all that is needed. That, plus overly aggressive farming practices, can even lead to a "Dust Bowl."

In the early decades of the 20th century, American farmers moved into the Great Plains as part of a drive to convert the region into an agricultural paradise. The farmers came from the East and the South. However, those regions typically receive double, triple, or quadruple the annual rainfall of the Panhandles. Furthermore, typical wind speeds in the East and South are lighter than in the high Plains.

In the summer of 1931 a drought began, not unlike other droughts in the Panhandles in other times. The difference was that this time all the soil was plowed and ready to go airborne. When the winds came, dust storms developed early and often: 14 dust storms in 1932 and 38 in 1933. The sky turned black, thousands of years of topsoil literally gone with the wind (see photograph). Dust blew and drifted like snow—except that dust doesn't melt and instead had to be swept out of homes and shoveled off all exposed areas. Crops died, year after year. Animals died of starvation. Children died from pneumonia triggered by dust inhalation. It was a meteorological and ecological disaster that one journalist dubbed the "Dust Bowl".

The drought lasted almost a decade, until late 1939. By that time, 25% of the population of the Panhandles had fled, many to California. This mass exodus became the inspiration for John Steinbeck's classic novel *The Grapes of Wrath*.

Today, however, "dust emission in this region . . . [is] at a historical minimum," according to a recent research study. Why? Wiser farming practices, including soil conservation efforts, and wetter conditions.

▨ Heat Bursts

Our next small-scale wind sounds like something from a science fiction movie script. You're living somewhere on the high plains, say the Oklahoma Panhandle. You go to bed on a warm night, with a rumble of thunder in the distance. Then, in the middle of the night, you wake up in an oven. The temperature has shot up to daytime levels since you went to bed, the air is parched dry, and the winds are howling, but there's no thunder. What force of nature could make the temperature soar and generate a gale in the middle of what seemed like a calm, peaceful night?

What you've just experienced is called a **heat burst**, and it can be easily explained using the concept of adiabatic warming and static stability we first learned in Chapter 3. When a thunderstorm forms (**FIGURE 12-11a**), unstable air rises; however, as the storm matures (Figure 12-11b), the updraft weakens, and the precipitation in it evaporates. This evaporative cooling causes the air to become cooler and denser than its surroundings, and it sinks as a downdraft. As the air sinks, it is compressed and warms at the dry adiabatic lapse rate. If the air begins its descent at relatively high altitudes, then by the time it reaches the surface it is much warmer and drier than other air at the surface (Figure 12-11c). Like a hot version of a microburst, the air "splashes" against the ground and rushes outward rapidly, causing a heat burst.

Heat bursts are most common at night during the late spring and early summer in the southern Great Plains. Heat bursts are associated with thunderstorms and typically occur during the night. Based on recent research in Oklahoma, an "average" heat burst raises the temperature 10.2° F (5.7° C), drops the dew point 14.4° F (8° C), triggers a maximum wind gust of 66 kilometers per hour (41 mph), and lasts only about 30 minutes. On May 22 and 23, 1996, a widespread heat burst across southwest Oklahoma led to scorching evening temperatures of more than 100° F (38° C) and $18 million in damage from wind gusts of up to 169 kilometers per hour (105 mph)!

Heat bursts occur on such short time and space scales that they are hard to capture in hourly weather observations. A heat burst was observed by the Sioux Falls' automated surface observation system (see Chapter 5) on August 3, 2008 (**FIGURE 12-12**). The temperature at 4 AM CDT was 74° F (23° C) with a dew point of 66° F (19° C). At about 4:10 AM CDT the temperature began to rapidly increase, reaching a maximum value of 101° F (38° C) at 4:26 AM CDT. During this time, the dew point dropped 22° F to 44° F (7° C). The rapid changes in temperature were accompanied by wind gusts of 50 to 60 mph (80 to 96 kilometers per hour). After the heat burst ended, the temperature returned to 74° F (23° C) by 4:45 AM.

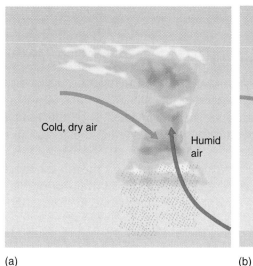
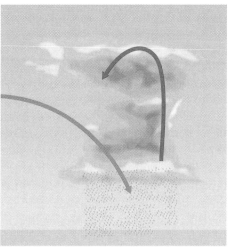
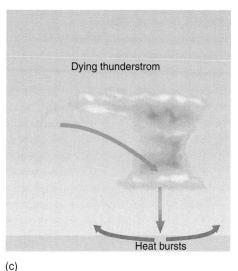

(a)   (b)   (c)

FIGURE 12-11 Schematic explanation of the development of a heat burst. A thunderstorm develops with an updraft (a) and then matures and develops a rain-cooled downdraft (b). The downdraft warms adiabatically and reaches the ground as a hot, dry wind called a "heat burst" (c). (Source: Oklahoma Climatological Survey & The Oklahoma Mesonetwork.)

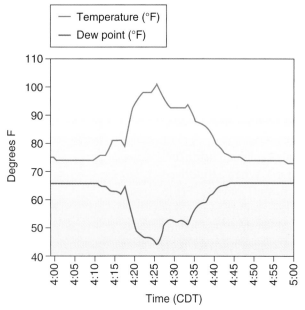

FIGURE 12-12 The temperature and dew point changes associated with a heat burst in Sioux Falls on August 3, 2008. (*Source*: NOAA.)

■ Chinooks

On the western edge of the Great Plains, mountains meet flatlands. When air moves down these mountain slopes, it goes down in a hurry. The peaks of the Rockies reach more than 3 kilometers (10,000 feet) in altitude; in contrast, the Colorado plains are only 1.5 kilometers high. Similarly, the highest Black Hills of western South Dakota are more than 2 kilometers (7000 feet) tall, but the nearby towns and cities are less than 1 kilometer (about 3000 feet) in elevation. As with sinking air in a heat burst, adiabatic compression of air moving down a mountain causes a warming and drying-out of the air. A dry, warm wind is thus created whenever the large-scale pressure gradient moves air down the slopes of the Rockies, or the Black Hills of western South Dakota, onto the flatlands. Plains residents call this dry, warm wind a **chinook**.

The chinook's nickname of "snow eater" arises because a chinook quickly warms a location above freezing while dropping its relative humidity into the single digits. Snow is able to melt and evaporate rapidly under these windy conditions. Not far from Rapid City, South Dakota, in 1943, a chinook rocketed the temperature up from −4° F (−20° C) to 45° F (7° C) in only 2 minutes, a world record! In other parts of the world, this type of wind is called a **foehn** (European Alps), a **puelche** (west slopes of the Andes Mountains in South America), a **koembang** (in Indonesia), or a **zonda** (east slopes of the Andes in Argentina).

On the Great Plains, these extreme changes are mostly a curiosity, significant only for farmers and the comparatively few residents of the area. We will see that when the same type of wind occurs on the heavily populated West Coast, it is as big a threat to life and property as El Niño or earthquakes.

THE WEST

The Western U.S. enjoys the richest array of small-scale winds because its geography is the most varied, from tall peaks to flat deserts. Many of the local wind patterns we have already studied, such as microbursts, mountain waves, and cold-air damming, are frequently observed in the West as well. Other winds are unique to the complex geography of the West. Once again, we can explain their existence as the combination of pressure gradients interacting with geography and topography.

■ Mountain/Valley Breezes and Windstorms

We've already discussed how temperature differences between ocean and coastline, or a Great Lake and its shores, can lead to pressure gradients that drive local wind circulations. The same thing can happen along the slopes of high mountains such as the Rockies. These small-scale winds are called **mountain and valley breezes** and closely resemble the ocean and land breezes of Chapter 6.

FIGURE 12-13 is a schematic of how these breezes develop. During the day, the thin air above the high mountainsides warms quickly. The warm air rises and creates local low pressure along the slopes. Air from the lower valleys moves in to replace it, creating an upslope breeze that becomes strongest around noon. This is the valley breeze.

At night, the high mountain slopes cool very quickly. This cold, dense air forms a local high-pressure area. The pressure gradient drives a gentle breeze down the slope into the valley that is strongest just before sunrise. This is the mountain breeze.

Mountain and valley breezes are usually gentle, just a few kilometers or miles per hour. Even so, they can have a profound impact on local weather and climate. In some bowl-shaped valleys in the western U.S., bitter-cold air sinks overnight during winter, creating a reverse treeline below which it is too cold for trees to survive! Mountain and valley breezes are also found throughout the world. For example, the **hira-oroshi** is a breeze that occurs along the mountainous shoreline of Lake Biwa west of Tokyo, Japan.

In mountainous regions with steep-sided, snow-covered large plateaus, the cool mountain breeze is anything but gentle. In some cases, wind gusts can exceed 160 kilometers per hour (100 mph). These more violent relatives of mountain breezes are called **katabatic winds.** They occur all over the world and are called the **bora** (Adriatic coast of Europe), the **mistral** (French Riviera), and the **fall wind** (Greenland and Antarctica). In the U.S., Colorado and the Columbia River valley experience katabatic winds. Their violence is a result of the strong pressure gradient built up by the chilling of air over the high snowy plateau and the steep slopes that allow the wind to rush downhill quickly.

Still other mountain winds affect the mountainous regions of the world. Any time there is a strong pressure gradient across the mountains, the wind tries to find a place to break through the barrier of rock. Low points in the mountains, called "gaps" or "passes," become wind tunnels in these cases. The **gap winds** that develop can easily exceed 160 kilometers per hour (100 mph) in the strongest cases. The **squamish** of British Columbia, the **levanter** of the Strait of Gibraltar between Spain and Africa, and the **tehuantepecer** on the Pacific Coast of Central America are all examples of gap winds.

In Boulder, Colorado, just northwest of Denver, downslope winds associated with mountain gravity waves can cause particularly ferocious **Boulder windstorms.** Winds well over 160 kilometers per hour (100 mph) have been observed in some cases. FIGURE 12-14 depicts the winds from the Groundhog Day Boulder windstorm of 1999. These windstorms cause considerable property damage; damage from Boulder's winds averages about a million dollars per year. In areas near Boulder, building codes have been changed to reduce the frequency and amount of damage.

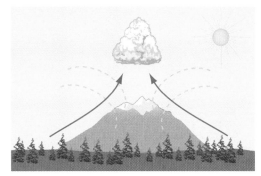

Valley breeze

Mountain breeze

FIGURE 12-13 Schematic explaining the mountain and valley breezes. Purple lines are isobars.

Dust Devils

A much more benign wind blows along the sands of the desert Southwest. There, intense daytime heating helps spin up thin, rotating columns of air called **dust devils** (FIGURE 12-15). Unlike its cousins the tornado and the waterspout, the dust devil appears to be a creature created

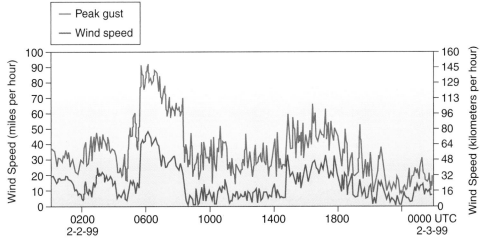

FIGURE 12-14 Winds in the Boulder, Colorado, windstorm of February 2, 1999. Notice how quickly the winds increase and decrease and how consistently strong they are for hours. (*Source*: University Corporation for Atmospheric Research.)

solely by solar heating. The Sun above desert bakes the ground until the surface air becomes unstable and rises, creating a local low-pressure center usually only a few meters (yards) across and 100 meters (330 feet) tall. As the little vortex begins to spin, dust and sand are drawn into the circulation and a dust devil is born. It usually lasts only a few minutes. Dust devils usually form between 11 AM and 2 PM local time.

Unlike waterspouts and tornadoes, dust devils do not require any clouds, showers, or thunderstorms above them; they form under clear skies in a hot Sun. Like waterspouts, the winds in a dust devil are much weaker than in tornadoes, and they rarely cause damage. There are exceptions, however. A June 2008 dust devil near Casper, Wyoming, collapsed a shed and killed one person. Close relatives of the dust devil are the **willy-willy** of Australia and the **simoom** whirlwind of the African and Arabian deserts. Dust devils have also been photographed on Mars by NASA's *Spirit* rover!

FIGURE 12-15 A dust devil observed in Arizona on June 10, 2005, by a team of researchers from NASA and the University of Arizona.

Lenticular Clouds

As our tour approaches the mountains of the Coastal Range, we encounter another small-scale wind and cloud feature that combines the beauty of the dust devil with the beast of a Boulder windstorm. It is called a **lenticular cloud** (FIGURE 12-16).

The name lenticular means "lens-shaped." These clouds hang over mountainous regions for hours, moving little if at all. To modern eyes, lenticular clouds look like the hovering motherships in space alien movies. In fact, the first modern "sighting" of a UFO occurred near Mount Rainier, Washington, and was probably a type of lenticular cloud. They are also seen in Alaska and Hawaii, and some of the most impressive lenticular clouds occur near the isolated Norwegian volcanic island of Jan Mayen northeast of Iceland.

FIGURE 12-16 A lenticular cloud at sunset on January 3, 1996, near Tehachapi, California, as photographed by former student Cynthia Stoneburner. This smooth, sculpted cloud is a sign of severe mountain-related turbulence nearby.

What does a lenticular cloud have to do with small-scale winds? Everything. When winds blow across high mountain ranges in certain circumstances, vigorous gravity waves develop downwind of the mountains. Air rising on the crest of the wave just past the mountain becomes saturated, forming the lenticular cloud (FIGURE 12-17). Because the wind and the mountain anchor the wave crest in the same place, the cloud is stationary—just as a river eddy downstream from a boulder remains in the same spot. The swirling pattern of winds around and over the mountain sculpts the lenticular cloud into unique and ever-changing shapes.

The beauty of the lenticular cloud sounds a warning alarm to pilots. The wind circulations beneath lenticular clouds are extremely turbulent (FIGURE 12-18), despite the seemingly calm, smooth appearance of the cloud. Aviators know to avoid these situations if at all possible.

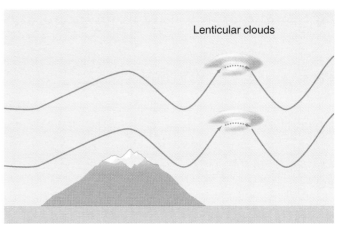

FIGURE 12-17 This schematic shows the wavelike path air takes as it flows over a mountain and a lenticular cloud forms.

Santa Ana Winds

As our tour begins its westward descent over the coastal mountains of California, we discover yet another downslope wind. This wind, the **Santa Ana wind**, combines the characteristics of its close kin the chinook with the damage potential of a Boulder windstorm.

The Santa Ana wind occurs when the pressure gradient caused by an anticyclone over the Rockies (FIGURE 12-19), in combination with friction, forces already-dry air from the mountainous West down the Coast Range in northern California or down the San Gabriel Mountains in southern California and all the way out into the ocean. As with the chinook, a Santa Ana causes the temperature to increase and the relative humidity to plummet because of adiabatic warming. This happens during September through April, but Santa Ana winds most often make the news in October (see below).

There are a few key differences between chinooks and Santa Anas. There is no snow for the Santa Ana to "eat" in fall in California. The trees and grasses of California are naturally dry in autumn. Also, more than 30 million people live in California, as opposed to less than 1 million in all of South Dakota. As a result, the Santa Ana wind turns some of America's largest metropolitan areas into bone-dry, roasting tinderboxes.

One spark, cigarette butt, or lightning strike later, regions affected by the Santa Ana wind become huge roaring fires. On October 20, 1991, a Santa Ana wind with speeds of more than 32 kilometers per hour (20 mph) dropped the relative humidity in the Oakland, California, vicinity to less than 10% for an entire day. The remains of a brush fire flared up, and the inferno was on. When the fire was finally extinguished, 25 people were dead, more than 2000 buildings were destroyed, and damage was assessed in the billions of dollars.

Twelve years after the Oakland conflagration, similar Santa Ana wind–related fires raged throughout southern California in October 2003, killing 24 people, injuring another 246, closing airports, burning 721,791 acres (2921 km²) and threatening major urban areas. The smoke from the fires spread over 1600 kilometers (1000 miles) out to sea. Another similar Santa Ana-related fire in southern California in October 2007 wreaked similar havoc (see Chapter 1).

Numerous winds around the world bring hot, dry conditions to regions downwind of mountains and deserts. The **berg wind** of South Africa, the **leveche** of Spain, the **khamsin** of Egypt, the **leste** of the Canary Islands,

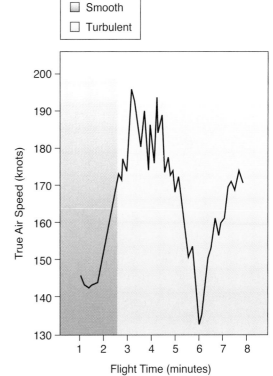

FIGURE 12-18 The airspeed of a plane flying downwind below the region of lenticular clouds. Notice the rapid transition from smooth flying to turbulent conditions with large and abrupt fluctuations in airspeed. (Adapted from Lester, P. *Turbulence.* Jeppesen, 1994.)

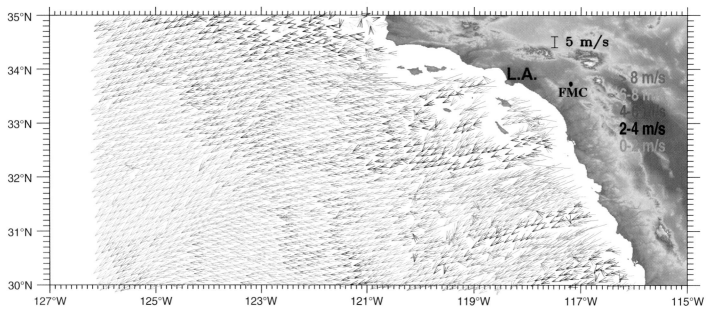

FIGURE 12-19 Santa Ana winds measured by QuikSCAT. Satellite-derived winds off the coast of Los Angeles California ("LA" on map)—associated with a Santa Ana event in January 2004. The fastest winds (red) are rushing down off the San Gabriel Mountains (yellow, brown, and white areas to the north and east of Los Angeles) and across the coastline, becoming warmer and drier as they descend. The colored arrows represent various ranges of wind speed (see legend to the right of the image). (Courtesy of JPL/NASA.)

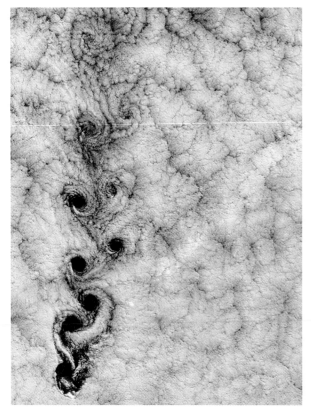

FIGURE 12-20 A visible satellite picture of a von Kármán vortex street downwind of Guadalupe Island just offshore Baja, California, on July 25, 1996.

and the **sirocco** of the Mediterranean all combine aspects of the Santa Ana wind with other types of circulations we have studied in this text, such as the dryline (see Chapter 9).

■ von Kármán Vortex Street

Before we conclude our American tour of winds, we peek out the airplane window at an amazing sight: a long interlocking chain of eddies rippling in the clouds downwind of an island (**FIGURE 12-20**). This chain of eddies is called a **von Kármán vortex street**, named for Caltech engineering professor Theodore von Kármán, who pioneered modern research on this and many other topics in turbulence and aerodynamics.

These vortices are caused when a mountain interrupts the smooth flow of wind past it and causes the airflow to be deflected mainly laterally, instead of vertically, as in the case of mountain waves or lenticular clouds. The wind closest to the mountain feels its frictional effect because of turbulence and slows, but the deflected wind on either side of the mountain proceeds downwind quickly. This creates horizontal wind shear that causes the deflected winds to roll up into interlocking pairs of vortices, one cyclonic and the other anticyclonic. Although the eddies may extend for hundreds of kilometers downwind, the von Kármán vortex street is a fundamentally small-scale process. It does not require the Coriolis force for its existence.

These vortices show up best in moist marine atmospheres with just enough low-level cloud cover to make them visible from the

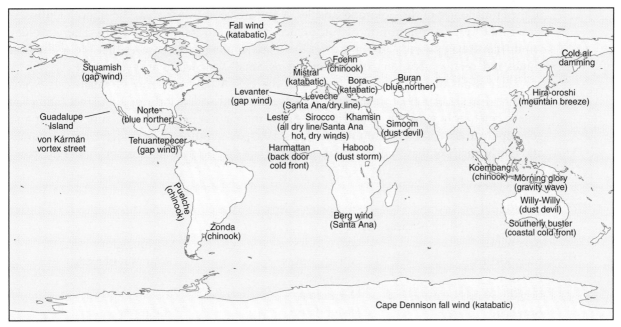

FIGURE 12-21 Selected small-scale winds from around the world. Their American counterparts are listed in parentheses below the names.

air or from satellite. Some of the most beautiful examples of von Kármán vortices are found along the coast of Baja California, as in Figure 12-20, and downwind of the Canary Islands west of Africa. They are not dangerous; they are stunning visual reminders of the fascinating pageant of small-scale winds that blow across our planet.

THE BIG PICTURE

Let's step back and survey small-scale winds. Geography is destiny for these winds. The Hoosiers of flat Indiana need never fear a Santa Ana, just as residents of Minot, North Dakota, will never experience a coastal front. However, each region of the country has small-scale winds. Their visible signatures are all around us, occurring at any time and place where the combination of pressure gradient and landscape is right.

Small-scale winds act locally, but they are found globally. This has profound consequences for weather. Worldwide, these winds connect the synoptic weather of extratropical cyclones and anticyclones (Chapter 10) and the mesoscale weather of thunderstorms (see Chapter 11) with the motions of the atmosphere at the smallest scales, and thus with the surface of the Earth itself. FIGURE 12-21 summarizes the global extent of small-scale winds, relating each overseas example with its closest American cousin.

PUTTING IT ALL TOGETHER

Summary

Winds occur on every imaginable scale. Small-scale winds are driven primarily by the PGF and are slowed by the effects of "friction." The main type of friction in a fluid such as the atmosphere is actually turbulence caused by swirling eddies of different sizes. These eddies are stirred up by the Sun, by wind blowing around obstacles, and by the atmosphere itself. This causes the characteristic gustiness of winds that is frequently observed.

An amazing variety of small-scale winds exists in the U.S. Some winds are gentle, such as those that slide mountain breezes down the sides of the Rockies. Other winds are violent, such as derechos and microbursts dropping out of thunderstorms or Boulder windstorms blowing out of the Front Range of the Colorado Rockies. A few are deceptive, such as the smooth lenticular cloud that hides a commotion of turbulence underneath its polished sides. There are dry winds that eat snow and cause fires, and there are rippling gravity waves that "draw" lines in clouds and, on rare occasions, spark windstorms. Small-scale winds are everywhere.

These kinds of winds are not limited to the U.S. A dazzling array of small-scale winds exists worldwide. For all of their diversity, however, small-scale winds on our planet (and even other planets) can generally be attributed to the interaction between a pressure gradient and geographic features such as mountains, plains, lakes, or coastlines. The colorful, unique names and rich histories of these local winds disguise the fact that they are all "brothers under the skin."

■ Key Terms

You should understand all of the following terms. Use the glossary and this chapter to improve your understanding of these terms.

Berg wind	Gap winds	Morning glory
Blue norther	Gravity waves	Mountain breezes
Bora	Haboobs	Norte
Boulder windstorms	Harmattan	Puelche
Bow echo	Heat bursts	Santa Ana wind
Buran	Hira-oroshi	Simoom
Chinook	Katabatic winds	Sirocco
Coastal front	Khamsin	Southerly buster
Cold-air damming	Koembang	Squamish
Derecho	Lake breezes	Tehuantepecer
Dust Bowl	Lenticular cloud	Turbulence
Dust devils	Leste	Valley breezes
Dust storms	Levanter	Viscosity
Eddies	Leveche	von Kármán vortex street
Fall wind	Microburst	Willy-willy
Foehn	Mistral	Zonda

■ Review Questions

1. A tiny swirl in the bathtub can look a lot like an extratropical cyclone. Using facts from this chapter, explain how they differ.

2. Why do meteorologists link turbulence with friction, when in fact there aren't any rough surfaces rubbing together in the atmosphere?

3. Go to your kitchen or bathroom sink and make three different sizes of eddies. What causes them, how big are they, and how long do they last?

4. On a U.S. map that includes elevations, find the highest mountain peaks in the entire eastern U.S. Does their location surprise you? How does their location affect cold-air damming events?

5. Would mountain or valley breezes be more intense on the northern or the southern slopes of a mountain? Why would there be a difference?

6. Based on your understanding of gap winds, explain why the streets of skyscraper-filled Manhattan can be so windy on a day when the winds in the suburbs around New York are light to moderate.

7. A real estate agent tries to sell you a house in Colorado that's near the bottom of a valley, ringed by tall mountains. The view is gorgeous. What meteorological advantages and disadvantages are there to purchasing a house in this location? How would your heating and cooling costs be different if the house were up near the mountain peaks?

8. It's a hot, sunny day in Dallas, Texas. A low is approaching from the west, however. What different kinds of small-scale winds could lead to turbulence for a plane taking off and ascending, or descending and landing, near Dallas?

9. Why are windmills and wind turbines placed on tall towers?

10. An old folklore saying goes, "The winds of the daytime wrestle and fight/Longer and stronger than those of the night." Based on what you've learned about turbulent eddies, explain why this is usually true.

11. The planet Mars has very high mountains and flat plains. It's very dry (few clouds, no thunderstorms) and dusty. Its axis tilts about the same as Earth's so that Mars's tropics receive a lot of sunshine all year long. Based solely on this information and facts found in this chapter, what kinds of small-scale winds would you expect to find on Mars?

12. It's the week before your wedding. You're in your seventh-story apartment watching television with your fiancée. Suddenly, the local TV weatherman comes on, shows the radar, and points out a "bow echo" approaching quickly. What kind of small-scale wind is coming? What should you do immediately? (This is a true story!)

13. True or false: The air rushing out of a thunderstorm feels cold because it comes from high up in the troposphere, where the air is very cold. Explain your answer.

14. A neighbor says, "I don't care about all those other watches and warnings; the only thing that scares me is a tornado." Tell the neighbor what other kinds of small-scale winds can cause damage almost as severe as a tornado, and why that is so.

Observation Activities

1. What is a common small-scale wind where you live? Describe the dominant forces that cause this wind.
2. Turbulence is an important energy transfer mechanism. Describe ways that are not listed in the book in which you can observe the presence and effects of turbulence.

This rain cloud icon is your clue to go to the *Meteorology* Web site at http://physicalscience.jbpub.com/ackerman/meteorology/. Through animations, quizzes, web exercises, and more, you can explore in further detail many fascinating topics in meteorology.

13 Weather Forecasting

AFTER COMPLETING THIS CHAPTER, YOU SHOULD BE ABLE TO
- List and describe the various ways that people forecast the weather
- Discuss why numerical weather forecasts are needed
- Describe and explain the significance of the steps of a numerical weather forecast
- Explain why numerical forecasts are not perfect and the reasons that they are getting better

INTRODUCTION

"What will you do after college?" This question inspires excitement, but also dread, in many students. Your future is wonderful to ponder but difficult to predict. So many different variables can affect your decisions. So many twists and turns may lie ahead!

Similarly, the question, "What will the weather be tomorrow?" has captivated, but also vexed, meteorologists for decades. Weather forecasting is one of the most fascinating aspects of meteorology. However, an accurate forecast requires a thorough knowledge of all the variables of the atmosphere and also an understanding of how the atmosphere "twists and turns" and changes over time. Until the last century, scientists often assumed that precise weather forecasting was impossible.

It's not surprising, then, that the founder of modern weather forecasting was someone who could not easily answer the question, "What will you do after college?" Lewis Fry Richardson (pictured) graduated from Cambridge University in England in 1903, but he avoided majoring in any one branch of science. After college, he changed from one job to another and from one science to another. When he was 26 years old, Richardson's mother noted that he was having a "rather vacant year with disappointments"—an early version of a slacker! A few years later, Richardson became interested in numerical approaches to weather forecasting. The rest is history: Richardson created, from scratch, the methods of modern weather forecasting that are now used worldwide. His remarkable story will guide us through the intricacies of computer-based forecasting.

Before we get to Richardson, however, we will survey the various ways that weather forecasting can be done without a computer. From the dawn of civilization until the 1950s, humans had to rely on a variety of imperfect forecasting techniques. This was true during World War II, when a handful of meteorologists had to make the most important forecast in human history: whether or not to launch the "D-Day" invasion of Europe (photo at left). Democracy or dictatorship? For a brief moment in history, the answer depended on an accurate weather forecast. The meteorologists in 1944 used methods that you can use today, and we learn about them in this chapter, too.

METHODS OF FORECASTING BY PEOPLE

Folklore

Long before computers and The Weather Channel, humans needed weather forecasts. Farmers and sailors in particular needed to know if storms were approaching. Over time, various **folklore forecasts**, often in the form of short rhymes, were devised and passed down through the generations. Although memorable, the folklore forecasts are of uneven quality—some good, others laughably bad.

Some of the earliest folklore forecasts in Western culture, such as the following, come from ancient mariners:

> Red sky at night, sailor's delight
> Red sky at morning, sailor take warning.

This saying is fairly accurate. A clear western sky at sunset allows the Sun to shine through the atmosphere, its light reddening as a result of Rayleigh scattering (see Chapter 5) and then reflecting off clouds in the eastern sky. Clouds to the east usually move away; recall from Chapter 6, 7, and 10 that storms in the middle latitudes generally travel to the east under the influence of jet-stream winds. The reverse is true in the morning, when the red sunlight shines on storm clouds approaching from the west. However, this folklore doesn't work at all in overcast conditions or at tropical latitudes (see Chapter 8), where weather often moves from east to west.

Another famous folklore forecast focuses on a different optical effect: the halo. What does the presence of a halo mean? According to the soothsayers,

> If a ring forms around the moon,
> 'Twill rain or snow soon.

This is another relatively accurate saying. High ice-crystal clouds such as cirrostratus often occur in advance of a mid-latitude cyclone's warm front. Cirrostratus clouds, in turn, cause halos. So the halo may be the first sign of an approaching warm front and cyclone and thus bad weather—especially at night. (Notice in Chapter 10 that halo-making cirrostratus clouds were reported in the vicinity of the *Edmund Fitzgerald* wreck site some 36 hours before the shipwreck.) Furthermore, this saying also applies if the halo forms around the Sun.

There are limitations to this saying's accuracy, however. This folklore forecast requires the Moon (or Sun) to be up and visible through thin high clouds, conditions that are somewhat hit or miss. Also, not all warm fronts produce extensive regions of high clouds, and not all high clouds are the result of a warm front. Nevertheless, it seems to work: A dedicated amateur meteorologist in Britain in the 1800s observed 150 halos in 6 years and noted that nearly all of the halos preceded rain by 3 days or fewer. (However, it often rains every few days in Britain, so this statistic is less impressive than it initially appears.)

One famous forecast that *doesn't* work in the United States is Groundhog Day, known as Candlemas Day in medieval Europe:

> If Candlemas Day is bright and clear
> There'll be two winters in that year;
> But if Candlemas Day is mild or brings rain,
> Winter is gone and will not come again.

What the weather is at one location in Europe or Pennsylvania on February 2 tells us very little about the weather for the rest of the winter season. The popularity of Groundhog Day in the U.S. has much more to do with clever marketing than it does with forecast accuracy.

Although the Groundhog Day festivities don't help with weather forecasting, some accurate folklore forecasts do involve animals and plants that are weather sensitive. One example is the saying:

> When spiders' webs in air do fly
> The spell will soon be very dry.

This forecast seems to be well founded. Here is one possible explanation for why it works: Spiders want to catch insects in their webs, and insects will often breed after a rain. However, spiders don't want to have their webs covered with water; a sopping-wet spider web is easily detected, making it a lousy trap for insects. So spiders spin webs at the end of a period of rain as the air dries out—for example, after a cold frontal passage in spring. As you know from Chapter 10, high-pressure systems follow cold fronts and often lead to a spell of dry weather. The spider doesn't know meteorology, but its actions do fit into a predictable pattern of weather changes that humans can anticipate with some success.

As meteorology developed into a scientific field in the 19th century, folklore forecasts were gradually replaced with their modern equivalent: rules of thumb based on observations of clouds, barometric pressure, and winds. We have already encountered one such example in Chapter 10: FitzRoy's rhyme about fast pressure rises leading to strong winds behind a low-pressure system. FIGURE 13-1 shows a modern example, known as a "decision tree," that uses concepts such as pressure gradients and the lifted index from Chapters 6 and 11 to predict the probability of high winds over west Texas. BOX 13-1 includes several much simpler rules that you can use to make your own weather forecast when you are away from the TV, radio, or computer.

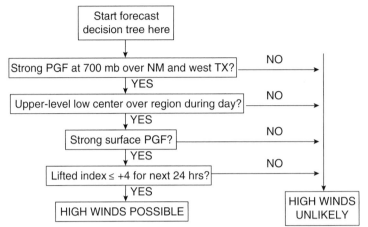

DECISION TREE FOR HIGH WINDS (sustained westerly component of 30 mph or greater) IN THE NORTHERN PART OF WEST TEXAS (mainly winter and spring for the first 12 and 24 hour fcst periods)

FIGURE 13-1 A modern rule of thumb "decision tree" for forecasting the occurrence of high winds in west Texas, adapted from work by National Weather Service meteorologists Greg Murdoch and Jim Deberry. The logic behind this decision tree is as follows: Strong pressure (or height) gradients cause strong winds. Meanwhile, low pressure aloft and a relatively low lifted index ensure that strong winds above the surface can be easily dragged down to the ground during the day, when solar heating stirs up the atmosphere into turbulent eddies. When all of these factors are present, high wind gusts are likely. (*Source*: Murdoch, G. and Deberry, J., Aspects of Significant West Wind Events Across West Texas and Southeast New Mexico. NOAA, 1999. Retrieved February 28, 2011, from http://www.srh.noaa.gov/maf/?n=research_sigwestwinds.)

Persistence and Climatology

Other forecasting methods rely on the continuity of weather from one day to the next or from one year to the next. A **persistence forecast** is simple: *The weather you are having now will be the weather you have later.* The accuracy of this forecast is very much dependent on where you are, the type of upper-troposphere winds that exist over your location, and how far in advance of a forecast you want.

For example, a 24-hour persistence forecast works pretty well in places like Hawaii or southern Florida. These locations are rarely under the direct influence of a strong jet stream. However, persistence can be a terrible forecast where the jet stream is strong. In these regions, growing low-pressure systems cause rapidly changing weather. Britain and the northeastern U.S. are locations where persistence forecasts for more than a few hours in the future are not very good. The reverse is true in situations where the upper-level winds are "stuck" in a "blocking" pattern (see Chapter 7) for several weeks. In these cases—for example, during a drought or the Flood of 1993 in the midwestern U.S.—a persistence forecast may be fairly accurate for days or even weeks at a time.

A **climatology forecast** relies on the observation that weather for a particular day at a location doesn't change much from one *year* to the next. As a result, a climatology forecast is this: the weather you will have will be the same as the long-term average of weather for that day or month. The most obvious climatology forecast for the Northern Hemisphere middle latitudes is "cold in December, warm in July." You don't need to be a meteorologist to make that forecast! Its success derives from the fact that weather, although changeable, is strongly determined by the tilt of the Earth and the global energy budget discussed in Chapter 2.

Climatology forecasts can be quite specific. A favorite is the "White Christmas" forecast: What is the percent chance that snow will be on the ground at a particular location on

Box 13-1 Personal Weather Forecasting

Would you like to be a weather forecaster? The following is a table that can help you forecast the weather based only on what you can see: No weather instruments are necessary! Both "typical" and less likely but "possible" forecasts are included. Many of them are actually forecasts based on the Norwegian cyclone model in Chapter 10. For more precise forecasts for specific regions, you need the guidance of numerical models of the atmosphere, which are discussed in this chapter.

Clouds/ Precipitation	Cloud Movement Toward	Surface Wind From	Typical Forecast	Possible Forecast
Cirrus shield covering the sky	E or NE	NE	Cloudy with a chance of precipitation within 2 days	Continued fair weather or major storm coming
Wispy cirrus	SE	NW	Fair weather and unseasonably cool	Bitter cold at night if clouds go away
Cirrocumulus in bands	E or NE	E or NE	Changing weather soon	Major intensifying storm coming
Nimbostratus and stratus, rain and fog	NE	E or SE	Turning partly cloudy and warmer, rain ending	May warm up even at night
(Same as above)	NE	NE	Windy with cold rain	Rain turning to ice or snow if surface air cold and dry enough
Towering cumulus to the west	E or NE	S or SW	Thunderstorms soon; then clearing and turning colder	Severe weather possible soon
A few puffy cumulus	Stationary	Light and variable	Partly cloudy; cold in winter, hot in summer	Scattered thunder- storms if very clear (cold, dry) above the clouds
Stratocumulus with flat bases	SE	NW and gusty	Winds dying down at sundown and cool	Increasing high clouds by morning if next storm approaches
Hazy and humid	Stationary	Calm	Hot; unseasonably warm at night	More of the same for a week or longer
Clear with new snow on the ground	None	Light and northerly	Rapidly dropping temperatures after sunset	Record low temperatures by morning
Cumulonimbus with continuous lightning in the distance	Anvil top pointed just to your right	Into the cloud and gusty	Thunderstorm with heavy rain soon	Severe weather imminent with hail, a tornado, or both
Lenticular cloud near mountains	Stationary	Across the mountains	Partly cloudy; high winds downslope of the mountains	UFO reports!

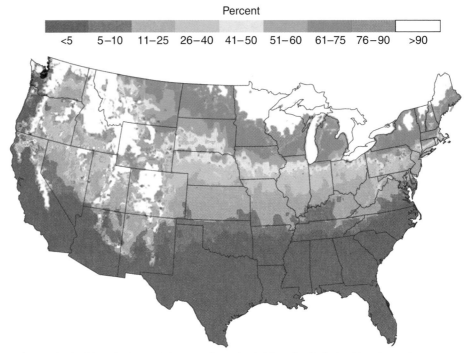

Percent

| <5 | 5–10 | 11–25 | 26–40 | 41–50 | 51–60 | 61–75 | 76–90 | >90 |

FIGURE 13-2 Climatological forecast probability of a "white Christmas" across the lower 48 U.S. (*Source*: NOAA.)

December 25 (**FIGURE 13-2**)? This forecast depends on the past 30 years of weather observations on Christmas Day across the U.S. It would change somewhat if more, fewer, or different years were used in the climatology.

Today's numerical forecast methods still use climatological statistics as a "reality check," as we will discuss later on in this chapter. The use of the statistics ensures that the computer models aren't going off the deep end, climatologically speaking.

▇ Trend and Analog

We know that persistence forecasts are doomed to failure, ultimately, because the weather does change. An approaching cyclone brings precipitation and falling barometric pressure. Thunderstorms sprout ahead of a cold front. A **trend forecast** says that weather will change based on the movement and approach of weather features such as an extratropical cyclone. However, a trend forecast assumes that although the weather features are moving they are not otherwise changing (i.e., are "steady-state") in speed, size, intensity, and direction of movement.

Benjamin Franklin pioneered the concept of trend forecasting in 1743. Clouds ruined an eclipse for Ben in Philadelphia, but his brother in Boston saw the eclipse clearly. The clouds did not reach Boston until many hours later. Franklin discovered that "nor'easter" cyclones frequently moved up the East Coast, a trend that could be used to forecast the weather (**FIGURE 13-3**).

Trend forecasts work well for a period of several hours. Forecasting for such a brief period is called **nowcasting**. This is because large-scale weather systems such as cyclones don't change very much over a short time period. However, in Chapter 10 we learned that cyclones have definite life cycles and can change remarkably in size, speed, and intensity from one day to the next. Therefore, the accuracy of trend forecasts declines quickly when they are made for longer than a few hours ahead. Modern nowcasting combines trends with high-resolution numerical forecast model information.

The **analog forecast** also acknowledges that weather changes, but unlike the trend method, it assumes that weather patterns can evolve with time. The key—and flawed—assumption for the analog forecast is that *history repeats itself,* meteorologically speaking. The analog forecaster's task is to locate the date in history when the weather is a nearly perfect match, or *analog,* to

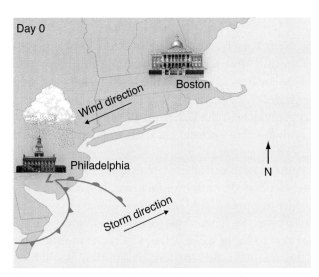
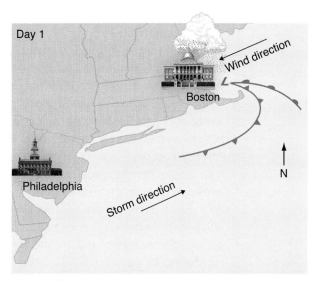

FIGURE 13-3 A trend forecast based on the assumption that a mid-latitude cyclone moves, unchanging, up the East Coast. (*Adapted from*: Gedzelman, S. D., *The Science and Wonders of the Atmosphere*. John Wiley & Sons: [1980] p. 14.)

today's weather. Then the analog forecast for tomorrow is simple: Whatever happened in the day after the analog will be the weather for tomorrow. The forecast for the day after tomorrow is whatever happened in the second day after the analog, and so forth.

Analog forecasting therefore requires many years of weather maps and an efficient way to compare one map with another. One approach, borrowed from personality testing in psychology, is to categorize the weather into a small number of **weather types**. If today's weather has a "type A personality" (**FIGURE 13-4**), then you can create a multiday forecast based on how the weather evolved in type A cases. If today's weather isn't type A, then you try to match it to

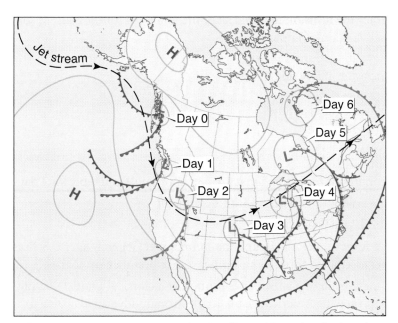

FIGURE 13-4 North American Weather "type A" used in the heyday of analog forecasting in the 1940s. The goal of analog forecasting is to identify a weather type that matches today's weather and to use the left-to-right sequence of weather in that type as a forecast for the weather for the next several days. Compare this weather type with the life cycle of the *Edmund Fitzgerald* cyclone in Chapter 10. (Adapted from Krick, I., and Fleming, R. *Sun, Sea, and Sky: Weather in Our World and in Our Lives.* Lippincott, 1954.)

one of the other types (six main types were identified in the 1940s and 1950s) and hope for a perfect match.

But does meteorological history repeat itself? Very rarely, although we saw in Chapter 10 that strong Great Lakes cyclones do recur around November 10, and type A does resemble the evolution of the *Fitzgerald* storm of 1975. For how long could the historical analog help you forecast the weather—a day? Possibly. A week? Unlikely. A month? Impossible. Even so, meteorologists used analog forecasting by types widely in the U.S. from 1935 until about 1950. Overblown claims for its long-term accuracy then led to a strong backlash among professional meteorologists.

Researchers have returned to weather typing via sophisticated statistical approaches. Today we know that certain weather patterns in widely separated locations—for example, Alaska and the U.S. East Coast—vary in relation to one another in a predictable way during global phenomena such as El Niño-Southern Oscillation (see Chapter 8). These long-distance relationships are called *teleconnections* and are used by forecasters today to make general forecasts months into the future. In short-term forecasting, an analog method called "pattern recognition" is still used by weather forecasters to supplement today's computerized methods. In the end, however, the complexities of weather, like human personalities and the twists and turns of human history, defy simple categorization.

A REAL LIFE-OR-DEATH FORECAST: D-DAY, JUNE 1944

It was the largest military invasion of all time. Nearly 160,000 Allied soldiers and support personnel, over 5000 ships and 13,000 airplanes assembled along the southern coast of England during World War II in the spring of 1944. Their goal was to cross the narrow, often stormy English Channel into Nazi-occupied France. The decision of when to launch the invasion was largely in the hands of meteorologists.

This forecast could not rely on folklore; it was one-of-a-kind. Ships, boats, aircraft, and paratroopers all required special weather conditions. The Allied commanders determined that suitable weather for the invasion required all of the following for the coast of southern England and the Normandy region of France:

1. Initial invasion around sunrise
2. Initial invasion at low tide
3. Nearly clear skies
4. At least 3 miles of visibility
5. Close to a full Moon
6. Relatively light winds
7. Nonstormy seas
8. Good conditions persisting for at least 36 hours and preferably for 4 days

And, to make the task of forecasting ever more difficult:

9. At least 2 days' advance forecast of these conditions.

To create this forecast, the Allied supreme commander General (later U.S. President) Dwight "Ike" Eisenhower assembled three teams of the world's best meteorologists:

- Caltech meteorology professor (and one-time classical pianist) Irving Krick and his protégés formed the American team. Krick strongly advocated analog forecasting.
- An "odd couple" led the British Royal Air Force team. Sverre Petterssen of the Bergen School was an expert in the science of air masses, cyclones, and upper air patterns. C. K. M. Douglas made his forecasts by looking at weather maps and guessing what would happen next by pattern recognition and his intuition.
- Cambridge engineer Geoffrey Wolfe and Rhodes scholar mathematician George Hogben headed the third team, from the British Royal Navy. Their expertise was the forecasting of waves on the English Channel.

The three forecast teams devised separate forecasts using their diverse methods. Over the phone, they hashed out their differences with their leader, British meteorologist James Stagg. Stagg molded the three forecasts into one consensus forecast and presented it to Eisenhower.

The immense Allied invasion force came together in England in late spring of 1944. Eisenhower asked his meteorologists a crucial initial question: What were the odds, month by month, that the weather required for the invasion would actually occur based on past experience? The meteorologists made a fateful climatology forecast. The odds were 24 to 1 in May, 13 to 1 in June, and 33 to 1 in July. The reasons for the long odds: Low tide with a full Moon is a fairly rare event. Also, persistent calm weather because of "blocking highs" is more common in western Europe in late spring than in summer.

Based on these odds and the need for more time to assemble the invasion force, Eisenhower settled on early June, probably June 5, as the "D-Day" of invasion. Because of the requirement of a low tide around dawn, the only alternative would be 2 weeks later. After that, the climatological odds for success would drop quickly.

However, just as May turned to June, placid weather turned stormy over Europe. A series of strong mid-latitude cyclones developed and deepened (**FIGURE 13-5a–c**), a winter-like pattern not seen in the Atlantic in June in the past 40 years! Fronts crossed the English Channel every day. Climatology, persistence, and even analog forecasts did not promise much insight into this weather pattern.

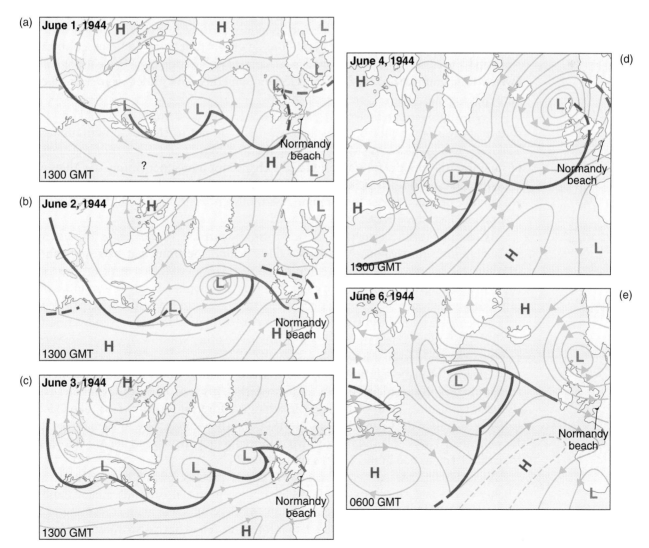

FIGURE 13-5 Allied surface weather maps for the North Atlantic at 1300 Greenwich time for June 1 (a), June 2 (b), June 3 (c), June 4 (d), and at 0600 Greenwich time on June 6 (e), the dawn of D-Day. Question marks indicate where the Allies did not have adequate weather observations. (Modified from Shaw, Roger H. and Innes, William editors. *Some Meteorological Aspects of the D-Day Invasion of Europe*, 1984. © American Meteorological Society. Reprinted with permission.)

What was the all-important forecast for June 5? The three teams of forecasters debated this point for a week. Krick's team consistently found historical analogs that called for acceptable weather that day. Petterssen's theories and Douglas's intuition just as consistently indicated deteriorating weather. By June 4, the Royal Navy forecasters also foresaw weather unfit for the invasion. Stagg presented Eisenhower with a bleak, windy forecast.

Early on June 4, Eisenhower canceled the invasion for the next day—even though the weather overhead was clear and calm! The forecasters were right; the 5th was a windy, rainy mess in the Channel, a potential disaster for the invaders. Meanwhile, the largest invasion force in history waited anxiously. Further delays meant lost time and lost lives.

As if on cue, the weather suddenly changed for the better. Just a few hours after the decision to cancel the June 5th invasion, one of Stagg's deputies telephoned him. He told Stagg, "A cold front has turned up from somewhere and is already through the Irish west coast" (Figure 13-5d). Defying the Norwegian model and trend and analog forecasts, several lows had merged into a giant over northern Britain. Their combined winds were pushing cold air and better weather rapidly southeastward toward the English Channel. Meanwhile, a large low over the Atlantic was slowing down.

Would there be a gap between the two storm systems? The teams hastily drew up new forecasts. The analog forecasters said the cold front would rush through the Channel rapidly, but why believe them when their consistent forecast for good weather on June 5th was a failure? However, both the Royal Air Force and Royal Navy teams concurred. For once, the different forecast approaches all converged on a single forecast. It was this: There *might* be *just enough* time of *marginally acceptable* weather for Eisenhower's force to cross the Channel successfully.

Eisenhower seized the initiative. Before dawn on June 5, he gave the go-ahead for the Normandy invasion. Ironically, rain and gales battered his location as he made the fateful decision. Twenty-four hours earlier, he had canceled the same invasion under clear skies. Such was his faith in his meteorologists' forecasting ability.

Meanwhile, German meteorologists also noticed the cold front that would give the Allies a chance at invasion on June 6th, but their warnings did not reach the German military leaders. The Nazis had spent a tense May expecting the invasion. Then rainy weather at the beginning of June, and the gloomy forecasts signaled a chance for a breather. Surely the Allies would not cross the Channel until later in the month! Field Marshal Rommel left the front to be with his wife on her birthday. Other high-ranking Nazi officers went hunting and to a Paris nightclub. The revised German forecast went unheeded.

As a result, the Allies caught the German army by surprise. June 6 was, as forecast, marginally acceptable invasion weather. Northwest winds behind the front that had swept through the previous day (Figure 13-5e) churned up rough seas. However, the element of surprise gave the Allies the edge against one of history's most formidable armies. Partly as a result, World War II in Europe ended earlier and more definitively than it would have otherwise. Sverre Petterssen hailed the forecast for the gutsy Allied invasion of Normandy as "meteorology's finest hour."

The forecasters' celebration was short lived. The other window of opportunity in June for D-Day was June 17 to 20. After the invasion, the Allies used this period to resupply their troops, now engaged in fierce fighting along the coast of France. What was the forecast? The Allied meteorologists all agreed this time: excellent weather. Instead, the English Channel experienced its worst windstorm in 40 years. The climatological forecast was for a less than 1% chance of gale-force winds in the Channel in June. Instead, gales blew along the entire length of the Channel for more than 3 days. It is said that this windstorm did more damage to the Allied forces in 4 days than the Germans had in the 2 weeks since D-Day!

A weather forecast had helped win World War II. Even so, it was obvious that there had to be a better way to make forecasts. Folklore, climatology, persistence, trends, analogs, and intuition were not enough. Even the Norwegian cyclone model couldn't explain everything.

There *was* a better way to make weather forecasts. It had been discovered and tried out over a quarter-century earlier, in wartime, in France, and then it had been forgotten. This method, numerical weather prediction, would revolutionize meteorology and turn weather forecasting into a true science. Its prophetic discoverer: Lewis Fry Richardson. We now turn to his story to understand how modern weather forecasts are made.

L. F. RICHARDSON AND THE DAWN OF NUMERICAL WEATHER FORECASTING

Richardson's inspiration came from the father of modern meteorology, Vilhelm Bjerknes. In the early 1900s Bjerknes proclaimed, "The problem of accurate [forecasting] that was solved for astronomy centuries ago must now be attacked in all earnest for meteorology." Astronomers can forecast eclipses years in advance. Why can't we forecast tomorrow's weather? Bjerknes' answer: Meteorologists don't yet possess the two basic ingredients for an accurate forecast. They are as follows:

1. The current conditions of the atmosphere over a wide area, known as **initial conditions**.
2. Mathematical calculation of the atmosphere's future conditions, using the physical laws that govern the atmosphere's changes. In a nearly exact form, they are called the **primitive equations** because they are so complicated that they cannot be solved with algebra and pencil and paper.

Bjerknes admitted that this was a "problem of huge dimensions . . . the calculations must require a preposterously long time." He turned his attention to air mass and cyclone studies instead, revolutionizing meteorology, but his grand plan for a science of weather forecasting was left undone.

Into this void stepped Richardson. During his "slacker" period, Richardson gained on-the-job experience reducing difficult equations to lots and lots of simple arithmetic formulas that could be solved with calculators. The technique of approximating tough real-world problems with numbers is called **numerical modeling**. The numerical formulas used are called a **model**, just as a realistic approximation of a train is called a "model railroad" (see **BOX 13-2** for a more in-depth explanation of this topic).

Then in 1911 a vision came to Richardson. In this "fantasy" he saw a vast circular room like a theater. In this room, people at desks representing every part of the globe were solving the primitive equations numerically. It was a kind of meteorological orchestra, with numbers replacing notes. The combined effort was not a symphony but instead a forecast.

This visual metaphor enthralled Richardson. He would approximate the difficult equations of the atmosphere with a numerical model. All Richardson needed were some data to use for initial conditions, the primitive equations, and "a preposterously long time" to run his numerical model and make a forecast. Or, because the forecast would take a very long time to do by hand, he planned to do a "hind"-cast of a weather situation that had already happened, as a test case for his approach.

World War I gave Richardson his opportunity. He was a member of the Quaker religion and therefore a pacifist, morally opposed to war. Married and in his mid 30s, Richardson could have easily dodged the draft in England. Instead, in 1916, he registered as a conscientious objector to the war. For the next 2 years, he voluntarily drove an ambulance in war-torn France, caring for the wounded while bombs exploded nearby.

During lulls in the fierce fighting, Richardson would fiddle with his equations and his methods of solving them. On a "heap of hay" he attempted the first-ever mathematical weather forecast as Bjerknes had envisioned it. The task was an immense undertaking for one person, so he limited his "forecast" to a prediction of wind and barometric pressure at a point in Germany for the morning of May 20, 1910, using data gathered across Europe earlier that same morning. Even with such a limited "forecast," it likely took Richardson up to 10 hours a week for 2 whole years to compute it!

What did Richardson get for all his troubles? A *really* bad forecast. According to his calculations, the pressure at one spot in Germany (**FIGURE 13-6**) was supposed to change 145 millibars in just 6 hours. That's the difference between a major hurricane and an intense high-pressure system! Because his "forecast" was for an actual day many years earlier, Richardson could go back and compare his result with reality. May 20, 1910, was tranquil, and the barometric pressure in Germany

Box 13-2 Modeling the Equations of the Air

The "primitive equations" of the atmosphere are actually very easy to say in words. They are the conservation principles of momentum, mass, energy, and moisture, combined with the Ideal Gas Law from chemistry. What is hard about these equations is solving them for the variables meteorologists care about—the wind or the temperature.

For example, the equation governing the change in the west-to-east wind u can be written using the abbreviations from Chapter 6 as follows:

$$\text{Change of } u \text{ at a point over time} = \text{Advection of } u + \text{PGF} + \text{CF} + \text{FF}$$

This equation is all tangled up. Here's why: What we want to solve for, u, isn't all by itself in one term like in a simple high-school algebra problem. The west–east wind is on the left-hand side. It's also hiding in the advection term. It affects the north–south wind in the Coriolis force term. Also it's embedded in the poorly understood effects of turbulent friction. This messy situation, in which what you're solving for is also part of the answer for it, is called "nonlinearity." There's no simple solution to this equation; it's the mathematical equivalent of trying to run while standing on your own shoelaces. This is one of many reasons why computers, not pencil and paper, are used to make forecasts.

But computers don't do algebra easily. Everything is 0s and 1s to them—numbers. A way must be found to translate that word equation into pure numbers. One method for doing so is called "finite-difference approximations." This is the technique that L. F. Richardson used in his first-ever numerical weather forecast.

A finite-difference approximation works a lot like a strobe light at a dance hall or disco. A strobe light flashes on and off rapidly; a dancer sees snapshots of his or her surroundings in the intermittent light. Similarly, a finite-difference approximation "sees" what's going on for an instant . . . and then sees nothing . . . and then sees things for another instant . . . and then nothing . . . and so on, as the model marches forward in time.

For example, the left-hand side of our equation can be written as this finite-difference approximation:

$$\text{Change of } u \text{ at a point over time is approximately equal to } \left(u_{\text{now}} - u_{\text{earlier}}\right)$$

$$\div \text{ the time elapsed between now and earlier}$$

Computers can subtract and divide at lightning speeds, so this kind of approximation suits them perfectly. The approximation lies in the fact that a "finite" or measurable amount of time may have elapsed between *now* and *earlier*. This isn't the same as the instantaneous change in u at a point, which is what the original primitive equations require, but it is close enough in many instances.

Similarly, the advection term involves changes in wind over distance, and the finite-difference approach replaces the change with the following:

$$\left(u_{\text{here}} - u_{\text{there}}\right) \div \text{distance from here to there}$$

In this case, the "strobe effect" is in space, not time; the model "sees" and calculates weather variables at some places but not others.

The great advantage of finite-differencing is that nasty math is reduced to lots and lots of arithmetic calculations. Computers can do these effortlessly, although the drudgery would drive a human being to distraction.

The big drawback is that a forecast based on the finite-difference approximation is *not* the same thing, exactly, as the original problem. It is a model of the real thing, just as a robot can be a model of a human being. Robots can short-circuit or blow up; humans can't. In other words, models can have special problems that the real-world situation doesn't.

In the case of modeling the atmosphere, the big difference between the model and reality is that the atmosphere is a fluid that is always everywhere in time and space. In the model, the finite-difference calculations are only made *here* and *there* and *now* and *earlier*. This strobe effect is the Achilles' heel of modeling; we explore it further in Box 13-4.

had changed a millibar at most, not 145. His forecast fantasy had flopped; his approach was apparently a failure.

Richardson put a good face on his failed test-case forecast and called it "a fairly correct deduction from . . . somewhat un-natural initial [conditions]." Meteorologists generally applauded his Herculean effort, but no one followed in his footsteps. As we saw earlier, World War II forecasts used every technique *except* Richardson's. By that time, Richardson himself had quit meteorology on moral grounds because the science was being used for military purposes.

Richardson's forecast was not a total failure. Richardson, as usual (BOX 13-3), was simply too many decades ahead of his time. As we'll see, a combination of meteorological and computer advances were needed to turn his failure into success. The invention of the electronic computer during World War II revived interest in numerical weather forecasting. The modern computer could shave the "preposterously long time" of forecast-making from years to hours or less. In 1950, a simplified computerized next-day forecast was made in 24 hours. This meant that forecasts could be made for real-life situations, not just test cases. The era of real-time forecasting by numerical modeling had begun! By 1966, advances in computers and meteorology theory finally allowed meteorologists to use Richardson's approach on an everyday basis.

Today, weather forecasts worldwide follow a numerical process that Richardson would instantly recognize as his own—with a few key improvements.[1] In the following section, we compare today's methods versus Richardson's to understand the process better. Along the way, we discover why Richardson's forecast failed.

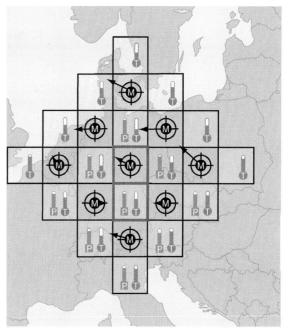

FIGURE 13-6 A pictorial representation of Richardson's famous forecast for May 20, 1910. The boxes on the map divide Europe into a grid that facilitates the numerical forecast. Boxes with an "M" in them are where Richardson's model used observations of momentum (i.e., wind); "P" and "T" refer to pressure and temperature observations, respectively. Richardson used this data to make his erroneous pressure forecast for the red box in the center near Munich, Germany. (Adapted from Brian Hayes, "The Weatherman," *American Scientist*, 89:[2001].)

THE NUMERICAL WEATHER PREDICTION PROCESS: THEN AND NOW

Step 1: Weather Observations

Richardson made a "forecast" for May 20, 1910, because it was one of the few days in history up to that time when there had been a coordinated set of upper-air observations. Even so, he wasn't working with much data—just a few weather balloon reports.

A numerical forecast is only as accurate as the observations that go into the forecast at the beginning of its run, the "initial conditions." Because weather moves from one place to another rapidly, tomorrow's weather is influenced by today's weather far upstream, and next week's weather can be affected by today's weather a continent away. For this reason, forecasters must have lots of data worldwide. Today we have global sources of data of many different types to give the forecast the best possible start.

FIGURE 13-7 shows the diverse types and the spatial distribution of weather data for a typical 6-hour period used by a global weather forecast model in May 2010. Surface observations, radiosondes, and satellite measurements supply most of the data used for model initial conditions. Weather satellites, in particular, now provide more than 99% of all data used in weather forecast models. Airlines are also now providing forecasters with over 100,000 observations per day of mostly upper-tropospheric data along major flight paths (FIGURE 13-8).

---

[1]The very latest numerical models actually compress and combine the steps that follow. To understand the overall approach, however, we keep each step distinct—as was the case for the earlier generations of numerical weather prediction models.

Box 13-3 L. F. Richardson, Pioneer and Prophet

The hero of this chapter, L. F. Richardson, left an eternal mark on science by doggedly pursuing simple questions with difficult solutions. Early in his career, Richardson learned numerical solution methods by studying "unsexy" problems. A prime example is the flow of heat through peat bogs (swamps). A Cambridge University friend was shocked: "*Dried peat*?" Richardson replied, "Just peat." Silly as it sounded, it was the ideal training ground for Richardson's trailblazing work on numerical weather prediction a few years later.

Around the time of his now-famous forecast, Richardson also became interested in turbulence (see Chapter 12)—another phenomenon that is easy to see but difficult to explain. Today, the onset of atmospheric turbulence is predicted using a special parameter known as the "Richardson number," in honor of his pioneering efforts on this subject.

Richardson's work on turbulence garnered interest among those wanting to study the movement of poison gas in the atmosphere. His pacifist beliefs offended, Richardson promptly quit meteorology. He went back to school and, at the age of 48, earned a second bachelor's degree in psychology. He studied human perception, gradually becoming interested in the psychology of war.

In the years during and after World War II, Richardson combined his mathematical prowess, his understanding of complex phenomena from meteorology, and his Quaker pacifism. He tried to quantify and predict (and thus avoid) war in much the same way that he had forecast the weather, with data and equations. The political scientists of his era were completely befuddled at this elderly meteorologist with his equations of war (below, a photograph from a scientific meeting in 1949; note the sleeping woman and the exasperated man in the row in front of Richardson!). Richardson encountered great difficulty getting any attention, or even a publisher, for his laborious calculations.

Yet Richardson was, as usual, on the right track. Decades after his death in 1953, scholars realized the worth of his approach. One short paper of his was rediscovered; it was titled "Could an arms-race end without fighting?" In the nuclear Cold War days of 1951, Richardson's equations said, "Yes, without a shot being fired," if one side outspent the other on armaments and the weaker nation bankrupted itself. Almost 40 years later, the Berlin Wall came down, some say, as an outspent Soviet Union relinquished power quietly in the face of American military buildups. Coincidence—or prophecy?

In the 1960s, a mathematician named Benoit Mandelbrot stumbled across an obscure paper of Richardson's. Its simple subject: How long is a coastline? What a simple question—that is, until you realize that the shorter a ruler you use, the longer a coastline is. Mandelbrot extended Richardson's work and discovered that the length of the coastline is intimately related to "fractal geometry," an aspect of chaos theory.

Meanwhile, in Boston, MIT meteorologist Ed Lorenz, using a numerical model based on Richardson's ideas, discovered "sensitive dependence on initial conditions." This, too, is a hallmark of chaos and is discussed at the end of this chapter. Together, Lorenz and Mandelbrot, intellectual heirs of Richardson, brought chaos theory to the forefront of 20th century science (not to mention the movie *Jurassic Park*).

L. F. Richardson was so far ahead of his time that he won no major scientific awards or prizes while he was alive. He was the antithesis of a scientific prodigy, doing his best work after the age of thirty. For much of his life he taught in obscure universities. He was not an inspiring lecturer (as the photo suggests). He did his research "on the side." He was a loner and a dreamer. Today, however, Richardson's name adorns meteorological prizes, peace-studies foundations, and fundamentals of mathematics and meteorology. This pioneer of modern meteorology was also one of the most creative and visionary scientists of the entire 20th century.

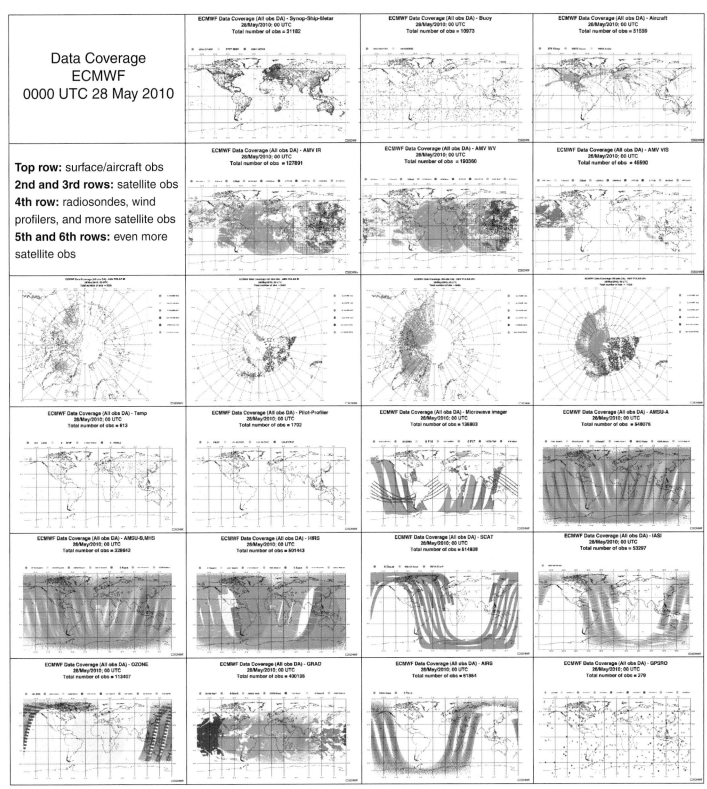

FIGURE 13-7 Weather data gathered for the 0000 UTC 28 May 2010 forecast made by the ECMWF numerical weather prediction model. (With thanks to the European Centre for Medium-Range Weather Forecasts [ECMWF] for allowing the use of its data.)

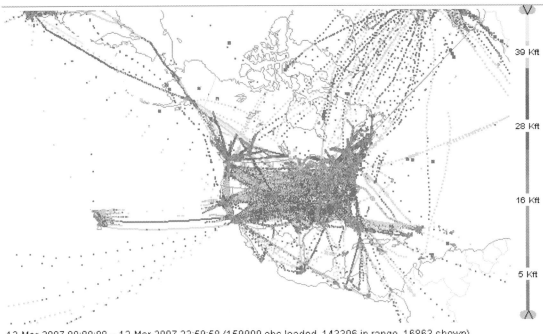

FIGURE 13-8 Graphic showing the distribution of nearly 160,000 aircraft weather data observations gathered by wide-bodied jets over North America on March 12, 2007. Blue colors indicate flight observations in the upper troposphere or lower stratosphere; red colors indicate observations nearer to the ground as the plane was taking off or landing. (Courtesy of Stan Benjamin and William Moninger, NOAA Earth Systems Research Laboratory/Global Systems Division.)

This vast and continuous data-collection process is overseen by the World Meteorological Organization and relayed worldwide. In the U.S., the National Centers for Environmental Prediction (NCEP) receive the data for use in their forecast models.

■ Step 2: Data Assimilation

To do their work, most numerical models look at the atmosphere as a series of boxes. In the middle of each box is a point for which the model actually calculates weather variables and makes forecasts. The result of this three-dimensional boxing-up of the atmosphere is known as the **grid**. The point in the middle is the **gridpoint**, and the distance between one point and another is called the **grid spacing**. FIGURE 13-9 illustrates this "gridding" of the atmosphere in a model. These models are called, appropriately enough, **gridpoint models**.

Gridpoint models of the atmosphere can get fussy when the data in the initial conditions are not obtained at exactly the location of the grid points. The observations were sparse and irregularly spaced in Richardson's test case. He needed data at the centers of the boxes in Figure 13-6, but instead, his observations were at randomly scattered cities. The process of creating an evenly spaced data set from irregularly spaced observations is called **interpolation**. It was a key part of Richardson's forecast and remains so today for some models.

Richardson then inserted the interpolated data into his numerical model. Observed data can be "bumpy" and out of balance (see Chapter 6).

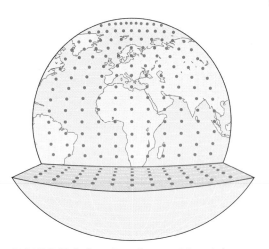

FIGURE 13-9 Dividing the world up into boxes for use with a gridpoint model. Forecasts and calculations are made only at the gridpoints, which are depicted as dots on this diagram. (*Source*: CSIRO.)

Observations are likely to include little gusts and swirls that do not reflect the large-scale weather. Putting this bumpy data into a forecast model is a lot like driving a sport-utility vehicle with no shock absorbers over a bumpy road. The variables predicted by the forecast model, just like the SUV, can bounce around out of control. As a result, the forecast can, like an SUV, "crash" and become useless. No one knew this in Richardson's day, however, and so he inserted out-of-balance data into his model and ended up with a really bad forecast.

Today, meteorologists know that a crucial step in forecasting is to take the unbalanced "shocks" out of the data. This is called **data initialization** and can be thought of as the mathematical equivalent of shock absorbers. The goal of data initialization is essentially to achieve balance in the data—in this case, to mutually balance wind and pressure, just as we discussed with respect to geostrophic balance in Chapter 6—before those data enter the model.

The multiple jobs of interpolating and balancing the data for use in numerical models are collectively called **data assimilation**. For those who like food more than SUVs, this step can be summarized metaphorically in the following way: Meteorologists use data assimilation to "cook" the raw data they feed into the numerical model so that the forecast doesn't get poisoned.

How important is data assimilation for making accurate forecasts? Irish meteorologist Peter Lynch re-ran Richardson's forecast on a modern computer, using the same initial data and the same equations. The only difference: Lynch did two runs, one without any data assimilation step (just like Richardson) and one in which he inserted a data assimilation step. The result? Lynch's assimilated forecast predicted a very accurate pressure change of about 1 millibar, not Richardson's ridiculous result of 145 mb! A snapshot of these two forecasts at the 1-hour mark is shown in FIGURES 13-10 and 13-11. Without data assimilation, the forecast is a big mess of

Pressure, 20 May 1910, 7Z + 01, Sea Level

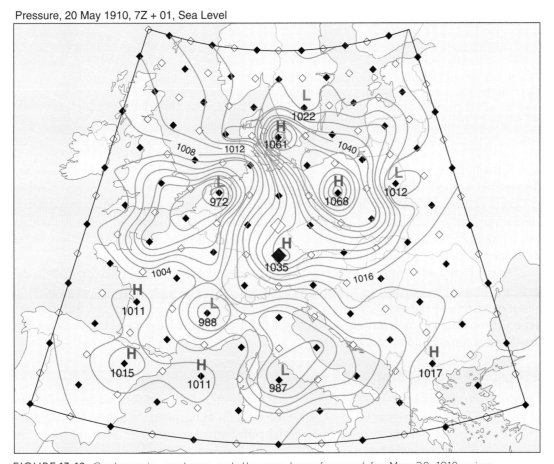

FIGURE 13-10 Garbage in, garbage out: the one-hour forecast for May 20, 1910 using Richardson's model with initial data that was unassimilated and out of balance. (The diamonds on the graph correspond to the gridpoints of Richardson's model, as shown in Figure 13-6.) Notice all of the unrealistic bull's-eyes of high and low pressure, causing a tight pressure gradient near Munich (center of diagram) that was not observed on that day. (Copyright © 2006 P. Lynch. Reprinted with permission of Cambridge University Press.)

Pressure, 20 May 1910, 7Z + 01, Sea Level

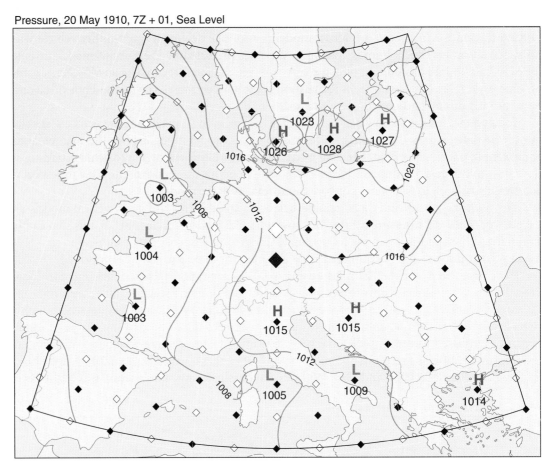

FIGURE 13-11 As in Figure 13-10, except that the initial data were balanced via data assimilation. The bull's-eyes have been eliminated and the pressure gradient near Munich (center of diagram) is small, as was actually observed. (Copyright © 2006 P. Lynch. Reprinted with permission of Cambridge University Press.)

unrealistic bull's-eyes of intense high and low pressure (Figure 13-10); with data assimilation, the weather map is realistic and accurate (Figure 13-11).

So, Richardson wasn't entirely mistaken; his forecast was, as he claimed, poisoned by "unnatural" initial conditions (although not quite in the way that he surmised). Also, Richardson was typically ahead of his time: the mathematical "shock absorbers" of data assimilation weren't developed in meteorology until about 40 years after his World War I-era forecast.

More recently, the successful assimilation of satellite data into numerical forecast models since the mid 1990s has led to marked improvements in forecasts for regions that suffer from a lack of surface or radiosonde observations, such as the Southern Hemisphere. Without data assimilation, satellite observations would be of much less benefit to weather forecasting, and weather forecasts would be less accurate.

Step 3: Forecast Model Integration

Once the data are "cooked," the model's formulas "ingest" it, and the real forecast begins. Millions of arithmetic calculations may be made to get forecasts for atmospheric variables at a later time. Then these forecasts are used as the new initial conditions for a forecast at a still later time. The whole process piggybacks on itself and marches forward in time. Mathematicians call this *integration,* and it is the same concept as integration in college calculus. This step puts together the two key ingredients of the forecast: the observed data and the model's formulas, which are approximations to the actual primitive equations describing the atmosphere.

Richardson solved his formulas by hand and obtained his erroneous result in this step. Today, computers have replaced Richardson's manual calculator. The advances in computing in the past 60 years have been phenomenal. Forecasts that took 24 hours on a 30-ton computer in 1950 can now be performed in less than 1 second on a cellphone!

However, today's models are also much more detailed than Richardson's, requiring far more computing power than a cellphone or personal computer. For example, a 24-hour global forecast for just five weather variables can take more than 1 trillion calculations.

The end result is that numerical weather forecast models always use the fastest supercomputers in the world, "pushing the envelope" of computing as much as any other science. For example, in 2010, the NWS installed new supercomputers at NCEP capable of making 69.7 trillion calculations *per second* (FIGURE 13-12). How far have we come since Richardson's day? It would take a person with a calculator over 2 million *years* to make as many calculations as the new NCEP supercomputers can perform in 1 *second*.

All of this computational power is required because of how the model simulates real life. In this step, the physical distance between the interpolated data points becomes very important. The smaller the grid spacing, the easier it is for the model to "see," or resolve, small-scale phenomena. This is called good, or fine, **resolution**. Wide spacing between grid-points is called poor, or coarse, resolution. FIGURE 13-13 shows examples of both coarse and fine resolution; notice that fine resolution simply implies more gridpoints in a given area than coarse resolution.

The downside is that fine resolution in a gridpoint model means more points at which the model must make forecasts. So, a forecast model with very fine resolution takes a very long time to compute, even on a supercomputer. Even worse, the forecast can "blow up" if the ratio of grid spacing versus time resolution is too small, even if the original data were silky-smooth. (Why? See BOX 13-4 or visit the text's Web site.) Richardson didn't know this; this limitation wasn't discovered by mathematicians until 1928. His forecast would have gone haywire even if he had employed meteorological "shock absorbers" and had spent

"Weather Forecast Model"

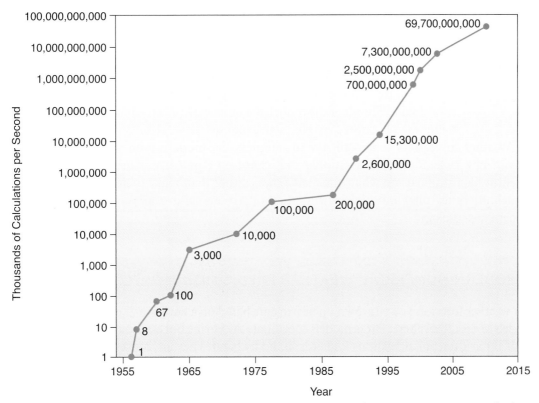

FIGURE 13-12 Logarithmic graph of the speed (in thousands of operations per second) of the computers used for forecasting at the National Center for Environmental Prediction (NCEP) since 1955. Every jump between horizontal lines is a factor of 10 increase in computing speed. (Data from Tim Vazquez.)

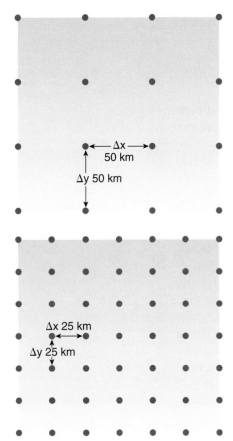

FIGURE 13-13 Examples of two different grid spacing used in current weather forecasting models. Each dot represents a gridpoint at which the model makes calculations. Notice that coarser grid spacing (top) means far fewer gridpoints per unit area than finer resolution (bottom). This reduces both the computer run time, but also the accuracy, of the coarser-grid model.

countless hours more to calculate the pressure in Germany for a still later time. Today, this fine-resolution requirement on gridpoint models is a major reason that forecast models take a long time to run, even on the most powerful supercomputers.

▣ Step 4: Forecast Tweaking and Broadcasting

Richardson made a "forecast" for one place in Germany. He translated his results into a sea-level pressure prediction. Then he broadcast his result in the form of a book. However, the time elapsed from forecast to broadcast was more than 4 years!

Today, forecast centers follow an approach similar to, but much faster than, Richardson's. Computers create forecasts of variables that forecasters care about: temperature, dew point, winds, and precipitation. Forecasters also look at other more complex variables. Atmospheric spin, also known as *vorticity* (see Chapter 10), turns out to be especially useful in forecasting. Some computer forecasts are for as short as 6 hours into the future, and others are for as long as 15 days ahead. Then the forecasts are transmitted worldwide in both pictures and words. Using this information, everyone from the NCEP's colleagues at local National Weather Service offices to your local TV weather personality adds his or her own "spin" to the information, creating a specific forecast tailored to the consumer's needs. (Read **BOX 13-5** to learn how weather information is used by meteorological "detectives.")

In this last step, some of the other forecasting methods—especially climatology and modern-day analog approaches—sneak in the "back door" of the modern forecasting process. They are used to make small improvements on the numerical forecast. The best-known of these approaches is Model Output Statistics, or **MOS** (pronounced "moss"), that is used by NCEP. MOS is a series of statistical relationships between a particular model's predictions and observed weather variables at a particular location. These complex statistical formulas are used to tweak the "raw" model output. It's basically a scientific version of experience, with MOS doing the equivalent of saying, "Ah, I recall that the model makes too warm a forecast in these situations, so fix the forecast so it's a little cooler than what the model said." The older the model, the more experience is built into MOS, and the better the MOS forecast is for a given model.

Weather forecasters today still try to outguess the models, even with the corrections provided by MOS. But these days, unlike in Richardson's time or in World War II, forecasters are most likely to focus on what "the *model* says" rather than make their own forecasts from scratch. Bjerknes' plan and Richardson's method, once a failure, are now the foundations of modern meteorology.

MODERN NUMERICAL WEATHER PREDICTION MODELS

The weather forecasts you've grown up hearing are based on a fairly small number of numerical models of the atmosphere. These models coordinate the boring but essential work of performing all of the arithmetic calculations Richardson did by hand. They are converted from arithmetic formulas into complicated computer programs composed of tens of thousands of lines of computer code. Hidden in all of the details is the stepwise scientific process we examined in the last section, particularly data assimilation and model integration.

Like other complicated pieces of equipment—a car, for example—these computer programs are given names and have their own unique characteristics. Professional forecasters swear that they even have identifiable quirks and "personalities." Now that we've talked about the numerical

Box 13-4 "Blowing Up" a Forecast Model

In Box 13-2, we identified the Achilles' heel of gridpoint models of the atmosphere. It is related to the distance between adjacent gridpoints, known as the "grid spacing," and the time elapsed between one forecast calculation and the next, known as the "time step."

The problem itself is the "strobe effect." If a strobe light at a dance flashes too slowly, dancers can't see where everyone is in between flashes. Dancers are moving too fast for the strobe, and they'll bump into each other and cause a big mess.

Similarly, if weather simulated in a forecast model moves too fast for the time step of the model, then the forecast will be a big mess. When this happens, meteorologists say the model "blows up" because the mess consists of larger and larger, totally unrealistic numbers for the weather variables the model is calculating.

Richardson didn't know this because the field of numerical modeling was in its infancy in the early 1920s. This problem hadn't been discovered yet! In 1928, three mathematicians named Courant, Friedrichs, and Lewy developed a criterion that, if violated, would lead to the "blowing up" of a finite-difference model. This "CFL" criterion is as follows:

Speed of fastest winds in model ≤ Grid spacing ÷ time step

The upshot of the CFL criterion is that a modeler cannot arbitrarily choose a horizontal grid spacing without also taking into account the time step of the model. If you want fine horizontal resolution to see small-scale weather, you *must* have fine time resolution too. Otherwise, the model "blows up" and the forecast crashes. For a horizontal resolution of 50 kilometers, a model usually needs a time step on the order of just 10 minutes.

What does it look like when a model "blows up"? Try it yourself. A simplified "barotropic" model is located on the text's Web site. Nice troughs and ridges turn into total messes when the horizontal grid spacing and time step don't satisfy the CFL criterion. The forecast is ruined. The model is useless.

In short, the CFL criterion is the "there's no such thing as a free lunch" rule translated into numerical weather forecasting. Meteorologists want fine resolution in both time and space, just like we want a free lunch. Both inevitably come at a price, however. For meteorologists, the price is high: their best models are required to make forecasts every few minutes. These minute-by-minute forecasts are not released to the public; imagine the information overload! These mini-forecasts are used by the model itself, and only by the model, to satisfy the CFL criterion. This wastes valuable computer time and vastly increases the computational requirements of numerical weather prediction. But it is a necessary chore. Without it, numerical weather forecasts would "blow up" in meteorologists' faces.

weather prediction process in general, we can look at the individual models that actually help make the forecasts for tomorrow and beyond (see **TABLE 13-1** for a summary).

Short-Range Forecast Models

The first truly modern numerical forecast model was called the "Limited Area Fine Mesh Model," or **LFM** for short. Meteorologists at NCEP's forerunner, the National Meteorological Center (NMC), developed it in the early 1970s. It closely followed Richardson's methods but for the U.S., not Europe.

The LFM was, as its name advertised, "limited." It was better than flipping a coin or reading the Farmer's Almanac. However, it made forecasts only for North America. Its longest forecast was for 48 hours from the present, and its grid spacing was a very coarse 160 kilometers (100 miles). The LFM, like a nearsighted person, could only "see" larger details and not the finer details of weather, even features several hundred miles across. As a result, it's not a surprise that the LFM didn't forecast the 1975 Edmund Fitzgerald cyclone perfectly (see Chapter 10). The LFM is no longer used by NCEP, but it was the forerunner of the later forecast models.

Box 13-5 CSI: Weather

Are you a "CSI" fan? The forensic scientists on these TV shows use highly skilled and specialized knowledge to solve crimes that mystify local law authorities. But did you know that forensic meteorologists can do the same thing using weather and climate information to unravel civil and criminal cases that would otherwise go unsolved?

Pam Knox, the Georgia Assistant State Climatologist and wife of your co-author, is also a forensic meteorologist (see photo). She has served as an expert witness in many court cases that involve the weather.

How does one become a forensic meteorologist? Knox, who earned a bachelor's degree in physics and a master's degree in meteorology, embarked on this side career while their son Evan was an infant. She applied for and obtained Certified Consulting Meteorologist (CCM) status from the American Meteorological Society. This rigorous application process includes a take-home test that lasted 3 months, plus periodic recertification. With her CCM designation and her educational training, Knox has the credentials that lawyers covet in their expert witnesses.

What does a forensic meteorologist actually study? "Any aspect of life that involves outdoor activities and has an economic or social cost can potentially be affected by adverse weather," Knox says. In over ten years of work as a Certified Consulting Meteorologist, she has worked on cases involving damage to homes due to storm surge and wind gusts caused by hurricanes, traffic accidents, and injuries and deaths due to hydroplaning caused by excess rainfall on busy roads and highways. She has also been involved in assessing humidity and light conditions for drug busts, train wrecks, and car crashes.

The case of a truck-train collision in southeast Georgia in May 2003 illustrates how a forensic meteorologist works. A logging truck stopped at an unmarked train crossing and then proceeded across the tracks directly into the path of a passenger train moving at 70 mph. The train hit the truck and derailed, killing the truck driver and one of the train engineers and injuring 26 others. In her analysis, Knox used satellite imagery to show that the driver of the truck, as he was stopped at the tracks looking toward the train, stared directly into the bright rising sun just as clouds parted. Surface weather observations revealed that early morning fog near the ground probably kept the truck driver from seeing the train racing down the track—even though the horrified train engineers could see him clearly from their higher vantage point. As a result of her meteorological detective work, Knox was able to create a plausible scenario for why the truck driver drove right into the path of a speeding train.

It might not seem quite as exciting as the TV shows. But someday bad weather may affect you and put you in the middle of a legal battle. Then, the work of forensic meteorologists could be just what you need to obtain justice.

In the early 1980s the Nested Grid Model (**NGM**) was implemented at NMC as an improvement to the LFM. Its name refers to the finer inner grid over the U.S. that was inside or "nested" within a more coarse-resolution grid. This was done because the U.S. was the region the NMC forecasters wanted to forecast most accurately; the coarser grid outside of the U.S. saved computing time. The NGM also focused more attention on the jet-stream winds; the LFM's ability to "see" in the vertical was as fuzzy as in the horizontal. More and smoother data were incorporated into the model calculations. The NGM was discontinued in March 2009.

In the 1990s, U.S. government scientists began using two new models, the **Eta** (now called the North American Mesoscale, or **NAM**) and the Rapid Update Cycle (**RUC**). These models incorporate continued improvements in resolution in all three dimensions. The RUC model is

TABLE 13-1 Summary of Numerical Weather Prediction Models Discussed in the Text

Model	First Used	Horizontal Resolution	Number of Vertical Grid Points	Makes Forecast for	Forecast Issued at (BCE or EST)	Longest Forecast	Best Used for
Gridpoint models							
Richardson's	1916–1918	200	5	Near Munich, Germany	1922, for a day in 1910	3 hours	History lessons
LFM	1971	160	7	North America	Twice a day	48 hours	Discontinued in 1993
NGM	1982	80–160	16	Northern Hemisphere, best for North America	Twice a day	48 hours	Discontinued in 2009
Eta	1993	12	60	North America	Every 6 hours	84 hours; 180 hours over smaller areas	Replaced by NAM in 2006
RUC (Replaced with Rapid Refresh running Advanced Research WRF in 2011)	1994	13	50	North America	Every hour	18 hours	Short-term forecasts over U.S.
NAM (NEMS NMM-B)	2006	12	60	North and Central America and adjoining oceans	Every 6 hours	84 hours; 180 hours over smaller areas	Wide variety of U.S. forecasts
Hurricane WRF	2007	9–27	42	Moves with tropical cyclones	Every 6 hours when storm is present	5 days	Tropical cyclones
UKMET Unified Model	1972	25	70	Globe	Twice a day	6 days	Short-to-medium-range forecasts
Spectral models							
GFS (AVN, MRF)	1980	23	64	Globe	Every 6 hours	16 days	Overseas and U.S. forecasts
ECMWF	1979	16	91	Globe	Twice a day	10 days	Wide variety of forecasts

The resolution, frequency of runs, and forecast length of each model have changed over time as computing power has increased. The information in the table is the most recent available at the time of printing, valid as of February 2011. (Based on information from Dr. Stephen Jascourt, UCAR/COMET.)

run every 3 hours, and the NAM is run four times every day. These are vast improvements on Richardson's pace of one forecast computed every 6 weeks! However, these models are limited to short-term forecasts, a few days at most.

The latest advance in U.S. numerical weather prediction is the Weather Research & Forecasting models, or **WRF** (pronounced "worf"), which is based on the Eta model. As the name implies, WRF is designed to allow both forecasters and researchers to use the same basic computer code to create a suite of different models tailored to different situations. Some of these WRF variants are described in Table 13-1.

Although American scientists took the early lead in numerical weather prediction, other countries such as Great Britain and Canada also have well-known models. For example, starting in the 1970s, the British developed the United Kingdom Meteorological Office (UKMO) model, known as the **UKMET** Unified Model.

Medium-Range Forecast Models

For forecasts of a week or so into the future, NCEP has relied on a slightly different forecast technique. So far, all of the models we've looked at have been gridpoint models that chop the atmosphere into boxes. It's as if the atmosphere were a football stadium, and the gridpoint model "saw" the sellout crowd only in terms of one person per row or aisle. However, as we learned in Chapter 10, the atmosphere, like crowds in football stadiums, can do "the wave." In fact, the atmosphere can be described completely in terms of different kinds of waves, such as Rossby waves (see Chapter 7) and gravity waves (see Chapter 12).

Spectral models use this concept as the basis for a very different forecast approach. Instead of calculating variables on points here and there, the spectral approach takes in the whole atmosphere at once and interprets its motions as the wigglings of waves. A primary advantage is that computers can compute waves more efficiently than data at lots of points on a map.

Spectral models, because they can run on a computer faster than gridpoint models, have been used to make longer range and more global forecasts than gridpoint models. The "Aviation" (**AVN**) or "Spectral" model was developed at NMC in 1980. Its forecasts were especially useful for transcontinental airline flights—hence the nickname "Aviation." A special version of the AVN, the **MRF**, made global forecasts for up to 15 days in advance. AVN and MRF were consolidated and renamed the Global Forecast System (**GFS**) in 2002, which now makes 16-day forecasts for the globe.

For decades, U.S. numerical weather prediction models were the best in the world. Today, the best model is probably the one created and run at the European Centre for Medium-Range Weather Forecasts, or **ECMWF**. The ECMWF model is a type of spectral model but with more waves (i.e., better resolution) than the GFS and special attention to data assimilation. It is used to make forecasts as long as 10 days in advance. Instead of creating a different medium-range model for each nation, the European countries have collaborated on one very good model.

In a nutshell, today's forecasts are based on science and very complicated computer programs. However, human meteorologists are still a very important part of the forecast process. They use the models' predictions as "guidance" for the forecast that gets issued to the public. Table 13-1 summarizes the characteristics of these models, and FIGURE 13-14 shows an example of visualized model output.

No single numerical weather prediction model is perfect. Each one is a little different, with differing strengths and weaknesses, and the data entering the model are also imperfect. These small differences can lead to large differences in forecasts. So, which forecast can we trust?

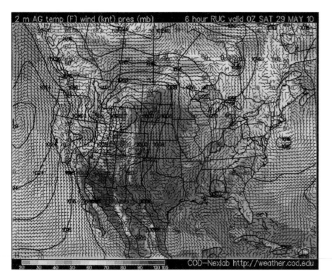

FIGURE 13-14 An example of a 6-hour forecast of surface temperature (shading, in °F), pressure (contours, in mb) and winds (wind barbs, in knots) from the RUC model, valid for 0000 UTC May 29, 2010. The RUC model computes these variables, and many others, at each of the locations shown with a wind barb—in fact, for legibility the winds from only every *sixth* gridpoint are depicted here!

In the 21st century, weather forecasting is increasingly using the concept of **ensemble forecasting**, which uses multiple runs of one model (or even combining runs from different models) to improve overall forecast accuracy and provide estimates of forecast confidence. We'll come back to this idea shortly. Ensemble forecasting requires more computer power than was available back in 1993, the year of the story in the next section: the successful forecast of the biggest storm of the 20th century in the eastern U.S.

A REAL LIFE-OR-DEATH FORECAST: THE "STORM OF THE CENTURY," MARCH 1993

The Medium-Range Forecast

The eastern U.S. basked in unseasonable warmth in early March of 1993. Cherry trees were on the verge of blossoming in Washington, DC. Thousands of college students headed south by car or plane toward the Gulf Coast beaches for spring break.

There was just one problem. By late Monday, March 8, medium-range forecasts at the NMC (renamed NCEP in 1995) began calling for an epic winter storm that weekend. Snow, wind, and bitter cold were on the horizon? Unbelievable!

Overseeing the forecast process at NMC in Washington was meteorologist Louis Uccellini. He knew only too well how unreliable the official forecasts could be. In 1979, the worst snowstorm in more than 50 years had surprised the nation's capital (see Figure 10-21). It dropped up to 50 centimeters (20 inches) of unforecast snow on Washington. The nation's snowbound leaders, who vote every year on the National Weather Service's budget, took note.

Uccellini studied that storm and pioneered efforts to understand why the forecast had "busted." In March of 1993, he was the one in charge of the nation's forecast and, by his own admission, "sweating bullets." Before the end of the week, President Clinton and two Cabinet secretaries would be getting personal briefings on the forecasts of an impending blizzard.

But on March 8, the only sign that the early spring weather would come crashing to a halt was in the output of numerical weather prediction models at NMC and overseas. The MRF forecast 5 days into the future called for an extratropical cyclone to develop very far to the south, over the Gulf of Mexico, and then curve up the East Coast and strengthen. The MRF suggested that by the morning of the 14th (**FIGURE 13-15a**) the storm would be a potent 983-mb low just east of Washington.

However, during the previous month, the MRF had made one wrong forecast after another. Strong cyclones popped up on the forecast maps but were no-shows in real life. Could the

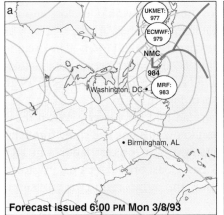 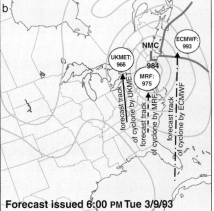 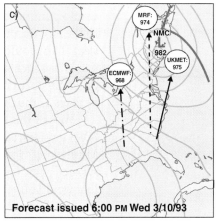

FIGURE 13-15 Medium-range forecasts of the "Storm of the Century" as of 6:00 AM EST on March 14, 1993, from three numerical models: the MRF, ECMWF, and the UKMET models. The left panel (a) shows the forecasts for the 14th issued on March 8; the middle panel (b) shows forecasts for the 14th issued on March 9; and the right panel (c) shows forecasts for the 14th issued on March 10. In each panel, numbers indicate the lowest predicted central pressure of the cyclone at that time. The L and fronts on each panel represent the NMC meteorologists' own forecasts based on a synthesis of the model projections. (Adapted from Shapiro and Gronas, editors. *The Life Cycles of Extratropical Cyclones.* American Meteorological Society, 1999.)

forecasters trust the MRF now? The NMC meteorologists skeptically accepted the MRF's guidance. The words "chance of snow" entered the extended forecast for the nation's capital on Tuesday.

In the days ahead, Uccellini's group compared the results from the world's best medium-range forecast models as they continually ingested new data and calculated new predictions. On Tuesday night (Figure 13-15b), the MRF continued to predict a Gulf low that would intensify and hug the East Coast. However, the ECMWF model painted a much less dire picture. It envisioned a weak 993-mb low far out to sea over the Atlantic. The UKMET forecast was for a monster 966-mb low over Lake Ontario! Which model was right? The NMC forecasters stuck with the MRF guidance but played it safe and kept the low at 984 mb, the same as the last forecast.

By the night of Wednesday the 10th, the forecast picture flip-flopped (Figure 13-15c). Now the European Centre model called for an intense low over Erie, Pennsylvania, not the weak Atlantic low it had predicted only a day earlier. This time it was the British model that called for a weaker low out to sea. The American MRF, in contrast, more or less stayed the course laid out in its earlier model runs. The NMC forecasters trusted the MRF's more consistent forecast. An intense winter storm would hit the East Coast that weekend—even if the computer models were disagreeing as vehemently as the human forecasters before D-Day!

By the next day, Uccellini's group at NMC had gained enough confidence to do something unprecedented: warn the nation several days in advance of a killer storm. The vague "chance of snow" gave way to official statements warning of "a storm of historic proportions forecast over the mid-Atlantic." This was a big risk, both professionally to the meteorologists and financially to the country. Millions of dollars in lost business would result if the weekend turned out to be nice and sunny.

The Short-Range Forecast for Washington, DC

The time for medium-range forecasts had ended. On March 11th, the first short-term storm forecasts from the higher-resolution models came out. Using them, NMC and local NWS meteorologists could make specific snowfall forecasts.

The LFM, NGM, Eta, and AVN models all predicted that a strong low (or lows) would move northeastward out of the Gulf of Mexico into Georgia. Then, as a classic "nor'easter" (see Chapter 10), it would deepen as it moved north to Chesapeake Bay.

The NMC meteorologists held their collective breath and predicted a huge 963-mb low—as intense in terms of pressure as a Category 3 hurricane—would pass just east of Washington on Saturday night (March 13). As Friday the 12th dawned, they stuck to that forecast. The cyclone would pummel the nation's capital initially, but the track of the cyclone would be close enough to the city for the low's warm air to invade the region and change the snow to freezing rain or sleet. This would keep snow accumulations down near the coast (FIGURE 13-16a). The cyclone would be much stronger than the surprise snowstorm back in 1979. However, because of a slightly different storm path, the overall snow totals this time would be about half as much as in 1979, perhaps as much as 30 cm (1 foot), the sum of the forecasts in Figure 13-16.

Was all this really going to happen? A local TV weatherman told his viewers, "When those NWS guys start using terms like 'historic proportions' . . . you know this one will come through." Meanwhile, Uccellini, aware of the models' disagreements and past forecast failures in DC, sweated a few more bullets. Washington and the entire U.S. East Coast braced for a record storm that would bury the Appalachian Mountains with feet of snow and dump a wintry mix of snow, sleet, and freezing rain on the large coastal cities.

The Short-Range Forecast for Birmingham, Alabama

Tough forecasts weren't confined to the East Coast that day. An intensifying Gulf low with bitter-cold air also means a chance of snow for the Gulf Coast states—even at spring break. In Birmingham, Alabama, a metropolitan area of 1 million nestled in the foothills of the southern Appalachians 400 kilometers (250 miles) from the Gulf, March snows aren't unheard of. But

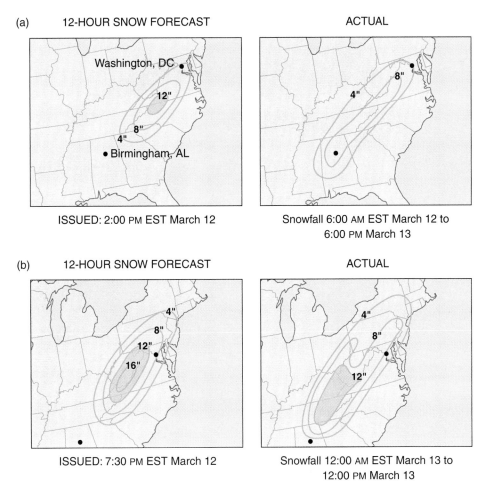

FIGURE 13-16 NMC's 12-hour snow forecasts for the "Storm of the Century" (left) and what actually happened (right) for two overlapping time periods on March 12 and 13, 1993. The region of 16 inches of snow in the lower left figure was the most snow ever forecast by NMC in a 12-hour period. (Adapted from Shapiro and Gronas, editors. The *Life Cycles of Extratropical Cyclones*. American Meteorological Society, 1999.)

when National Weather Service meteorologists in Birmingham started calling for 6 inches or more of snow on Friday, March 12, afternoon and evening, eyebrows went up and panicked residents cleaned out supermarket shelves.

Six inches of snow? This was even more than the NMC models suggested (Figure 13-16a). The Birmingham forecasters relied on their local experience and knowledge of the city's hilly topography, too small-scale for the forecast models to "see." They had a hunch that, despite the blooming trees and flowers and warmth, the conditions were just right for a climatology-defying snowstorm in Birmingham.

Even so, there was no proof that the storm would come together. Proof would come shortly.

■ The Storm of the Century Appears

On the morning of Friday the 12th, meteorologists' eyes turned to the Texas Gulf Coast. There, all of the ingredients for cyclone growth discussed in Chapter 10—upper-level divergence; temperature gradients; even warm, moist water—came together in a once-in-a-century mix. The storm exploded, Uccellini said later, "like an atom bomb." The cyclone engulfed the Gulf. By afternoon, oil rigs off the coast of Louisiana were reporting hurricane-force winds. In its first 24 hours of life, the cyclone's central pressure would drop almost 30 mb. By midnight, the storm had the characteristics of a major cyclone: strong fronts, a massive "comma cloud," and a powerful squall line (**FIGURE 13-17**).

The incredible pressure gradient caused by the dropping pressure led to the first widespread Southern blizzard on record—snow combined with cold and howling winds. The atmosphere

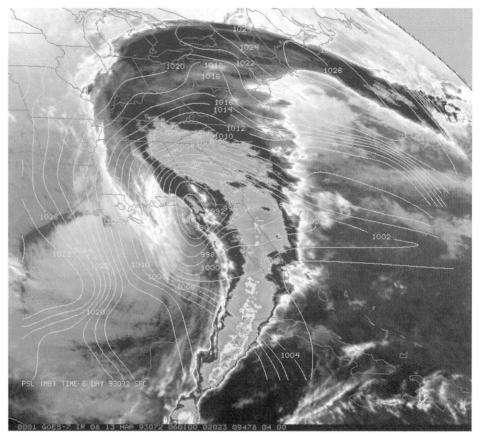

FIGURE 13-17 An infrared satellite image of the "Storm of the Century" making landfall over the Florida Panhandle at 1:00 AM. Easter time on March 13, 1993. Estimated surface pressures are shown with yellow lines; yellow-and-red shaded regions indicate tall thunderstorm clouds.

became so unstable that thunderstorms developed in the cold air, dumping several inches of snow every hour. In Birmingham, a radio transmitter on top of 300-meter (1000-foot) Red Mountain was struck twelve times by eerie green lightning during "thundersnow."

Nearly 50 University of Wisconsin students heading for Panama City, Florida, during spring break were stranded in Birmingham when their bus skidded off a mountainous road. Boasted one, "Don't they have sand or salt around here? Back home this is nothing." The thundersnow raged on.

A Perfect Forecast

The numerical models hadn't foreseen the cyclone's explosive development over the Gulf (although the out-of-date LFM had been giving hints). NMC forecasters now revised the models' predicted central pressures downward. The models were often accused of going overboard with their doomsday forecasts, but on the 13th, nature itself was "over the top!"

What did this mean for the East Coast? The Eta and NGM models now indicated that the cyclone would move inland over Virginia with a central pressure well below 960 mb. This storm track would turn all snow to rain over the East Coast cities. The AVN, however, insisted on the same overall scenario that had been based on recent guidance from the very similar MRF model: a 960-mb cyclone over Chesapeake Bay. The forecasters knew that the AVN had a reputation for forecasting East Coast cyclone tracks accurately. NMC thus ignored the "Armageddon" predictions of the Eta and NGM and based their forecast on guidance from the AVN. They were right. At 7:00 p.m. on the 13th, the storm of the century was centered over Chesapeake Bay, where barometers measured its central pressure at a record-low 960 mb, exactly as NMC had predicted. *It was a perfect forecast of a once-in-a-lifetime event.*

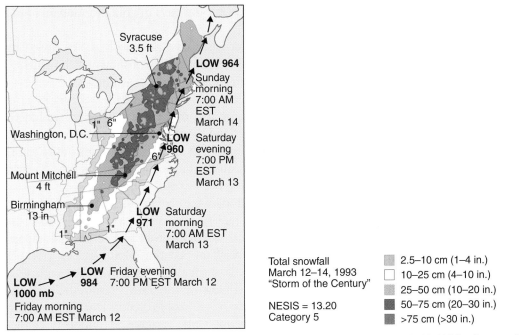

FIGURE 13-18 The path of the "Storm of the Century" on March 12 through 14, 1993, along with the total amount of snowfall reported across the southern and eastern U.S. (Courtesy of NOAA.)

Nowcasting in DC

National Weather Service forecasters in Washington, DC, using the NMC forecasts and their local knowledge, had given 1 to 2 days of notice of impending winter storm and blizzard conditions. On the morning and afternoon of the 13th, bands of thunderstorms of snow and sleet rotated around the cyclone into the Washington vicinity. The computer models couldn't "see" these narrow bands, and neither could older weather radars. However, Doppler radar (see Chapter 5) could. The local forecasters used the radar to make very short-term forecasts of precipitation type, intensity, and winds. Instead of a vague "heavy snow today" forecast, Washington residents received hour-by-hour nowcasts of the tiniest details of a cyclone the size of half a continent.

The Aftermath

The "Storm of the Century" buried the eastern U.S. in enough snow that, if melted, would cover New York State with 30 centimeters (1 foot) of water. The amount and extent of the snowfall was unprecedented in American history (**FIGURE 13-18**).

In Birmingham, the official total of 33 centimeters (13 inches) achieved the rare "hat trick" of snowfall records. It was the most snow ever recorded in the city in 1 day, 1 month, and even 1 whole year! In the hills around Birmingham, retired National Weather Service forecaster J. B. Elliott carefully measured 43 centimeters (17 inches) of snow. The forecast had been an underestimate, but it gave residents the warning they needed that a major storm was on the way.

The deep white blanket of snow and an intense anticyclone behind the low caused Birmingham temperatures to drop to −17° C, or 2° F—an all-time March record (**FIGURE 13-19**). Meanwhile, half the state of Alabama was without power and therefore heat in most cases. "Prairie Home Companion" storyteller Garrison Keillor, by chance in Birmingham to do his live public-radio variety

FIGURE 13-19 The aftermath of the "Storm of the Century" in the hills of Birmingham, Alabama. Wet, wind-driven snow is defying gravity and hanging off the roof of the co-author's boyhood home. The deep blue sky at top left is a sign of the rapidly sinking and drying air of the intense anticyclone that followed the storm. The record snow led to massive power outages that cut off heat for days to tens of thousands of homes in Alabama, including the Knoxes' house.

FIGURE 13-20 The March 1993 "Storm of the Century" buried East Coast cities with heavy, but not record-breaking, snowfall. Nevertheless, the cumulative impact of the storm across the entire region ranks it as the #1 worst Northeast snowstorm in history.

show, sang, "Oh Susannah, now don't you cry for me/I'm going to Alabama for a couple of days to *ski*."

Washington (**FIGURE 13-20**) and other East Coast cities such as New York City fared just as predicted. They generally received about a foot of snow, which turned to ice and slush as the low and its warmer air moved up the coastline.

Some East Coast residents grumbled that this wasn't the century's worst storm; they'd seen worse. They didn't realize that much worse weather had indeed occurred, as forecast, just a few miles inland. Three million people lost electric power nationally. Storm winds gusting up to 232 kilometers per hour (144 mph) shut down airports across the East. This caused the worst aviation delays in world history until the terrorist attacks on September 11, 2001.

Over a decade later, Louis Uccellini and colleague Paul Kocin created the Northeast Snowfall Impact Scale (NESIS) to rate and rank snowstorms in the Northeast since the late 1800s, based on storms' meteorological severity as well as their societal impact. Sure enough, the March 1993 storm easily topped the NESIS list, as a NESIS Category 5 storm (refer to Figure 13-18). It was truly the "Storm of the Century."

Other regions were also dealt a cruel blow by the unexpected ferocity of the cyclone's winds and thunderstorms. Forty-seven people died in Florida as a result of tornadoes and storm–surge-like flooding. Cuba suffered $1 billion in coastal flood damages. Severe weather battered Mexico's Yucatan peninsula. Forty-eight people died at sea in this truly "perfect storm."

However, the $2 billion in damages and up to 270 deaths attributed to the storm in the U.S. were a fraction of what might have been. What if the year had been 1939 instead of 1993? Folklore, persistence, climatology, trends, and analogs— the methods of forecasters in 1939—could not have forecast such an extreme event as the Storm of the Century. Today's numerical models of the atmosphere, Richardson's fantasy come true, could and did make the right forecast—but only when coupled with the wisdom of the human meteorologists.

At NMC, Louis Uccellini could stop sweating bullets. Numerical weather prediction had saved the day. Afterward, he observed, "We [meteorologists] attained a level of credibility that we never had before and we haven't lost since. People took action based on our forecasts and they take action today . . . that's the future."

WHY FORECASTS STILL GO WRONG TODAY

Even with forecasting successes like the Storm of the Century, meteorologists are not satisfied. "We always want to push the limits," Louis Uccellini says. The reason is that numerical weather prediction isn't perfect. As we saw earlier, even in the best of forecasts there are times and regions where the models are wrong. Even when tomorrow's forecast is right, it's fair to ask this: Why couldn't we have known it weeks, not days, in advance?

The limits of prediction today have their roots in Richardson's original forecast. Richardson's model, as we've seen, had several limitations:

1. He didn't have much data to work with.
2. The forecast was for a small area of the globe.
3. The forecast was for much less than 1 day.
4. The complicated nature of the equations forced him to make approximations.
5. His initial surface data were out of balance and made a mess of the forecast.
6. His model's wide grid spacing would have made a mess of his forecast if the initial data hadn't done it already.
7. The forecast itself inspired little confidence because of its unrealistic results.

These limitations directly relate to today's numerical forecast models. We now examine a few of today's limits of prediction.

Imperfect Data

Important meteorological features still evade detection, especially over the oceans. The model results are only as good as the data in its initial conditions. So if the observational network of surface stations, radiosondes, satellites, and airplanes happens to miss a key region of the atmosphere, say near the jet stream in the vicinity of a developing oceanic cyclone, then the forecasts for that storm will be poor. Imperfect data were a reason for Richardson's forecast failure and explain some forecast "busts" that still happen in the modern era of numerical weather prediction, including the surprise blizzard that shut down the federal government (once again) in Washington, DC, on January 25, 2000.

Faulty "Vision" and "Fudges"

Today's forecasts also involve an inevitable tradeoff (Box 13-4) between horizontal resolution and the length of the forecast. This is because fine resolution means lots of points at which to make calculations (refer to Figure 13-13). This requires a lot of computer time. A forecast well into the future also requires lots of computer time because each day the model looks further into the future requires millions or billions more calculations.

So today's models are still not able to "see" many small-scale phenomena, from thunderstorms to many of the topics discussed in Chapter 12, not to mention clouds, raindrops, and snowflakes. To compensate for this fuzzy "vision" of models, the computer code includes crude approximations of what's not being seen. These are called **parameterizations** (FIGURE 13-21). Although much science goes into them, these approximations are nowhere close to capturing the complicated reality of the phenomena. This is because, as we learned in Chapter 12, the smallest-scale phenomena are often the most daunting to understand. Therefore, it is not an insult to meteorologists' abilities to say that parameterizations are "fudges" of the actual phenomena.

Forecast models have to parameterize or "fudge" all the small-scale phenomena listed previously. Worse yet, the atmosphere's interactions with other spheres, such as the ocean or the land, also have to be approximated—usually very poorly. This seemingly trivial part of modeling turns out to be a critical area for forecast improvement today.

1) Incoming Solar Radiation
2) Scattering by Aerosols and Molecules
3) Absorption by the Atmosphere
4) Reflection/Absorption by Clouds
5) Emission of Longwave Radiation from Earth's Surface
6) Condensation
7) Turbulence
8) Reflection/Absorption at Earth's Surface
9) Snow
10) Soil Water/Snow Melt
11) Snow/Ice/Water Cover
12) Topography
13) Evaporation
14) Vegetation
15) Soil Properties
16) Rain (Cooling)
17) Surface Roughness
18) Sensible Heat Flux
19) Deep Convection (Warming)
20) Emission of Longwave Radiation from Clouds

The COMET Program

FIGURE 13-21 Twenty different meteorological processes in this figure require parameterization in forecast models because they are too small, too complicated, or both for the models to compute precisely. (Copyright, University Corporation for Atmospheric Research, COMET Program.)

Would Richardson have believed it? Of course. His 1922 forecast book includes entire exhaustive sections devoted to the proper handling of the effects of plants, soil moisture, and turbulence on his forecast.

Rapid improvements in computing speed are helping, however. For example, during the 2010s, the best high-resolution forecast models will be able to "see" thunderstorms. This will allow the meteorologists who run the models to turn off the imperfect parameterizations for thunderstorms in those models and let the models generate their own "real" thunderstorms—something that was unthinkable just a decade or two ago. This should lead to sizable improvements in weather forecasts, especially with regard to precipitation.

◼ Chaos

Why settle for a 15-day forecast? Assume you own the world's fastest supercomputer in the year 2020. If the trend in Figure 13-12 holds, by the year 2020 your computer should be able to do quadrillions of calculations each second! With that much computing power, why not make a forecast for each day out to 1 month in advance? Surprisingly, you would not necessarily get a better forecast. Brute-force numerical weather forecasting, with extremely fine resolution, has its limits.

The reason for these limits is a curious property of complex, evolving systems like the atmosphere. It is called "sensitive dependence on initial conditions" and is a hallmark of what's popularly known as **chaos theory**. Chaos in the atmosphere does not mean that everything is a mess. Instead, it means that the atmosphere—both in real life and in a computer model— may react *very* differently to initial conditions that are only *slightly* different. This property of the atmosphere was uncovered by MIT meteorologist Ed Lorenz in one of the most celebrated discoveries of 20th-century science.

Because we don't know the atmospheric conditions perfectly at any time, chaos means that the resemblance between a model's forecast and reality will be less and less with each passing day. (The same is true for different runs of the same model with slightly different initial conditions.) Meteorologists believe that a 2-week forecast is the eternal limit for a forecast done Richardson's way. No amount of computer improvements, parameterization advances, or complaining will change this limit.

ENSEMBLES, THE FUTURE OF FORECASTING

Is chaos then the eternal roadblock for numerical weather prediction? Not quite. In the past two decades, meteorologists have figured out how to use chaos theory to give us something Richardson sorely needed—a better forecast based on many different model runs. It also gives meteorologists a way to quantify the uncertainty in a forecast, that is, an estimate of confidence in a particular forecast. This new and improved forecasting method is called **ensemble forecasting**, and it works for both short- and long-range forecasts (see Chapter 16). Ensembles are the future of weather forecasting.

Ensemble forecasting is the meteorological equivalent of "seeking a second opinion" from a doctor or a lawyer. If every doctor you visit says you need to go to the hospital, you're likely very ill. If only one doctor says you need to go to the hospital and the rest give you a clean bill of health, the odds are much higher that you are healthy. In ensemble forecasting, meteorologists get lots of "second opinions" from different model runs, and then they use the overall results from all the simulations to create the best possible forecast that takes into account the widest possible range of possibilities.

Here's a recipe for ensemble forecasting. Make a numerical forecast for a certain day. However, there is no guarantee that the data used by the model for the initial conditions were measured precisely and accurately. For example, radiosonde measurements of temperature have an error range of about $0.6°$ C ($1°$ F). So make a different forecast using slightly different initial conditions that are within the error range of the observing instruments. Then make yet another forecast with yet another set of slightly different initial conditions. Repeat for many different sets of initial conditions. Then compare all of the different forecasts, which are called the "ensemble." If most or all of the different forecasts agree, then there is a high degree of confidence that their prediction will become reality. If the different forecasts give wildly different results, then there is lower confidence in whatever forecast is eventually chosen. In the latter case, forecasters also get a "feel" for what the possibilities are for a rare weather event, such as a severe cyclone.

Past: Ensembles and Extratropical Cyclone "Lothar," 1999

On the day after Christmas Day 1999, a severe extratropical cyclone named "Lothar" swept across the English Channel with a central pressure as low as 974 mb. Extreme wind gusts close to 210 kilometers per hour (130 mph) blew through Paris and across much of France, Germany, and Switzerland, causing damage in the billions of Euros.

Could any numerical model anticipate such a swift and severe storm as Lothar? The top row in FIGURE 13-22 suggests that the answer is "no." The official forecasts of the state-of-the-art ECMWF model reveal little hint of the deep cyclone that was observed, the shaded region in the panel labeled "Actual."

However, the logic behind ensemble forecasting says that a single model run is not enough to base a forecast on. The 50 forecasts shown in the bottom five rows of Figure 13-22 are all different simulations of this same storm using the same ECMWF model at the same resolution as the forecast in the first image in the top row. The other difference is that each of the 50 model runs was begun with slightly different initial data that account for the imperfections in our weather observing systems. Of these 50 model runs, about 20, or 40%, of the runs show signs that very low pressure would develop over western Europe.

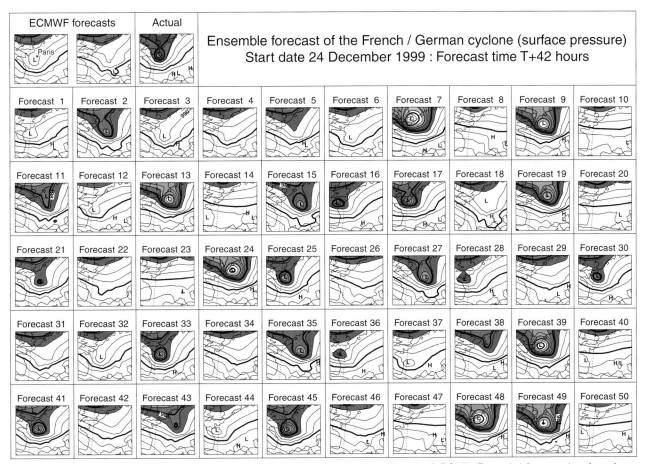

FIGURE 13-22 Top row: Lower-resolution (left) and high-resolution (center) ECMWF model forecasts of surface pressure versus the observed analysis (right) during a severe extratropical cyclone on December 26, 1999, over western Europe. Bottom five rows: Ensemble forecasts of the same storm using 50 ECMWF model runs. Compare the "Actual" conditions to each panel forecast. Notice that about 40% of the ensemble model runs closely resemble what actually happened. This indicates that the chance of a severe storm was unusually high. (With thanks to the European Centre for Medium-Range Weather Forecasts [ECMWF] for allowing the use of its data.)

With this kind of knowledge, forecasters could immediately recognize that the risk for an epic storm was unusually high, even if one or two model runs showed no hint of it. Ensemble forecasting, therefore, creates the ability to make what are called *probabilistic forecasts,* in which the forecast itself comes with a "percent chance" attached. In the case of Lothar, the ensemble forecasting approach would have given about a 40% chance of an extreme wind event for western Europe.

Present: Ensembles Go Global

Ensemble forecasting requires a sophisticated understanding of the atmosphere and computer models and exceptional amounts of computing power. Until the past decade or so, it wasn't feasible to do ensemble runs in real-time weather forecasting—the ensembles simply took too long to run to give results in a reasonable amount of time. Once again, advances in computing have made the impossible possible. Today, ensemble forecasts are made at most of the world's forecasting centers. NCEP runs a global ensemble based on the GFS model four times a day with 20 "members" (i.e., 20 different forecast runs) apiece. Other NCEP models are also run in ensemble mode for shorter range ensemble forecasts.

The ability to make probabilistic forecasts is only one advantage of ensemble forecasting. Another involves data assimilation. Studies have shown that the best methods of data assimilation use information from ensembles. We've already seen in this chapter how important data assimilation is to forecast accuracy. So it makes sense that if ensembles help improve data assimilation, then ensembles are the way to go.

How good can ensemble forecasts be? FIGURE 13-23 shows a 10-day forecast of 500-mb height patterns for the evening of February 6, 2010, based on the average of all the members of the NCEP ensemble runs. The orange and red shadings indicate regions where forecast confidence was high—in other words, locations where nearly all of the ensemble members agreed on what the 500-mb

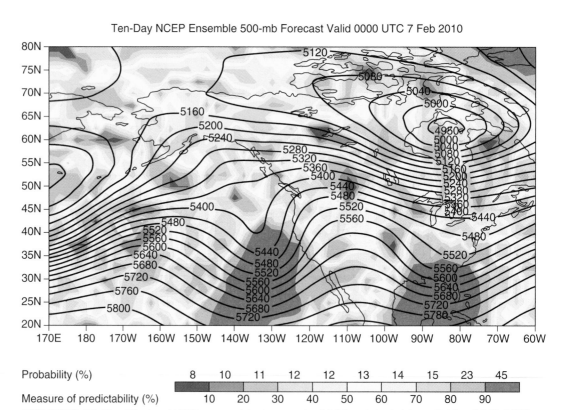

FIGURE 13-23 The 10-day NCEP ensemble forecast of 500-mb heights valid for 0000 UTC on February 7, 2010, made at 0000 UTC on January 28, 2010. Orange and red shaded regions are those in which confidence in the forecast was high; blue shaded regions are where confidence was low. Notice the high confidence in the deep trough over the eastern U.S. (lower right). Ten days after this forecast was made, a deep 500-mb trough and surface low-pressure system did in fact form over the East Coast as forecast, causing a historic snowstorm. (Courtesy of NOAA.)

heights would be. On January 27, there was high confidence that a deep low-pressure trough would be over the eastern U.S. on February 6. Sure enough, on February 6 the record snowstorm dubbed "Snowmageddon" by President Obama (see Chapter 9) hit the East Coast. Although the storm was even stronger than foreseen by the ensembles in late January, the ability to predict a snowstorm 10 days in advance would have seemed like science fiction in L. F. Richardson's day.

Future: Models Talking to Models

During World War I, running a model of the atmosphere for a few points at one time was at the very limit of human ability and took years to do. In 1950, it took a 30-ton computer 24 hours to make one 24-hour forecast for one continent. Today, cell phones can make the same forecast in less than a second, and supercomputers can spit out dozens of ensemble "members" for the globe in mere minutes. Now that forecast models can be solved with such blinding speed, what does this portend for the future of forecasting?

One possibility, which is already becoming reality, is that forecasting will become a densely interconnected web of models that "talk" to each other and provide information about a wide variety of physical processes—not just weather—at a wide variety of scales.

FIGURE 13-24 tries to capture this brave new world of modeling. At the coarsest resolution in time and space at the top right, global ensemble forecasts generate data that use the initial conditions developed by global forecast models such as the GFS. These global ensembles then feed data to higher resolution ensemble models designed for certain regions, not the globe. These ensembles use initial data from single-run regional models, which are partly based on data from the global models. The regional models feed information into even higher resolution local models. The regional and local models provide forecast information for other models dealing with air pollution at very small scales (see Chapter 15). Meanwhile, information from the global model and global ensembles drives other computer models that make forecasts of hurricanes, ocean waves, and storm surge (see Chapter 8). At far right, all of these model results go through various "post-processing," including MOS to create realistic forecasts for specific locations.

It may sound a bit like science fiction, but this scenario of the future is already becoming reality at NCEP and other forecast centers worldwide. In just over a half-century, weather forecasting has

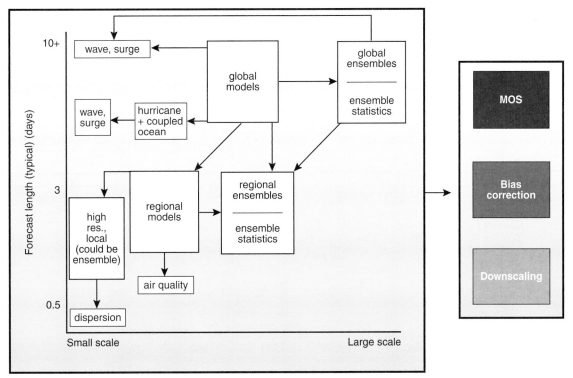

FIGURE 13-24 A schematic showing how different forecast models developed for differing time and space scales and purposes "talk" to each other. (Copyright, University Corporation for Atmospheric Research, COMET Program.)

evolved from meteorologists talking with other meteorologists while looking at weather maps, to computer models "talking" to other computer models and generating graphics and animations of future weather. In Chapter 16, we examine the rapidly expanding field of climate forecasting.

THE PROPER PERSPECTIVE

Why is it that the task of weather forecasting seems to be so complex and error-prone compared with predicting, say, a solar eclipse? L. F. Richardson explained it first and best: "The [forecasting] scheme is complicated because the atmosphere is complicated." Far from being a failure and a waste of time, modern weather forecasting is a triumph.

In just a half century, numerical weather prediction has turned weather forecasting into a true science. FIGURE 13-25 charts this progress in terms of decreasing forecast errors from the 1940s to the late 1990s. Errors in NMC/NCEP forecasts for North American weather at 500 mb plummeted from the days of human forecasters in the 1950s to nearly "perfect" computerized forecasts by the late 1990s. A 3-day forecast in 1990 was as accurate as a 1.5-day forecast in 1975.

More recent improvements in forecasting are depicted in FIGURE 13-26. This bar graph shows the day at which the GFS forecast for the Northern Hemisphere became no better than a coin toss from the late 1980s to the late 2000s. In 1993, a 6-day GFS forecast was just barely better than a coin toss. But by 2008, skill existed in the GFS forecast out beyond 7.5 days (Figure 13-26 top). As in Figure 13-25, a good rule of thumb seems to be that *weather forecast skill extends 1.5 days further into the future every 15 years*. The ECMWF model is even better, with skill out beyond 8 days for Northern Hemisphere forecasts (Figure 13-26 bottom). In February 2010, the ECMWF model broke the 10-day barrier for skillful forecasts for the very first time.

These improvements are the result of better computers, better science, and new generations of enhanced computer models of the atmosphere. However, challenges remain. Precipitation forecasts are harder because of the spotty nature of rain and snow, but even so today's forecast of 2.5 centimeters (1 inch) of rain 2 days from now is as accurate as a 1-day forecast of 1 inch of rain made 20 years ago. Similar improvements have been made for medium-range forecasts as well. As a consequence, today's media place much more emphasis on and confidence in multiday weather forecasts than they did just a decade ago.

The benefits to society of weather forecasting are enormous. A recent study estimated the value of weather forecasts to American households at over $30 billion annually, compared with only

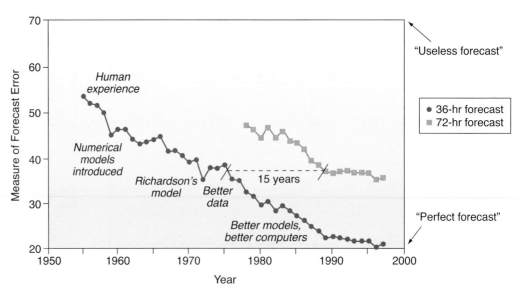

FIGURE 13-25 Errors in NMC/National Center for Environmental Prediction model North American forecasts for 36-hour forecasts at 500 mb. Notice the steady decline in errors since the introduction of numerical models in the late 1950s.

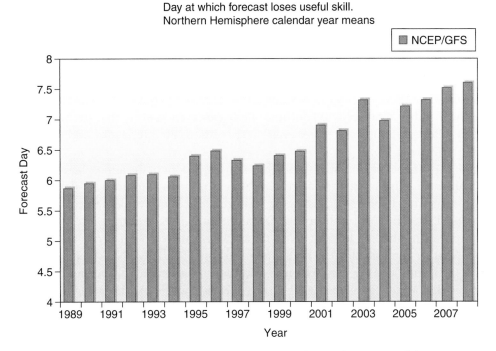

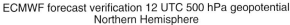

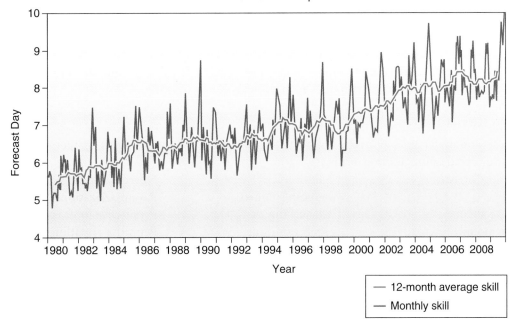

FIGURE 13-26 Improvements in the days of skill of forecasts for the NCEP GFS model (top) and the ECMWF model (bottom). Both models now provide skillful forecasts out to more than one week in advance. (Part a courtesy of Dr. Louis W. Uccellini, Director of the National Centers for Environmental Prediction. Part b, with thanks to the European Centre for Medium-Range Weather Forecasts [ECMWF] for allowing the use of its data.)

about $5 billion per year in government spending on weather forecasting—a six-to-one ratio of benefits to costs. Furthermore, as forecasts keep improving, the benefits will keep on growing.

These improvements in weather forecasting are steady but generally unspectacular, except in the case of a Storm of the Century. In your lifetime, you will grow accustomed to—and spoiled by—better forecasts and more accurate longer-range weather forecasts. As you do, please try to remember that weather forecasting was a confusion of competing unreliable methods as recently as World War II. Thanks to the work of L. F. Richardson and his successors, weather forecasting has become a modern quantitative science.

PUTTING IT ALL TOGETHER

Summary

People can forecast the weather in a wide variety of ways. They can recall folklore rhymes, or they can watch the skies. They can assume that today's weather will persist into tomorrow or that the trend of weather will continue. They can even assume that the weather will be typical from a climatological perspective. These methods are hit or miss and inadequate for most modern needs.

Numerical weather prediction is the solving of the equations of the atmosphere—a "model"—on a computer. Lewis F. Richardson created the first numerical forecast model. Although his "forecast" turned out to be wrong, his overall approach was sound. Modern forecast models ingest observational data, take out imbalances via data assimilation, and then integrate the models forward in time to obtain a forecast. With these models, meteorologists can make precise and accurate weather forecasts for the first time in history. The LFM, NGM, NAM, RUC, WRF, UKMET, GFS, and ECMWF are nicknames of modern numerical forecast models. Statistical relationships known collectively as MOS help turn forecast model predictions into realistic forecasts for particular locations based on climatological experience.

Modern weather forecasting today fuses advanced computer modeling with human insight. Together, they save lives and protect property through increasingly accurate predictions, as in the "Storm of the Century" in March 1993. Ensemble forecasting allows meteorologists to get many "second opinions" on which to base even better forecasts farther out into the future and to quantify the confidence level of forecasts. Limits exist on how good forecasts can become, however. Imperfect data, imperfect knowledge of how the atmosphere works, limits on computing power, and chaos make forecasts go wrong. Even so, modern weather forecasting is one of the great achievements of modern meteorology and all of science. It will continue to improve during your lifetime through new techniques such as ensemble forecasting.

Key Terms

You should understand all of the following terms. Use the glossary and this chapter to improve your understanding of these terms.

Analog forecast	Gridpoint	Numerical modeling
AVN	Gridpoint models	Parameterizations
Chaos theory	Grid spacing	Persistence forecast
Climatology forecast	Initial conditions	Primitive equations
Data assimilation	Interpolation	Resolution
Data initialization	LFM	RUC
ECMWF	Model	Spectral models
Ensemble forecasting	MOS	Trend forecast
Eta	MRF	UKMET
Folklore forecasts	NAM	Weather types
GFS	NGM	WRF
Grid	Nowcasting	

Review Questions

1. In Britain it was once said, "If it rains on St. Swithin's Day (July 15) it means 40 more days of rain." What kind of forecast is this? Do you trust it? Why or why not? Can you think of a region in the world where it actually could rain 40 days in a row?

2. What is the difference between a persistence forecast and a trend forecast?

3. You travel to a region where the north–south temperature gradient is very strong. Do you think a persistence forecast will be reliable? (Hint: Review the concept of the thermal wind relationship in Chapter 6.)

4. How might global warming, if real, affect the usefulness of climatology forecasts in the future?

5. Some meteorologists say that chaos theory proves that analog forecasts cannot work. Why do they say this? (Hint: Think about the small differences between one weather map and another.)

6. In Chapter 10, we learned that on November 9, 1975, a small but potent extratropical cyclone was centered over Wichita, Kansas, and headed northeast toward the lower peninsula of Michigan. Would persistence have been a good forecast for the Lake Superior region? What about a climatology forecast? Would a trend forecast have accurately anticipated the storm's future strength and path?

7. Is a wrong answer always a failure in science? Explain how Richardson's forecast woes helped lead to much more accurate weather forecasts today.

8. If the November 1975 "Fitzgerald" cyclone (see Chapter 10) happened today, do you think it would be forecast better than it was by the LFM model? Answer with reference to the details in Table 13-1.

9. You are an aviation forecaster and you need a good forecast for an airplane flying over Japan in the next 6 hours. Which American forecast model would you look at and why?

10. NCEP is having a picnic for its forecasters and their families in Washington, DC, in a week. Which one of their own models would the NCEP forecasters look at to decide whether the picnic will be rained out? Which overseas model would they look at first? Why does this overseas model make the most accurate forecasts?

11. A radiosonde launching site costs roughly $100,000 a year to operate. Why don't we lobby the government to shut down all the radiosonde sites and spend the money instead on multimillion-dollar supercomputers?

12. Which type of data-gathering method provides more than 99% of all data used in numerical weather prediction?

13. Why is it that different models can give widely varying predictions for the same time and the same region, as shown in Figure 13-15?

14. Why can't today's numerical models make forecasts for individual thunderstorms 5 days ahead of time?

15. A grandparent says, "Weather forecasts always were terrible, always will be terrible." After reading this chapter, do you agree? How could you use the examples of D-Day and the "Storm of the Century" to describe the advances of modern weather forecasting?

■ Observation Activities

1. Collect three folklore forecasts from family or friends. Can you explain these sayings based on material covered in this book?

2. For 1 or 2 weeks keep track of the forecasts from a local weather source. Set up a table with five columns: date, observed temperature, normal temperature, 24-hour predicted temperature, and 6-day predicted temperature. Over the course of 2 weeks, fill in this table. Which turns out to be more accurate: the 24-hour forecast or the 6-day forecast? Why?

3. What is a memorable weather forecast "bust" for you? Why do you remember this bust and why do you think the forecaster erred?

4. Visit http://wxchallenge.com and follow the forecasts and results of this collegiate weather forecasting contest. How well do climatology and persistence forecasts do in this contest? Which forecast models are used in the contest? Participate in the contest if your college or university is registered to participate (see http://wxchallenge.com/info/teams.php for a list, then contact the local manager at your school).

This rain cloud icon is your clue to go to the *Meteorology* Web site at http://physicalscience.jbpub.com/ackerman/meteorology/. Through animations, quizzes, web exercises, and more, you can explore in further detail many fascinating topics in meteorology.

14 Past and Present Climates

AFTER COMPLETING THIS CHAPTER, YOU SHOULD BE ABLE TO
- Identify the main climate zones around the world
- Explain how scientists study past climates
- List the natural processes that can affect climate, and describe how each one can lead to climate change

INTRODUCTION

Los Angeles residents do not expect to have snow at Christmas. No polar bears are found roaming the deserts of Arizona, and diamondback rattlesnakes do not inhabit arctic ice fields. Tropical rainforests teem with life, but life is much sparser in the great deserts such as the Sahara. Summer is the off-season for hotels and restaurants in Florida, but it is the busiest time of year for the same businesses in Maine and on Cape Cod. *Life adapts to climate.* Climate, the overall weather that prevails from year to year in an area, profoundly affects the distribution and abundance of life forms and the activities of people.

Today, Antarctica has a frigid climate where humans can survive only because of our ability to build shelter from the weather elements. A recent analysis of temperatures by Dr. Susan Solomon (pictured), a senior scientist at the National Oceanic and Atmospheric Administration, indicates that an unusually cold Antarctic autumn contributed to the death of Captain Robert F. Scott and his four comrades on their 1500-kilometer (900-mile) trek back from the South Pole in March 1912. Temperatures were 10° to 20° C colder than expected during the race to the South Pole. The cold weather cut in half the distance the explorers could travel in a day. A blizzard trapped them in a tent, where they froze to death 18 kilometers (11 miles) from a supply depot.

Susan Solomon.

Scott's party had lost the race to the South Pole to Roald Amundsen, who reached the pole a month before. However, the Scott expedition revealed that Antarctica once basked in warmth. Among the 16 kilograms (35 pounds) of rocks the expedition collected were fossils of *Glossopteris*, a seed fern. This fossil is scientific evidence that the current ice-covered continent was once fertile.

Climates usually remain the same during a person's lifetime, but climates have also changed remarkably during the Earth's lifetime. This chapter will define climate and describes how and why climate changes over time.

DEFINING CLIMATE

Climate is to weather what a friend's personality is to his or her mood. In other words, weather captures the atmosphere's short-term ups and downs, whereas climate sums up its long-term behavior. More precisely, **climate** can be defined as the collective state of the atmosphere for a given place over a specified interval of time. There are three parts to this definition:

1. *Location,* because climate can be defined for a globe, a continent, a region, or a city. In this chapter, we concentrate on regional and global-scale climates.
2. *Time,* because climate must be defined over a specified interval. In Chapter 3 we used 30-year averages, whereas in studying Earth's history we may use averages of a century or longer.
3. *Averages and extremes of variables* such as temperature, precipitation, pressure, and winds. In Chapter 3 we focused on temperature, but in this chapter we look at both temperature and precipitation.

We have already made use of climate to understand the atmosphere. For example, in Chapter 3 we compared the 30-year averages of monthly temperature of different regions to understand the controls of the seasonal temperature cycle. In Chapters 11 and 12 we used climate data to study the preferred locations and frequency of thunderstorms and windstorms. In this chapter we learn how to classify various climates and explain climates of the distant past.

CLIMATE CONTROLS

The five basic factors that affect climate are very similar to the temperature controls we studied in Chapter 3. They are latitude, elevation, topography, proximity to large bodies of water, and prevailing atmospheric circulation. Latitude determines solar energy input. Elevation influences air temperature and whether precipitation falls as snow or rain. Mountain barriers upwind and downwind can affect precipitation of a region as well as temperature. Topography also affects the distribution of cloud patterns and thus solar energy reaching the surface. The thermal properties of water moderate the temperature of regions downwind of the region. Atmospheric circulation is somewhat less regular than other factors, but consistent large-scale circulation patterns, such as the position of the subtropical high-pressure belts or the Intertropical Convergence Zone (ITCZ), exert a systematic impact on the climate of a region. These surface and atmospheric features produce variations in temperature and precipitation that create different climate patterns.

Climate can be modified by natural events, like volcanic eruptions, and changes in the amount of solar energy, are natural. Changes can also be caused by human activities, such as building cities and adding greenhouse gases to the atmosphere.

CLASSIFYING TODAY'S CLIMATE ZONES

Are you a "type A" personality? Have you ever taken a personality test that categorized you as an "ENFP"? Pop psychology tries to categorize the complexities of human personalities into a few types, which are usually represented by one or more letters. Scientists in all disciplines use classification for similar reasons: to make sense of complex natural systems.

Climatologists use this approach with climate. There are many different regional climates across the world. To make sense of this variability, climatologists use classification schemes that resemble the personality categories of psychology. The difference with climate classification is that the categories identify regions, not people, that have similar characteristics.

A challenge in designing a climate classification scheme is that climates, like personalities, do not have clear dividing lines. We often classify the climate of a location using descriptive terms for temperature and precipitation: for example, hot and dry, warm and wet, or cold and dry. We choose temperature and precipitation because these two weather parameters are extremely important to life. Temperature and precipitation together determine the environmental conditions under which certain plants and animals flourish while others perish.

Between 1918 and 1936 Vladimir Köppen (pronounced KEPP-in) devised the climate classification scheme that is most widely used today. He used vegetation and temperature as natural indications of the climate of a region. Improvements have been made to the original **Köppen scheme**, most recently in 2007 and is referred to as the Köppen-Geiger map.

The current Köppen-Geiger classification scheme has six main groups, each designated with a letter: Tropical Moist (A), Dry (B), Moist with Mild Winters (C), Moist with Severe Winters (D), and Polar (E). Some groups are described by two- and three-letter designations that are reminiscent of personality classifications. The second letter usually refers to whether and when a dry season occurs, and the third letter denotes differences in temperatures. FIGURE 14-1 organizes and summarizes these climate zones by their temperature and precipitation characteristics. The dominant climate classification over land is arid B (about 30%) followed by cold D (about 25%) and tropical A (19.0%). Temperature zones (C) only occupy about 13%, about the same as polar E zones.

Understandably, the Köppen-Geiger climate zones are closely related to both geography and the global circulation of the atmosphere. FIGURE 14-2 shows the global distribution of the Köppen-Geiger climate zones, and FIGURE 14-3 depicts this distribution in relation to the large-

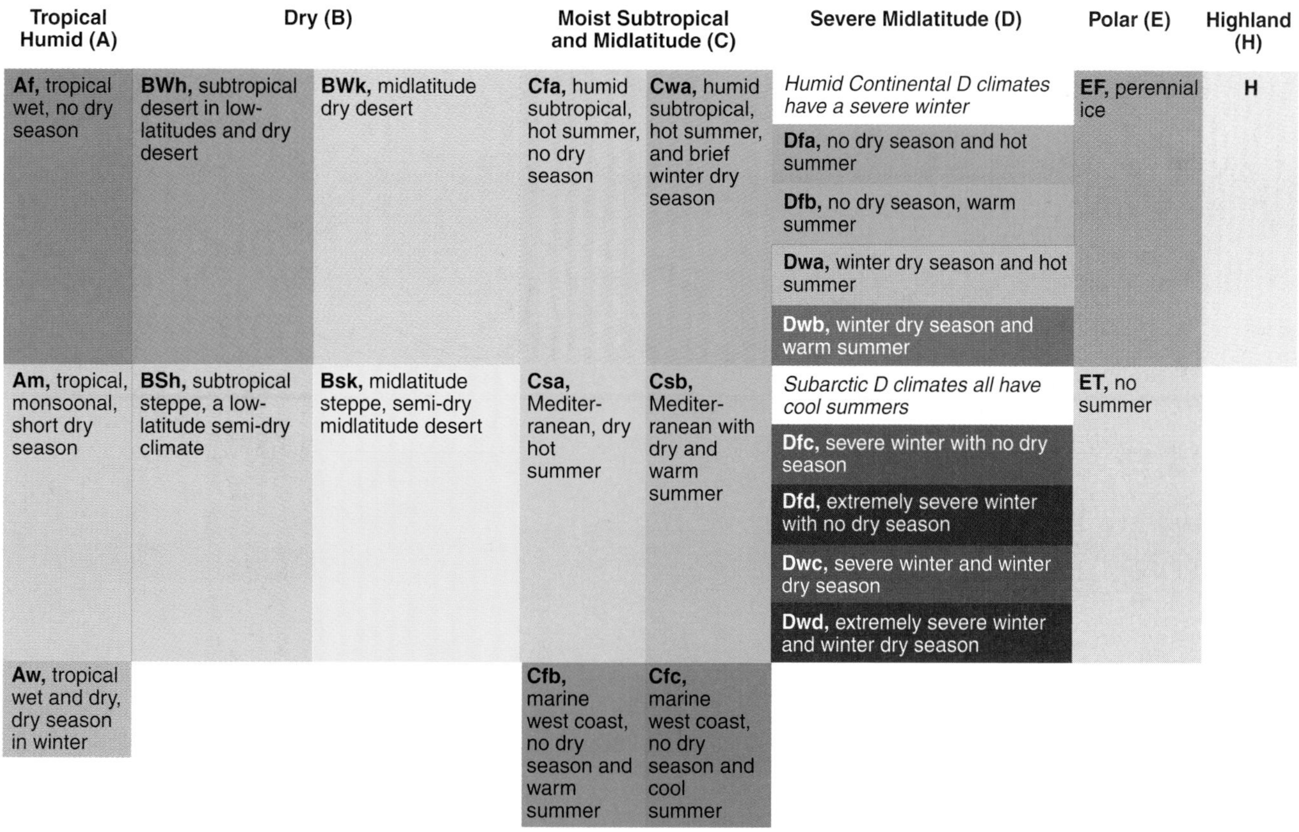

> "Discovering Climate Types" to test your knowledge of the various climate types.

Tropical Humid (A)	Dry (B)		Moist Subtropical and Midlatitude (C)		Severe Midlatitude (D)	Polar (E)	Highland (H)
Af, tropical wet, no dry season	**BWh,** subtropical desert in low-latitudes and dry desert	**BWk,** midlatitude dry desert	**Cfa,** humid subtropical, hot summer, no dry season	**Cwa,** humid subtropical, hot summer, and brief winter dry season	*Humid Continental D climates have a severe winter* **Dfa,** no dry season and hot summer **Dfb,** no dry season, warm summer **Dwa,** winter dry season and hot summer **Dwb,** winter dry season and warm summer	**EF,** perennial ice	H
Am, tropical, monsoonal, short dry season	**BSh,** subtropical steppe, a low-latitude semi-dry climate	**Bsk,** midlatitude steppe, semi-dry midlatitude desert	**Csa,** Mediter-ranean, dry hot summer	**Csb,** Mediter-ranean with dry and warm summer	*Subarctic D climates all have cool summers* **Dfc,** severe winter with no dry season **Dfd,** extremely severe winter with no dry season **Dwc,** severe winter and winter dry season **Dwd,** extremely severe winter and winter dry season	**ET,** no summer	
Aw, tropical wet and dry, dry season in winter			**Cfb,** marine west coast, no dry season and warm summer	**Cfc,** marine west coast, no dry season and cool summer			

FIGURE 14-1 Overview of the main climatic groups with respect to temperature and precipitation. In general, temperature decreases from left to right of the figure.

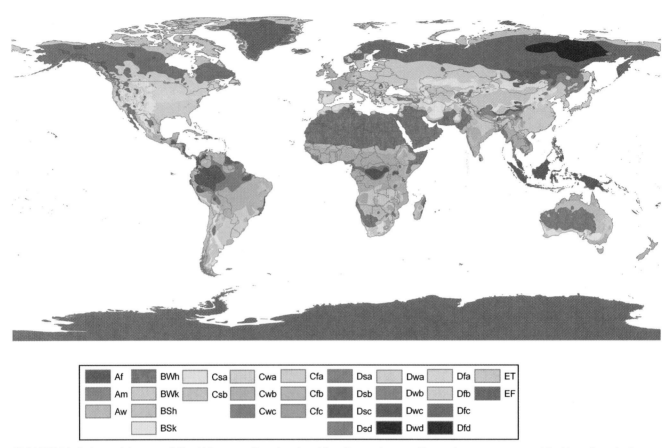

FIGURE 14-2 A world map of the Köppen climate classification scheme, the colors correspond to the climate types defined in Figure 14-1. (*Reproduced from*: Peel MC, Finlayson BL, and McMahon TA [2007], Updated world map of the Köppen-Geiger climate classification, *Hydrology and Earth System Sciences*, 11, 1633–1644.)

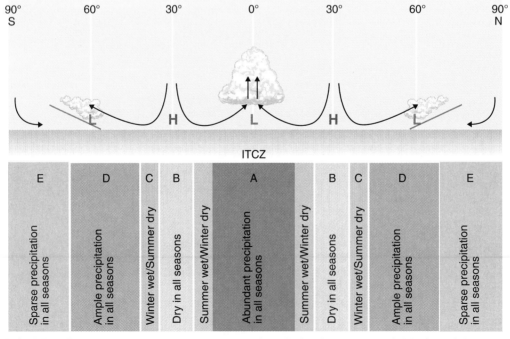

FIGURE 14-3 A cross-section of pressure, wind, and cloud patterns in latitude, relating the Köppen climate classes to the global scale atmospheric circulation patterns discussed in Chapter 7.

scale wind patterns we studied in Chapter 7, such as the ITCZ. Next, we discuss the characteristics of each zone. We use plots of climate data called **climographs** to depict the characteristic monthly mean temperature and precipitation of each of these climates.

Tropical Humid Climates

The mean monthly temperature of **tropical humid climates** (**A**) is high, at least 18.3° C (65° F). The range of the annual temperature is small, usually less than 10° C (18° F). As a result, killing frosts are absent in A type climate regions. The diurnal variation in temperature in A climate regions is often larger than the annual variation.

Although A climates have abundant rainfall, typically more than 100 centimeters (39 inches) per year, they can have different precipitation patterns. A-type climate zones are therefore subdivided into three subtypes: tropical wet climates (**Af**), tropical wet-and-dry climates (**Aw**), and tropical monsoon (**Am**) climates. Examples of these climates are Iquitos, Peru (**Af**), Pirenopolis, Brazil (**Aw**), and Rochambeau, French Guiana (**Am**). FIGURE 14-4 shows the annual temperature and precipitation patterns of these cities.

The tropical wet climates (Af) have temperatures that are distributed fairly uniformly throughout the year. Precipitation each month averages between 17.5 and 25.4 centimeters (6.9 and 10 inches). Af climates also have a diurnal precipitation pattern, with most thunderstorms occurring in the afternoon, triggered by solar heating of the surface. Vegetation in Af climates is very lush, as in the tropical rainforests of Brazil and the Congo. Analysis of Figure 14-2 shows that Af regimes often border Aw climates.

Tropical wet-and-dry climates (Aw) have wet and dry seasons. Summers are wet, and winters are dry in Aw climates. This seasonal rainfall is linked to the seasonal migration of the ITCZ. There is a cool season in Aw climates, which occurs during winter. The vegetation of Aw climates is usually savanna or tropical grasslands with scattered deciduous trees, as in the grasslands of Africa.

Tropical monsoon climates (Am) are climates that have a short dry season (Figure 14-4). Monthly average temperatures of Am climates are uniform throughout the year. These climates tend to occur in regions that have seasonal onshore winds that supply ample moist air. Orographic lifting can also help to enhance the precipitation of Am regimes.

Dry Climates

Dry climate zones (**B**) are located in regions where potential evaporation and transpiration exceeds precipitation. There is more land area with B climates than of any other single type

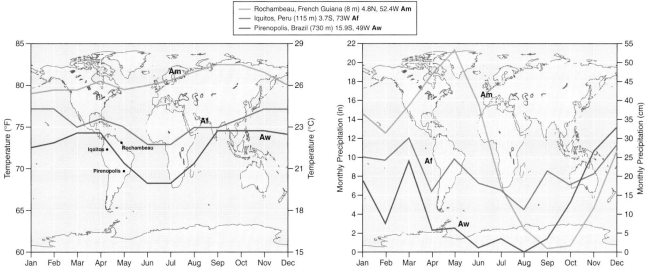

FIGURE 14-4 Monthly temperature and precipitation for three tropical climate regimes: Iquitos, Peru (Af), Pirenopolis, Brazil (Aw), and Rochambeau, French Guiana (Am). The altitude, latitude, and longitude of each station are given in the legend.

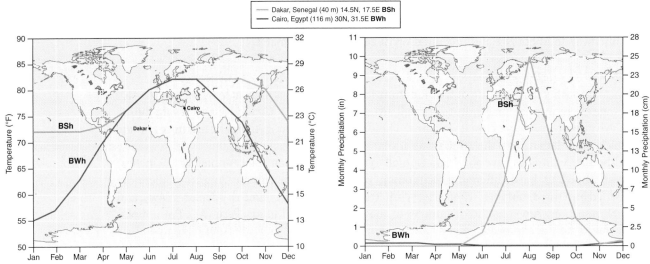

FIGURE 14-5 Monthly temperature and precipitation for the dry subtropical climates of Dakar, Senegal (BSh), and Cairo, Egypt (BWh). (Data from CIRA/Colorado State University and NOAA.)

of climate. The descending branch of the Hadley cell or a rain shadow caused by mountain barriers causes the lack of precipitation in many of the B climate zones.

There are two subtypes of the B climates: steppe or semiarid (**BS**) and arid or desert (**BW**). Inspection of the climate zone map in Figure 14-2 indicates that the BS climates are transition zones, situated between humid climates and desert climates.

Dry climates are those for which evaporation exceeds precipitation. Dry climates span the tropics and the poles. **BSh** and **BWh** are warm dry climates, typical of tropical regions. **FIGURE 14-5** illustrates the temperature and precipitation of Dakar, Senegal, and Cairo, Egypt, which are examples of BSh and BWh climates, respectively. The precipitation peak over Dakar results from the seasonal migration of the ITCZ (see Chapter 7). The annual mean precipitation of Cairo (BWh) is much lower than that of Dakar (BSh), and its annual temperature range is larger.

Hollywood's portrayal of deserts is usually one of hot sweltering days, an intense Sun, and large sand dunes. The cities of Dakar and Cairo tend to conform to this image, but not all dry climates are hot, tropical deserts. **BSk** and **BWk** climates, such as San Diego, California, and Santa Cruz, Argentina, are cool and dry climates at nontropical latitudes (**FIGURE 14-6**). **BWh** and **BSh**

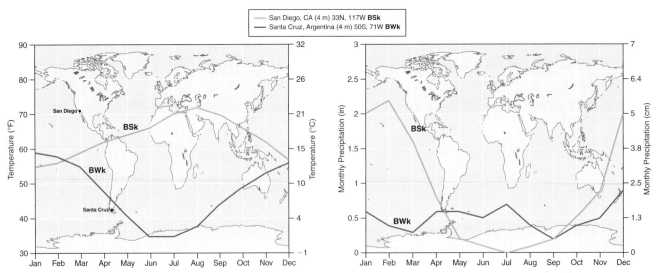

FIGURE 14-6 Temperature and precipitation for dry climate regimes of the middle latitudes at San Diego, California (BSk), and Santa Crux, Argentina (BWk). December and January are summer months in the Southern Hemisphere.

climates have a mean annual temperature of 18° C (64° F) or higher, whereas **BWk** and **BSk** have a mean annual temperature below 18° C. **BSk** and **BWk** climates also usually have more precipitation and less evaporation than their warmer counterparts, BSh and BWh.

The difference between these two climates is in total annual precipitation. BSk climates usually have more precipitation than BWk climates. BWk climates are often located in the rain shadows of large mountain ranges or the interior of continents. BSk climates are mid-latitude steppe regions and have annual temperatures similar to those in BWk regions.

Moist Subtropical and Mid-Latitude Climates

Moist subtropical and **mid-latitude climates** (**C**) are characterized by humid and mild winters. Although temperatures vary appreciably from day to day, the average temperature of the coolest month is below 18° C (64° F) and above 0° C (32° F). Geographically, the subtropics lie between the tropics and the middle latitudes; however, subtropical climates also often lie in the mid-latitude regions. This is where the largest annual temperature ranges are observed because tropical and polar air masses govern the weather at different times of the year.

In the tropics, seasons are distinguished by wet and dry cycles; in the middle latitudes, seasons are distinguished by annual variations in temperature. In the tropical regions, plants become dormant with a lack of precipitation. In subtropical climates, plants go dormant because of low temperatures.

There are three major subgroups of subtropical climates: the marine west coast (**Cfb** or **Cfc**), the humid subtropical (**Cfa** or **Cwa**), and the Mediterranean (**Csa** or **Csb**). We examine these next.

Marine West Coast (Cfb, Cfc)

Summers and winters of marine West Coast climates are usually mild with no dry season, although precipitation can vary throughout the year. The Cfb regime has a warm summer, and the Cfc has a cool summer. Cfb and Cfc climates are usually near the coast. Examples of these climates are those of Bergen, Norway (Cfb), and Reykjavik, Iceland (Cfc) (**FIGURE 14-7**).

The characteristic temperature and precipitation of marine west coast climates are determined by the advection of air over ocean currents to the land. This moderates the annual range in temperature. When cool water is upwind, the summer high temperatures are held down. Warmer ocean currents upwind lead to milder winter temperatures. The coldest month of the year has an average temperature above freezing, making snowfall rare.

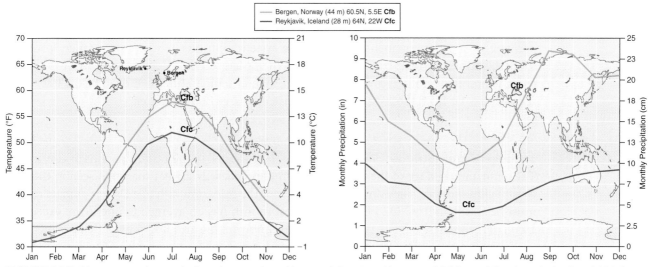

FIGURE 14-7 Marine west coast climates such as those at Bergen, Norway (Cfb), and Reykjavik, Iceland (Cfc), have mild temperatures throughout the year with no dry season. Cfb have warm summers and Cfc have cool summers. This climate regime does not always have to lie on the west coast of a continent, as the name would imply.

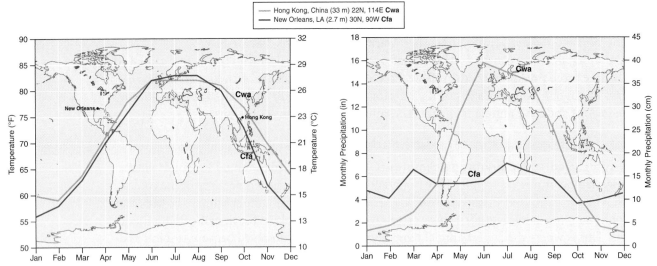

FIGURE 14-8 Humid subtropical climates such as those at New Orleans, Louisiana (Cfa), and Hong Kong, China (Cwa), are hot and wet. The Cwa regime has a drier winter than summer, whereas the Cfa is wet all year.

The name of this climate type suggests that these climates lie along the west coasts of continents. This is because cool ocean currents predominate along west coasts in the subtropics (see Chapter 8). However, these climates are also found along southeastern Australia and southeastern Africa.

Humid Subtropical (Cfa, Cwa)

Humid subtropical climate locations, such as New Orleans or Hong Kong, have hot summers (**FIGURE 14-8**). Summer daytime high temperatures typical of this regime are in the 27° C to 32° C (80° F to 90° F) range. The humid conditions keep the low temperatures in the evening from getting very cool. Winter temperatures are mild. Although mean temperatures may be above freezing in winter, it is not uncommon for daily minimum temperatures to drop below 0° C (32° F). Precipitation in humid subtropical climates is plentiful, 75 to 250 centimeters (30–100 inches) per year. Summer precipitation is usually associated with convection; extratropical cyclones (see Chapter 10) provide the winter precipitation. **Cfa** climates have similar precipitation amounts throughout the year, whereas **Cwa** regions have a definite seasonal precipitation cycle, with a brief dry period in the winter.

Mediterranean (Csa, Csb)

A dry summer and a wet winter characterize the "Mediterranean" climate of Lisbon, Portugal, and Santiago, Chile (**FIGURE 14-9**). The lack of precipitation in summer is associated with the presence of a semipermanent high-pressure system. Summer temperatures range from hot to mild, and winter temperatures are mild. When Mediterranean climates are located along a coast, winter temperatures are also very mild. Winter temperatures can drop below freezing if the region is far from the moderating influence of a large body of water.

Severe Mid-Latitude Climates

The **severe mid-latitude climates (D)** tend to be located in the eastern regions of continents. The temperature range of the D climate regimes is generally greater than that seen in the west coast C climate types. The average temperature of the coldest month of a D-type climate regime must be less than 0° C (32° F). The warmest month has an average temperature exceeding 10° C (50° F). These climate types usually have snow on the ground for extended periods.

There are two basic D climate types: humid continental and subarctic. The second letter **f** indicates that the climate has no dry season, whereas a second letter of **w** indicates a dry season in winter.

For D climates, a third letter of a, b, or c indicates a hot summer, a warm summer, or a cool summer, respectively. A hot summer climate (**Dfa** or **Dwa**) has a warmest month of greater than

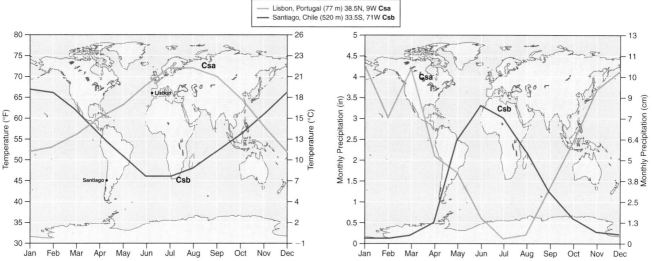

FIGURE 14-9 Mediterranean climates at Lisbon, Portugal (Csa), and Santiago, Chile (Csb), have a distinct dry summer with maximum precipitation in the winter. June, July, and August are winter months in Chile.

22° C (72° F) with at least 4 months above 10° C (50° F). Finally, a D type climate with d as a third letter indicates an extremely severe winter with a cool summer. Next we examine some types of D climates.

Humid Continental (Dfa, Dfb, Dwa, Dwb)
Humid continental climates have a large range in temperature; each subgroup has severe winters and cool-to-warm summers. The climates in the subgroup denoted by an f (e.g., **Dfa**, **Dfb**) do not have a dry season. Examples of the **Dfb** and **Dwb** climates are Fargo, North Dakota (Dfb), and Vladivostok, Russia (Dwb) (**FIGURE 14-10**). Both cities have a large annual temperature range. Vladivostok has a strong summertime maximum in precipitation, whereas Fargo's range in monthly average precipitation is smaller.

Subarctic (Dfc, Dfd, Dwc, Dwd)
Subarctic climates have a very large range in annual temperature. Winters are very long and cold, and summers are brief and cool. Fairbanks, Alaska (Dfc), and Verkhoyansk, Siberia (Dfd), are examples of subarctic climate regimes (**FIGURE 14-11**). Both have very cold winters with monthly average temperatures below freezing for at least 5 months. Precipitation is greater in summer than winter for both cities, but it is never more than a few inches per month. The poleward displacement

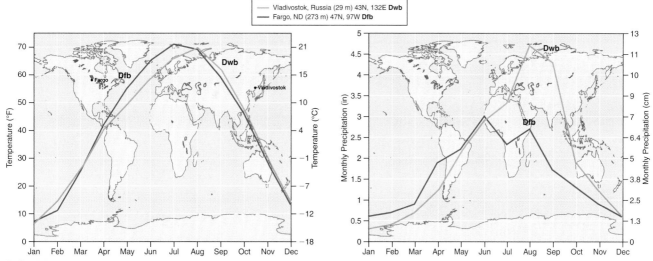

FIGURE 14-10 Climographs for Vladivostok, Russia (Dwb), and Fargo, North Dakota (Dfb). The existence of a winter dry season at Vladivostok separates these two climate types.

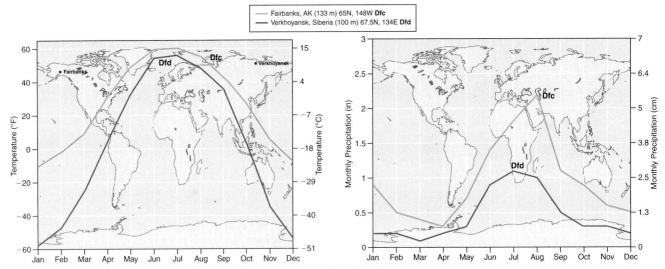

FIGURE 14-11 Fairbanks, Alaska (Dfc), and Verkhoyansk, Siberia (Dfd), are examples of subarctic climates. Both cities have a very large annual range in temperature and small amounts of precipitation.

of the mid-latitude cyclones leads to this maximum precipitation in summer. These dry, severe conditions are best suited to coniferous forests, known as **boreal forests** in North America and **taiga** in Asia.

Polar Climates

Polar climates (**E**) occur poleward of the Arctic and Antarctic circles. Polar climates are extremely cold and have little precipitation (**FIGURE 14-12**). The mean temperatures of polar climates are less than 10° C (50° F) for all months. This cut-off temperature is the minimum temperature for tree growth. Annual precipitation, which is mostly frozen, is less than 25 centimeters (10 inches) of melted water. However, these regions are not considered deserts because precipitation still exceeds evaporation. The climates have a marked seasonal temperature cycle that corresponds directly to the solar input.

A distinction is made between two polar climate types: tundra (**ET**) and ice caps (**EF**). This distinction is made based on the warmest month being warmer (ET) or colder (EF) than 0° C (32° F). Greenland and the Antarctica Plateau are examples of EF climates. EF climate zones have essentially no vegetation, whereas tundra occupies ET climate zones. The vegetation of tundra is primarily mosses, lichens, flowering plants, and some woody shrubs and small trees.

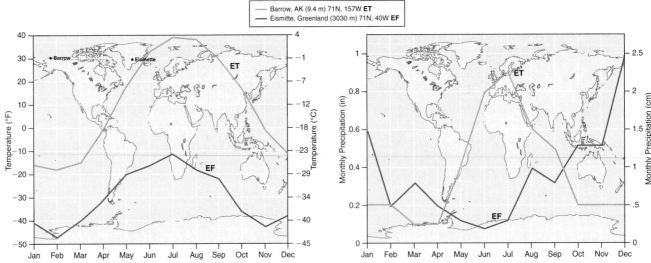

FIGURE 14-12 Polar climates such as those at Barrow, Alaska (ET), and Eismitte, Greenland (EF), have long cold winters and are typically poleward of 70°.

ET regions have a layer below the surface that is perennially frozen, a condition referred to as **permafrost**. During the summer, enough energy is received so that the top layer of soil thaws. This causes the tundra to become wet and swampy. About a meter below the surface the ground is still frozen. This frozen layer may extend downward for hundreds of meters.

PAST CLIMATES: THE CLUES

Now that we have classified today's climates, it is natural to ask: Have these climates always been the same? To answer this question, climatologists study a fascinating variety of clues from the Earth's past. However, the study of past climate does not simply yield knowledge for its own sake. A fuller understanding of past climates enables scientists to better predict future climates and assess the impact of humans on climate, as we will see in Chapter 15.

Past climates can be divided into two main, unequal categories. **Historical climate** refers to the climate of the past several thousand years, during which humans have kept some sort of record of climate conditions. The instrumental record covers the period during which scientific instruments have been used to quantify climate, an era that began around 1600. *Historical data* consists of nonquantitative records of weather, such as diaries. **Paleoclimate** ("paleo-" is a Greek root that means "ancient") refers to climate conditions that existed in the billions of years before the dawn of human civilization. Paleoclimatology is the study of climates of the distant past.

Describing today's climates is rather straightforward because of the large number of observations we have available for analysis. Determining past climates is more challenging. Paleoclimatologists use environmental records to infer past climate conditions. To determine climates of the distant past, paleoclimatologists seek remains from the period that reflect the climate at the time of their creation. By mapping the distribution of these climate remains, such as ancient coal and salt deposits, glacial material, and the distribution of plants and animals sensitive to climate types (such as hippopotami and palm trees), paleoclimatologists seek to discover climate changes of the past and the causes of these changes.

Determining the global and regional climates of the distant past is like solving a mystery. Scientists have to think like detectives, gathering evidence wherever and however it can be found. They look for clues hidden in libraries, trees, ice, land, and oceans that help to show what past climates were like and what caused them. Next we examine some sources of information that paleoclimatologists use to study past climates. We start with the climate clues that are closest to our own era and work backward in time; FIGURE 14-13 summarizes these clues, the eras that they tell us about, and how detailed in time their information is.

Historical Data

Historical documents can provide valuable clues about past climates. Farmers' logs, travelers' diaries, newspaper articles, and other written records often include descriptions of weather conditions. Holy men of the Shinto faith have recorded observations of when Lake Suwa in Japan freezes over in winter and when the ice starts breaking apart. These observations provide proxy records of seasonal temperatures for this region for more than 500 years. Similar records have been kept at other lakes across the continents, and these data indicate that lakes and rivers in the Northern Hemisphere are freezing later and thawing sooner. On average, the freezing now occurs 9 days later, and the thaw occurs about 10 days earlier than they did 150 years ago.

Paintings, drawings, and other works of art can also provide clues to past climates. The Sahara Desert of today covers most of North Africa with desolate sands. Cave paintings in and around the Sahara Desert (FIGURE 14-14) that date from between 10,000 and 2000 years before the present depict giraffes, elephants, and crocodiles. These paintings suggest a much wetter climate than today; one that supports the fauna and flora required by these painted images.

Tree Rings

Climatic conditions and weather variations influence tree growth. The diameter of a tree trunk increases as the tree grows. In regions with distinct growing seasons, this growth appears as concentric

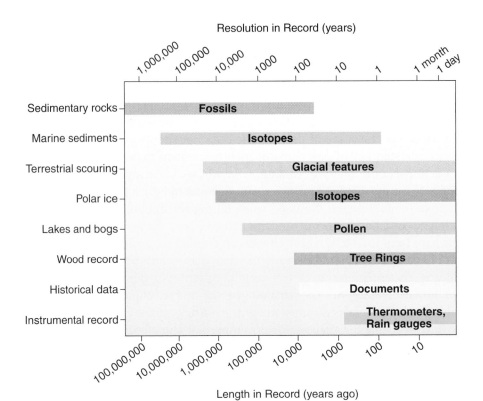

FIGURE 14-13 Climate clues and the lengths of time for which they provide evidence for past climate conditions.

FIGURE 14-14 Cave and rock paintings in the Sahara Desert and Saudi Arabia reflect a climate that once supported enough green vegetation to support herds of wilds animals. This example is from the Tadrart Acacus mountain range in the Sahara desert, near Libya.

rings (**FIGURE 14-15**). Trees generally produce one ring a year and indirectly record environmental conditions each year.

The width of tree rings can be used to gather information about climates from several thousand years ago up to recent times. The width of each ring indicates how fast the tree grew during a particular time period. It is a function of available water, temperature, and solar radiation. Thick rings indicate favorable growing conditions, whereas thin rings suggest poor growing seasons. Some tree species are more susceptible to temperature variations, and others are more sensitive to variations in water availability. The tree rings of different species are helpful in determining what caused the growth spurt or suppressed it. Fire-scarred trees also can provide precise dates of forest fires. Charcoal records in sediments are also used to reconstruct the fire history of an area.

The science of studying tree rings to ascertain a climatic condition is known as **dendrochronology**. This method of research relates tree ring width to modern-day precipitation, which turns the width of tree rings into a yardstick for estimating precipitation for periods when no human observations of precipitation are available.

For example, **FIGURE 14-16** shows the amount of precipitation for Iowa as determined indirectly using dendrochronology. This analysis indicates that the 1930s, a period known as the "Dust Bowl," were very dry (see Chapter 12). For these years, dendrochronologists can compare the tree ring widths with historical measurements of precipitation made by meteorologists. This gives them confidence in interpreting the widths of the tree rings in terms of precipitation amounts. According to the tree ring data from Iowa, very dry times also occurred in 1700 and 1820—periods for which we have little in the way of historical precipitation data. However, by using the relationships between tree ring width and observed precipitation in modern times, dendrochronologists can estimate with confidence the precipitation for these periods as well. In this way, dendrochronology allows climatologists to infer climate conditions long before the dawn of modern scientific instruments.

FIGURE 14-15 The growth of a tree trunk appears as rings in regions with distinct growing seasons. The width of each ring indicates how fast the tree grew during a particular year. This ponderosa pine was a sapling in 1597. The other dates in the figure denote fire scars.

Pollen Records

Trees leave climate clues in addition to the width of their growth rings. Trees also produce pollen that can accumulate in a given location, such as a lake. Although this might not be good if you have certain allergies, pollen is useful to paleoclimatologists. This is because pollen degrades slowly and each species of pollen is identifiable under a microscope by its distinctive shape. Pollen grains that are washed or blown into lakes can accumulate in sediments and provide a record of past vegetation, providing clues to the regional climate. Although pollen grains do not give the same detailed year-to-year information that tree rings do, they are valuable in extending our understanding of climate backward for tens of thousands of years by using **radiocarbon dating** of the pollen grains (**BOX 14-1**).

A pollen record extracted from a northern Minnesota bog provides evidence of a changing climate over the past 11,000 years (**FIGURE 14-17**). The oldest (deepest) layers with pollen indicate

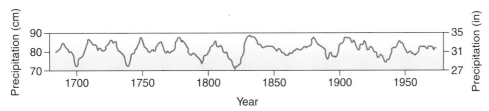

FIGURE 14-16 Measuring and counting tree rings can provide clues about past climates. The graph shows a plot of annual precipitation in Iowa derived from analysis of tree rings. Notice the dry periods in the decades around 1700, 1740, 1820, 1890, and 1930. (Duvick, D. N. and T. J. Blasing, 1981. A dendroclimatic reconstruction of annual precipitation amounts in Iowa since 1680. *Water Resources Research*, 17:1183–1189.)

Box 14-1 Dating Ancient Climates

How do we know how old rocks and fossils are? Radioactivity is used to date ancient climate. Some elements undergo radioactive decay in predictable ways. We can get an idea of the age of a sample of soil or rock or pollen grains by measuring how much of these elements is present in them.

A few heavy elements will spontaneously disintegrate into simpler elements, losing one or more neutrons in the process. The original element is referred to as the parent isotope, and the simpler element is called the daughter isotope. Atoms of the same element that have different numbers of neutrons in their nuclei are called isotopes. For example, carbon (C) has three isotopes: ^{12}C, ^{13}C, and ^{14}C. For a particular element, the rate of decay is constant and is expressed in terms of the *half-life*—the time it takes for a given amount of the parent isotope to decay to half that amount. The age of rocks and minerals can be determined by measuring the amount of daughter isotope relative to the amount of the parent isotope. The ratio of the amount of the daughter isotope produced to that of the parent isotope remaining provides an accurate dating method.

All living things contain carbon. There is about one ^{14}C atom for every 100 billion ^{12}C atoms in a living creature. Once a plant or animal dies, the ^{14}C begins to decay, with a half-life of 5760 years. The percentage of ^{14}C that remains is determined by how long ago the plant or animal died. Radiocarbon dating, as this method is referred to, is good for samples up to 50,000 years old and has an uncertainty of about 15%.

Uranium is used in dating inanimate objects, such as rocks. Uranium-238 decays into lead-206 with a half-life of 4.5 billion years. Lead-206 is created only from the decay of uranium. Thus, the ratio of uranium-238 to lead-206 tells us how old an object is (in the absence of any human-produced radioactivity). If there are equal amounts of uranium-238 and lead-206, then the object is 4.5 billion years old. This is true for the oldest Earth and Moon rocks, and this is how we know the age of the Earth and the Moon.

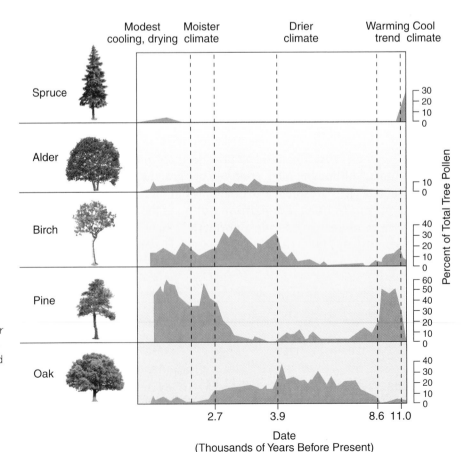

FIGURE 14-17 Tree pollen layered in lakes and bogs provides evidence of changing climate conditions. Pines prefer warm, moist climates, whereas oak trees do well in drier climates. The percentage of pollen from a given species provides evidence for paleoclimatologists to determine previous climate conditions. The pollen samples used to construct this diagram come from a bog in northwestern Minnesota and provide climate clues for the past 11,000 years. (Adapted from T. Graedel and P. Crutzen. *Atmosphere, Climate, and Change.* Scientific American Library, 1997.)

that spruce trees, which require cool climates, were the dominant species. The pollen record indicates that the spruce trees were replaced by pine and birch, which require a much warmer climate. About 8600 years ago the climate changed again, the dominant tree changed to oak, which require drier conditions than pine. Oak and birch were the dominant trees between about 3900 and 2700 years ago. As the climate again became warmer and moister, the oak declined and the pine became the dominant species. Therefore, pollen records provide evidence that climate has changed repeatedly in parts of the United States during the past several thousand years.

Air Bubbles and Dust in Ice Sheets

Clues to climate can be buried in ice just as in lake sediments. Bubbles of air trapped in ice provide windows to the past for atmospheric chemists and climatologists. Air bubbles get trapped in **glaciers** and **ice sheets** (BOX 14-2) as snow gets compressed. Glaciers that exist today can hold gas bubbles tens or hundreds of thousands of years old. These trapped bubbles provide a record of the concentration of atmospheric gases such as carbon dioxide (CO_2) and methane (CH_4) over the past several hundred thousand years.

FIGURE 14-18 shows the concentration of atmospheric carbon dioxide and methane obtained from a nearly 2-mile long ice core cut from an ice sheet in Antarctica. A core this long has bubbles that date back to more than 650,000 years ago. Also shown on this figure are estimates of temperature changes during this period. Notice the 100,000 year periodicity in the temperature and the two greenhouse gas concentrations. The warmer temperatures are apparently related to higher concentrations of carbon dioxide and methane, but cause and effect are very difficult to separate. We explore this relationship between temperature and "greenhouse gas" amounts further in Chapter 15. Explaining the observed rapid cooling after each warm period remains a major goal of paleoclimatologists. Evidence suggests that North Atlantic circulations play a role in the cooling. As discussed in Chapter 8, warm waters of the North Atlantic transfer heat to the atmosphere and then sink and return to the equator at deep ocean depths (see Figure 8-6). As the climate warms, this conveyor belt of heat transfer is shut down, which leads to a rapid and widespread cooling. The ice core measurements indicate concentrations of CO_2 and CH_4 are higher today than any period in the past 650,000 years.

Analysis of six ice cores from the rapidly shrinking ice fields of Mount Kilimanjaro, Tanzania, indicates that these tropical glaciers formed approximately 11,700 years ago. The cores also suggest that the area around Kilimanjaro was much wetter 9500 years ago. At that time, Lake Chad covered approximately 350,000 square kilometers—much larger than its current size of 17,000 square kilometers. The cores also indicate that catastrophic droughts occurred about 8300, 5200, and 4000 years ago. These conclusions are based on the amount of an isotope of oxygen (depleted values suggesting drier climates), methane (high amounts suggest wetlands), and visible dust layers (indicative of drought). Today, these ice fields are rapidly melting (FIGURE 14-19) and are predicted to disappear in the next 25 years.

Dust in ice sheets can be caused by climate-changing volcanoes (see Chapter 3) or by dry, windy conditions that lead to soil erosion. The soil can be transported by small-scale winds in the form of dust storms (see Chapter 12). For example, dust storms from the Sahara Desert (see Box 8-3) inject dust into the global circulation. Global-scale winds can carry this dust as far away as the poles, where the dust can then be detected in ice cores. Highly acidic dust containing sulfuric acid is a clear indication of volcanic activity, as we see later in this chapter.

Marine Sediments

Materials have been deposited in layers on the ocean floor for millions of years. The deeper the layer, the older is the material. These deposits can include soil from wind erosion or floods, ash from volcanic eruptions, and shells of animals. In ocean sediments, the shells of animals are primarily calcium carbonate ($CaCO_3$), a compound that makes up limestone. The calcium carbonate is very useful for tracking past climates by the relative amounts of different versions of oxygen atoms, the "O" in $CaCO_3$. Different versions of the same element, such as oxygen, are called **isotopes**, and they have different number of neutrons and thus different atomic weights.

Box 14-2 Glaciers and Icebergs

Glaciers form on land when the accumulation of ice and snow in winter exceeds summertime melting. An *ice sheet* is a mass of glacier ice with an area of more than 50,000 km² (20,000 mi²). As the snow accumulates on the ice, the ice crystals are crushed under the pressure of the weight of the ice above. Trapped air is expelled, forming bubbles. Eventually, this forms larger ice crystals and the glacial ice compacts and has a blue appearance. This blue color arises from the fact that ice weakly absorbs red light but scatters blue light back to the eye.

The ice sheet on the continent of Antarctic formed from the accumulation of snowfall of thousands of years. When the ice gets to be about 30 meters thick, its weight causes it to flow downhill, although it is a solid. How fast it flows is a function of how steep the land is and the size of the glacier. The speed can range from a few centimeters to 10 meters per day. If the glacier ends in the ocean, it floats out to sea and forms ledges called *ice shelves*. Great blocks of ice can break away before they have a chance to melt and become *icebergs*. This process is called *calving*.

The accompanying figure is a satellite image of a calving process that occurred in Antarctica during the spring of 2000. This iceberg is approximately 298 kilometers (185 miles) long and 37 kilometers (23 miles) wide—a surface area about twice that of the state of Delaware! It is about 400 meters (a quarter-mile) thick, with more than 90% lying below the water.

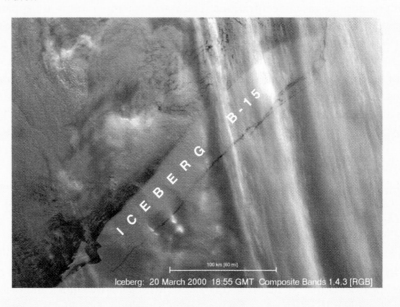

Iceberg: 20 March 2000 18:55 GMT Composite Bands 1,4,3 [RGB]

An enormous part of the iceberg, called B-15A, trapped sea ice in the Ross Sea, making delivery of fuel and supplies to the science outpost at McMurdo Station difficult. By December 2001, the increased sea ice nearly isolated large Adelie penguin colonies, making it difficult for the birds to return from their feeding grounds in the open sea during the breeding season. In October 2003, the iceberg broke up enough to let open sea water reach the Adelie penguin colony. By late November 2006, several large pieces were seen off the coast of New Zealand—one was still over a mile long. Pieces of the iceberg are still floating in the seas around Antarctica.

These Antarctica ice shelves have not only calved large icebergs; some have recently disintegrated. The Larsen Ice Shelf is along the east coast of the Antarctic Peninsula in the northwest part of the Weddell Sea. Parts of that ice shelf disintegrated in February 2002. The Wilkins Ice Shelf, which is on the western side of the Antarctic Peninsula, began retreating in 1990, after being stable for most of the last century. It began to disintegrate in 2008 and is all but gone now. Scientists are still studying the cause of this disintegration, but these breakups are likely a combination of the warming temperature trends of the Antarctica Peninsula and ocean swells generated by distant storms over the ocean, perhaps as far away as Alaska.

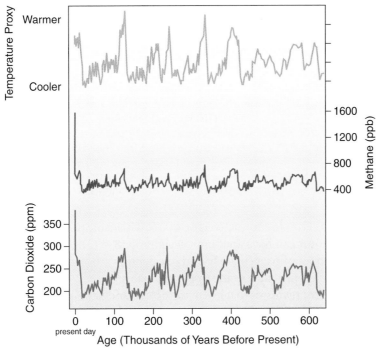

FIGURE 14-18 Concentration of atmospheric carbon dioxide and methane over the past 650,000 years determined from analysis of the chemistry of air bubbles trapped in an ice core cut from an ice sheet in Antarctica. Variations in these gases are correlated with changes in temperature. (Adapted from Brook, Edward J., *Science* 310 [2005]: 1285–1286.)

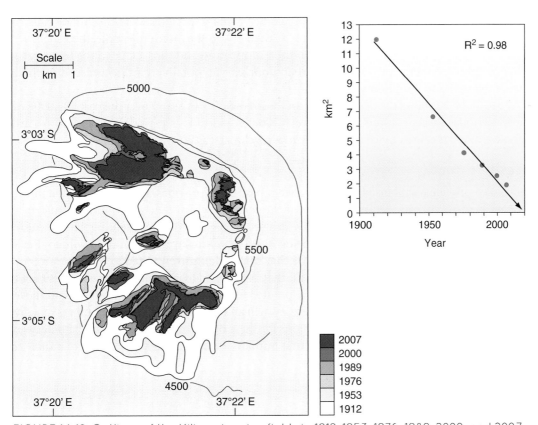

FIGURE 14-19 Outlines of the Kilimanjaro ice fields in 1912, 1953, 1976, 1989, 2000, and 2007. The upper right graph shows the areal coverage as a function of year. As of 2007, 85% of the ice that was on the mountain in 1912 is now gone. A value of R^2 near one indicates a high correlation between decreasing ice coverage with time. (Thompson, L. G., H. H. Brecher, E. Mosley-Thompson, D. R. Hardy, and B. G. Mark. 2009. Glacier loss on Kilimanjaro continues unabated. *Proceedings of the National Academy of Sciences*.)

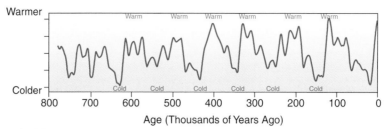

FIGURE 14-20 Variations in average temperature as determined from the ratio of ^{18}O to ^{16}O measured from fossil shells over the last 800,000 years. Periods as warm as today occur infrequently. Warm and cold periods recur approximately every 100,000 years. (Adapted from Imbrie, J. and J. Z. Imbrie, *Science* 202 [1980]: 943–953.)

Most oxygen atoms have an atomic weight of 16. This isotope oxygen is denoted as ^{16}O. Other isotope oxygen atoms can have an atomic weight of 18 (^{18}O). ^{16}O is much more common than ^{18}O. These two isotopes of oxygen are found in calcium carbonate in ocean sediments. The ratio of ^{18}O to ^{16}O ($^{18}O/^{16}O$) provides a clue to past climates.

Foraminifera are microorganisms that live in the oceans and have hard shells made of calcium-containing compounds, including calcium carbonate. The relative amount of isotopes ^{18}O to ^{16}O in the shells of these marine protozoans is related to the amount of continental ice that was present when they were alive. The proportion of ^{18}O to ^{16}O in ocean water is partly controlled by the volume of water in continental ice sheets. Because ^{16}O is lighter than ^{18}O, it can evaporate from water more easily. So, the precipitation that forms from ocean water and falls on glaciers has relatively more of the lighter isotope in it. The lighter water molecules tend to accumulate in snow and ice that form the glaciers. If more glacial ice accumulates, more of the ^{16}O isotope is retained in the ice sheets.

As a result, during colder climatic periods there is a higher concentration of ^{18}O in ocean water than during warmer periods. During colder climatic periods, as foraminifera construct their shells, they incorporate relatively more ^{18}O than ^{16}O. As the foraminifera die, their shells settle on the ocean floor and provide a record of the isotope ratio and thus a clue to the temperature. When we pull sediment cores from the ocean floor, we can obtain an indirect record of climate during the past 2 to 3 million years!

FIGURE 14-20 shows the departures from the average $^{18}O/^{16}O$ ratio over the past 300,000 years. Warm periods alternate approximately every 100,000 years. This periodic oscillation in Earth's climate is a major clue as to what mechanisms alter climate, as we see shortly.

Fossil Records

Fossils also provide useful insight into the distant past. They provide a means to track life through the ages because they are an integral part of the rocks in which they are found. The age of the rocks can be dated (refer to Box 14-1), providing evidence for climate extending back hundreds of millions of years. Fossil ages can also be deduced from the layer of ground in which they lie, if that layer has conspicuous characteristics that can be definitively dated.

Fossils reveal ancient animal and plant life that can be used to infer climate characteristics of the past. For example, tropical plants often have pointed tips so that the moisture can drip off the leaf. Plant fossils that have pointed leaves indicate a warm and moist climate. Large numbers of a given fossil also indicate favorable climate conditions for these organisms. In this way, climatologists can infer the climate of Earth long before the first humans walked on the planet—a remarkable feat of science!

PAST CLIMATES: THE CHANGE MECHANISMS

As detectives, climatologists are not content to know the "what" of past climate. Paleoclimate research is a "who-dun-it." In other words, what mechanisms have forced climate to change as it has over the entire span of Earth's history (**TABLE 14-1**)? How do these mechanisms cause these changes, and over what periods of time do they operate? In the following we examine several

TABLE 14-1 **Earth's Geological Periods**

Era	Period	Epoch	MYBP (Approximate Beginning of Period)	Features	Climate Features	Mean Global Temperature Present Cold Warm	Mean Global Precipitation Present Dry Wet
Cenozoic	Quaternary	Holocene	0.01	Age of Mammals	Little Ice Age (1450–1850 AD); Medieval Climatic Optimum (900–1200 AD)		
		Pleistocene	1.6		The Ice Age		
	Tertiary	Pilocene	5.3				
		Miocene	23.8	Earliest hominids			
		Oligocene	36.7				
		Eocene	57.8				
		Paleocene	65		Beginning of a cooling period, leading to glaciers in Antarctica		
Mesozoic	Cretaceous		144	Mass extinction at the end of the Mesozoic Era; Age of Dinosaurs	Intraglacial climates: no glaciers during the Mesozoic Era		
	Jurassic		208	Breakup of Pangaea			
	Triassic		245	Increasing volcanic activity			
Paleozoic	Permian		286	Final assembly of Pangaea; mass extinction at end of period	Glaciers in southern Africa, South America, and Australia recede, desert regions expand		
	Carboniferous		360	Extensive coal formation	Hot and humid		
	Devonian		408	First amphibians	Warm and dry		
	Silurian		440				
	Ordovician		505	Beginning formation of Pangaea; Primitive fish			
	Cambrian		570		Widespread ice age		
Pre-Cambrian			4600		Evidence of ice sheets		

Source: Adapted from Bradley, R., *Quaternary Paleoclimatology*, Allen and Unwin, Boston, 1985.

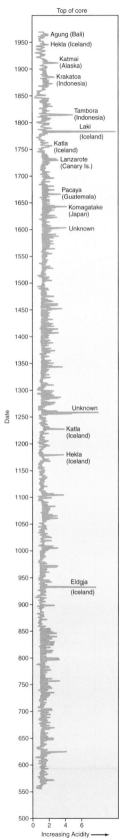

Volcanic Eruptions in a Greenland Ice Core

of the most prominent naturally occurring climate change mechanisms, in approximate order from those that act quickest to those that act on timescales unimaginably long to humans. (In Chapter 15 we take up the subject of climate change caused by humans themselves.)

Volcanic Eruptions

Previously in this text we have discussed how volcanic eruptions have affected local, regional, and even global climate. Locally, the ash may immediately reduce sunlight reaching the surface and cause cooling. Globally, sulfuric acid droplets launched into the stratosphere by the eruption can reflect sunlight and cool much of the globe for a period of several years (see Chapter 3). A series of eruptions therefore has the potential to affect climate on the time scale of a decade or more.

Because volcanoes emit sulfur compounds that turn into sulfuric acid in the atmosphere, sharp increases in surface acidity are an indication that fallout from a volcanic eruption was deposited onto the Earth's surface. This allows paleoclimatologists to detect volcanic eruptions in ice cores (**FIGURE 14-21**). By comparing the timeline of acidity with the timing of known volcanic eruptions, many of the "spikes" of high acidity in this figure can be definitively traced to specific eruptions. Some of the sharpest spikes in Figure 14-21 are for relatively minor eruptions that occurred near Greenland, but large eruptions such as Tambora in 1815 can spread significant volcanic debris worldwide. While the 2010 eruptions of Eyjafjallajökull, Iceland, caused massive aircraft flight delays, it did not impact global conditions as the ash remained in the troposphere.

Volcanic climate change is also relevant to prehistoric climates tens of millions of years ago, when immense eruptions covered swaths of continents with lava thousands of meters deep.

Asteroid Impacts

Once thought to be a figment of science fiction writers' imaginations, it is now an accepted scientific fact that objects from outer space can and do collide with planets. The collision of the Shoemaker-Levy comet with Jupiter in 1994 provided spectacular evidence that, to paraphrase the poet John Donne, no planet is an island and that extraterrestrial objects can affect a planet's atmosphere. The climate effects of an asteroid impact are probably similar, but much more pronounced and devastating, than those of a large volcanic eruption. Depending on its exact location, a major asteroid impact and the debris ejected by the impact can cause extended darkness, global fires, acid rain, ozone loss, and even mile-high tsunamis. The result of these extraterrestrial catastrophes can be a cooling of global climate and widespread extinctions, which in turn can cause long-term changes in climate.

Sixty-five million years ago, the age of the dinosaurs ended abruptly, coinciding with the extinction of approximately 75% of the total number of living species. The father-and-son science team of Luis and Walter Alvarez proposed in the 1980s that this extinction was caused by the indirect effects of an asteroid impact on Earth. In 1990 evidence of just such an impact occurred about 65 million years ago was found near the Yucatán Peninsula (**FIGURE 14-22**). Named for a local village, the Chicxulub crater is a 189-kilometer (112-mile) wide impact crater visible in gravity and magnetic field data. The crater size is consistent with a 10- to 20-kilometer (6- to 12-mile) wide asteroid. Remnants of the asteroid have been found in sediments worldwide, confirming its global influence.

The Chicxulub impact was probably not a once-in-history event. There are indications in the fossil record of other extinction events that occur every 26 million years or so (**FIGURE 14-23**). Scientists are coming to the realization that on very long time scales the history of Earth, including its climate, may be periodically upset by asteroid and comet impacts.

FIGURE 14-21 The annual acidity of layers of an ice core from central Greenland for the years 553 to 1972. The Icelandic fissure eruptions of Laki in 1783 and Eldgja in 934 stand out, largely because of their proximity to Greenland. In terms of global climate impact, the two most significant eruptions during this period were Krakatoa in 1883 and Tambora in 1815, which led to the "year without a summer" a hemisphere away in North America and Europe. (Modified from Graedel, T. and P. Crutzen. *Atmosphere, Climate and Change.* Henry Holt & Company, 1997.)

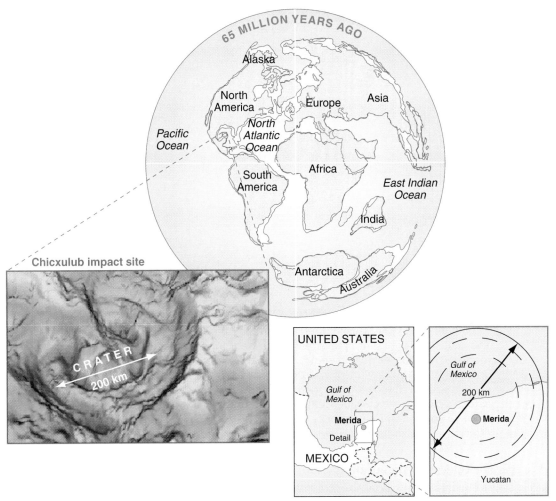

FIGURE 14-22 A collision with a large meteor is believed to have caused a climate change that resulted in the extinction of 70% to 90% of all species at the end of the Mesozoic Era. The top map indicates the landmasses and the point of contact at the time of the collision. The outlines of the Chicxulub Crater are seen today in gravity and magnetic field data of the Yucatán Peninsula region (inset). The blue regions in the inset are low density rock that was pulverized during the impact.

Solar Variability

Less dramatic than asteroid impacts, but probably just as influential to climate, are variations in the amount of energy the Earth receives from the Sun. Since the invention of the telescope in the 1600s, observers have recorded variations in the numbers of dark spots—"sunspots"—on the Sun's surface. These variations normally follow a regular cycle with peaks 11 years apart (**FIGURE 14-24**). NASA and NOAA satellites now monitor these sunspots and the radiative energy output of the sun. This cycle is coincident with an oscillation in solar energy output of a few watts per square meter. The Sun's output is slightly higher during periods with large numbers of sunspots.

It is difficult to prove a direct relationship between the sunspot cycle and climate of the troposphere. However, some atmospheric scientists theorize that the stratosphere acts as an amplifier of the small variations in solar output and causes a discernible 11-year cycle in climate.

As shown in Figure 14-24, between the years 1645 and 1715 the number of sunspots was dramatically lower than observed before or since. This period is known as the **Maunder Minimum**. It is hypothesized that the reduction in solar energy output during this period could have cooled the Earth.

The historical record supplies some evidence supporting this hypothesis. The period between about 1400 and 1850 is called the **Little Ice Age** in Europe, although it was not a true

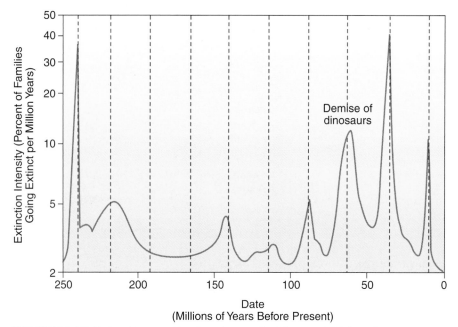

FIGURE 14-23 Extinctions in the fossil record of the past 250 million years. The solid line represents the rate of disappearance of species families in the fossil record per million years. The arrows are spaced 26 million years apart. Notice how well the arrows correspond to the major extinction events, such as the demise of the dinosaurs. This may be an indication that asteroid impacts occur cyclically every 26 million years, caused by as-yet-unknown astronomical patterns. (Adapted from T. Graedel and P. Crutzen. *Atmosphere, Climate, and Change.* Scientific American Library, 1997.)

ice age. The coldest portion of this period, accompanied by the greatest advance of mountain glaciers, occurred around 1750. In geological terms, the Little Ice Age started and ended very quickly. Around 1570, Europe was 1° C to 2° C cooler than it is today. The Thames River in London froze over 11 times in the 17th century, but it has not frozen over in the last 100 years. Canals in Holland also routinely froze solid during this Little Ice Age, as depicted in paintings by the Dutch artists Brueghel, Avercamp, and van der Neer. Access to Greenland was largely cut off by ice between 1410 and the 1720s.

Another explanation for the Little Ice Age is a slowing of warm ocean currents that flow toward Europe. A **Medieval Warm Period** occurred just prior to the Little Ice Age. During this time, large amounts of fresh water flowing into the North Atlantic might have interrupted the ocean circulation patterns.

More recently, the Sun was completely spotless on over 70% of days in 2008 to 2009, the deepest solar minimum observed in almost a century. Despite the lack of sunspots, NASA

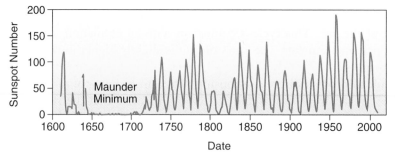

FIGURE 14-24 The yearly average number of sunspots observed since Galileo discovered them in 1610. Notice the regular interval of 11 years between peaks in the numbers of sunspots. Between 1645 and 1715, a period known as the "Maunder Minimum," almost no sunspots were observed. (*Source*: NASA.)

reported that global annual temperatures in 2009 were the second *warmest* in the past 130 years. We explore the reasons why in Chapter 16. Variations in solar output may also affect climate on the time scales of decades to centuries. In addition, there is the **weak Sun paradox** that concerns the earliest billion years of the Earth's atmosphere. At the very beginning of Earth's history, the Sun was young and its output should have been only 70% to 80% of its current output, according to the best theories of astrophysicists. If the Earth's climate is sensitive today to the small changes in solar output during the sunspot cycles, then surely the climate of early Earth should have been profoundly cooler with such a faint Sun. However, the evidence from paleoclimatology is that the Earth's temperature billions of years ago was almost exactly the same as now. There is currently no satisfactory explanation for this puzzling contradiction, although greenhouse gases may have also played a role.

Variations of the Earth's Orbit: Milankovitch Cycles

On time scales of sufficient length, the shape of the Earth's orbit around the Sun and the tilt of its axis are not constant. Instead, they oscillate with periods that are tens of thousands of years in length. In the 1860s and 1870s, a Scottish janitor named James Croll devised a theory explaining how changes in the tilt and orbit of the Earth could cause climate change. Throughout the early 1900s, Serbian engineer Milutin Milankovitch refined and expanded Croll's theory. In the theory's current form, the **Milankovitch cycles** (FIGURE 14-25) that describe the variation of the Earth's orbit are as follows:

"Orbital Cycles and Climate": explore how changing the incoming solar energy can change Earth's climate.

1. **Precession:** The Earth wobbles on its axis once every 27,000 years, similar to a spinning top. This alters the relationship between the solstices and the distance from the Earth to the Sun. For example, 11,000 years ago the Northern Hemisphere summer solstice occurred at perihelion, when the Earth is closest to the Sun. (This is almost the exact opposite of the case today; see Chapter 2.) This "synching up" of summertime with perihelion made the differences between winter and summer more pronounced 11,000 years ago than they are today. More generally, the solstices and equinoxes move slowly forward through the calendar with each passing year, a phenomenon known as precession.
2. **Obliquity:** The tilt of the Earth also changes slightly, with a dominant cycle every 41,000 years. The change in angle of inclination is only about 1° from the present tilt, from about 22° to 24.5°. However, as we learned in Chapter 2, the Earth's tilt is a critical factor in climate. For example, these small changes in tilt lead to solar radiation changes of up to 15% in the high latitudes. "Oblique" means neither parallel nor perpendicular, so changes of the Earth's angle with respect to the Sun often go by the name "obliquity."
3. **Eccentricity:** The shape of the Earth's orbit becomes more or less elliptical on time scales of about 100,000 years. At present, the maximum difference between Sun and Earth distances during the year is only 3%, but over the past several hundred thousand years, this number has been as small as 1% (a nearly circular orbit) and as large as 11% (a more elliptical orbit). This orbital variation gets its name because the deviation of an ellipse from a perfect circular shape is known in geometry as "eccentricity."

These three orbital variations take place simultaneously. Like overlapping musical tones, the cycles of orbital variations create overtones and resonances that are not quite the same as the original cycles. The result is that Earth's climate is affected by these Milankovitch cycles on four different periods: 19,000, 23,000, 41,000, and 100,000 years.

The cold periods experienced 20,000, 60,000, and 160,000 years ago as shown in Figure 14-18 probably resulted from the combined effect of Earth's orbital variations. The cold episodes are separated by lengths of time that closely match the expected periods of climate change resulting from Milankovitch cycles. However, the match is not quite perfect, and it is difficult to explain how variations of a few percent in solar energy on Earth can lead to the 10° C (18° F) variations in global temperature inferred from paleoclimate data. Research continues on this subject.

◼ Plate Tectonics

On the time scale of millions of years, not even the location of the continents can be considered constant. Continental movements are extremely slow in everyday terms—1 to 10 centimeters (up to a few inches) per year. Over millions of years, continents have migrated from the poles to the tropics and the tropics to the poles.

Because the location and latitude of land affects climate, the motions of the continents undoubtedly affect climate on very long time scales. If a continent moves poleward, its climate cools as its solar energy gains decrease. In addition, wherever continents collide mountain ranges are created, which further alters climate.

The theory of continental movement, now known as **plate tectonics**, was pioneered by meteorologist Alfred Wegener, the son-in-law of Vladimir Köppen. Wegener wrote to his future wife in December 1910, "Doesn't the east coast of South America fit exactly against the west coast of Africa, as if they had once been joined [**FIGURE 14-26a**]? This is an idea I'll have to pursue." Wegener's idea was ridiculed by the scientists of his day, but a half-century later the research of paleoscientists confirmed its essential truth. Where the continents fit together in the distant past, they also match in rock formation, glacier flow patterns, fossil record, and continuity in mountain ranges.

Approximately 300 million years ago the continents of today were joined in one supercontinent referred to as **Pangaea** (or Pangae) (Figure 14-26b). Approximately 160 to 230 million years ago Pangaea began to drift apart (Figure 14-26c), eventually forming Laurasia and Gondwanaland. Laurasia consisted of what are today Asia, Europe, and North America. Gondwanaland was comprised of South America, Africa, India, Australia, and Antarctica. As the continents continued

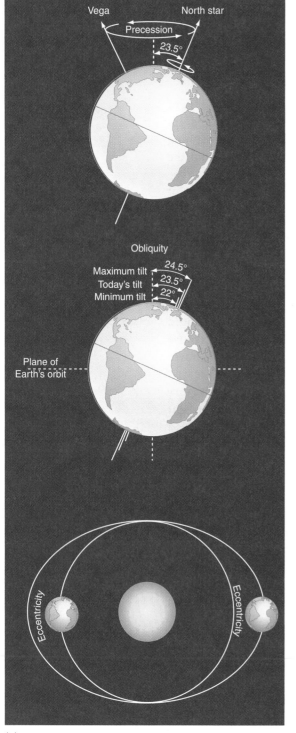

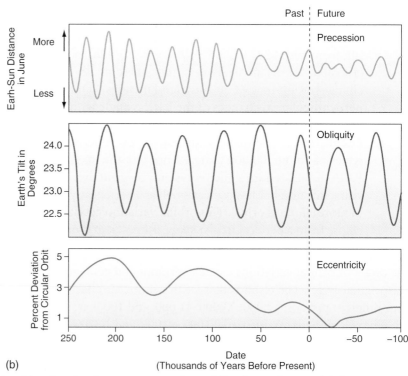

FIGURE 14-25 (a) The three orbital variations that lead to climate change: precession, obliquity, and eccentricity (exaggerated in the drawing). (b) The cycles of these orbital variations during the past, present, and future. (Part b adapted from T. Graedel and P. Crutzen, Atmosphere, climate, and change. *Scientific American Library*, 1997.)

to drift apart, some land masses collided, forming today's mountain ranges, including the Himalayas and Rocky Mountains.

Notice in Figure 14-26 that Pangaea was largely a tropical supercontinent, whereas today much of the world's land is located at high latitudes. The trend toward mountainous high-latitude land is a "recent" phenomenon, occurring during the past tens of millions of years. This may explain the most notable climate development of the Earth's recent past: the ice ages.

An **ice age** is a period of global cooling that leads to the creation of vast ice sheets across the Earth's land masses. We are currently in an ice age that began approximately 10 million years ago. Before that, the Earth was comparatively warm for hundreds of millions of years (Table 14-1).

Why are there ice ages? It is possible that the increasing concentration of land at high latitudes, combined with its increasing elevation, have progressively cooled the Earth over long periods. It is far easier to create and sustain glaciers in high-latitude mountains and continents than on water. For this reason, scientists think that plate tectonics may have directly contributed to the formation of continental ice sheets.

Just 20,000 years ago, North America was buried under 1.6 kilometers (1 mile) or more of ice, the edge of which extended as far south as what is now St. Louis, Missouri (**FIGURE 14-27**). Why has the ice retreated since then? One possibility is that the Milankovitch cycles that led to more pronounced seasonality 10,000 years ago also helped melt the ice sheets during the warm Northern Hemisphere summertime. In this way two different climate change mechanisms can combine to create a complex history of global climate.

Changes in Ocean Circulation Patterns

We close with a climate change mechanism that can be both rapid and long lasting. Today, surface ocean water in certain regions of the North Atlantic and Antarctic oceans sinks because it is cold and salty enough to be denser than the water beneath it. Oceanographers have discovered that the location and strength of this sinking water drives ocean circulations across the globe. However, past changes in the composition of the ocean waters, as well as changes in the locations of the continents, have altered the location of sinking ocean waters. Oceanographers contend that, as a result, ocean circulations in the past have been dramatically different than today's. Because of the influence of ocean circulation on climate, past climates may also have been dramatically different.

One possible example of the effect of ocean circulation on climate occurred at the end of the last advance of ice across the Northern Hemisphere. Probably because of changes in the Earth's orbital patterns, global temperatures warmed about 13,000 years ago. This caused the glaciers covering North America to melt rapidly, raising sea levels at a rate of 1 centimeter (about 0.4 inch) per year. Then temperatures over North America and Europe suddenly cooled for almost 1000 years. This relatively cold period existed between 12,000 and 11,000 years ago (**FIGURE 14-28**) and is known as the **Younger Dryas** (named for the reappearance of a polar wildflower, Dryas octopetala, in Europe).

A possible cause for the cooling during the Younger Dryas is the melting of the North American Laurentide ice sheet. The massive floods of fresh water from the melting glaciers may have rushed out of the

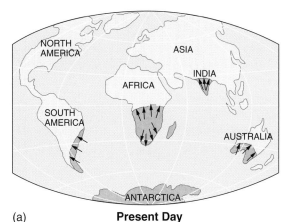

(a)   **Present Day**

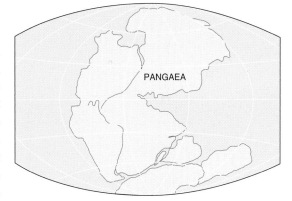

(b)   **Jurassic**
          135 million years ago

Permian
(c)   225 million years ago

FIGURE 14-26 Map (a) shows the continents as they appear today. About 135 million years ago the plates were moving apart (b). Approximately 300 million years ago the continents of today were joined in one supercontinent referred to as Pangaea (c). The shapes of the continents fit together like a jigsaw puzzle. In addition to the jigsaw pieces locking together, the patterns on the surface must match. Similarly, patterns on the Earth's surface must match to support the theory that the continents were truly combined. Notice how evidence found on several continents for ice sheets in the distant past tends to confirm the theory of a joined-together supercontinent.

FIGURE 14-27 The extent of the ice sheets over North America during the Pleistocene. The arrows indicate the direction of advance of the sheets.

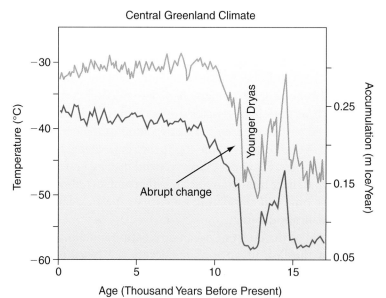

FIGURE 14-28 The temperature (orange) and ice accumulation (blue) in central Greenland on a yearly basis from the present (left) to 18,000 years ago (right), as inferred from ice cores. Notice the rapid spikes of warming and cooling near the end of the last Ice Age, between 10,000 and 15,000 years ago. The very cold episode about 12,000 years ago is the "Younger Dryas." Climate scientists believe these spikes are related to changes in ocean circulation related to the melting of the ice sheet. (*Source*: NOAA.)

St. Lawrence River Valley and into the North Atlantic. This could have caused North Atlantic surface ocean water to become much less salty and less dense. If the surface water in the North Atlantic became too light to sink, this could have shut down the "normal" circulation of the North Atlantic for hundreds of years. The change in vertical ocean motions could, in turn, have changed circulation patterns throughout the oceans of the Northern Hemisphere and even the world. In particular, a weakened northward transport of heat by warm ocean currents could have caused the cooling associated with the Younger Dryas in Europe and North America.

During the past 15 years, new research on the Greenland ice sheets has revealed abrupt climate changes associated with the Younger Dryas, on the time scale of just a few years. For example, at the end of the Younger Dryas the annual mean temperature warmed a whopping 15° C (27° F), and the annual precipitation doubled in less than a decade (Figure 14.28)!

WHY STUDY CLIMATES OF THE PAST?

What ever happened to the dinosaurs? This is a question on the mind of countless children and many grown-ups as well. To solve this puzzle requires us to understand the environment these creatures lived in and how that environment might have rapidly changed. Acting like detectives, scientists have collected various types of proxy data to research the answer to this and many other questions about Earth's past. Of course there is uncertainty in these proxy methods. Thus, scientists look for consistencies in the evidence to support their interpretation. They take many samples of various proxy data sets and look for similarities and discrepancies. The challenge is then to explain how these data sets are consistent with a given explanation. For example, a puzzle of the dinosaur extinction being caused by the Chicxulub meteorite is that the impact occurred 300,000 years prior to the mass extinctions. How this early impact could have led to global extinctions is still debated. Postulations that long-lasting volcanic activity, in what is now India, contributed to the extinction have been debated and generally concluded to have not played a major role. Understanding the role climate change in the extinction of the dinosaurs may help us to explain observations of the many mass extinctions that have occurred during Earth's history.

We also study past climate changes to help us understand the impact of the environment on human civilizations. For example, the ice cores on Kilimanjaro indicate widespread drought in tropical Africa about 4200 years ago that lasted for 300 years. These data are consistent with archeological evidence of the demise of the ancient Akkadian city of Shekhna, in what is now Syria. A population of 10,000 people deserted the city during this time of widespread drought. While conclusions have been reached, however, there is still more research to do. Although there was evidence of a volcanic eruption during this time, it is unlikely the volcano caused the drought. So, questions still remain: "How can a drought persist for three centuries?" and "Could it happen again?"

Instrument observations of our atmosphere only cover the last 150 years. During this time, as we saw in Chapter 3, we have observed changes in temperature and precipitation distributions. We have a good understanding of these observed changes. What these observations do not tell us is if these changes are typical. Understanding climate change from a scientific point of view requires us to gain insight on how these observations link to past climates. One lesson from the paleoclimate studies is that Earth's climate is always changing—sometimes very abruptly. Thus, by studying past climatic variations, scientists seek to gain clues on what might lie in the future. Today, climate predictions are accomplished using computer models (see Chapter 16). One way to assess the performance of these models is to compare how well they simulate changing climates as indicated in the paleoclimatic records. Past and present climate data provide a clearer understanding of our current climate and how climate changes.

This concludes our tour of natural processes that affect climate. However, we have not accounted for 100% of the observed changes in Earth's climate. Other factors, including natural variability, the ability of climate to oscillate without being forced to, may explain the changes in climate that are not driven by the mechanisms we have discussed here. This is particularly relevant to observations during the past 30 years indicating that Earth's climate is warming.

PUTTING IT ALL TOGETHER

▇ Summary

Climate varies from place to place and over time. The five basic climate controls are latitude, elevation, topography, proximity to large bodies of water, and prevailing atmospheric circulation.

Climate classifications are developed to organize the complex and varied climates of the world. The Köppen-based climate classification is the most widely used scheme for describing today's climates. The most recent Köppen-Geiger scheme has six main groups, each designated with a letter: Tropical Moist (A), Dry (B), Moist with Mild Winters (C), Moist with Severe Winters (D), and Polar (E).

Through a variety of scientific evidence, climatologists have been able to determine the past climates of Earth. This evidence includes historical records, tree rings, pollen deposits, air and dust trapped in ice sheets, sediments found on the ocean floor, and fossils. This evidence indicates that during most of the past 500 million years the Earth enjoyed a warmer and more congenial climate than we experience today. This mild climate has been repeatedly punctuated by cooler periods, including ice ages.

Scientists have proposed several different mechanisms by which Earth's climate can change naturally on a variety of time scales. Emissions from volcanic eruptions can cool the Earth for a few years. Asteroids and comets can crash into the Earth, causing mass extinctions and the dawn of new eras in Earth's biological and climatic histories. The Sun's energy output can vary over short and very long time scales. The Earth's orbit also varies on time scales of tens of thousands of years, altering the distribution and amount of solar energy reaching Earth. The motions of the continents over millions of years have probably triggered eras of worldwide warmth as well as ice ages. Changes in the ocean circulation can help cause abrupt climate change during time periods of a decade or less.

For the past 20,000 years, records of climate are rather complete. Over this time span the Earth's climate changed from a period of extreme glaciation to a warmer interglacial period. Alternating episodes of relatively cooler and warmer temperatures have characterized the past several thousand years, including the Younger Dryas about 11,000 years ago and the more recent Little Ice Age (1400 to 1850) in and near Europe. Changes in ocean circulation and solar output may explain these oscillations in climate. Recent increases in global temperature may be a result of human activity, which we examine in the next chapter.

▇ Key Terms

You should understand all the following terms. Use the glossary and this chapter to improve your understanding of these terms.

Boreal forests	Historical climate	Maunder Minimum
Climate	Humid continental	Medieval Warm Period
Climographs	Ice ages	Milankovitch cycles
Dendrochronology	Ice sheets	Moist subtropical and
Dry climates zones (B)	Isotopes	mid-latitude climates (C)
Eccentricity	Köppen scheme	Obliquity
Glacier	Little Ice Age	Paleoclimate

Pangaea

Permafrost

Plate tectonics

Polar climates (E)

Precession

Radiocarbon dating

Severe mid-latitude
climates (D)

Subarctic climate

Taiga

Tropical humid climates (A)

Weak Sun paradox

Younger Dryas

Review Questions

1. What is the difference between weather and climate?

2. Describe the climate where you live in terms of the Köppen climate classification scheme. Using Figure 14-2, find a location on a different continent that has the same classification as your hometown.

3. What features separate a tropical climate from the climate of the poles?

4. What type of parameters should be included in a climate classification scheme that would be useful to city planners? To farmers?

5. Develop a climate classification scheme based on precipitation type and amount that would be useful to people who want to avoid regions with lots of snow or ice storms.

6. Consider how you might design a climate classification scheme that is based on factors other than temperature and precipitation. For example, how could you use the frequency of air mass types to categorize climate?

7. In a million years from today, human activities from our time might be studied to determine the climate of our age. Describe how things you do today could leave behind a useful signature of today's climate.

8. Explain why the deepest layers of ice on a glacier are the oldest.

9. What was the Little Ice Age? Was it really an ice age? Why or why not?

10. Why are lower sea levels associated with ice ages?

11. Over a century ago, trees cut down from old-growth forests in the upper midwestern United States routinely sank in the cold waters of Lake Superior while being floated downstream to furniture makers. The cold lake waters have preserved these logs, and today furniture makers are raising them from the depths to make custom furniture. Can you think of a way to use these trees for paleoclimate purposes? How could you get around the problem of not knowing exactly when the trees were cut down?

12. What is plate tectonics? Why is it important in climate change studies?

13. What climate change mechanisms might cause a mass extinction in the near future?

14. Discuss how a mile-high ice sheet would affect climate conditions because of its height, distribution across a continent, and temperature.

15. Based on the number of craters on the Moon, we expect to find 400 large craters (50 kilometers wide) on Earth. However, only about 160 known impact craters, with only about 14 craters larger than 50 kilometers wide, have been found on Earth so far. Why do we find so few craters on Earth?

16. What is the Maunder Minimum? When did it happen, and how might it be connected to climate change in the past? Answer the same questions with regard to the Younger Dryas.

17. Use Figure 14-25 and the discussion in the text to explain how Milankovitch cycles should affect the Earth's climate in the future—for example, the next 50,000 years.

18. Following are the recent average temperatures of the Northern Hemisphere and Southern Hemisphere for winter, summer, and the year. The annual range is given as well as the differences between the Hemispheres. Discuss the causes of these observed differences.

	Winter	Summer	Year	Annual Range
Northern Hemisphere	8.1° C (46.6° F)	22.4° C (72.3° F)	15.2° C (59.4° F)	14.3° C (25.7° F)
Southern Hemisphere	9.7° C (49.5° F)	17.0° C (62.6° F)	13.3° C (55.9° F)	7.3° C (13.1° F)
Difference	−1.6° C (−2.9° F)	5.4° C (9.7° F)	1.9° C (3.5° F)	7.0° C (12.6° F)

■ Observation Activities

1. Climate classification is for a large region and may not apply on the local scale. Get temperature and precipitation data for a city near you from a state climatologist Web site (see http://cdo.ncdc.noaa .gov/cgi-bin/climatenormals/climatenormals.pl). Compare these data with the climate classification discussed in this chapter, and argue whether the classification is appropriate or not.

2. Microclimate is the climate of a small area, such as a woodland, lake, or even near a fence. Look for evidence of different microclimates where you live. Describe these microclimates and the factors that determined these features.

3. What kind of vegetation, if any, did your location have 18,000 years ago at the peak of the last glacial maximum (see www.esd.ornl.gov/projects/qen/NAL2215.gif). Look out your window and try to visualize what your area looked like back then.

This rain cloud icon is your clue to go to the *Meteorology* Web site at http://physicalscience.jbpub.com/ ackerman/meteorology/. Through animations, quizzes, web exercises, and more, you can explore in further detail many fascinating topics in meteorology.

Human Influences on Climate

15

AFTER COMPLETING THIS CHAPTER, YOU SHOULD BE ABLE TO
- Name the major pollutants in our atmosphere and describe their impacts
- Describe evidence that global warming is occurring
- Explain the roles of feedback mechanisms in climate change
- Relate acid rain and the ozone hole to human-caused changes to the atmosphere
- Discuss global warming and the roles of land-surface changes, clouds, and the oceans in climate change

INTRODUCTION

John Ruskin taught what he saw, and by the end of the 19th century he didn't like what he saw. A professor of fine arts at Oxford University in England, Ruskin once said, "I have come to the conclusion that it is not Art that I loved but Nature." The Industrial Revolution fouled the landscapes Ruskin adored with clouds of smoke and soot, however.

Ruskin's knowledge of the atmosphere was based on observations of storms and sunsets in diaries and sketchbooks that spanned a period of 50 years. In 1884, certain that his data reflected a fundamental change in the atmosphere, Ruskin gave a lecture entitled, "The Storm-Cloud of the Nineteenth Century." In this lecture he called the 1880s a "period which will assuredly be recognized in future meteorological history as one of phenomena hitherto unrecorded in the courses of nature." He described, probably for the first time, the phenomenon later known as acid rain. Ruskin also captured the impact of humans on climate in a clever turn of phrase by charging that "the empire of England, on which formerly the Sun never set, has become one on which he never rises." Later scholars agreed that Ruskin bore witness to an actual change in the atmosphere; he first saw the darkening "plague-cloud" when the burning of sooty coal increased dramatically across Europe.

A century later, climate scientists following Ruskin's lead have announced publicly that humans are very likely responsible for global climate change. In this chapter we explore how scientists have reached these conclusions. Are the scientists brave and accurate prophets in the mold of John Ruskin, or are they falsely "crying wolf"? This chapter is intended to supply you with the knowledge to develop informed opinions concerning the often-contentious topics of pollution, acid rain, the ozone hole, and global warming. We begin by discussing how a small change can start a process that leads to a much bigger change.

FEEDBACK MECHANISMS

As we have seen in previous chapters, climate is the result of an interplay of different mechanisms. To understand climate change, we need to understand how and when different mechanisms lead to these changes.

Feedbacks occur when one change leads to some other change, which can act to either reinforce or inhibit the original change. A **positive feedback mechanism** is one that enhances an existing trend of a change in climate. It is through positive feedback mechanisms that small changes can lead to large ones.

You can learn about positive feedback by experimenting with a microphone and an amplifier with a stereo speaker attached. If you stick the microphone into the speaker, here is what happens. First, the sounds that the microphone detects are amplified and emitted by the speaker. Then the microphone picks up these amplified sounds, which are then amplified even more and emitted by the speaker. This cycle repeats continuously. In just a few seconds, a deafening high-pitched sound develops even in a room that was initially quiet. You have probably heard this annoying sound at a concert, large lecture course, or another event that used microphones.

Sound engineers have long called this effect "feedback." This name arises because the sound into the microphone is repeatedly "fed" back into the amplifier, getting louder and louder with each cycle from the microphone to the speaker. The "positive" part of the name means that the amplification of the sound grows with time.

Positive feedbacks can occur in climate and contribute to rapid climate change, just as positive feedback with sound equipment leads to a rapid change in noise level. For example, the amount of water vapor in the atmosphere is a positive feedback mechanism involved in climate change. As the air temperature warms, there is increased evaporation from surface waters, resulting in higher atmospheric water content. As the atmosphere warms, more water vapor can exist in the atmosphere. This causes the atmosphere to warm even further, causing more water vapor to evaporate into the atmosphere, enhancing the warming. Although human emissions of carbon dioxide are attributed to the current trends in global warming, it is the water vapor feedback that causes most of the warming.

Another example of a positive feedback mechanism is the **ice-albedo temperature feedback**. Ice sheets affect climate by reflecting more sunlight than other types of surfaces, such as bare ground and areas covered by vegetation. This reduces the amount of the Sun's energy that can warm the planet's surface. Other things being equal, the more ice, the cooler the Earth is. Ice-covered areas can expand when the atmosphere along the margins of established ice regions cool. The increased area of ice reflects even more sunlight and further reduces the amount of solar energy absorbed by the surface. This reduction in solar energy gains causes further cooling and results in the formation of more ice and so on. This is a positive feedback loop: More ice causes a cooling, which reduces the temperature so that more ice develops, leading to a further cooling (**FIGURE 15-1**).

Also, a retreat of an ice sheet is also a positive feedback because it would cause a warming, as a result of the lower albedo, and lead to a warming and a further retreat of the ice sheet. Positive feedbacks can cause "less, less, less" as well as "more, more, more." A classic example of this in human culture is a stock market crash, in which financial worries amplify and cause an accelerating trend of dropping stock prices.

A **negative feedback mechanism**, in contrast, damps out an existing trend of climate change. A simple example of a negative feedback mechanism involves the influence of carbon dioxide (CO_2) concentration on plant photosynthetic rates. In the process of photosynthesis, plants use carbon dioxide and water to make sugar. An environment rich in carbon dioxide accelerates the growth of many plant species. This is a negative feedback: increasing carbon dioxide concentrations allow plants to grow faster and thereby increase the overall photosynthetic rate, which then removes increased amounts of carbon dioxide from the atmosphere. So, by a negative feedback loop more carbon dioxide results in a decrease of carbon dioxide,

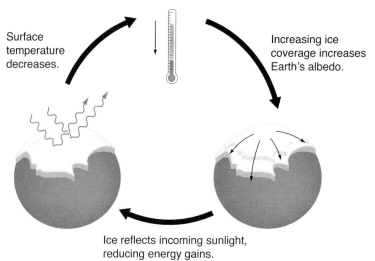

Surface temperature decreases.

Increasing ice coverage increases Earth's albedo.

Ice reflects incoming sunlight, reducing energy gains.

FIGURE 15-1 Ice-albedo temperature feedback explains how a small change in surface temperature can grow into a larger change.

and the end result is an equilibrium instead of an accelerating trend as in the case of a positive feedback.

Of course, there are other limiting factors in a plant's ability to respond to an enriched carbon dioxide environment, such as a lack of water and nutrients. Also, insects and other pests might also enjoy the warmer environment associated with high concentrations of carbon dioxide, which can reduce the number of plants. Feedback processes can quickly become complicated even in the simplest climate change scenario.

HUMAN IMPACTS

In the previous chapter we described current climate conditions and past climate characteristics. We briefly discussed the tools used to detect climates of the distant past and clearly demonstrated that climate changes. Detecting these changes is one challenge; another is attribution. Scientists work hard at attributing the most likely causes for an observed change. In this section, we discuss some changes that are easily attributed to human activities. Then we move on to observations that are consistent with a world that is warming. The final chapter provides evidence, and our level of confidence, that these observed global changes result from human activity.

Air Pollution

Air pollution is composed of airborne solid and liquid **particulates** called aerosols as well as gases that, when in high concentrations, seriously affect the lives of people and animals, harm plants, or threaten ecosystems. Pollutants can arise from human activities (anthropogenic sources) or from natural sources such as dust storms and volcanic eruptions. In previous chapters we focused on the natural sources, so here we examine anthropogenic air pollution.

Carbon monoxide (CO), a colorless and odorless gas, is a prime example of a harmful pollutant. Carbon monoxide forms when fossil fuels are not completely burned during combustion—for example, in car engines or in forest fires (FIGURE 15-2). It is very toxic because it disrupts how red blood cells absorb oxygen. As a result, inhalation of carbon monoxide reduces the body's ability to provide oxygen to the body. In cities, the carbon monoxide levels can approach unsafe levels in confined areas, such as garages and tunnels. Carbon

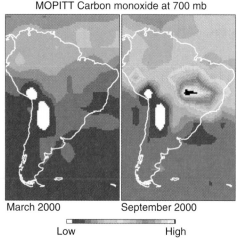

MOPITT Carbon monoxide at 700 mb

March 2000   September 2000

Low   High

FIGURE 15-2 A research satellite image of carbon monoxide in the lower troposphere over South America in connection with deforestation and burning of forests.

monoxide is also very dangerous in the home, where it can be produced by heating devices that are not operating properly.

Lead is a particulate released into the atmosphere when, for example, treated gasoline is burned in car engines. It settles out of the atmosphere and is easily absorbed into the human body. Even small amounts of lead cause brain damage and lower IQs in infants. For this reason, unleaded gasoline was developed and replaced leaded fuels in American automobiles by the 1980s, reducing lead emissions dramatically. Some possible societal benefits of reduced lead emissions are explored in **BOX 15-1**.

Pollutants can also develop from harmless chemicals that are emitted directly into the atmosphere but then become noxious gases or particulates after chemically combining with other atmospheric constituents. Human activities produce oxides of sulfur, in particular **sulfur dioxide** (SO_2) and **sulfur trioxide** (SO_3). These sulfur oxide compounds are released into the atmosphere primarily through the burning of fossil fuels that contain sulfur. Levels of sulfur dioxide and sulfur trioxide can become large when the activities concentrate the compounds over a small region that allows the pollutants to reach high levels, as in urban and industrial areas. Sulfur dioxide is a highly corrosive gas that irritates the human respiratory system. Sulfur trioxide is an important pollutant because it readily combines with water vapor to form droplets of **sulfuric acid** (H_2SO_4), creating acid rain, which we discuss in detail in the next section.

Oxides of nitrogen, in particular **nitric oxide** (**NO**) and **nitrogen dioxide** (NO_2), are two important air pollutants. Nitric oxide is a by-product of high temperature combustion, such as in automobile engines and electric power generation. Indeed, atmospheric concentrations of nitrogen dioxide in urban areas are well correlated with the density of vehicular traffic. Nitric oxide is a very reactive gas and quickly forms nitrogen dioxide. Nitrogen dioxide also is a toxic gas that is emitted by automobile engines. High concentrations of nitrogen dioxide give polluted air its reddish brown color. In high concentrations, oxides of nitrogen cause serious pulmonary problems.

Hydrocarbons are compounds made of hydrogen and carbon atoms. Examples of hydrocarbons are methane, butane, and propane, which can occur as either a gas or particulate. Hydrocarbons are also called **volatile organic compounds** (**VOCs**). During the day, hydrocarbons can combine with nitrogen oxides and oxygen to produce photochemical **smog**.

The main component of **photochemical smog** is ozone (O_3). **Ozone** is a chemically active molecule and is considered a corrosive gas. In the presence of sunlight, oxides of nitrogen from engine exhaust and hydrocarbons react to form a noxious mixture of aerosols and gases. This mixture includes ozone, formaldehyde, and PAN (peroxyacetyl nitrates). Exposure to high concentrations of ozone irritates the eyes, nose, and throat and causes coughing, chest pain, and shortness of breath. Ozone also aggravates diseases such as asthma and bronchitis.

TABLE 15-1 summarizes recent trends in emissions of major pollutants for the United States as a whole. Although the Clean Air Acts of 1970 and 1990 have led to remarkable improvements in most directly emitted pollutants, tropospheric ozone has emerged as a growing threat to health. Ozone alerts have become a commonplace occurrence in many urban regions of the United States, and gasolines specially formulated to reduce ozone pollution are sold (at a higher price) in these regions.

Air pollution is not just a modern-day problem. Cave dwellers undoubtedly had to address local air-quality problems such as smoke from fires. The Hopi Indians had to deal with sulfur dioxide emitted from burning coal to make pottery. England has long had a problem with coal burning. Air quality was so poor that it has been estimated that in 18th-century England half the children died before their second birthday.

TABLE 15-1 Trends in Air Pollutant Concentrations in the United States from 1990 to 2007

Pollutant	Lead	Carbon Monoxide	Sulfur Dioxide	Particulates	Nitrogen Dioxide	Smog
% Change	−80%	−67%	−54%	−19%	−35%	−10%

Box 15-1 Taking an Unleaded Bite Out of Crime?

Could a healthier environment be reducing crime in the U.S.? At first glance, this seems unlikely for at least three reasons: (1) Since 2003, a majority of Americans surveyed annually in a Gallup Poll have said that crime is going up, not down, in the U.S. (2) It is a basic tenet of criminology that crime is related to the economy, so the economic downturn in the late 2000s should have led to increased crime, and (3) people cause crimes, not the environment, right? But what if all this conventional wisdom is wrong?

In May 2010, the U.S. Federal Bureau of Investigation released preliminary statistics showing that in 2009 violent crime dropped 5.5% versus 2008. This was the third consecutive yearly decrease in an accelerating trend toward fewer violent crimes in the United States. Murder was down 7.2% in 2009; robbery was down 8.1%. The total number of violent crimes in 2009 was the lowest in at least 2 decades, despite a 20% increase in U.S. population over the same period. Nonviolent crimes such as arson and motor vehicle theft decreased at double-digit rates.

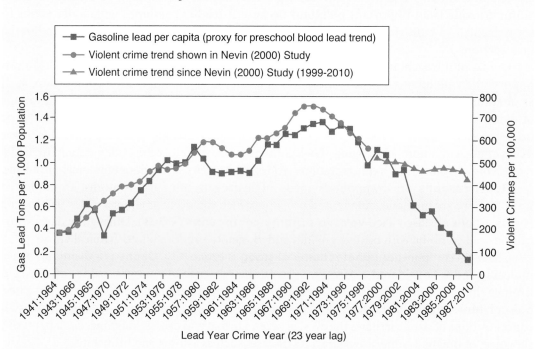

(Adapted from Nevin, Rick, *Environmental Research* 104 [2007]: 315–336.)

Experts in criminology struggled to explain how crime could plummet in 2009 during the most severe economic recession since the Great Depression. Big-city police chiefs credited increased vigilance by officers and the public. But crime decreased at similarly large rates in large metropolises and small towns alike. (Nevertheless, as of October 2009, 74% of Americans believed that crime was *increasing* in the U.S., according to Gallup's annual Crime Poll.)

Meanwhile, juvenile arrest rates have dropped by 33% nationally between 1997 and 2008. Again, criminologists are at a loss to explain why juvenile reformatories that were full in the 1990s are nearly empty today.

Could the reduction in crime be due to something even more basic than better police methods or expanded after-school activities for juveniles? Are high school- and college-age Americans somehow different today than they were 20 years ago, when violent crime was almost twice as common as it is today?

There is at least one undeniable difference. According to a 2007 study by economist Rick Nevin, a child in the 1990s had up to *10 times less* lead in his or her bloodstream than a preschooler in the mid 1970s.

This huge drop in blood lead levels is because of the near-total phase-out of leaded gasoline in the U.S. in 1986, which became a total ban in 1996. Prior to the phase-out and ban, leaded gas had been used in millions of vehicles on U.S. roads for over a half century. The Environmental Protection Agency has estimated that 200,000 tons of lead entered the atmosphere in the U.S. annually in the early 1970s because of leaded fuels. These lead emissions, plus lead in paints, ended up in the bodies of children who breathed in the fumes or nibbled at the paints. This made American children increasingly "leaded" from the 1940s until the 1980s, with a drastic decline in blood lead levels since.

How does lead relate to crime? It is well-established that lead poisoning causes violent behavior. A long-term medical study published in 2008 found that for every 5 microgram per deciliter increase in lead in the blood of a six-year-old child in Cincinnati, Ohio, in the late 1970s and early 1980s, the risk of that child being arrested for a violent crime as a young adult in the late 1990s and 2000s increased by a whopping 50%!

Therefore, it is plausible that the remarkable reduction in environmental lead levels discussed in this chapter could have a similarly large impact on crime rates, as children become teenagers and young adults. Economist Jessica Wolpaw Reyes at Amherst College observes that "it is certainly possible that lead is playing a role in these declines in violent crime. We are entering the years of criminal maturity for the 'fully unleaded cohorts.' This could potentially lead to drastically lower violent crime rates."

Furthermore, a lead-crime linkage has been found in statistical analyses for both the U.S. and other nations. Nevin's research demonstrated that 65% to 90% or more of the variation in violent crime in nine different countries was explained by lead. The figure, from Nevin's work, shows lead levels plotted versus U.S. violent crime levels 23 years later. The relatively close match in the two curves is not proof of a causal connection between childhood lead levels and U.S. crime rates when the kids become young adults, but it is intriguing.

Is the lead-crime connection the secret behind the puzzling drop in crime in the U.S.? No one knows for sure. Follow the story in the coming years to see what happens. If crime continues to decrease in parallel with blood lead levels, and independent of economic trends, then the answer may be "yes."

Aerosols and Clouds

Water is crucial to climate, so it is no surprise that water plays a critical role in global warming discussions. Water vapor and clouds play a vital role in preserving the balance between incoming and outgoing energy in the Earth. Water vapor is a strong greenhouse gas. Water vapor is transparent to solar radiation while being extremely effective at absorbing terrestrial radiation emitted from the Earth's surface.

Human activities add little water vapor to the troposphere, at least directly. Water vapor increases are a result of internal controls of the climate system. As we learned in Chapter 4, the equilibrium between air and water is such that saturation occurs at higher vapor pressures for warmer temperatures than for colder temperatures. Therefore, more atmospheric water vapor may be present in a globally warmed atmosphere. Increased water vapor concentration can warm the atmosphere further, leading to a further increase in atmospheric water vapor in a positive feedback loop. From this perspective, a strong case can be made for global warming resulting from increased greenhouse gases.

However, increased water vapor amounts may also enhance cloud cover that can induce a different global climate change. It is difficult to predict the effect of changes in cloudiness on climate. Different cloud types affect the climate in different ways. High, thin cirrus clouds can lead to a warming, similar to the greenhouse warming. Thick stratocumulus clouds cause a cooling by reflecting large amounts of solar energy back to space. Satellite observations indicate that the average global cloud distribution causes a net cooling of the planet.

Aerosols can also affect climate by causing changes in cloud radiative properties. Aerosols serve as cloud condensation nuclei (see Chapter 4). When more nuclei are present, more droplets

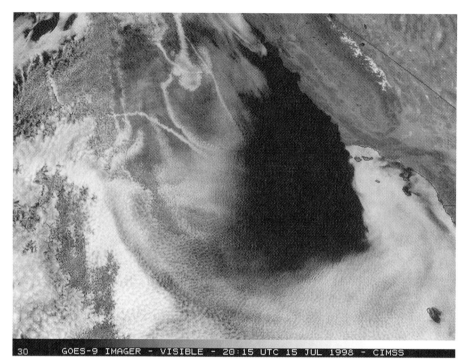

30   GOES-9 IMAGER - VISIBLE - 20:15 UTC 15 JUL 1998 - CIMSS

FIGURE 15-3 A visible image from a geostationary satellite of stratus clouds off the coast of California. The ragged lines in the upper portions of the figure are clouds whose reflectance has been enhanced by increased aerosols from ships moving below the clouds. This suggests that human activities can increase the brightness of clouds and thereby cause a cooling of the planet.

FIGURE 15-4 Contrails, the straight-line clouds over Lake Superior and Lake Michigan. Contrails form when hot, humid air from jet exhaust mixes with environmental air of low vapor pressure and low temperature.

will form in the cloud, and the droplets will be smaller. This makes the cloud more reflective (see Chapter 5), further reducing the solar energy reaching the surface.

Observations of cloud droplets over the Atlantic Ocean downwind of northeast North America indicate that these clouds tend to be composed of drops that are smaller than similar clouds that are in a pristine environment. Evidence of this is seen in satellite images of **ship tracks** (FIGURE 15-3). Pollutants from the ship engines rise upward into the cloud and serve as cloud condensation nuclei (CCN). This increases the number of cloud droplets in the cloud, making the cloud appear brighter. Smaller particles in high concentrations also make it less likely the cloud will yield precipitation. The effect of aerosols on the radiation budget through their impact on clouds is called an indirect aerosol effect on climate.

In addition to changing cloud properties, human activities may directly change cloud amount through the creation of clouds. For example, airplane-induced clouds may have a regional effect on climate.

Contrails

The white condensation trails left behind jet aircraft are called **contrails** (condensation trails). The satellite image in FIGURE 15-4 shows multiple contrails over Lake Superior and the upper peninsula of Michigan. Contrails form when hot, humid air from jet exhaust mixes with environmental air of low vapor pressure and low temperature. The mixing is a result of turbulence generated by the engine exhaust.

Cloud formation by a mixing process is similar to the cloud you see when you exhale in cold air and "see your breath." FIGURE 15-5 represents how saturation vapor pressure varies as a function of temperature. The blue line is the saturation vapor pressure for ice as a function of temperature (see Chapter 4). Air parcels in the region labeled saturated will form a cloud. Imagine two parcels of air, A and B, as located on the diagram. Both parcels are unsaturated. If B represents the engine exhaust, then as it mixes with the environment (parcel A) its temperature and corresponding vapor pressure will follow the dashed line. Where this dashed line intersects the blue line, the parcel becomes saturated and a cloud forms.

If you pay attention to contrail formation and duration, you will notice that they sometimes rapidly dissipate, but other times they will spread horizontally into an extensive thin cirrus layer. How long a contrail remains intact depends on the humidity structure and winds of the upper troposphere. If the atmosphere is near saturation, the contrail may exist for several hours. However, if the atmosphere is dry, the contrail will dissipate as soon as it mixes with the environment.

Contrails are a concern in climate studies because increased jet aircraft traffic may result in an increase in cloud cover over certain regions. It has been estimated that in certain heavy air-traffic corridors such as over Chicago, cloud cover has increased by as much as 20%. An increase in cloud amount changes the region's radiation balance. For example, solar energy reaching the surface may be reduced, resulting in surface cooling. Clouds can also reduce the terrestrial energy losses of the planet by absorbing energy emitted by the surface, resulting in a warming.

In the days following the September 11, 2001, terrorist attacks, all U.S. commercial jets were grounded, and for the first time in decades, American skies were briefly contrail-free. This 3-day period has been used to study the impact of contrails on surface temperature. The analysis suggests that contrails decrease the diurnal temperature range (see Chapter 3), largely by reducing the daytime maximum temperature.

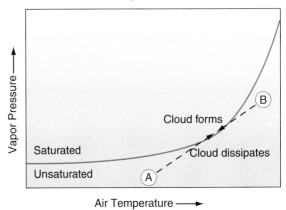

FIGURE 15-5 Saturation vapor pressure varies as a function of temperature. The blue line is the saturation vapor pressure for ice as a function of temperature (see Chapter 4). Air parcels in the region labeled saturated will form a cloud. Points A and B represent two unsaturated parcels, which if mixed can result in a saturated parcel.

"Growing a Contrail"

Acid Deposition

Acid deposition refers to the falling of acids and acid-forming compounds from the atmosphere to Earth's surface. Air pollution from industrial areas can become acidic and be carried downwind for many miles. When these acids settle on the ground, they can damage plants and aquatic life. This settling may occur as dry particles (dry deposition) or as rain, snow, or fog (wet deposition). When in the form of rain, the acid deposition is referred to as **acid rain**.

The **pH scale** (FIGURE 15-6) represents the acidity of a solution. pH levels range from 0 to 14, with 0 being extremely acidic. A pH of 7 is neutral. The scale is logarithmic, so a change in pH from 7 to 6 or from 5 to 4 represents a 10-fold increase in acidity.

Normal precipitation has a slightly acidic pH value of approximately 5.5. This is because some atmospheric carbon dioxide dissolves in the water drops as they form and grow. Acid rain forms because of the increased levels of sulfur dioxide and oxides of nitrogen that enter the atmosphere as a result of burning of high-sulfur-content coal. These gases dissolve in the cloud drops, making the precipitation more acidic and lowering the pH from 5.5 to between 4 and 4.5. This is an increase in acidity of a factor of 10 or more.

Industrial sources and petroleum-powered vehicles emit massive quantities of sulfur dioxide and nitrogen oxides into the atmosphere. The United States alone emits approximately 40 million tons per year. These chemicals undergo complex changes when in the atmosphere. Some get dissolved in raindrops, snow, or fog particles and produce weak solutions of sulfuric and nitric acids. As these acids fall to the ground, they can accumulate in lakes and can affect ecosystems.

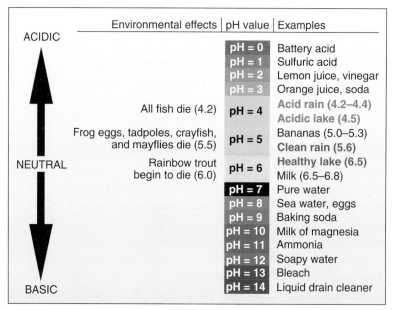

FIGURE 15-6 The pH scale is used to represent the acidity of a solution. pH levels range from 0 to 14, with 0 being extremely acidic and 7 being neutral. The scale is logarithmic, so a change in pH from 7 to 6 or from 5 to 4 represents a 10-fold increase in acidity. (Courtesy of EPA.)

For example, there has been a decline in the health of coniferous forests in the Appalachian Mountains from North Carolina to New England that may be related to acid rain.

In North America, acid rain is primarily a concern in the Northeast and Canada, downwind of sources of sulfur dioxide and nitrogen oxides. **FIGURE 15-7** depicts the pH level of precipitation measured over the United States. Notice the low pH values east of the Mississippi, with the lowest pH values (most acidic) occurring near Ohio. The situation is improving with the 1990 Clean Air Act amendments, which mandate reductions in sulfur and nitrogen acid-forming

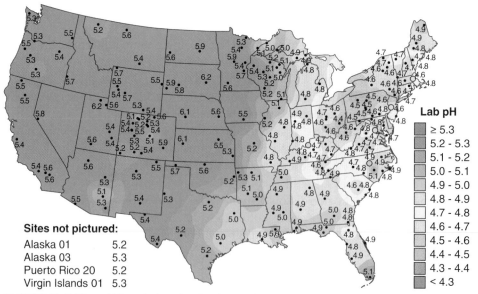

FIGURE 15-7 The pH level of water bodies in North America measured in 2007. This contour map was developed using the pH measurements at the specific sampling locations shown. The isolines of pH level are interpolated between data points. Brown shaded regions indicate acid rain regions where the pH is less than 5.0.

compounds. (We discuss how these reductions were made and attempts to model similar reductions in carbon dioxide emissions today in Chapter 16.) However, the problem is even more severe in Norway and Sweden, where an estimated 6500 lakes are essentially lifeless.

Acid rain is an example of how human activities that emit gases into the atmosphere can affect the environment. An even more dramatic example is the stratospheric ozone hole, which we discuss next.

Stratospheric Ozone Hole

As we learned in Chapters 1 and 2, ozone is produced in the stratosphere by the combination of three oxygen (O) atoms under the influence of sunlight. The concentration of atmospheric ozone is small, approximately three molecules of ozone for every 10 million air molecules. Through absorption of ultraviolet (UV) radiation, ozone plays a fundamental role in the radiation budget and the dynamics of life on Earth. Absorption of UV energy causes a heating that produces the increasing temperature with altitude, a characteristic feature of the stratosphere. Ozone absorption of UV also keeps this harmful radiation from reaching the surface. Reduction in amounts of ozone can lead to increased amounts of biologically damaging UV at the surface. Increased amounts of UV can lead to incidents of cataracts and deadly skin cancers known as melanoma.

FIGURE 15-8 shows the observed ozone over Antarctica from ground-based and satellite-based measurements. Note the sudden change in the amount of ozone after about 1975. By the middle of the 1990s, the total ozone was less than half its value during the 1970s.

Observations of the daily minimum total column ozone amounts during the past several decades show that values are lowest over the Southern Hemisphere in that hemisphere's spring season (**FIGURE 15-9**). During winter, the amounts of ozone over the South Pole region remain fairly constant. A decline in ozone is seen in September, and a minimum amount of ozone is observed in October. After October, ozone levels begin to increase. Why does this minimum occur today? Why did it get worse so quickly, and why does it happen in October?

In Chapter 2 (see Box 2-1) we explained ozone depletion as the result of the release of chlorofluorocarbons (CFCs) into the atmosphere during the middle decades of the 20th century. This is true, but the reason for the dramatic ozone losses seen over Antarctica is a complex interaction

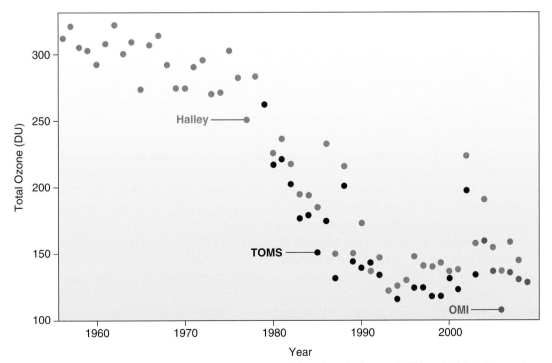

FIGURE 15-8 The observed ozone minimum over Antarctica between 1955 and 2009. The red circles represent the measurements from the ground, and the black and blue circles represent the observations from space from two different instruments. (Modified from GSFC/NASA.)

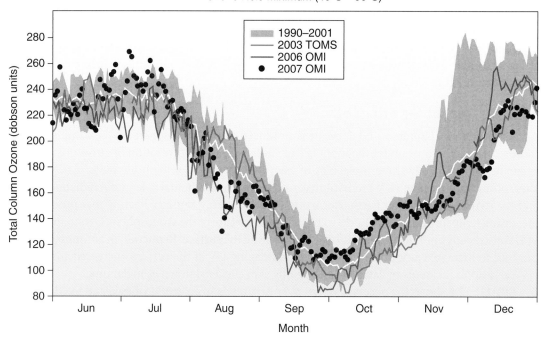

FIGURE 15-9 The daily minimum ozone values between 40°S and the South Pole as a function of year. The grey shaded regions represent the range of measurements between 1990 and 2001, whereas the dots and red and blue lines represent more recent observations from space. (Modified from GSFC/NASA.)

between clouds and atmospheric chemistry, including but not limited to the CFC-ozone interactions depicted in Chapter 2. Here we tell the rest of the story.

The winter atmosphere above Antarctica is very cold. These cold temperatures result from the high altitude of the Antarctic continent and the resulting energy losses caused by longwave radiation losses at the surface. Temperatures in the stratosphere can be less than −90° C (−130° F). In addition, the cold temperatures result in a temperature gradient between the South Pole and the Southern Hemisphere middle latitudes. Recalling the thermal wind law from Chapter 6, these temperature gradients lead to a belt of strong westerly stratospheric winds that encircle the South Pole region. These strong winds, referred to as the polar vortex, prevent the transport of warm equatorial air to the polar latitudes. This isolation of the south polar regions from the middle latitudes and tropical regions helps keep the stratospheric air very cold. The isolation from outside air also forces the polar vortex to "stew in its own juices" in terms of its chemical makeup.

These extremely cold temperatures cause water vapor and some nitrogen compounds to condense and form unique types of clouds. These **polar stratospheric clouds** (**PSCs**), also known as nacreous clouds, are composed of ice and frozen nitrogen compounds and form in air temperatures colder than approximately −80° C (−112° F). PSCs begin to form during June and dissipate in October, the Antarctic spring.

In the wintertime, chemical reactions on the surface of the particles composing PSCs result in chemical reactions that remove the chlorine from the atmospheric compounds. When the Sun returns to the Antarctic stratosphere in the spring, sunlight splits the chlorine molecules into highly reactive chlorine atoms and ozone is rapidly depleted (as discussed in Chapter 2). Destruction is so rapid over the South Pole region in the Southern Hemisphere springtime that it has been termed a "hole in the ozone layer," and it is seen every October.

Why does the ozone hole appear only over the South Pole? Stratospheric clouds composed of ice particles are common over Antarctica but are rarer over the Arctic regions. This is because the Arctic stratosphere does not normally get as cold as the air over Antarctica because of Rossby waves (see Chapter 7) in the stratosphere that prevent the development of an isolated vortex

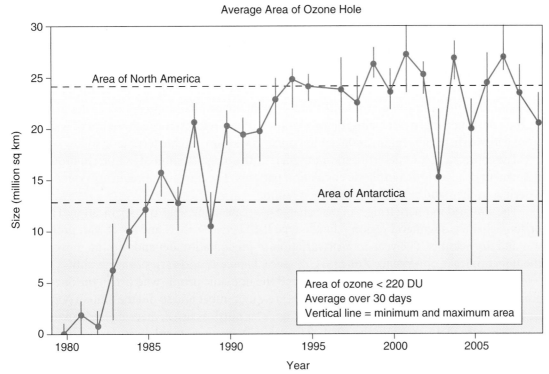

FIGURE 15-10 The annual average size of the ozone hole, in million of square kilometers, as a function of year between 1979 and 2009. The ozone hole is defined as the area for which ozone is less than 220 DU (Dobson units). The circles represent the average for the year, and the vertical bars are the minimum and maximum measured area of the ozone hole during that year. (Modified from GSFC/NASA.)

over the North Pole. As a result, although ozone depletion of 15% to 20% has been observed in certain regions of the Arctic stratosphere, the development of a concentrated "ozone hole" has not been observed as of yet.

The Antarctic ozone hole varies in size during each year and also on a year-to-year basis. This is demonstrated by plotting the size of the annual average size of ozone hole from 1979 to 2009 (**FIGURE 15-10**). This plot shows that the size of the ozone hole has varied from nearly zero in 1979 to an area larger than North America in 2000. The vertical lines about each circle represent the range of values in that year and indicate the large variations throughout the year. The small size of the ozone hole in 2002 resulted from unusually high temperatures of the Antarctic stratosphere that year, which prevented the formation of PSCs.

As mentioned in Chapter 1, international governmental agreements to limit CFC use should allow the ozone layer to repair itself fully over time. The Montreal Protocol in 1987, followed by increasingly stringent agreements, has nearly eliminated CFC use. As a result, CFC levels in the atmosphere are now decreasing. This is not a "quick fix," however. The long lifetimes of these molecules and the cooling of the stratosphere due to greenhouse gas increases (which may promote PSC development) may delay a full recovery of the ozone layer until perhaps the end of this century.

To summarize, the stratospheric ozone hole is the end result of the innocent release of tiny amounts of chlorine-containing chemicals into the troposphere, which over a period of decades found their way into the stratosphere. Once in the stratosphere, a series of chemical reactions on cloud surfaces led to a catastrophic decrease in the amount of ozone over Antarctica. This indicates that climate change is complicated and that when studying climate change it is wise to expect the unexpected and to anticipate unforeseen consequences.

Changing Land Surfaces

As discussed in Chapters 2 and 3, the Earth's surface has a direct effect on regional weather and climate. Large-scale changes in land use, such as urbanization, deforestation, and agriculture, affect the surface of Earth. Changes in the surface can result in a change in the energy budget of the region, which can modify the local climate. Two examples of how human activities modify the regional climate are desertification and urbanization.

Desertification

Desertification refers to the spreading of a desert region because of a combination of climate change and human impacts on the land. Practices that leave the land susceptible to desertification include the following: overgrazing, deforestation without reforestation, diversion of water away from a formerly fertile region, and farming on land with unsuitable terrain or soil.

The consequences of desertification can include the magnification of long-term drought and permanent changes in atmospheric circulation patterns. The human consequences can be severe: a decline in the standard of living and even famine.

The semiarid southern fringes of the Sahara Desert are vulnerable to desertification. The Sahel, or sub-Sahara, is a semiarid region (climate type Bsh) between 14°N and 18°N with pronounced wet and dry seasons. The year-to-year variation in precipitation depends on the movement of the Intertropical Convergence Zone (ITCZ) (see Chapter 7) and varies considerably.

In the early 1960s, rainfall was plentiful and the nomadic people who live in the Sahel found ample grazing lands for cattle and goats. Herds grew in number and so did the human population. In 1968 the ITCZ did not bring the needed rains as far north as usual, marking the beginning of a severe drought that lasted into the 1980s. The drought, in concert with overgrazing, turned large regions of pasture into a wasteland. The drought peaked in 1973 when rainfall totals were only half the long-term average. The Sahara Desert moved southward into the Sahel, and a famine ensued that took the lives of more than 100,000 people and affected more than 2 million. Rainfall eventually returned to the region but in smaller quantities than were experienced in the 1950s and 1960s. Lake Chad, in West Africa, is now only 5% as large as it was just 35 years ago!

It is possible that this change in regional precipitation is a natural variation. Indeed, the southern boundary of the Sahara shifts position north and south by 100 kilometers (60 miles) in different years. However, the drought may also be enhanced by a biogeophysical positive feedback mechanism. The Sahel lies below the descending branch of the Hadley cell during its dry season. Sinking air warms and is thus an energy gain for the atmosphere. With a reduction in rainfall, there is less vegetation. The reduced vegetation results in an increase in surface albedo, reducing the energy gains of the surface. To make up for reduced energy gains from the surface, the atmosphere subsides, warms, and dries out. This warming and drying associated with the sinking motion of the atmosphere further enhances desert conditions. Thus, the reduced precipitation of a semidesert region is a positive feedback mechanism in which reduced precipitation leads to changes in the surface budget that result in sinking air, further reducing precipitation.

An even more pronounced example of human-caused desertification has occurred in recent decades in the vicinity of the Aral Sea in central Asia. The government of the Soviet Union diverted the rivers feeding this vast inland lake to provide water for the growing of cotton. Deprived of its sources of water, the Aral Sea began shrinking rapidly (FIGURE 15-11). Evaporation left minerals behind, and the shrinking lake exposed gritty, dusty shorelines and sea floors.

In 1960, the Aral Sea was one lake with a surface area of 68,000 square kilometers (26,000 sq mi). By 2007, the Aral Sea was 10% of its original size and divided into three lakes. By the year 2009, the southeastern lake was gone. As the Aral Sea has shrunk, both the climate and the culture of the region have changed for the worse. The evaporated waters exposed some 40,000 sq km of dry sand and salt and formed a new desert called the Aral Karakum Desert. Winds picked up this dust and created an Asian "Dust Bowl" (see Chapter 12) across a once-flourishing region. The dust can cause health problems for local populations, and replacing the lake with dry soil results in colder winters and hotter summers for the region. The Aral Sea fishing industry collapsed because the lake grew increasingly salty and inhospitable to aquatic life. Meanwhile, the local climate has changed, shortening the growing season and forcing local agriculture to require even more diversion of water from the Aral Sea, further worsening the situation. Fortunately, there are now efforts by Kazakhstan to replenish the North Aral Sea. A dam was completed in 2005, raising water levels and lowering the salinity of the lake.

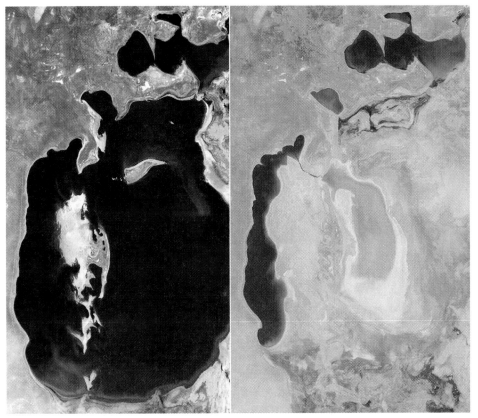

July - September, 1989   October 5, 2008

FIGURE 15-11 The disappearance of the Aral Sea, as seen by satellite during the period from 1989 to 2008.

Urban Heat Islands

The **urban heat island** effect refers to the increased temperatures of urban areas compared with a city's rural surroundings. The urban heat island is a well-documented example of inadvertent modification of climate by human activities. It is a classic example of how changing the energy balance of a region can affect the regional climate.

On average, the city is warmer than the countryside because of differences between the energy gains and losses of each region. A number of factors can contribute to the relative warmth of cities, such as heat from industrial activity, the thermal properties of buildings, and the evaporation of water. For example, the heat produced by heating and cooling city buildings and running planes, trains, buses, and automobiles contributes to the warmer city temperatures. Heat generated by these objects eventually makes its way into the atmosphere, adding as much as one third of the heat received from solar energy.

The thermal properties of buildings and roads are also important in defining the urban heat island. Asphalt, brick, and concrete retain heat better than do natural surfaces. Buildings, roads, and other structures add heat to the air throughout the night and thus reduce the nighttime cooling of the air so that the maximum temperature difference between the city and surroundings occurs during the night. The canyon shape of the tall buildings and the narrow space between them magnifies the long-wave energy gains. During the day, solar energy is trapped by multiple reflections off the many closely spaced, tall buildings, reducing heat losses by long-wave radiation (FIGURE 15-12). Pollution in the city's air also modifies the absorption of long-wave and short-wave radiation of the atmosphere.

Evaporation of water may also play a role in defining the magnitude of the urban heat island. During the day, the solar energy absorbed near the ground in rural areas evaporates water from the vegetation and soil. Thus, although there is a net solar energy gain, heating is reduced to some degree by evaporative cooling during evapotranspiration. In cities, where there is less vegetation, the buildings, streets, and sidewalks absorb the majority of solar energy input and warm up rapidly.

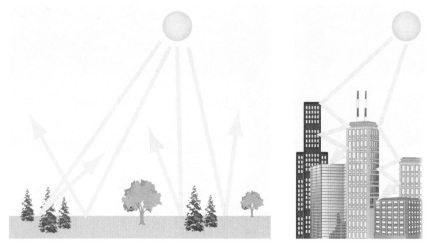

FIGURE 15-12 Concrete roadways and buildings reduce evapotranspiration and absorb more solar radiation than surrounding rural regions. As a result, urban areas are often warmer than rural areas.

The urban heat island is evident in statistical tables of surface air temperatures. The warmer temperatures of urban areas are also apparent in cloud-free satellite images. **FIGURE 15-13a** is a satellite image of infrared radiative energy exiting the atmosphere with a wavelength of 11 microns, which is directly correlated to surface temperature, for the Baltimore, Maryland, region. Urban heat islands appear in this image as bright areas that are warmer than the more rural regions around them. Compare the satellite image of temperature with the map of developed land in the Baltimore area (Figure 15-13b). Marked differences in air temperature are some of the most important contrasts between urban and rural areas, as demonstrated by the satellite observations. The intensity of the heating varies across a city, with the highest temperatures found near the city core. For instance, under clear skies and light winds, temperatures in central London during the spring reached a minimum of 11° C (52° F), whereas in the suburbs they dropped to 5° C (41° F), a difference of 6° C (11° F) (**FIGURE 15-14**).

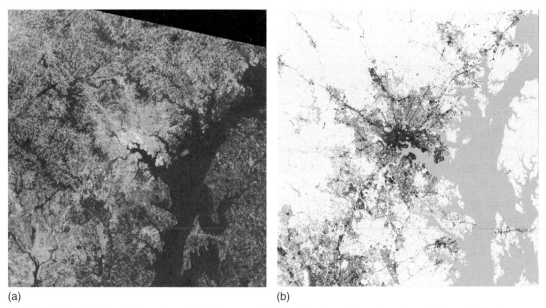

(a)   (b)

FIGURE 15-13 (a) An infrared satellite image demonstrating the urban heat island effect around Baltimore, Maryland. (b) The developed landscape around Baltimore; red areas have a high concentration of cement and asphalt (impervious surfaces), whereas white areas are primarily covered by plants.

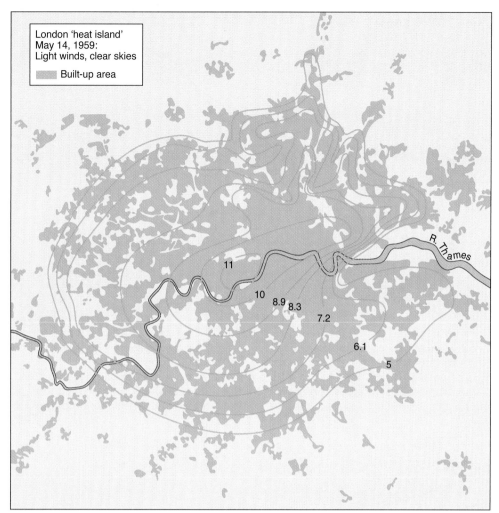

FIGURE 15-14 The minimum air temperature across the city of London, England and the surrounding landscape on May 14, 1959, under clear skies and light winds. Minimum temperatures in central London were 11° C, whereas in the suburbs the minimum temperature was only 4° C. This heat island effect was mapped by T. J. Chandler about 50 years ago. (Adapted from Chandler T. J. *The Climate of London.* Hutchinson, 1965.)

FIGURE 15-15 shows the magnitude of the urban heat island as a temperature difference between a city and its surroundings, for clear sky and light wind conditions, plotted as a function of city population. The different slopes of the relationship differ for cities in different parts of the world. This difference results from differences in the urban characteristics, the characteristics of the rural surroundings, and the prevailing climate. More typically, heat island intensities are in the range of 1° C to 3° C (about 2° F to 5° F) degrees, even for large cities, indicating the importance of wind and cloud effects. Seasonally, heat islands in temperate climates are a maximum during summer and fall when weather conditions are more favorable. Scientists are confirming that the urban heat island affects not just temperature, but also precipitation (**BOX 15-2**).

We can reduce the magnitude of the urban heat island through various mitigation strategies. Examples include planting trees and vegetation to shade the surface and provide evaporation cooling and increasing the reflectivity of surfaces such as roofs to reduce the amount of solar radiation absorbed.

Now that we have investigated several different ways in which humans have changed climate on local and regional scales, we are ready to look at the science of global warming. Climatologists are aware of the impact of the urban heat island and its potential to bias observations to warmer climates, and they account for this in their climatologies.

"Urban Heat Island," use satellite images to identify urban heat islands.

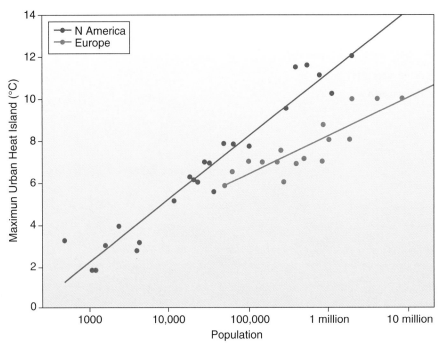

FIGURE 15-15 The magnitude of the urban heat island, expressed as a temperature difference between a city and its surrounding, for clear sky and light wind conditions, plotted as a function of the city's population. The slope of the relationships differs for cities in different parts of the world depending in part on the urban characteristics, the characteristics of the rural surrounding, and the prevailing climate. (Courtesy of T. R. Oke, The University of British Columbia.)

Box 15-2 Can a City Make It Rain?

The urban heat island isn't just about heat, it turns out. New research is confirming old suspicions that it can have profound impacts on precipitation, too.

Since the early to mid 1900s, meteorologists have thought that regions downwind of large cities received more than their expected share of rainfall. However, conclusive evidence has been hard to gather. For example, research on just one city doesn't eliminate the possibility that any precipitation anomalies are unique to that one city. More research and the advent of space-based satellites that measure precipitation have overcome this hurdle, however. The cities of New York, Atlanta, Phoenix, Houston, Tokyo, and Mexico City, among others, have been shown to affect precipitation amounts—especially downwind of the city centers. For example, a careful radar-based study of the Atlanta metropolitan area revealed that the eastern (downwind) suburbs of Atlanta received up to 30% more nighttime rainfall than surrounding areas.

How does a city make it rain? This is a topic of current research. The bubble of warm air over the city center and the frictional drag on wind by city buildings can set up local three-dimensional circulations similar to the sea and land breezes we discussed in Chapter 6. Computer modeling studies suggest that these circulations favor rising motion and downwind thunderstorm development, as shown in the figure. Another possible "trigger" for an urban precipitation effect is that the aerosols in the polluted air of a city may act as cloud condensation nuclei (see Chapter 4), causing more rainfall. The eventual explanation may include some of both triggers, and other factors as well.

As cities expand in size during the 21st century, the impact of urban areas on local weather and climate will grow. Over and above the more famous effect of global warming, these urban effects will be yet another aspect of anthropogenic climate change in our lifetimes.

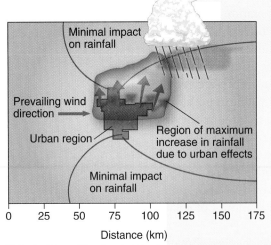

(Adapted from Shepherd, J.M., et al., Agronomy Monograph 55, [2010]: 1–29 and J. Aitkenhead-Peterson and A. Volder [ed.]. *Urban Ecosystem Ecology.* American Society of Agronomy, 2010.)

EVIDENCE OF GLOBAL WARMING

Because of the potential impact on society, global warming is a frequent topic in news reports. Sometimes these reports suggest that scientists are arguing whether greenhouse gases will change the climate. In scientific discussions, the issue is not whether greenhouse gases have increased or if the climate has changed. The evidence for climate change is all around us.

Over the past 2 decades, the global average surface temperature has increased noticeably (FIGURE 15-16). Although a few warm winters and hot summers in one location or region do not mean global warming, the observed warming trend over the last 2 decades may indicate a significant global change. As we see in this chapter, the observed warming is consistent with other observed changes:

- There is a widespread retreat of nonpolar glaciers.
- The Arctic sea ice has thinned by 40% in recent decades in late summer to early autumn and decreased in extent by 10% to 15% since the 1950s in spring and summer.
- Northern Hemisphere snow cover has decreased in area by 10% since the 1960s.
- The global mean sea level has increased at an average annual rate of 1 to 2 mm during the 20th century.
- The growing season has lengthened by about 1 to 4 days per decade during the last 40 years in the Northern Hemisphere, especially at higher latitudes.
- The duration of ice cover on lakes decreased by about 2 weeks over the 20th century in mid and high latitudes of the Northern Hemisphere.

Is this temperature change a natural fluctuation in our climate, or is it a result of human activities in the last 150 years or so? This question is a focus of the last chapter of this book and is one of the most important scientific questions in the world today.

Global warming is part of two larger stories: how climate changes, according to the ongoing research of atmospheric scientists, and how humans have changed the Earth's atmosphere. To place the subject of global warming in its proper context, we first examine these topics

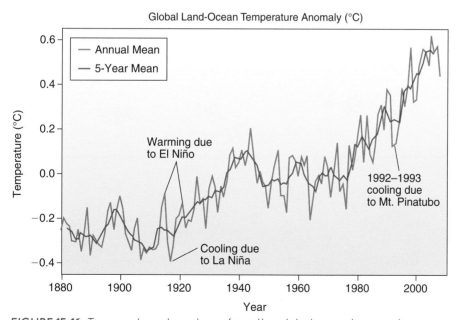

FIGURE 15-16 Temperature departures from the global mean temperature over land since 1880. The reference years are from 1951 to 1980. Notice the trend of increasing temperature over the last 2 decades. Note the shorter-term effects of the ocean temperature phenomena El Niño and La Niña, which have periodically affected global temperatures throughout this period. The episodic cooling effect of the Mt. Pinatubo volcanic explosion is also indicated. (*Source*: NASA/GISS.)

in some detail. We see that climate change is exceptionally complex and that there is ample evidence that humans can and do make large changes to our atmosphere.

Climate includes the interactions of the atmosphere with the land, the oceans, and living things. Because of this, scientific conclusions about global change must include complex computer models of climate that attempt to integrate all of the interactions in predictions of future climate. We examine the computer-model forecasts of global warming in Chapter 16. Later here we summarize some of the scientific complexities that lurk within these models.

Atmosphere

As we learned in Chapters 1 and 2, the Earth's atmosphere contains gases that are transparent to solar radiation but absorb large amounts of terrestrial infrared radiation. Carbon dioxide, methane (CH_4), water vapor, and CFCs all are "greenhouse gases" that act to trap heat in the Earth's lower atmosphere, just as a greenhouse keeps plants warmer than if they were exposed to the elements. Without the greenhouse effect, the Earth would be too cold for most life to exist.

However, human civilization has significantly changed the amount of greenhouse gases in the atmosphere during the past century. Since the beginning of the Industrial Revolution in the 19th century, concentrations worldwide have increased by 25%. Today, the concentration of carbon dioxide is increasing at a rate of about 0.5% per year. Methane has doubled over the same period because of human activities such as coal mining and agriculture, leaky natural-gas lines, and cattle. CFCs are a human creation and did not even exist in the atmosphere until after 1930.

Based on the radiation concepts from Chapter 2, it would seem obvious that the larger the amount of greenhouse gases, the more energy trapped in the lower atmosphere—and as a result a higher global temperature. Of course this warming has not been uniform across the globe. Some regions are warming faster than other regions of the globe. FIGURE 15-17 shows the observed temperature differences between 1955 and 2005. Alaska and Siberia are two regions that are warming the fastest, and this is affecting the permafrost in those regions. Thawing permafrost can release large amounts of carbon dioxide, methane, and nitrous oxide, also known as laughing gas, which is also a greenhouse gas. The Antarctic Peninsula is also warming rapidly. This, in part, contributes to the breakup of ice sheets we discussed in Chapter 14.

A change in the diurnal cycle of temperature has also been detected. Although many land areas are showing an increase in both the minimum and maximum daily temperatures, the minimum temperature has increased about twice as fast as maximum temperature. This has resulted in a decrease in the diurnal temperature range.

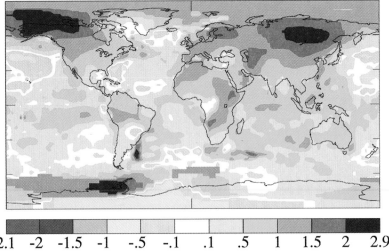

FIGURE 15-17 The annual temperature changes of the past 50 years. The scale represents the difference between observed temperatures in 2005 and those observed in 1955. Positive values indicate a warming. The global average temperature of 2005 is 0.59° C warmer than the global average temperature 50 years earlier. (Courtesy of GISS/NASA.)

Oceans

The interactions between the atmosphere and the oceans in a globally warming world are also complex and poorly understood. As we learned in Chapter 4, water absorbs energy by warming up, changing phase, or both. Because the Earth's surface is 70% water, the oceans can absorb a considerable amount of energy trapped by increasing greenhouse gases. Scientists believe this has slowed the warming of the atmosphere during the past several decades.

But the oceans themselves have also been warming. Like air and other fluids, water expands as its temperature increases (i.e., its density goes down as temperature rises). As climate change increases ocean temperatures, initially at the surface and over centuries at depth, the water will expand, contributing to sea level rise due to thermal expansion. During the 20th century, sea level rose about 15 to 20 centimeters (roughly 1.5–2.0 mm/year). Satellite measurements indicate that over the past decade the rate of sea level rise has increased to approximately 3 mm/year (FIGURE 15-18). Thermal expansion of the ocean alone cannot explain this rise. Another contributor to sea level rise is the melting of glaciers and the ice sheets of Greenland and Antarctica. The melting of sea ice does not directly contribute to a rise in sea level, as it is already floating in the ocean. (In the same way, melting ice in a drink doesn't cause the glass to overflow.) However, melting sea ice contributes indirectly as smaller amounts of sea ice allow for more solar energy to penetrate the ocean and warm the waters.

Warming water and melting land ice have raised global mean sea level 45 millimeters (1.7 inches) from 1993 to 2008. Thermal expansion and melting of land ice each contribute about half of this observed sea level rise, but the rise is not constant across the globe. There are variations across the ocean basin because of wind patterns and ocean currents.

So far we have thought of the oceans as simply a big bathtub that absorbs heat from the atmosphere and melted ice from land. On the contrary, the oceans are dynamic and have complex three-dimensional circulations, as we learned in Chapter 8. When this is taken into account, the role of the oceans in climate change becomes far more complicated. For example, vertical motions (convection) in the ocean are very sensitive to temperature. If global warming changes the amount and location of convection, climate change could accelerate rapidly. For example, during the Younger Dryas period, about 10,000 years ago (see Chapter 14), melting ice sheets apparently

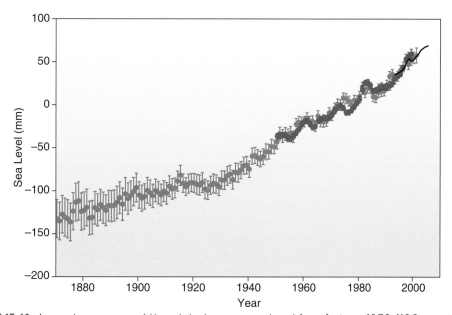

FIGURE 15-18 Annual averages of the global mean sea level (mm) since 1870 (100 mm is about 3.9 inches). The red and blue points represent observations from tide gauges, while the black line is from satellite measurements. The vertical lines with each point represent errors of the measurement. (*Source*: Intergovernmental Panel on Climate Change, "Climate Change 2007," The Fourth Assessment Report, Working Group 1, Figure 5.13; Red curve from: Church and White, 2006; Blue curve from: Holgate and Woodworth, 2004; Black curve from: Leuliette et al., 2004.)

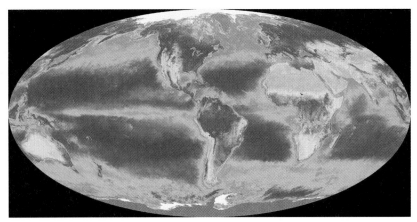

FIGURE 15-19 A composite satellite image from the NASA SeaWiFS project depicting 3 years of data on land vegetation amount and ocean chlorophyll concentrations, which indicate the locations of plankton in the ocean. The green, yellow, and red areas in the oceans are biologically active.

interfered with oceanic convection and led to a rapid cooling of the climate of the Northern Hemisphere that lasted hundreds of years. Could atmospheric warming lead to a similar disruption of climate as usual? This depends on the details of temperature and precipitation changes caused by global warming and is an extremely difficult question to answer.

The oceans are, like the land surface, teeming with life. This underwater biosphere may turn out to hold the crucial clues for global climate change. The warmer ocean temperatures impact those species that prefer cooler temperatures. Coral are marine animals that are sensitive to the temperature of the waters where they live. Warm temperatures can result in coral bleaching, a whitening of corals when the plant is stressed. The recent warming of the oceans has resulted in the death of about a quarter of the world's coral reefs.

Chapter 1 discussed the carbon cycle and the role of ocean as storage of carbon. The absorption of anthropogenic carbon emission by the ocean is changing the chemical equilibrium of the ocean. The increased CO_2 dissolved in the ocean decreases the pH, making it more acidic and impacting marine ecosystems. Changes in carbon cycle and its impact on biological activity are difficult to quantify at the global scale. To improve our global knowledge of life on land and sea, an extensive series of international satellite campaigns is in progress and is observing the biosphere in unprecedented detail (**FIGURE 15-19**). However, the problems ahead are challenging and improved understanding may take years or even decades.

Cryosphere

The **cryosphere** describes the portions of the Earth's surface with ice. Chapter 14 discussed observations of shrinking mountain glaciers. These ice masses are only one component of a changing Earth's cryosphere—changes linked to a warmer climate. There have also been changes in the amount of sea ice. **FIGURE 15-20** shows the observed monthly ice extent over the Arctic for the month of September between 1979 and 2009. There is a decline in the extent of the ice of about 11% per decade. The amount of the ice cover loss was particularly pronounced along the Eurasian coast in 2005 and 2007 when, by the end of the summer, there was an ice-free Northern sea route. Observations also indicate that the sea ice thickness has declined.

There are two factors that contribute to these changes in sea ice. The first is the large warming of this region as shown in Figure 15-17. Circulation patterns also play a role in this decreasing trend. An increase in storms has resulted in winds that push the ice away from shore and out toward the ocean, which helps to melt the ice faster.

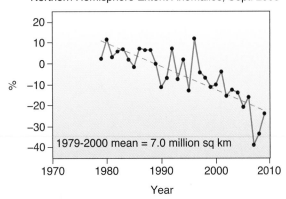

Northern Hemisphere Extent Anomalies, Sept. 2009

1979-2000 mean = 7.0 million sq km

FIGURE 15-20 Monthly change in September ice extent for 1979 to 2009. The dotted line is the trend, showing a decline of the amount of ice of about 11% per decade. (*Source*: National Snow and Ice Data Center.)

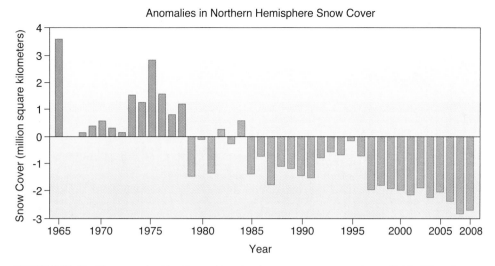

FIGURE 15-21 Changes in Northern Hemisphere snow cover since 1965. (Courtesy of Riccardo Pravettoni, UNEP/GRID-Arendal.)

The large ice sheet on Greenland is a remnant of the last ice age. While Greenland is about a quarter of the size of the United States, the Greenland ice sheet is over 3 kilometers (two miles) thick in some places and contains enough water that if it were to all melt into the oceans would raise the global sea level about seven meters (23 feet). While scientists do not anticipate the Greenland ice sheet to completely disappear, it is rapidly melting. Coastal glaciers are slipping into the Atlantic while puddles of water lie on the surface of the ice during summer, increasing the melt rate.

With increased temperature in the Northern Hemisphere, we might expect a decrease in snow cover. Indeed, the average snow cover in the Northern Hemisphere has decreased (**FIGURE 15-21**), especially in the spring and summer. A decrease in snow cover is a positive feedback mechanism. As snow cover disappears earlier in the spring, the large amounts of energy that would have melted the snow can now directly warm the soil, further warming atmospheric temperatures (see Chapter 3).

The changes in snow cover have broad impacts on ecosystems and society. Clearly, reduced snow cover negatively impacts the economy of regions that rely on wintertime tourism, such as skiing and snowmobiling. As another example, the earlier snow melt across the western United States results in an earlier runoff from the mountain rivers that supply water to agriculture and for recreational activities. The reduced snowfall amount combined with the earlier melting and runoff results in longer summertime periods with a low river flow.

Records over the past century of various lakes shows that the durations of ice cover on lakes is shortening. Fall freezes are occurring later and spring thaws earlier. This is demonstrated in the ice cover of Lake Mendota in Madison, WI (**FIGURE 15-22**), one of the most studied lakes in the world. The reduced ice cover means there is greater evaporation from the open water, which contributes to lower lake levels. This also leads to warmer waters in the lake, which affects the lake's ecosystem, as fish species prefer particular water temperatures.

Biosphere

The **biosphere** comprises all of Earth's living organisms. The biosphere plays an important role in the climate, regulating the carbon cycle, the hydrologic cycle, and the heat budget of the planet. Climate changes can be linked to changing land and vegetation patterns and vice versa. Changes in the atmosphere and surface in turn affect the biosphere. Here we look at two examples of this relationship.

Snow cover regulates surface temperature in two ways. First, the high albedo reduces the incoming solar radiation. This keeps the soil from warming until the snow melts, which limit biological activity in the frozen soil. As discussed in Chapter 2, snow is also a good insulator, which keeps snow covered soil from getting too cold during winter. Thus, in Arctic regions the temperature range of the soil is smaller under snow cover than snow-free conditions. These properties have resulted in a fragile ecosystem called the tundra. The increased frequency of snow thaw that results from a warmer climate has implications for animals, plants, and birds that live in

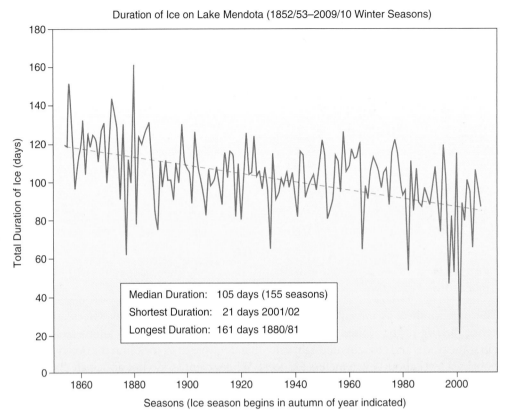

FIGURE 15-22 The total duration of ice, in days, on Lake Mendota in Madison, Wisconsin. (*Source*: Wisconsin State Climatology Office.)

the tundra. For example, some sandpipers that breed in the Arctic require snow-free patches at the time of breeding so that the chicks hatch at the same time the insects emerge. Change in snow cover can impact this timing, while also exposing nests to predators that prefer snow-free conditions.

If you do any gardening, you know that frost determines the growing range for many plants. So changes in the temperature, particularly a reduction in nighttime minimum temperatures, may influence where plants can be grown. The warmer temperatures mean that spring is arriving sooner. Apples and grapes are blooming about 8 days earlier in the Northeast than just 50 years ago. In Washington, D.C., 89 of 100 plants studied were blooming on average 4.5 days earlier now than in the 1970s.

If you are involved with gardening, you probably are also aware of the hardy zones listed on seed packets. The U.S. Department of Agriculture developed the zones and first published them in 1960. A hardiness zone provides information on the type of plants capable of surviving certain climatic conditions. These conditions include the ability of a plant to survive a minimum temperature range. The climate zones are determined from temperature records kept by the National Climatic Data Center, or NCDC. Recognizing that climate varies over long time periods, every 10 years the NCDC computes a revised 30-year average temperature and extreme temperatures for the U.S. Between 1961 and 1990 and 1971 and 2000, the 30-year average minimum temperatures in winter increased at nearly all locations in the continental U.S. (**FIGURE 15-23**). This suggests that when the USDA next revises the hardy zone map, the zones in your area may change.

■ The Intergovernmental Panel on Climate Change Assessment

The **Intergovernmental Panel on Climate Change** (**IPCC**) was formed by the World Meteorological Organization (WMO) and the United Nations Environment Programme (UNEP) in 1988. It is a group comprised of over 1000 scientists and experts from more than 100 countries. This group does not conduct research about climate change when they meet. Instead, the group writes reports that describe our current knowledge about climate change, based on published scientific literature. These reports are used by government officials throughout the world to aid in their

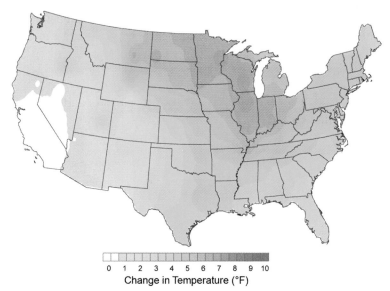

FIGURE 15-23 The trend in minimum temperatures in winter for the continental U.S. Temperatures are rising faster in winter than in any other season, especially in many key agricultural regions. (Courtesy of NOAA/NCDC.)

decision making. In October 2007, the IPCC was awarded the Nobel Peace Prize jointly with Former U.S. Vice President Al Gore for their efforts to build up and share knowledge about climate change.

The IPCC report in 2007 concluded that global warming is "unequivocal"—in other words, unquestionable. Furthermore, the IPCC report states that the observed increase in the globally averaged temperatures since the mid 20th century is "very likely" due to increased greenhouse gas concentrations resulting from human activities. The phrase "very likely" in IPCC parlance means that there is at least a 90% chance that the temperature rises that have already been observed are due to humans.

In this chapter, we have demonstrated how observed changes in climate are consistent with expectations from the theory of global warming. Now that we know what has happened, we can speculate what will happen in the future. Chapter 16 examines the young science of climate forecasting, concluding with a detailed look at predictions of future climate change.

PUTTING IT ALL TOGETHER

Summary

Recent observations indicate that global surface temperatures are warming. This warming has coincided with rapidly increasing amounts of atmospheric greenhouse gases. These gases play a crucial role in Earth's climate by affecting the energy budget. Human activities are increasing the concentrations of greenhouse gases.

Humans have been altering weather and climate on local and regional scales for centuries. Pollution in the form of lead, carbon monoxide, and tropospheric ozone causes health problems and photochemical smog. Oxides of nitrogen and sulfur are primarily to blame for the development of acid rain, which kills trees and aquatic life. The release of CFCs, in combination with stratospheric wind and cloud patterns and chemical reactions, led to the ozone hole over Antarctica. In addition, by diverting rivers and building cities, humans have changed local and regional precipitation and temperature patterns markedly. These examples reveal that feedback processes within the atmosphere and connecting the atmosphere with the Earth's surface can cause relatively small initial changes in the atmosphere to amplify into significant climate changes.

Scientists are still learning how the atmosphere interacts with water (water vapor, clouds, and the oceans), especially in the situation of rising temperatures as a result of increasing amounts of carbon dioxide and other "greenhouse gases." Evidence from the atmosphere, oceans, cryosphere, and biosphere supports the conclusion of the IPCC that global warming has already occurred. The IPCC has also concluded that the chance that these changes in global temperature have been caused by humans is at least 90%.

Key Terms

You should understand all of the following terms. Use the glossary and this chapter to improve your understanding of these terms.

Acid deposition
Acid rain
Air pollution
Biosphere
Carbon monoxide (CO)
Contrails
Cryosphere
Desertification
Feedback
Hydrocarbons
Ice-albedo temperature feedback
Intergovernmental Panel on Climate Change (IPCC)

Lead
Negative feedback mechanism
Nitric oxide (NO)
Nitrogen dioxide (NO$_2$)
Ozone (O$_3$)
Particulates
pH scale
Photochemical smog
Polar stratospheric clouds (PSCs)
Positive feedback mechanism

Ship tracks
Smog
Sulfur dioxide (SO$_2$)
Sulfur trioxide (SO$_3$)
Sulfuric acid (H$_2$SO$_4$)
Urban heat island
Volatile organic compounds (VOCs)

Review Questions

1. Why do we use unleaded gasoline in cars today?
2. A Fortune 500 company saw its stock drop rapidly in value because a few worries about its financial health were reported publicly; the drop in stock value generated more worries that were then confirmed, leading to still more investor concern and a massive sell-off of its stock. Is this an example of a positive feedback mechanism or a negative feedback mechanism? Why?
3. How is photochemical smog formed?
4. Overall, is pollution increasing or decreasing across the United States? Why is this change occurring?
5. Do atmospheric scientists expect the ozone hole to get better or worse during the next century? What do they base their predictions on?
6. The city of Los Angeles diverted water from Tulare Lake in California, once the largest freshwater lake west of the Mississippi River, to satisfy the city's increasing need for water. The lake dried up completely. How do you think the climate in the vicinity of the lake might have changed as a result?
7. Why are the average temperatures of cities often greater than those of the surrounding rural region?
8. Can you think of other differences in weather between the city and its rural surroundings?
9. Describe how increased amounts of water vapor could help accelerate global warming. Describe how increased amounts of low clouds could help slow down global warming. How might increased amounts of high clouds affect global warming, and why?
10. Describe how a slight increase in the horizontal extent of glaciers could cause glaciers to expand.

Observation Activities

1. This chapter discussed various potential changes we may experience with a warming planet. How do these potential changes impact the location where you live?
2. Newspapers sometimes include editorials on global warming—some argue that warming has not occurred; others say it has. What evidence do these editorials present? Is this evidence fairly presented?

This rain cloud icon is your clue to go to the *Meteorology* Web site at http://physicalscience.jbpub.com/ackerman/meteorology/. Through animations, quizzes, web exercises, and more, you can explore in further detail many fascinating topics in meteorology.

Climate Forecasting

16

AFTER COMPLETING THIS CHAPTER, YOU SHOULD BE ABLE TO
- Explain how monthly and seasonal climate forecasts are made
- Interpret a monthly/seasonal climate forecast for your area
- Explain the role of global climate models in long-range climate forecasting
- Describe some of the key results of long-range climate forecasts based on increasing greenhouse gas concentrations
- List some possible societal responses to forecasts of global climate change

INTRODUCTION

It's the middle of the night in late June in the Andes Mountains of South America, and farmers are making . . . a *climate forecast!* Since about the year 1600, potato farmers from Peru and Bolivia have ascended to high ridges and mountaintops in late June. They look to the northeast sky near dawn for the appearance of the Pleiades, a star cluster. The farmers make the following prediction: If the Pleiades are bright with lots of stars visible, then there will be abundant rainfall during the rainy season from October to March, some 3 to 9 months later. This means a good potato harvest the next fall. But if the Pleiades are dim and few in number, the farmers predict drier than normal conditions and a poor potato harvest. In such cases, the farmers often alter their normal planting schedule to make the best of a bad situation.

We learned in Chapter 13 that weather forecasts based on conditions on a particular day, such as the famous Groundhog Day forecast, are usually worthless. We also learned that even the most accurate modern weather forecasts are limited to no more than about 2 weeks into the future. So how could a 3- to 9-month climate forecast based on stargazing have any validity? And yet scientific research by anthropologist Ben Orlove and climate researchers John Chiang and Mark Cane has shown that the Andean potato farmers' forecasts are quite accurate. How is this possible?

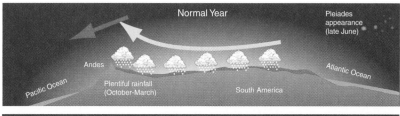

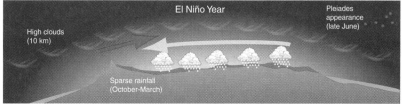

(Modified from Benjamin Orlove, John Chiang, Mark Cane, "Ethnoclimatology in the Andes," *American Scientist* 90:[2002].)

The answer lies in the El Niño phenomenon that we studied in Chapter 8 (see figure). During El Niño years, thunderstorms form over the unusually warm waters of the central and eastern Pacific Ocean, offshore of Peru. The thin cirrus clouds associated with the thunderstorms anvils are blown to the east and obscure the starlight of the Pleiades, making them appear dim. Furthermore, during El Niño years the lower tropospheric winds are west to east. This causes dry downslope winds instead of the normal, wetter upslope winds. Therefore, dim Pleiades = El Niño = dry weather for many months = poor potato harvest!

The potato farmers didn't know all this science back in 1600, or even today. But they did know about poor and plentiful harvests. The farmers carefully observed the night sky, and they made a connection between the two that worked and has been verified by modern science.

Keep the Andean potato farmers' story in mind as we survey the state-of-the-art of climate forecasting in this chapter. Today's monthly and seasonal climate forecasts, like the farmers', rely on established relationships between climate and key climate forcing mechanisms—in particular, El Niño-Southern Oscillation. The modern climate forecasts, like the farmers', are also limited to predicting changes versus normal conditions instead of giving precise numbers like a weather forecast.

Today's long-range climate forecasts stretch decades and even centuries into the future. They use ensemble model techniques to peer beyond today's climate relationships into a world profoundly changed by increasing greenhouse gases. It turns out that these predictions have grave implications for high-altitude locations such as the Andes. Warming temperatures could cause extinctions of plants and animals and eliminate glaciers that have served as renewable sources of freshwater for untold generations.

Such massive changes are well beyond the ability of even the shrewdest stargazing farmer to predict—but are the consensus of the world's climate modeling experts. Will modern civilization heed the "global warming" forecast and adjust its practices to make the best of a bad situation, just as the Andean potato farmers do when the Pleiades are dim? We close this chapter by looking at what can be done to avert the worst consequences of global warming.

WEATHER FORECASTS VS. CLIMATE FORECASTS

There is a crucial distinction to be made between weather and climate forecasts. Weather forecasts can provide precise numbers for high and low temperatures and precipitation on an hourly basis. In contrast, climate forecasts are often *probabilistic*, that is, given in terms of the *chance* of a location to be warmer/colder or wetter/drier versus normal conditions over the whole period of the forecast. Climate forecasts are also not the same as the climatology forecasts we saw in Chapter 13. Whereas climatology forecasts are simple long-term averages, today's climate forecasts attempt to predict the future *deviations* from these long-term averages.

The "fuzziness" in climate forecasts is the inevitable result of chaos. As we learned in Chapter 13, chaos in the atmosphere imposes a 2-week limit on precise forecasts. Ensemble forecasting, which was invented for climate forecasts and now is used extensively in weather forecasting, allows climate forecasters to use multiple runs of a computer model to gain a sense of the possibilities for future climate. In this way climate forecasts can be made, even though pinpoint forecast numbers are impossible to obtain. As we'll see, ensemble forecasting is at the heart of today's climate forecasting, particularly at longer ranges.

MONTHLY/SEASONAL CLIMATE FORECASTS

In addition to weather forecasts, the forecast centers of the world also issue **monthly** and **seasonal climate forecasts** for up to about 1 year in advance. As their names imply, these forecasts are for conditions to be expected for, say, the next month or for upcoming seasons (e.g., forecasts issued in fall for winter, spring, summer and the next fall).

Making Monthly/Seasonal Climate Forecasts

Monthly forecasts use the longest range weather forecasts from numerical weather prediction models. For example, at NCEP, the 30-day climate forecast is based partly on the 2-week forecast from the GFS model (see Chapter 13 for more details). Although we learned in Chapter 13 that the weather forecast skill for this model is no better than chance after about 1 week, this is partly because the timing of weather systems is wrong. However, for a monthly *climate* forecast, exact timing is not as important as the overall pattern.

Other information that is incorporated into the 30-day forecast comes from ensemble runs of the NCEP Climate Forecast System model (**CFS**). CFS is a version of the GFS that has been augmented for seasonal climate forecasting, for example, with a more realistic treatment of the ocean (more details on this when we discuss global climate models). The 30-day forecast is also modified based on existing patterns of ocean temperatures, such as El Niño, and soil moisture patterns. In essence, this is a persistence forecast of ocean and soil conditions based on statistical relationships of how these factors affect climate in a certain region. The final 30-day forecast is therefore a blend of statistics and model results.

At seasonal timescales, the influence of ocean temperature anomalies such as El Niño or La Niña on the atmosphere is probably the single most crucial forecast component. This is especially true for forecasts of Northern Hemisphere winter and spring. Why? Because the occurrence and location of tropical thunderstorms help drive the Hadley Cell and in turn affect the location of the jet streams—leading to the global atmospheric response called ENSO. In a probabilistic sense, the presence of El Niño or La Niña "loads the dice" toward a particular climatic outcome in some parts of the world, making certain outcomes more likely. Even so, as with a dice game there is no certainty in the outcome.

Seasonal forecasts also take into account the climatic impacts of other oscillations, such as the North Atlantic Oscillation and the Pacific Decadal Oscillation (see Chapter 8). Trend and analog forecasting methods are also employed. They are the statistical cousins of their weather-forecasting counterparts in Chapter 13. Seasonal forecasts also use results from the CFS and statistical modeling methods.

One problem with seasonal climate forecasts is this: Climatologists understand the effects of ENSO on climate much better than they can *forecast* the onset of an El Niño or La Niña event. This is because we still don't fully understand the atmospheric and oceanic processes that cause El Niño and La Niña. And so while seasonal forecasts during El Niño or La Niña can be quite accurate, it is much harder to anticipate their development and therefore the climatic consequences several months later.

For example, in FIGURE 16-1, the 40 different ensemble forecasts of ocean temperature anomalies in the central Pacific for 2010 are compared with the actual observations of ocean temperatures through April 2010. The model forecasts generally called for a continuation of

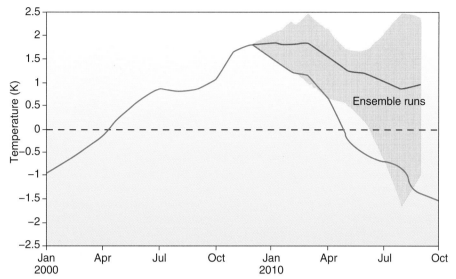

FIGURE 16-1 December 2009 ensemble model predictions of equatorial Pacific Ocean temperature anomalies in 2010 (beige region; average in thick blue line) from the NCEP Climate Prediction Center. The actual observations of equatorial Pacific ocean temperature anomalies are shown in the red line. Only one out of the 40 ensemble runs came close to predicting the rapid waning of El Niño (unusually warm) conditions in early 2010. (Courtesy of NOAA.)

unusually warm El Niño conditions. Instead, the observed ocean temperatures cooled more rapidly than forecast, even by the coldest ensemble forecast. The fact that the vast majority of ensemble runs did not accurately forecast the cooling implies that more research is needed to understand ENSO before we can have better seasonal forecasts.

Interpreting Monthly/Seasonal Climate Forecasts

Monthly and seasonal climate forecasts look very different than weather forecasts. There are no low- and high-pressure systems, no fronts, and no specific numbers for high and low temperatures. Instead, the maps show chances of above- or below-normal climate. What do these maps mean?

Three classes of conditions are forecast on these maps: A, for above normal; B, for below normal; and EC, for equal chances of either above or below normal. In the case of equal chances, the percentages are roughly 33% apiece for above normal, below normal, and normal conditions. Where the odds favor A or B and are above 33%, the percentage chance is labeled and contoured, and the region is shaded.

FIGURE 16-2 shows a typical seasonal climate forecast for the U.S., with temperature on the left and precipitation on the right. For summer 2010, the forecast was for warmer than normal temperatures in most of Alaska, much of the West, and the Southeast and cooler than normal temperatures in the Great Plains and upper Midwest. However, the highest confidence for unusual warmth was only 50% over parts of Nevada, Utah, and Arizona. And so while the odds favored summer warmth in those locations in 2010, it was far from a "sure thing."

Figure 16-2 also calls for a drier than normal Pacific Northwest and an unusually wet Great Plains and Gulf Coast in summer 2010. However, the odds were just slightly tipped toward those possibilities. For large parts of the U.S., the forecast was for "EC." In other words, for much of the nation the precipitation forecast was a coin flip.

How accurate are these types of climate forecasts? FIGURE 16-3 depicts the skill of the NCEP 90-day temperature forecast. In some years, the climate forecasts have great accuracy, particularly in predicting which regions will be warmer or colder than normal, as during the

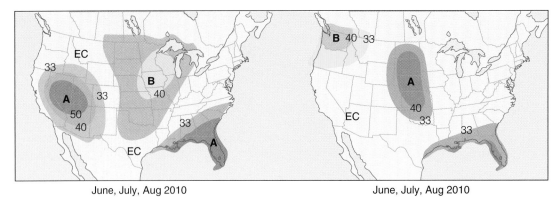

June, July, Aug 2010   June, July, Aug 2010

FIGURE 16-2 Three-month seasonal forecasts of temperature (left) and precipitation (right) for North America for summer 2010, from the NCEP Climate Prediction Center. At left, orange/red regions are those with the highest chances (33% to 50%) of above-normal temperatures; blue areas have the highest chances (33% to 40%) of below-normal temperatures. At right, green regions have the highest chances (33% to 40%) of above-normal precipitation; brown areas have the highest chances (33% to 40%) of below-normal precipitation. White regions labeled "EC" have equal chances of normal, below-normal, or above-normal conditions. (Courtesy of NOAA.)

El Niños of 1997–1998 and 2003–2004. In other years, there is little skill. Overall, however, there *is* skill in 90-day forecasts—they are better than a coin flip. Precipitation forecasts (not shown), however, have marginal skill. Both temperature and precipitation forecasts are generally at their most reliable for winters during strong El Niño or La Niña events. Precipitation climate forecasts during these events can be as highly accurate as temperature climate forecasts.

Now that we have toured the world of monthly/seasonal climate forecasting and learned of its successes and struggles, we can turn to long-range climate forecasting. In particular, we'll examine the predictions of global climate change that have become a focal point of both modern science and international politics.

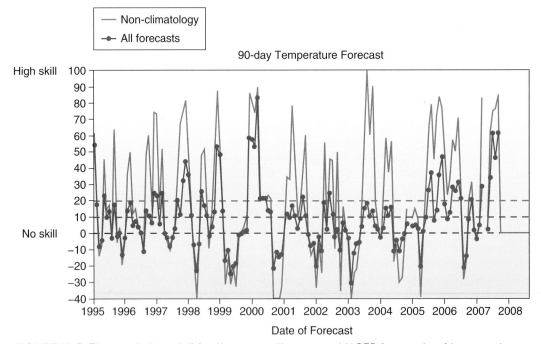

FIGURE 16-3 The predictive skill for three-month seasonal NCEP forecasts of temperature for 1995 to 2007. The red line is for regions where the forecast was for something other than climatology; the blue line is for all forecasts. The higher the line above zero, the more skill in the forecast. Compare this figure with the weather forecast skill graphics in Chapter 13—it's much tougher to make an accurate climate forecast! (*Source*: Climate Prediction Center/NOAA.)

LONG-RANGE CLIMATE FORECASTS

Seasonal forecasts are dominated by ENSO partly because an El Niño or La Niña event lasts roughly a year or so. As we expand our focus to make forecasts for decades or centuries, however, longer term climate change mechanisms become more important than ENSO. These include the natural climate change mechanisms, such as volcanoes, that we studied in Chapter 14. Also included is the human-caused, or anthropogenic, climate change mechanism of increasing "greenhouse gases" in the atmosphere.

Will global climate change in the future? If so, what will be causing the change? To answer these questions, climate researchers use complex computer programs known as **global climate models** (**GCM**s). These GCMs, close cousins of the numerical weather prediction models studied in Chapter 13, in turn produce the predictions of "global warming." How do GCMs lead to climate forecasts? Let's find out.

Making Long-Range Climate Forecasts with GCMs

A GCM calculates the variables that describe global climate using mathematical equations derived from physical principles such as conservation of energy (see Chapter 2) and Newton's laws of motion (see Chapter 6), as shown schematically in **FIGURE 16-4**. The roots of today's GCMs can be traced directly to the methods first developed by L. F. Richardson for numerical weather forecasting (see Chapter 13). They contain many of the same parameterizations (see Figure 13-21) as weather forecasting models. There are several key differences, however.

A crucial difference is that most weather forecasts have focused exclusively on simulating the atmosphere, although this is changing. Climate models, however, must also accurately simulate the oceans, the biosphere, and cryosphere in order to capture the various interactions. This *coupling* of the atmosphere and other spheres is suggested by the arrows in Figure 16-4. In reality, coupled modeling is extremely complicated and remains an unsolved problem in climate modeling.

Explore the inter-relationship between the biosphere and climate in a very simple model in "Daisyworld"

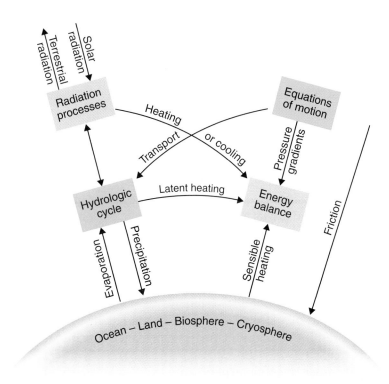

FIGURE 16-4 A schematic of the different processes simulated in a global climate model. These interactions are expressed in a GCM as mathematical expressions in thousands of lines of computer code.

Because of the computational challenges of coupled models, the horizontal and vertical resolution of GCMs is much coarser than in weather forecast models.

Another difference is that GCMs have often been used as research tools to test the effect of changing a particular parameter, rather than simulating precise climate evolution (which isn't possible anyway, given chaos). This approach is called **sensitivity testing** and is colloquially referred to as giving the model's climate a "big kick" and letting the model run until it reaches a new equilibrium. This type of GCM experiment also goes by the name of "**equilibrium run.**" This is the thinking behind simulations of global climate in which carbon dioxide is doubled at the beginning of the simulation, and the climate of the GCM then responds to this "big kick" by becoming warmer than in a **control run** that simulates present-day climate. A more realistic but more computer-intensive **transient run** adds a little more carbon dioxide to the GCM's atmosphere each year, similar to what happens in nature.

A third difference is that today's numerical weather forecasts can be proven right or wrong almost immediately. Even seasonal climate forecasts can be evaluated for accuracy within a year's time. But GCM predictions of climate a century from now cannot be proven right or wrong in any reasonable amount of time.

▨ Verifying Long-Range Climate Forecasts

How, then, do we know that GCM predictions are reliable? One check on GCM validity is to simulate the climates of the past and compare the simulations with data from the era in question. The CLIMAP project in the 1970s, for example, used GCMs to simulate the climate of the last Ice Age maximum 18,000 years ago. Climate researchers then compared the simulations with paleoclimate data from that same period. The models consistently predicted that the tropical oceans had been cooler than the data seemed to suggest—until 20 years later, when renewed research with Andean ice cores revealed that the models had been correct all along. Instead, the data had been flawed!

Another such "litmus test" for GCMs is the Industrial Revolution, from about 1860–2000. As we have seen, the global temperature during this period rose, then fell a bit during the middle part of the 20th century, and then rose steadily from about 1980 through 2000 (and beyond). Doubters of "global warming" cited the mid-century cooling as evidence that greenhouse gases could not be affecting the temperature. If global warming were true, how could carbon dioxide concentrations go up steadily while the global temperature went up and down? Often, the doubters invoked natural climate change mechanisms to explain the temperature variations.

In response, climate researchers ran GCMs for the period 1861–2000 for two different scenarios. In one simulation (**FIGURE 16-5**), only natural climate change mechanisms were included, such as solar energy variations and volcanic activity. In the other (**FIGURE 16-6**), the GCM included anthropogenic forcings such as increasing greenhouse gases and sulfate aerosols released by the burning of coal. These aerosols, released in great abundance before air pollution became a global concern (see Chapter 15), tend to cool the planet.

The GCM results for natural climate change do not reproduce the observed global temperature changes of the 20th century. But the GCM results that incorporate both greenhouse gases and sulfate aerosol effects closely mirror observed temperature during 1860–2000. The shaded region is the spread of model results, akin to an ensemble forecast. The observed temperatures are close to the ensemble results for the entire 140 years of the GCM runs. (More recently, the inability of a deep solar minimum in the late 2000s to reverse global warming reinforced the GCM result that natural climate change mechanisms are not responsible for recent temperature trends.)

These and many other scientific experiments for a wide variety of situations and eras have bolstered climate researchers' faith in the validity of GCM results. The GCMs, despite their imperfections, are able to reproduce changes in both past and current climate. This requires us to take very seriously the GCMs' consistent projections of global climate change in the future.

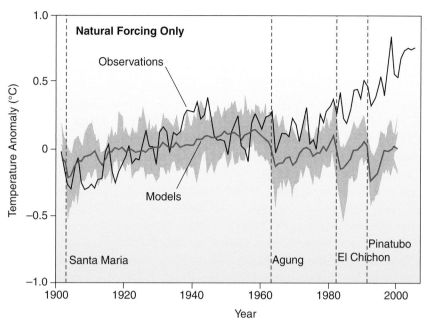

FIGURE 16-5 The observed changes in global temperature (thick line) over the past century. The thin lines represent ensemble runs of simulated changes using different GCMs that assume only natural forcing, such as volcanoes. Major volcanic eruptions during the 20th century are labeled. Note that the average of the model results (thick blue line) does not agree with the observations, particularly since 1980. (*Source*: Intergovernmental Panel on Climate Change, "Climate Change 2007," The Fourth Assessment Report, Working Group 1, Figure TS23.)

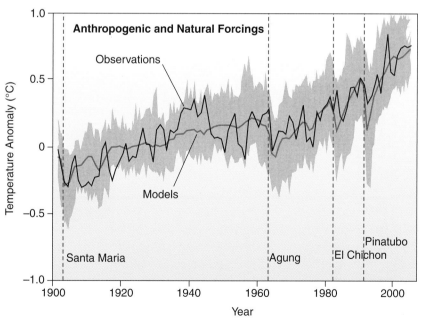

FIGURE 16-6 As in Figure 16-5, but in this case the ensemble runs of GCMs include anthropogenic effects on climate as well as natural mechanisms of climate change. Note the much better agreement between the average of the model results (thick red line) and the observations compared with Figure 16-5. (*Source*: Intergovernmental Panel on Climate Change, "Climate Change 2007," The Fourth Assessment Report, Working Group 1, Figure TS23.)

■ A GCM Genealogy

Before we can understand the latest predictions of global climate change from GCMs, however, we first have to learn their names. As with weather forecasting models, GCMs are often named after the institution where they were created.

The first atmospheric GCM (originally called a "general circulation model") was created by meteorologist Norman Phillips at the Institute for Advanced Studies at Princeton University in 1955. In contrast to forecast models, many of the early models were developed by researchers in academia and at government research laboratories, not in forecast centers. Interestingly, many of the early pathbreakers in GCM research also broke cultural barriers (FIGURE 16-7). For example, Japanese-born scientists who came to the U.S. in the years after World War II played a crucial role in early GCM development.

As FIGURE 16-8 shows, the early history of atmospheric GCMs sprouted from the work at the NOAA/Geophysical Fluid Dynamics Laboratory (**GFDL**) at Princeton; **UCLA**; and the National Center for Atmospheric Research (**NCAR**) in Boulder, Colorado. Most current leading GCMs, such as the **ECHAM** model from the Max Planck Institute for Meteorology (**MPIM**) in Germany, the **Hadley Centre** in the United Kingdom, and the NASA/Goddard Institute for Space Studies (**GISS**) model in the U.S., are descended from these early GCMs. Even more recently, GCMs in Australia (**CSIRO**), Canada (**CCCma**), and Japan (**CCSR/NIES**) have been created. Each model has its own unique approaches to the unsolved problems of climate modeling, for example, in the way that certain poorly understood processes are parameterized.

These are models that are used to make the most famous climate predictions of all: forecasts of global climate change, better known colloquially as "global warming." Global warming and GCMs became household words beginning in June 1988. That's when James Hansen, longtime director of the NASA/GISS in New York City, testified to the U.S. Senate that there was less than

FIGURE 16-7 Climate modelers Akira Kasahara (left) and Warren Washington (right) at the National Center for Atmospheric Research in Boulder, Colorado, in 1975.

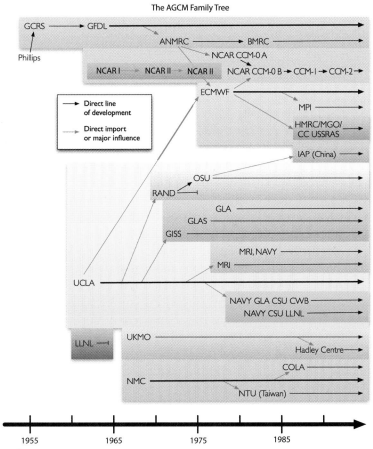

FIGURE 16-8 A "family tree" of atmospheric-only global climate models. The acronyms in this graphic stand for various GCMs, and many of them are defined in the text. (Reproduced from Paul N. Edwards, *A Vast Machine: Computer Models, Climate Data, and the Politics of Global Warming* [Cambridge: MIT Press, 2010], p. 168. Graphic by Trevor Burnham. A clickable version of the AGCM Family Tree, with acronym expansions and information about each lab and its models, may be found at http://pne.people. si.umich.edu/vastmachine/agcm.html.)

a 1% chance that the heat wave afflicting the U.S. at that time was due to natural, nonhuman causes (FIGURE 16-9). The predictions of future global climate change he presented to Congress were based on results from an early transient run of the GISS model.

The most famous of all climate predictions appear in the IPCC reports, which use results from the GCMs above in a coordinated effort to make long-range climate forecasts. Next, we look at IPCC and these climate forecasts below.

FORECASTS OF GLOBAL CLIMATE CHANGE

We learned about the Intergovernmental Panel on Climate Change, or IPCC, in Chapter 15. Since 1988, IPCC has issued four reports on climate change. At the heart of these reports are sophisticated modeling experiments using the world's best GCMs in an attempt to simulate future climate. Differences in the models allow for an ensemble approach, in which a range of results are obtained. In addition, for the third and fourth IPCC reports released in 2001

FIGURE 16-9 Dr. James Hansen, director of the NASA/ Goddard Institute for Space Studies (GISS), testifies to the U.S. Senate about global warming on June 23, 1988. Dr. Jerry Mahlman, climate modeler and then-director of the NOAA Geophysical Fluid Dynamics Laboratory (GFDL) at Princeton, is at right.

and 2007, respectively, a range of scenarios of greenhouse gas emissions was inserted into the model simulations. These different scenarios were based on differing societal decisions regarding the use of fossil fuels. We now examine those scenarios in detail.

■ IPCC Scenarios of Greenhouse Gas Emissions

For the GCMs to be able to simulate future climate, modelers have to make some assumptions about what will be affecting the climate. They assume that natural climate change mechanisms will operate as they have in the past. But what about the human component, in particular the addition of greenhouse gases? Will we continue to release these gases at the current rate? Will the amounts increase with growing world population and industrialization? Predicting what people will do is an even tougher challenge than simulating climate!

To give the GCMs some data to ingest, scientists affiliated with IPCC created four main "storylines" from which to develop various scenarios for future greenhouse gas emissions. These storylines, which are a little like science-fiction "alternate timelines," are detailed in **TABLE 16-1**.

These storylines vary depending on whether the nations of the world will join forces to protect the environment (A1 and B1) or if instead regions and nations will tend to go it alone with their own solutions (A2 and B2). The storylines also vary in terms of economic and population growth. The A1 storyline, in which rapid economic growth is counterbalanced by a leveling-off of population in the mid-21st century, also has its variant scenarios or "family members" in which fossil fuels are used intensively (A1FI) or are used in conjunction with nonfossil fuels (A1B) or are de-emphasized in favor of nonfossil fuels (A1T).

From a greenhouse gas perspective, the most polluting is the A1FI scenario; the least polluting are the B1 and A1T scenarios. The blended A1B scenario represents a "middle of the road" guess at future emissions.

Explore the "Global Carbon Budget— 1960 to 2100"

This spread of scenarios for greenhouse gas inputs into the environment has the same effect as ensemble forecasting. When inserted into a GCM, the various scenarios lead to a range of results for global climate change. When the scenarios are inserted into the 23 GCMs used in the IPCC report, the range of results is expanded even further.

This broad range of results, in turn, gives us a glimpse into the future in a probabilistic sense. We can't know for sure what the world will do with greenhouse gases. No one GCM is perfect either.

TABLE 16-1 IPCC Scenarios for 21st-Century Greenhouse Gas Emissions (Data from IPCC.)

Storyline	Family Member	Global or Local/ Regional Solutions?	Economic Growth	Population Growth	Other Choices	Gigatons of CO_2 Emissions from 1990–2100
A1		Global	Very rapid	Low; peak in mid 21st century	Rapid introduction of new, efficient technology	Depends on exact scenario; see below
	A1T	Global	Very rapid	Low; peak in mid 21st century	Non-fossil fuel emphasis	~1000–1200
	A1B	Global	Very rapid	Low; peak in mid 21st century	Blend of fossil and non-fossil fuels	1350–2100
	A1FI	Global	Very rapid	Low; peak in mid 21st century	Fossil fuel emphasis	2100–2600 (most)
A2		Local/ regional	Slow	High	Slow technological advances	1350–2000
B1		Global	Rapid in information & service industries	Low; peak in mid 21st century	Use of new, efficient technologies	700–1400 (least)
B2		Local/ regional	Intermediate	Moderate	Slower adoption of new technology than in A1 and B1	1100–1700

But by inserting a spread of scenarios into nearly two dozen GCMs and solving the equations for future climate, we can get a clearer sense of the possible range of outcomes.

Forecasts of Surface Temperature Changes

The four IPCC reports have been remarkably consistent in predicting a globally warmed world in the 21st century. The first IPCC report in 1990 predicted a global surface temperature increase of 0.3° C per decade during the 21st century, or 3° C in a century, assuming a "business as usual" scenario. By the time of the third report in 2001, the IPCC projected a temperature rise of 1.4° C to 5.8° C from 1990 to 2100. The third report placed even greater faith in the GCM results, as the science of global climate modeling matured. The range of results from the GCMs (**FIGURE 16-10**), like an ensemble weather forecast, lent greater confidence to the "big picture" consensus. All models predicted significant warming—some models (NCAR) a little less than others, some (CCSR/NIES) considerably more.

The 23 GCMs and 40 scenarios of greenhouse gas emissions employed in the fourth IPCC report in 2007 refined these earlier predictions of global warming. On the left-hand side in **FIGURE 16-11**, the average of the GCM results for three different IPCC scenarios is plotted. The colored bars to the right of this graph show the spread of model results for all the major scenarios in Table 16-1.

Again, as in previous IPCC reports, all scenarios inserted into all the GCMs lead to warming of the global average temperature. The A1FI scenario, with the greatest input of carbon dioxide into the atmosphere by 2100, leads to the greatest warming. The B1 scenario adds the least CO_2 and leads to the least warming. It should be stressed that the GCMs are not "rigged" to cause warming if higher values of CO_2 are inputted. The results in Figure 16-11 are the culmination of billions of calculations within each model for each scenario, based on our best understanding of how the climate system works.

On the right-hand side of Figure 16-11, the spatial pattern of warming due to greenhouse gas increases is shown for three scenarios and two different time periods during the 21st century. Regardless of the scenario, by 2090–2099, the entire globe will experience warming. The greatest warming occurs in the high latitudes of the Arctic in all simulations, consistent with the warming that has already been observed (see Chapter 15). In the A2 scenario, this high-latitude warming

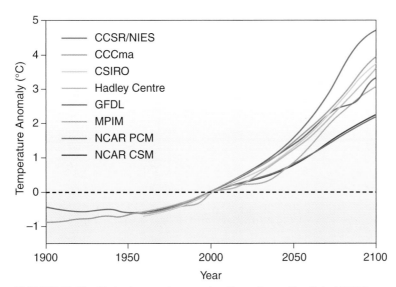

FIGURE 16-10 Global warming projections from the third IPCC report in 2001. Different GCM results are plotted. Note that the range of results for 2100 (roughly 2° C to 5° C warmer than 2000) is consistent with more recent projections. (Adapted from Robert A. Rhode, "Global Warming Projections," *Global Warming Act*. [http://www.globalwarmingart.com/wiki/File:Global_Warming_Predictions_png].)

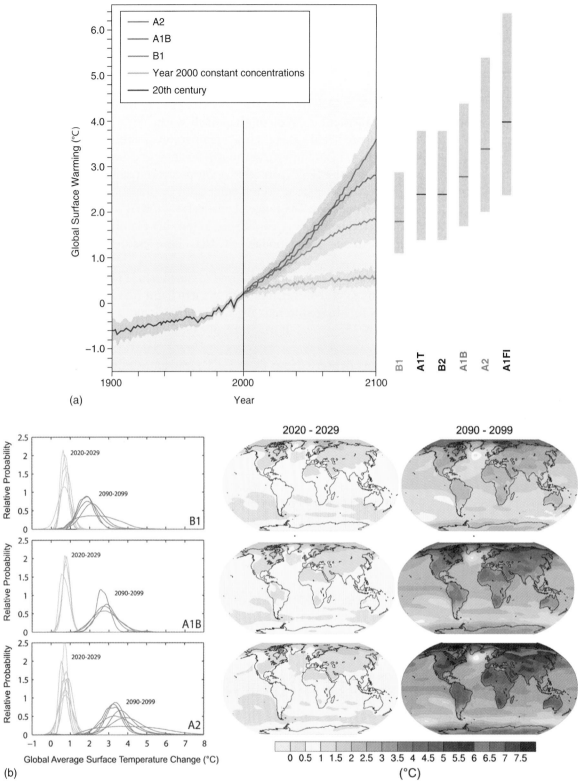

FIGURE 16-11 (a) Projections of global warming for various IPCC scenarios of fossil fuel use during the 21st century (Table 16-1). Three scenarios, plus a hypothetical case in which carbon dioxide concentrations were halted at 2000 levels, are plotted at left, along with observed temperatures during the 20th century. Shaded regions indicate uncertainty in the observed temperature record, or the range of ensemble run results in the case of the models. The averages and range of model results for all six scenarios are plotted in bar-graph form at right. (b) Patterns of global warming for two different periods in the 21st century for three different IPCC scenarios. At left, the global average surface temperature change is plotted for 2020 to 2029 and 2090 to 2099 in terms of probability. At right, the spatial pattern of the warming, as determined by the average of the model runs, is shown for the two time periods. Red and magenta regions at right indicate the largest temperature increases. The plots indicate that temperatures are projected to increase throughout the 21st century, and the largest increases are projected to occur in the high latitudes of the Northern Hemisphere. (*Source*: The Fourth Assessment Report "Climate Change 2007," Intergovernmental Panel on Climate Change, Working Group 1, Figure SPM6.)

exceeds 7° C in some locations. But even in the most benign B1 scenario, the GCMs predict a warming by 2100 of about 2° C over the world's continents. Even this is more than twice the amount of warming that was observed in the 20th century and that has led to the consequences detailed in Chapter 15.

The bottom-line IPCC projections of global temperature change are this: *The average surface temperature of the Earth is likely to increase by 1.1° C to 6.4° C (2° F to 11.5° F) by the end of the 21st century, with a best estimate of 1.8° C to 4.0° C (3.2° F to 7.2° F).* These projections bracket the original 3° C warming prediction in the first IPCC report in 1990. But the latest IPCC results are even more robust, due to the latest advances in climate modeling science and the work of thousands of scientists on the subject since 1990. As a result, IPCC now assesses the chances of this warming occurring at between 66% and 90%. *All scientific signs point toward a much warmer world in the 21st century.*

Forecasts of Precipitation Changes

A globally warmed world would also be a world in which the water cycle is revved up, according to IPCC projections. The ultimate physical reason for this is the relationship between temperature and saturation vapor pressure (see Chapter 4)—in a warmer world, the atmosphere can "hold" more water vapor. As a result, thunderstorms could be more common and more vigorous in the deep tropics. But these storms would also cause strong sinking of air and drier conditions in the descending branch of the Hadley Cell (see Chapter 7) in the subtropics. As seen in FIGURE 16-12, precipitation amounts in the late 21st century for the "middle of the road" scenario may increase 20% or more near the equator and decrease 20% or more in the subtropics. High latitudes are predicted to be much wetter in winter, largely because of the presence of warmer and moister air than at present.

Possible Impacts of Global Climate Change

It's tempting to say "so what" if the planet gets warmer—we'll all wear shorts in winter, or if it's wetter, we'll just buy an umbrella! "The devil lies in the details," however; in the case of global climate change, the crisis lies in the *consequences* of the predicted temperature and precipitation changes.

Global Impacts

FIGURE 16-13 depicts just a few of the possible impacts of global climate change for the "middle of the road" A1B scenario. As the temperature curve climbs, so does the severity of the consequences, according to IPCC. During the early 21st century, droughts become more frequent as the

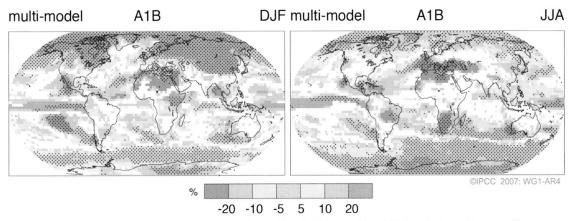

FIGURE 16-12 Precipitation changes simulated by GCMs for the A1B "middle of the road" scenario for winter (left) and summer (right) seasons in 2090 to 2099 versus 1980 to 1999 averages. Red regions would experience a decrease in precipitation; green and blue regions would experience an increase in precipitation. (*Source*: Intergovernmental Panel on Climate Change, "Climate Change 2007," *The Fourth Assessment Report*, Working Group 1, Figure SPM7.)

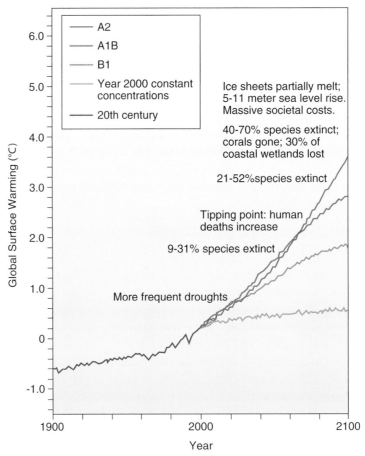

FIGURE 16-13 Similar to the global warming timeline in Figure 16-11a, except that the predicted impacts of global warming are labeled for various amounts of temperature rise.

hydrologic cycle becomes more vigorous. Warmer winters melt glaciers and snowpack. Then, by about 2050, the changes in temperature and precipitation cause habitats to disappear out from under plants and animals. Extinctions ensue.

Shortly after 2050, when the global average temperature has warmed 2° C versus the present, the impacts start affecting human populations in ways that can no longer be ignored. Heat waves, droughts, and floods take unprecedented human tolls. By century's end, most of the world's tropical rain forests and corals are gone, with extinctions of up to about half of all species.

And yet all this could be a prelude to even worse outcomes. If the warming reaches 5° C versus current conditions, partial melting of the Greenland and West Antarctic ice sheets could lead to massive coastal flooding that would reshape the world's coastlines and wreak havoc on the world's governments and economies.

These consequences sound like the plot of a disaster movie. To make them even more tangible for most of our readers, we now examine in detail a few possible impacts for the United States based on both IPCC and other projections.

Impacts on U.S. Coastal Areas
More than 50% of the U.S. population lives in coastal regions, and these areas are some of the fastest growing in the nation. Along with this growth comes increased vulnerability to natural hazards. Global climate change poses at least two hazards to many coastal areas: sea level rise and changes in hurricane frequency and intensity.

During the next century, sea level is expected to keep on rising (see Chapter 15). The latest IPCC report projects a rise of 21 to 48 cm (8 to 19 inches) for the A1B scenario. This is a conservative

FIGURE 16-14 A preliminary assessment of the potential of shoreline erosion of the United States. The degree of erosion is associated with changes in sea level, the frequency of storms, and the condition of the shoreline. (Courtesy of U.S. Global Change Research Program [www.globalchange.gov].)

forecast, based partly on the undisputed fact that water expands as it warms. Additional sea level rises on the order of several meters could occur if significant melting of the Greenland and West Antarctic ice sheets were to occur, similar to what occurred during the warm Last Interglacial period about 125,000 years ago.

An increase in global sea level is expected to have a large impact on coastal erosion. Low-lying coastal regions are particularly vulnerable. As **FIGURE 16-14** shows, the Atlantic and Gulf Coasts are particularly susceptible to erosion (as is northern Alaska). These regions have gentle slopes, and therefore, a rise in sea level causes a large change in the shoreline. The Pacific Coast may also be affected. For example, beach and cliff erosion increases during El Niño years when storms hit the coast. A long-term rise in sea level, if accompanied by more storms in a globally changed climate, could enhance erosion on the West Coast.

A different threat to coastal regions of the Atlantic and Gulf Coasts is hurricanes. There has been intense debate and controversy regarding whether hurricane frequency and intensity has been, or will be, affected by a rise in global temperature. One impediment to resolving this debate is the resolution of GCMs. As noted earlier, GCMs have coarser resolution than weather forecast models. This means that GCMs cannot accurately simulate the smaller scale features of intense hurricanes.

Recent research has partly circumvented this problem. Scientists at GFDL took the average of an ensemble of 18 GCM "global warming" runs and inserted this ensemble average into a high-resolution hurricane model. The results of this modeling experiment were: a near-doubling of the frequency of intense Category 4 and Category 5 hurricanes in the Atlantic, Gulf, and Caribbean by the end of the 21st century (**FIGURE 16-15**).

Modeled Category 4 & 5 Hurricane Tracks

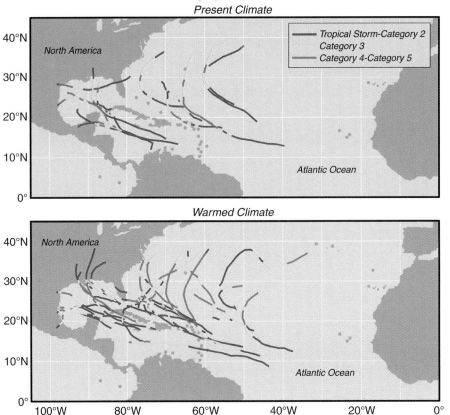

FIGURE 16-15 The results of an 18-model ensemble simulation of major Atlantic hurricane tracks (in yellow and red) in a control run representing present-day climate (top) and in a globally warmed climate (bottom). This simulation used the A1B scenario for the late 21st century. Note the increased number of major hurricanes in the globally warmed simulation. (Courtesy of NOAA.)

This research is not the last word on the subject of hurricanes and global warming. However, it raises the disturbing possibility of more frequent Hurricane Katrinas in the Gulf and Atlantic during the 21st century. The implications for damage along the East and Gulf Coasts due to storm surge and wind are sobering.

Impacts on U.S. Forests

How could a globally warmed world affect our nation's forests? An increase in atmospheric carbon dioxide levels generally leads to higher photosynthesis rates. This could increase forest growth and productivity. However, the increases may be limited by other conditions, such as the availability of moisture and nutrients. As discussed in Chapter 14, tree species have adapted to natural threats, such as fire, insects, droughts, and windstorms. Global climate change will likely force trees to adapt, migrate, or die.

FIGURE 16-16 depicts the forest types of the present day in the eastern half of the United States. These forests have adapted to their environments' temperature and moisture conditions. In the next century, however, these forests may face rapid changes to their environment. How will they respond?

In a warmer climate, it seems logical that tree species that prefer cooler conditions will shift poleward (or upward, in mountainous regions). However, trees cannot pull up their roots and migrate. How a given species adapts to environmental changes depends on its growth rate and its ability to disperse seeds. As temperatures rise, it's a race between the rate of change of the environment and the tree's ability to propagate into cooler conditions.

It's also possible for a species to run out of territory. For example, a species may propagate up to the highest slopes of a mountain range and then run out of habitat when even the highest slopes become too warm for it. In such cases, extinction is likely.

Dominant Forest Types

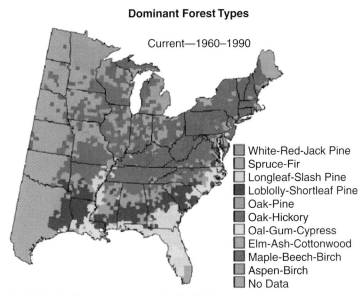

FIGURE 16-16 The current (1960–1990) forest types across the eastern United States. The distributions are determined by temperature, moisture, and soil conditions. (Courtesy of U.S. Global Change Research Program [www.globalchange.gov].)

In addition, climate change may also affect the frequency of fires, storms, droughts, and outbreaks of diseases and pests. Some or all of these changes in trees' natural hazards may put species at even more risk. In short, trees (and other plants and animals) are under multiple stresses in a globally warmed world.

FIGURE 16-17 presents two projections of forest distributions in the eastern U.S. by the end of the 21st century. The two projections are based on climate forecasts from the Hadley Centre and Canadian GCMs, respectively. The contrast between the forecast forests and the present-day forests is stark. If the projections are accurate, pine forests will decline in the Southeast, whereas oak forests will expand northward. The maple–beech–birch forest types will no longer exist in the eastern U.S. by 2100, according to these analyses.

Dominant Forest Types

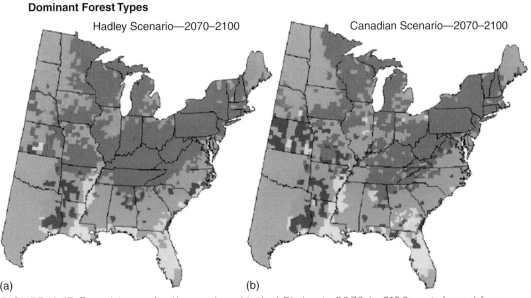

FIGURE 16-17 Forest types for the eastern United States in 2070 to 2100, as inferred from climate projections using the (a) Hadley Centre and (b) Canadian GCMs. The color coding for forest type is the same as in Figure 16-16. (Courtesy of U.S. Global Change Research Program [www.globalchange.gov].)

Are these predictions alarmist? They are not any more radical than the actual changes in forests that took place during the end of the last ice age, as revealed by research in past climates (see Chapter 14). For example, northern spruce forests like those in Canada today were found in the mountains of the Southeast about 10,000 years ago. The difference this time is that humans are causing the climate change. This time the changes may be so rapid that our biosphere is unable to respond.

SOME POSSIBLE SOCIETAL RESPONSES TO GLOBAL CLIMATE CHANGE FORECASTS

Given the severity of the possible impacts of global warming, a natural question to ask is this: "What can we do in response?"

A possible "role model" for our response is the reaction to the stratospheric ozone hole. In the case of the ozone hole, the global ban on CFC use will lead to the eventual removal of chlorine compounds from the stratosphere during the 21st century. In the absence of CFCs, the ozone layer will repair itself naturally.

Unfortunately, there is no similar healing process for global warming. Natural processes that reduce carbon dioxide in the atmosphere act on timescales that are far too long to be helpful. Except perhaps for some controversial "geoengineering" proposals, the goal of societal responses to global warming is to *limit the increases* in greenhouse gases; there is no way to *reduce* their concentrations as the world has done with CFCs.

There are two main types of societal responses. **Mitigation** is a response that seeks to limit the increases in greenhouse gases so that the worst impacts of global warming do not occur. A complementary societal response, **adaptation**, refers to steps taken to reduce our vulnerability to climate change and to take advantage of any benefits that may occur. In this chapter, we focus mostly on mitigation.

On an individual level, lifestyle changes can limit one's contribution to greenhouse gas concentrations. **BOX 16-1** outlines some of these individual responses. However, the results of the IPCC scenarios indicate that globally coordinated activities are more effective than local or regional solutions. Therefore, we now examine a few potential global responses to climate change.

Kyoto Protocol

The "Earth Summit" in Rio de Janeiro, Brazil, in 1992 formulated the UN Framework Convention on Climate Change. This international treaty proposed greenhouse gas reductions for the first time. In 1997, an update to this treaty was created at another meeting in Kyoto, Japan. This updated treaty, which has become known as the "**Kyoto Protocol**," took effect in 2005.

What does the Kyoto Protocol require? Industrialized nations that agree to or "ratify" the updated treaty promise to limit their greenhouse gas production to at least 5% below the levels each nation emitted in the year 1990. The goal is to keep the global temperature from climbing high enough to trigger the worst consequences shown in the top right of Figure 16-13.

A total of 187 nations or other governmental bodies has ratified the Kyoto Protocol. Among industrialized nations, only the U.S. and Australia were holdouts. Australia finally ratified in December 2007. In the U.S., concerns over the economic impact of the Kyoto Protocol, as well as doubts over the reality of global warming in some sectors of the U.S., have prevented ratification.

The Kyoto Protocol is far from perfect. Like the initial Montreal Protocol for CFCs, the environmental impact of the agreement is less than the political benefit of achieving international cooperation. The reductions in global temperature rise due to the Kyoto accord through 2100 are small, frustrating some environmentalists. However, the benefits in the 22nd and 23rd centuries are much greater. Meanwhile, the economic costs of ratcheting back greenhouse gas emissions to 1990 levels could be large—or they could be offset by avoidance of costly global warming consequences and the development of "green" industries. Rapidly developing nations such as China and India are not required to reduce greenhouse gas emissions, despite their large and growing use of fossil fuels.

Box 16-1 What Can I Do About Global Warming?

Societal responses to global warming can start with individual action. But which actions are the most effective at limiting greenhouse gas emissions? Here are some of the best possibilities, most of which have additional advantages beyond climate change mitigation:

- **Reuse and recycle products**. A significant portion of fossil fuel use is devoted to creating, wrapping, and shipping new products.
- **Limit air conditioning and heating use** by changing the thermostat, adding insulation, or both. This reduces the amount of fossil fuels that are burned at power plants to create electricity and/or heat.
- **Change incandescent light bulbs and replace them with compact fluorescent bulbs**. This, too, reduces the amount of fossil fuels burned by power plants. It is said that if each U.S. family changed just one light bulb, it would have the same impact on greenhouse gas emissions as removing 7.5 million cars from American roadways.
- **Use fuel-efficient vehicles and use them wisely**. Fight the car-ad mania for reckless driving. Even in a fuel-efficient vehicle, driving at 55 mph increases your fuel efficiency noticeably versus driving at 70 mph (your co-author has documented this with his Toyota Corolla). And, it reduces your chance of a fatal accident!
- **Find opportunities to carpool, use public transportation, walk, or bike to your destinations**. A resident of New York City emits less than one third as much greenhouse gases as a resident of other American cities, partly because New Yorkers use subways and walk more.
- **Use less hot water**. Fossil fuels are ultimately used to heat water, so your long, hot showers are affecting the environment! Shorter, cooler showers are better. (If you travel to a foreign country, you'll quickly find out that Americans are some of the few people on the planet who can and do indulge in long, hot showers every day.)
- **Kill the vampires**. Unplugging so-called "vampire appliances" that draw energy even when they are turned off can reduce your electricity bill and reduce energy and fossil fuel use.

These actions can reduce your "carbon footprint" by up to several tons of carbon dioxide per year. However, this is only about 30% of the average per capita footprint for Americans, which is close to 20 tons of CO_2 per year—over 4 times the amount released by an average resident of China and over 15 times more than released by an average resident of India. This means that the majority of Americans' greenhouse gas emissions are *institutionalized*: hidden deeply in our everyday cultural and business practices that are difficult for individuals to change. Thus, individual action has to be coupled to broader societal responses if carbon dioxide levels are to be kept below levels that will lead to severe climate consequences, according to the IPCC projections.

One more response you may not have thought of: you can be part of the scientific research on global warming! At http://climateprediction.net you can join thousands of people around the globe who run the Hadley Centre climate model in background on their computers (see the dots on the figure on the next page for their locations). It is billed as "the world's largest forecasting experiment for the 21st century." The climateprediction.net team (mostly atmospheric scientists from Oxford University) use the collective unused computing power of thousands of PCs to run more ensemble climate experiments than can be done on even the world's fastest supercomputers.

Over 2.5 million model simulations have been completed through this grassroots project. The climateprediction.net website also contains excellent information on the science of global warming and GCMs. You'll understand climate modeling and climate change better once you are part of the action!

(continued)

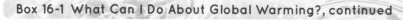

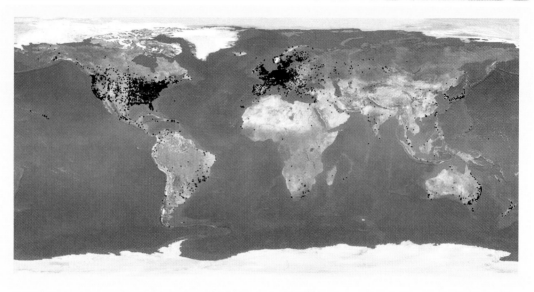

A final limitation of the Kyoto Protocol is time: It expires in 2012. As with the Montreal Protocol for CFCs, significant improvements will require new and more stringent subsequent agreements. However, the December 2009 Copenhagen conference designed to extend the Kyoto accord did not lead to such agreements.

In the final analysis, the biggest single difference between the ozone hole and global warming that makes solutions problematic is that the world simply was not nearly as dependent on CFCs as it currently is on fossil fuel burning. CFCs only propelled spray cans and cooled refrigerators; fossil fuels are currently the power source for the world. There can be no realistic ban on fossil fuel use as there is on CFC use. Even a limit on fossil fuel use is much harder to agree to and implement.

Carbon Trading and Taxing

Another societal response to global warming imitates the approach to a different environmental problem: acid rain. To limit sulfur dioxide emissions, the U.S. government created a trading system for sulfur emissions among the industries emitting SO_2 (e.g., energy companies burning coal). This approach successfully reduced SO_2 emissions and acid rain in the U.S. and is now being explored as a way to limit greenhouse gas emissions as well.

This response is referred to as **carbon trading** or "**cap-and-trade**." A simple analogy helps to explain cap-and-trade. You may have been in (or attended) a high school or college graduation where seating was limited. Lots of friends and relatives want to attend such events, but the auditorium may not have enough seats. The authorities forbid overflow crowds for safety reasons. So the authorities place a cap on how many people can attend the ceremony. Only as many tickets are printed as the number of seats in the auditorium. Graduates receive a certain limited number of tickets apiece. Some graduates end up with more tickets than they need; others desperately want more tickets so that more of their friends and family can attend. Those with too many tickets give their tickets to those who need tickets. In the end, it's a equitable win–win situation. The auditorium is safe, not filled beyond capacity. The authorities are happy. And most, if not all, of the people who wanted to attend the graduation are able to do so.

How does this analogy explain carbon trading? Instead of seats in an auditorium, it's the ability to emit carbon dioxide that authorities would regulate. Some businesses will emit less than their allotted share; others will need to emit more. Through a trading system, the cap on overall emissions could be maintained—in analogy with the total attendance in the auditorium. The trading option would still allow various businesses to burn fossil fuels as needed, instead of mandating inflexible across-the-board limits on all businesses. Furthermore, by running the entire trading system as a market similar to commodities trading, governments could set prices on the ability to emit carbon dioxide and reap monetary rewards from carbon trading. The profits from carbon trading could then be set aside to make other societal changes that would limit global warming.

However, carbon trading has its critics. The cap may be set too high to make a discernible difference in greenhouse gas concentrations. The market may function in unpredictable ways with unforeseen negative consequences, not unlike the recent events on Wall Street. The financial benefits may be smaller than expected or may be diverted from their original purposes. Nevertheless, as of the writing of this textbook, carbon trading was growing in popularity, particularly in Europe, to achieve the greenhouse gas emissions requirements of the Kyoto Protocol.

A simpler but controversial method of limiting carbon emissions than cap-and-trade is a **carbon tax** that is applied at the point at which carbon enters the economy. One hypothetical example of a carbon tax would be a 10-cent-per-gallon tax on gasoline in the U.S. to compensate for the societal impacts of carbon dioxide via global warming. A carbon tax provides certainty about what the price of carbon-based fuels will be. This is in contrast to carbon trading, in which the wheeling-and-dealing of the market sets the price. Carbon trading, however, provides certainty about how much emissions are allowed, unlike a carbon tax. To date, most nations have been unable to implement a carbon tax, although some nations have enacted less comprehensive energy taxes.

Geoengineering

Geoengineering is a more radical response to global climate change. Its underlying philosophy is this: "Technology got us into this bind; more technology will get us out of it." Visionary proponents of geoengineering have proposed a number of possible technological partial fixes to global warming. The common thread in these approaches is that they all propose to alter basic aspects of the climate system to limit global warming.

FIGURE 16-18 depicts a variety of geoengineering strategies. They range from ambitious responses to "Hail Mary" long-shot approaches. Most can be understood in the context of basic concepts covered in this textbook, however.

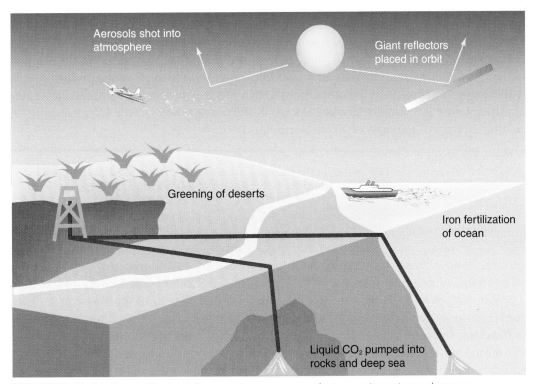

FIGURE 16-18 A schematic depicting various proposed geoengineering schemes.

For example, the concepts of energy and energy balance (Chapters 2 and 3) help explain several geoengineering proposals. These proposals assume that greenhouse gases will continue to increase and trap more and more heat in the Earth's lower atmosphere. To counteract this, geoengineers propose rebalancing the Earth's energy budget by carefully limiting the amount of sunlight reaching Earth. This could be done by increasing the Earth's albedo. One proposal calls for space-based mirrors to reflect sunlight away from Earth. Another proposal calls for intentionally injecting sulfur aerosols into the air to replicate the cooling that the Earth experienced in the mid 20th century due to sulfur pollution. Any mistakes or unintended consequences of such tinkering with solar energy inputs could be catastrophic, however.

Other geoengineering proposals attempt to rebalance the energy budget by physically removing CO_2 from the atmosphere. Widespread forestation of current desert areas would take up some of the additional carbon dioxide that we put into the air via fossil fuel burning. Some nations are capturing carbon dioxide and pumping it into vast underground storage areas. A few geoengineers have even proposed dumping iron into the world's oceans, fertilizing them, and potentially leading to a massive drawdown of CO_2 from the atmosphere into the ocean due to massive phytoplankton blooms. The ecological drawbacks of such plans make them less palatable than other approaches to mitigating global warming.

Geoengineering approaches are not a cure-all. Many climate scientists believe that we should explore these options, but should *not* implement them at this time. If the IPCC projections come true, however, there may be a point at which even "Hail Mary" technological options to stave off climate change will be necessary.

PERSPECTIVES ON GLOBAL CLIMATE CHANGE

Global climate change has become a lightning-rod political issue, especially in the United States. The stakes of global warming have focused a spotlight on a small scientific field that is often misunderstood and is unused to such public scrutiny.

To those unfamiliar with computer modeling of the atmosphere, the inevitable uncertainties of GCM results sound like sloppy science performed by biased participants. Scientists who find their research results distorted or misconstrued in public forums are tempted to assume that they are in a "gotcha" battle with disingenuous questioners of their methods.

Lost amid the sound and fury of Internet blogs is a simple fact. The experts in global climate change around the world, using the best scientific tools available to study the problem, believe the IPCC projections are likely to come true. And they are profoundly worried about the future of the planet. James Hansen, director of NASA/GISS whose testimony to the U.S. Senate in 1988 first aroused public awareness of global warming, said in 2010 that

> The predominant moral issue of the 21st century, almost surely, will be climate change, comparable to Nazism faced by Churchill in the 20th century and slavery faced by Lincoln in the 19th century. Our fossil fuel addiction, if unabated, threatens our children and grandchildren, and most species on the planet.

Hansen has even taken the extraordinary step (as a private citizen, not as director of NASA/GISS) of advocating and participating in civil resistance to force governments to act on global warming. Many climate scientists would not go this far. But it is a measure of the depth of feeling on this subject by at least one of the experts in global climate modeling.

What can we believe? The complex computer schemes behind global climate change projections may seem like voodoo. The first time a sky-savvy potato farmer suggested that the next year's potato harvest could be predicted by looking at the stars, his listeners may have thought the same thing. The stakes were high for the Andeans: Poor harvests could mean famine. They placed their faith in a method that, though it may have seemed odd at first, turned out to be reliable.

What should we do? Although science can provide a sense of future scenarios and their likelihood, any response must be a social and political decision, not a scientific one. The Andeans' observations and inferences of the dim Pleiades were, as it turns out, sound science—but the societal response was up to the farmers themselves.

The stakes for us today are even higher than they were for the Andean farmers, and the timeline for consequences is even longer: not just the next growing season, but decades and centuries into the future. Our responses may have to go against the grain of "business as usual" on individual, corporate, national and international levels. Change is difficult for human societies, at least at first. But as we have seen, anthropogenic *climate* change is already occurring and is expected to ramp up in the coming decades.

The Andeans trusted their experts' climate forecasts, acted on them, and preserved their way of life with some adaptations. Will we trust our climate experts and act upon their forecasts? The next chapter of this story will be written by you.

CLOSURE

We have come to the end of this text, but hopefully it is only the beginning of a lifetime of exploring the atmosphere on your part. Fears of global warming notwithstanding, meteorology is a fascinating subject that is simultaneously mysterious and literally right in front of you. We hope that this book has opened your eyes to the many ways we observe the atmosphere and how these observations are at the heart of the science of meteorology (FIGURE 16-19). Keep looking up at the sky! We also hope that you will hear each new weather and climate forecast with a deeper appreciation of the scientific effort that goes into each one. Our deepest wish is that you use what you have learned from this book to explain what you see and hear and read about in the vast world of weather and climate.

FIGURE 16-19 This is not an astronaut's photograph of Earth! Instead, it is a digital image based on a combination of high-resolution satellite data of land, sea, topography, ice, clouds, and even city lights. The resulting composite is even more informative than a photograph, and just as beautiful.

PUTTING IT ALL TOGETHER

Summary

Climate forecasts are different than weather forecasts. Instead of precise numbers and forecasts for a particular day and time, climate forecasts cover larger blocks of time and give chances of changes versus normal climate.

Climate forecasts up to about one year in advance tend to rely on statistical relationships between climate and processes such as El Niño-Southern Oscillation that are known to affect climate. Longer range climate forecasts rely on global climate models, or GCMs, to generate computer-based projections of future climate. These models are cousins of the numerical weather prediction models used in weather forecasting. However, unlike weather forecast models, GCMs incorporate sophisticated interactions between the atmosphere, ocean, land, and ice.

Research using GCMs began in the 1950s. Today dozens of GCMs produce climate predictions at research centers across the globe. The results of GCMs have been verified versus both past and current climate conditions.

The most famous climate forecasts are associated with projections of "global warming." The Intergovernmental Panel on Climate Change (IPCC) has coordinated and issued climate forecasts for the next century based on expectations that greenhouse gas concentrations will increase. The most recent IPCC report projects that global average temperatures at the surface will increase from 1.1° C to 6.4° C (2° F to 11.5° F) by the end of the 21st century, with a best estimate of 1.8° C to 4.0° C (3.2° F to 7.2° F). The forecast range reflects the spread in results of the ensemble of GCM runs as well as different scenarios for fossil fuel use in the 21st century. The precipitation forecast is for more droughts as well as more floods during the 21st century. This is largely because of a more vigorous hydrologic cycle driven by high temperatures and more evaporation.

Climate scientists and others are concerned about global warming because of the possible consequences of these temperature and precipitation changes. Heat waves, droughts, floods, intense hurricanes and species extinctions could all become more common during the 21st century. The consequences could be most severe for high-latitude, high-altitude, and coastal regions. A globally warmed world would also be a globally changed world, and in most cases not for the better.

In response, nearly all of the governments of the world (with the exception of the United States) have ratified the Kyoto Protocol. This treaty requires industrialized nations to reduce their greenhouse gas emissions to less than 1990 levels. Even so, the Kyoto Protocol by itself will not reduce warming during the 21st century by a significant amount. It will take future, more stringent international agreements to avert the consequences of global warming during the next century.

Other ways to limit greenhouse gas emissions include cap-and-trade, which is a market-based solution that establishes a cap on overall emissions rather than a fixed limit on emissions by each business, or a carbon tax. Yet another possible response is geoengineering the Earth's climate system, for example, by reflecting sunlight or reducing CO_2 in the atmosphere. Geoengineering is a radical option that probably should not be implemented now, but it may be needed in the future.

Global warming has become a "hot potato," politically speaking. The overheated political rhetoric obscures a basic point: The world's experts on global climate change, using the best scientific tools available, predict that the planet will warm during the next century—and that the consequences could be severe. Decisions about what to do in response will be made not by scientists, but by people, politicians, and nations.

Key Terms

You should understand all of the following terms. Use the glossary and this chapter to improve your understanding of these terms.

Adaptation	CCCma	CSIRO
Cap-and-trade	CCSR/NIES	ECHAM
Carbon tax	CFS	Equilibrium run
Carbon trading	Control run	Geoengineering

GFDL	Kyoto Protocol	Seasonal climate forecasts
GISS	Mitigation	Sensitivity testing
Global climate models (GCMs)	Monthly climate forecasts	Transient run
	MPIM	UCLA
Hadley Centre	NCAR	

■ Review Questions

1. What is one difference between a weather forecast and a climate forecast?
2. How can ocean temperature anomalies in the equatorial Pacific lead to accurate climate forecasts for locations thousands of miles away?
3. What is a global climate model (GCM)? When and where were GCMs pioneered?
4. Distinguish between control, equilibrium, and transient runs of a global climate model.
5. Which of the following are GCMs: GFDL, GFS, GISS, IPCC, NAM, and NCAR.
6. Why does the IPCC report contain results for many different scenarios such as "A1B"? What do these scenarios represent?
7. What is the IPCC projection for the amount of temperature change by 2100, and how likely is it that this will occur?
8. True/False: "Global warming" would affect only temperature, not precipitation amounts.
9. Name at least two ways that coastal regions could be impacted by global warming.
10. Why are high-altitude regions at particular risk from global warming?
11. Why is global warming a more difficult problem to address than the stratospheric ozone hole?
12. Discuss the pros and cons of the following societal responses to global warming: (a) Kyoto Protocol, (b) cap-and-trade, (c) carbon tax, (d) geoengineering, and (e) doing nothing.
13. What are some ways in which geoengineering proposals such as placing mirrors in orbit over Earth could go wrong? What could be some of the climatic consequences?
14. Who will ultimately decide the societal responses to global warming?

■ Observation Activities

1. Visit http://www.ncdc.noaa.gov/climate-monitoring/, and check on the state of the climate in both the U.S. and the globe. What are the temperature anomalies for a particular month or year? Have there been notable extremes? Notice that the pattern in the U.S. is not always the same as for the globe as a whole!
2. Determine the phase of ENSO at http://www.cpc.ncep.noaa.gov/products/precip/CWlink/MJO/enso.shtml, and then compare it with the seasonal forecasts found at http://www.cpc.ncep.noaa.gov/products/forecasts/. Based on what you have learned in Chapter 8 and this chapter, can you make a connection between the ENSO phase and the seasonal forecast?
3. Visit http://climateprediction.net, and find out how you can use your own personal computer to produce predictions of the Earth's climate up to 2100 and to test the accuracy of climate models.

This rain cloud icon is your clue to go to the *Meteorology* Web site at http://physicalscience.jbpub.com/ackerman/meteorology/. Through animations, quizzes, web exercises, and more, you can explore in further detail many fascinating topics in meteorology.

Appendix

MAP OF NORTH AMERICA

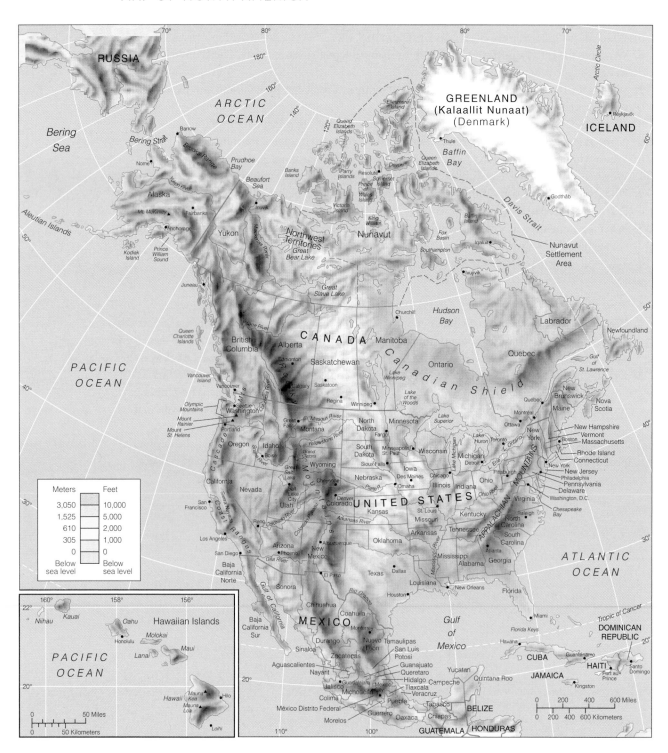

UNITS OF MEASUREMENT

International System of Units (SI)

Measurement	Unit
Length	meter (m)
Mass	kilogram (kg)
Temperature	kelvin (K)
Time	seconds (s)
Force	Newton ($N = kg\ m\ s^{-2}$)
Pressure	pascal ($Pa = N\ m^{-2}$)
Energy	Joule ($J = N\ m$)
Power	Watt ($W = J\ s^{-1}$)

SOME USEFUL CONVERSIONS

Length

1 meter = 3.281 feet = 39.37 inches
1 kilometer = 0.62 miles
1 mile = 1.61 kilometers
1 inch = 2.54 centimeters
1° latitude = 111 kilometers = 69 miles

Speed

1 knot = 1 nautical mile/hour = 1.15 miles/hour = 0.51 meters/second
1 meter/second = 2.24 miles/hour = 3.60 kilometers/hour = 1.94 knots
1 mile/hour = 0.45 meters/second = 1.61 kilometers/hour = 0.87 knots

Temperature

Celsius (°C) = (°F − 32)/1.8 = K − 273.15
Fahrenheit (°F) = 1.8 × (°C) + 32
kelvin (K) = °C + 273.15

Pressure

1 millibar (mb) = 100 pascals = 1 hectopascal = 0.02953 inches of Hg
1 inch of Hg = 33.865 millibars
1 standard atmosphere = 1013.25 millibars = 29.92 inches of Hg = 760 millimeters of Hg

ANNOTATED BIBLIOGRAPHY

Want to learn more about the topics covered in this textbook? Here is an annotated list of some of our favorite books and movies that relate to various portions of our book and that you can understand (and critique) with the knowledge gained from our book. For a list of great online resources, visit the Web site that accompanies this book, via **www.jblearning.com.**

▨ Books Relating To Material In:

All Chapters of *Meteorology, Third Edition*
Lester, Peter F., 2007: *Aviation Weather* **(hardcover, 3rd edition), 466 pages. Publisher: Jeppesen. ISBN: 088487446X.**
If you're interested in flying, this is the best book around on the subject of aviation meteorology. Lester covers introductory material in a very understandable manner and investigates each aviation hazard (e.g., turbulence, icing) in detail. The graphics are excellent.

Rauber, Robert A., John E. Walsh, and Donna J. Charlevoix, 2008: *Severe & Hazardous Weather: An Introduction to High Impact Meteorology* **(paperback, 3rd edition), 642 pages. Publisher: Kendall/Hunt. ISBN: 9870757550430.**
This highly detailed and informative textbook focuses on hazardous weather and covers many aspects of introductory meteorology in more detail than can be covered in a single semester. This is an excellent "next step up" book for readers of this text.

Stull, Roland B., 2005: *Meteorology for Scientists and Engineers* **(paperback, 2nd edition), 502 pages. Publisher: Brooks/Cole. ISBN: 0534372147.**
This highly useful, problem-oriented text covers all the topics of introductory meteorology from a mathematical perspective—but without any calculus. It is ideal if you want to get a deeper quantitative understanding of a particular topic in introductory meteorology.

Wallace, J. Michael, and Peter V. Hobbs, 2006: *Atmospheric Science: An Introductory Survey* **(hardcover, 2nd edition), 483 pages. Publisher: Academic Press. ISBN: 012732951X.**
The first edition of Wallace and Hobbs was the definitive junior–senior textbook for a generation or two of college meteorology majors. The long-awaited second edition is completely rewritten and expanded by the authors, professors of atmospheric science at the University of Washington, but in some cases, it's at a level beyond typical undergraduates. Nevertheless, if you want more "meat" in your meteorology than an introductory text can provide, you should turn to this book.

Chapter 1
Monmonier, Mark, 2000: *Air Apparent: How Meteorologists Learned to Map, Predict, and Dramatize Weather* **(paperback), 324 pages. Publisher: University of Chicago Press. ISBN: 0226534235.**
Weather maps are central to our ability to understand and predict the weather. *Air Apparent* chronicles the history of the weather map, highlighting key events, technology, and the people who attempted to unravel the mysteries of the atmosphere and develop methods of forecasting the weather.

Chapter 3
Solomon, Susan, 2002: *The Coldest March: Scott's Fatal Antarctic Expedition* **(paperback), 416 pages. Publisher: Yale University Press. ISBN: 0300099215.**
Award-winning meteorologist Susan Solomon took a break from her research on the Antarctic ozone hole to pen this account of a real-life historical drama. British explorer Robert Scott and four companions died in March 1912 on an Antarctic expedition, and Scott has been known ever since as a failure. Solomon examines weather records during Scott's exploration to vindicate Scott and show that his exploration was beset by extreme, life-threatening cold.

Chapter 4

Pretor-Pinney, Gavin, 2007: *The Cloudspotter's Guide: The Science, History, and Culture of Clouds* (paperback), 320 pages. Publisher: Perigee Trade. ISBN: 0399533451.
The science of clouds is interspersed with stories of how the weather has played an important role in history, art, and pop culture. Although not always scientifically accurate, Pretor-Pinney provides fascinating reading for anyone who has ever been accused of having their head up in the clouds.

Chapter 5

Bohren, Craig, 2001: *Clouds in a Glass of Beer: Simple Experiments in Atmospheric Physics* (paperback), 218 pages. Publisher: Dover Publications. ISBN: 0486417387.
This book and its sequel, *What Light Through Yonder Window Breaks?*, are very readable excursions into the world of atmospheric optics. Bohren, an emeritus professor of meteorology at Penn State, makes the science both accessible and fun via everyday examples. It has good hands-on experiment suggestions!

Lee, Raymond L., Jr., and Alistair B. Fraser, 2001: *The Rainbow Bridge: Rainbows in Art, Myth, and Science* (hardcover), 393 pages. Publisher: Penn State Press. ISBN: 0271019778.
Lee, a math professor, and Fraser, a meteorology professor, team forces to give rainbows their definitive treatment in this elegant book. This book is similar to Emanuel's *Divine Wind* in terms of interdisciplinary scope (e.g., reproductions of paintings of rainbows in art) and depth of scientific discussion.

Chapters 6 and 7

DeBlieu, Jan, 2006: *Wind: How the Flow of Air Has Shaped Myth, Life, and the Land* (paperback), 304 pages. Publisher: Counterpoint. ISBN: 1593760949.
This is a personal and poetic account of wind from the viewpoint of a writer, not a scientist. However, DeBlieu interweaves science with the role of wind in society. *Wind* won the John Burroughs Medal for natural history writing.

Chapter 8

Davis, Mark, 2002: *Late Victorian Holocausts: El Niño Famines and the Making of the Third World* (paperback), 470 pages. Publisher: Verso. ISBN: 1859843824.
Davis, the celebrated author of *City of Quartz* about Los Angeles, turns his focus to the interaction between climate oscillations and empire in this book. It is a remarkable synthesis of science and social criticism, chronicling the deaths of tens of millions in the emerging Third World due to a combination of ENSO-related droughts and the rise of imperial governments.

Emanuel, Kerry A., 2005: *Divine Wind: The History and Science of Hurricanes* (hardcover), 285 pages. Publisher: Oxford University Press. ISBN: 0195149416.
This is a handsome tome with the sophisticated visual look of a "coffee-table" book and the detail of a monograph on hurricanes. Emanuel, a professor of meteorology at MIT, reveals new details in both the history and the science of these storms.

Larson, Erik, 2000: *Isaac's Storm: A Man, a Time, and the Deadliest Hurricane in History* (paperback), 336 pages. Publisher: Vintage Press. ISBN: 0375708278.
On September 8th, 1900, an intense hurricane slammed without warning into the city of Galveston, TX, leaving thousands dead in its wake. Using telegrams, photographs, and first-hand accounts, Larson spins an intense and spell-binding tale that interweaves the story of one of America's greatest disasters with the politics and arrogance of an age when "the hubris of men led them to believe they could disregard even nature itself." One of the best weather-related books of the past two decades.

Pinet, Paul, 2008: *Invitation to Oceanography* (paperback, 5th edition), 576 pages. Publisher: Jones & Bartlett. ISBN: 0763759937.
One of the best introductory oceanography textbooks, Pinet's book is designed to be student friendly but includes plenty of detail for those wanting to learn more about the oceans.

Chapter 9

Kocin, Paul J., and Louis W. Uccellini, 2004: *Northeast Snowstorms* **(hardcover, two volumes), 818 pages. Publisher: American Meteorological Society. ISBN: 1878220640.**
This is the definitive analysis of "nor'easter" snowstorms by a famed forecaster and the director of NCEP. The *Wall Street Journal* named it one of the five best weather books.

Chapter 10

Stonehouse, Frederick, 2006: *The Wreck of the* **Edmund Fitzgerald (paperback), 277 pages. Publisher: Avery Color Studios. ISBN: 1892384337.**
One of the best of the many books on this 1975 shipwreck, Stonehouse's regional best-selling book covers the many contentious aspects of the shipwreck admirably. Stonehouse teaches Great Lakes maritime history at Northern Michigan University in addition to writing books on a wide variety of Great Lakes subjects.

Junger, Sebastian, 2009: *The Perfect Storm: A True Story of Men Against the Sea* **(paperback), 234 pages. Publisher: W.W. Norton & Company. ISBN: 0393337014.**
Junger's nonlinear meditation on the *Andrea Gail* shipwreck during the Halloween 1991 hybrid hurricane–extratropical cyclone became a runaway bestseller and a hit movie. The book captures the world of the Gloucester, MA, fishing industry, and its clinical digressions on unusual topics such as drowning are mesmerizing.

Chapter 11

Barry, John M., 1998: *Rising Tide: The Great Mississippi Flood of 1927 and How It Changed America* **(paperback), 422 pages. Publisher: Simon & Schuster. ISBN: 064176362X.**
The award-winning *Rising Tide* is a deep and rewarding account of the epic flood of 1927 and how it shaped American culture from the flood-ravaged South all the way to the White House. It's a classic of American history as well as disaster history.

England, Gary, 1997: *Weathering the Storm: Tornadoes, Television, and Turmoil* **(paperback), 264 pages. Publisher: University of Oklahoma Press. ISBN: 0806129425.**
A long-time Oklahoma TV meteorologist, Gary England, is known for being the first person to use Doppler radar to provide tornado warnings to the public. In this book, he tells his story, from a childhood interest in weather to experiences with violent storms and the trials of being a broadcast meteorologist.

Felknor, Peter, 2004: *The Tri-State Tornado: The Story of America's Greatest Tornado Disaster* **(Paperback), 140 pages. Publisher: iUniverse, Inc. ISBN: 0595311881.**
Tornado warnings were virtually unheard of in 1925 when a massive tornado tore a 200+ mile path across Missouri, Illinois, and Indiana, killing nearly 700 people and injuring thousands more. Felknor uses survivor accounts to bring to life the story of the worst tornado disaster in U.S. history.

Grazulis, Thomas P., 2003: *The Tornado: Nature's Ultimate Windstorm* **(paperback), 304 pages. Publisher: University of Oklahoma Press. ISBN: 080613587.**
Grazulis, the pre-eminent tornado archivist, uses his expertise to advantage in this readable and comprehensive account of tornadoes. He has stories and statistics about twisters that you cannot find anywhere else.

Uman, Martin A., 1986: *All About Lightning* **(paperback), 167 pages. Publisher: Dover Publications. ISBN: 048625237X.**
An oldie but a goodie, this Dover classic book addresses nearly all of your questions about the mysterious phenomenon of lightning. Uman, a longtime expert on lightning at the University of Florida, is an authoritative and very accessible "answer man."

Chapter 12

Mass, Cliff, 2008: *The Weather of the Pacific Northwest* **(paperback), 281 pages. Publisher: University of Washington Press. ISBN: 0295988479.**
Mass, a longtime professor of atmospheric science at the University of Washington, has penned this recent and first-ever comprehensive study of the unique weather of the Pacific Northwest.

Chapter 13

Cox, John D., 2002: *Storm Watchers: The Turbulent History of Weather Prediction from Franklin's Kite to El Niño* **(hardcover), 252 pages. Publisher: Wiley. ISBN: 047138108X.**
Storm Watchers takes you through the tumultuous history of weather prediction, from Benjamin Franklin's kite experiments to modern numerical weather models. Cox presents a captivating series of biographies detailing the contributions of the scientists who paved the way to our current understanding of the atmosphere.

Fine, Gary Alan, 2007: *Authors of the Storm: Meteorologists and the Culture of Prediction.* **Publisher: University of Chicago Press. ISBN: 0226249522.**
An insightful rendering of meteorologists' "idioculture" and work constraints in National Weather Service (NWS) offices. The study's primary setting is the Chicago office, but Fine conducted additional fieldwork at two other regional offices and at two national centers, the Storm Prediction Center and the National Center for Environmental Prediction.

Gleick, James, 2008: *Chaos: Making a New Science* **(paperback, 20th anniversary edition), 384 pages. Publisher: Penguin. ISBN: 0143113453.**
A bestseller in 1988, this book tells the fascinating multidisciplinary story of how chaos theory came into being. Gleick highlights the key role of meteorologists such as Ed Lorenz in pioneering our understanding of chaos in the natural world.

Harper, Kristine C., 2008: *Weather by the Numbers: The Genesis of Modern Meteorology* **(hardcover), 328 pages. Publisher: MIT Press. ISBN: 0262083787.**
This book is the story of the birth of numerical weather prediction, as told by a history professor at Florida State University who is also trained as a meteorologist. It received a 2008 Choice Award given by the Atmospheric Science Librarians International.

Petterssen, Sverre, 2001: *Weathering the Storm: Sverre Petterssen, the D-Day Forecast, and the Rise of Modern Meteorology* **(hardcover, edited by J. R. Fleming), 329 pages. Publisher: American Meteorological Society. ISBN: 1878220330.**
What was it like to be at the center of the meteorological revolution in the early to mid 20th century? Sverre Petterssen was there for most of it, and recounts his experiences candidly in this autobiography.

Vasquez, Tim, 2008: *The Weather Map Handbook* **(paperback, 2nd edition), 169 pages. Publisher: Weather Graphics Technologies. ISBN: 097068407X.**
Vasquez, a well-known storm chaser, former U.S. Air Force forecaster, and developer of meteorological software, gives a very practical and down-to-earth description of weather maps in this book. This and other Vasquez publications (see http://www.weathergraphics.com for a list) fill the gap between academic texts on "synoptic meteorology" and the tried-and-true (but rarely published) practices of everyday weather forecasters.

Chapter 14

Alley, Richard B., 2002: *The Two-Mile Time Machine: Ice Cores, Abrupt Climate Change, and Our Future* **(paperback), 240 pages. Publisher: Princeton University Press. ISBN: 0691102961.**
This Phi Beta Kappa Award-winning book by an acclaimed geoscientist explains the recent revolution in our understanding of climate based on data from ice cores. Quick changes in climate once thought impossible are now known to have occurred repeatedly in the past. Hailed as "wonderfully accessible, information-packed science reading" by the American Library Association.

Alvarez, Walter, 2008: *T. rex and the Crater of Doom* **(paperback), 216 pages. Publisher: Princeton University Press. ISBN: 0691131031.**
Why did the dinosaurs die off? This engagingly written book is by the UC-Berkeley geology professor who pioneered the hypothesis that the murder weapon was extraterrestrial—a comet or asteroid. Especially appealing to weather/climate fans is the explanation of how the impact

annihilated the dinosaurs through a combination of consequences, including climate change. The tale of how this hypothesis was proved is a classic example of how science actually works. One of the very best books on this list.

Chapter 15

Ruddiman, William F., 2010: *Plows, Plagues & Petroleum: How Humans Took Control of Climate* (paperback), 240 pages. Publisher: Princeton University Press. ISBN: 0691146349.

Ruddiman, a noted paleoclimatologist, makes a controversial but fascinating case that humans have been altering climate for thousands of years, since the rise of agriculture. He includes an excellent discussion of natural climate change related to Milankovitch cycles, too.

Chapter 16

Hansen, James, 2010: *Storms of My Grandchildren: The Truth About the Coming Climate Catastrophe and Our Last Chance to Save Humanity* (paperback), 320 pages. Publisher: Bloomsbury. ISBN: 1608195023.

This book by the longtime director of the NASA/Goddard Institute for Space Studies pulls no punches, from the subtitle onward. Based on global climate modeling results, Hansen concludes that global warming will be a disaster for human civilization. He discusses the politics of global warming as well as possible societal solutions.

Mann, Michael E., and Lee R. Kump, 2009: *Dire Predictions: Understanding Global Warming* (paperback), 210 pages. Publisher: Dorling Kindersley Limited. ISBN: 9780136044352.

The IPCC reports on global warming are tough reading, so Mann and Kump wrote a visuals-heavy interpretation of the 2007 IPCC report. It's an excellent, student-friendly way to get the gist of the IPCC report without spending days paging through the IPCC report online.

Philander, S. George, 2000: *Is the Temperature Rising? The Uncertain Science of Global Warming* (paperback), 240 pages. Publisher: Princeton University Press. ISBN: 0691050341.

Philander, a Princeton professor and expert on climate, gives a clear and scholarly account of the subject of global warming, including the role of the ocean. The difficulty of the topic is reduced with the simplicity of the author's basic approach, evidenced in chapter titles such as "Why the Peak of a Mountain is Cold" and "Why Summer is Warmer than Winter."

■ Movies

The Day After Tomorrow (2004, 124 minutes)

Climate change fills the big screen in *The Day After Tomorrow,* and the first few minutes are faithful to the science. Ice core research, abrupt climate change, and the role of the oceanic conveyor belt are all presented with surprising fidelity for a major motion picture. Then Hollywood takes over and the movie loses all scientific credibility. Low point: when "flash-freezing" is "explained" as occurring when air sinks so rapidly that adiabatic warming cannot take place. Readers of this book, throw your popcorn at the screen at that point!

Deep Impact (1998, 120 minutes)

What would happen if a comet hit Earth? *Deep Impact* surveys both the science and the societal implications with the best cast and crew of any movie on this list (at least five Oscar winners). The comet impact sequence has generally sound science, with just a few liberties taken.

Groundhog Day (1993, 101 minutes)

The best of all TV-meteorologist movies, this U.S. Film Registry selection explores the philosophical possibilities of being trapped in the same day, repeatedly, in Punxsutawney, PA, the capital of Groundhog Day. According to literary critic Stanley Fish, who rated *Groundhog Day* as one of the ten best American films ever:

> The miracle is that as the movie becomes more serious, it becomes funnier. The comedy and the philosophy (how shall one live?) do not sit side by side, but inhabit each other in a unity that is incredibly satisfying.

As you're watching, think about how chaos might affect what happens in Punxsutawney on Groundhog Day as the lead character (Bill Murray) consciously alters his own actions.

Koyaanisqatsi (1982, 87 minutes)

This nearly dialogue-free film by Godfrey Reggio illustrates "life out of balance" (the Hopi word "Koyaanisqatsi") with stunning imagery of the natural and human worlds, much of it from the U.S. Southwest. For meteorology buffs, the time-lapse photography of clouds and an inferior mirage are unsurpassed.

National Geographic's *Storm of the Century* (1998, 53 minutes)

In March of 1993, a storm of epic proportions ravaged the East Coast, spawning blizzards, thunderstorms, tornadoes, hurricane-force winds, and storm surges. This video chronicles the storm and the forecasters who correctly anticipated it through storm footage and first-hand accounts by the meteorologists who issued warnings as the storm wrought havoc from the Gulf Coast of Florida to Canada.

Tornado Video Classics (series of three DVDs)

Self-produced by tornado guru Tom Grazulis, this set of three DVDs is the definitive collection of tornado videos from the dawn of modern weather photography forward. You won't see these videos on TV, but *TVC* is required viewing for any serious severe weather fan. It is available at http://www.tornadoproject.com.

Twister (1996, 113 minutes)

The one movie that all weather buffs have seen, *Twister* is a fictionalized version of the mid-1990s VORTEX scientific experiment to understand tornadogenesis in the Great Plains. A classic "effects-driven" blockbuster that grossed a quarter-billion dollars worldwide, it has an excellent cast, expensive visual effects that aren't any more scientifically accurate than *The Wizard of Oz*'s, and bad writing with some science gaffes. The special effects earned an Oscar nomination, but *Twister* also won a Razzie Award for being the worst written high-grossing film.

The Wizard of Oz (1939, 101 minutes)

This beloved Oscar-winning classic also contains remarkably sound science in its depiction of a Kansas tornado. (True, tornadoes don't repeatedly move from side to side, but this error was repeated in *Twister* 57 years later.) Pay attention to the implicit weather safety lesson: Dorothy's trip to Oz begins when she suffers a head injury because she was standing next to a window as the tornado arrived!

Glossary

The number in bold following a glossary term's definition is the chapter in which the term first appears as a key term.

Absolutely unstable atmosphere Describes an atmospheric layer with an environmental lapse rate greater than 10° C per kilometer (the dry adiabatic lapse rate). (**3**)

Acceleration The rate at which velocity changes over time. (**2**)

Accretion The growth of an ice particle by collision and coalescence of supercooled water droplets. (**4**)

Acid deposition The falling of acids and acid-forming compounds from the atmosphere to the Earth's surface. (**15**)

Acid rain Acid deposition that occurs in the form of rain. (**15**)

Active sensors Remote sensing instruments that determine atmospheric conditions by sending out energy and measuring the energy that comes back to them. (**5**)

Adaptation Refers to the ability of a system to adjust to climate change (including climate variability and extremes) to limit potential damage, to take advantage of opportunities, or to cope with the consequences. (**16**)

Adiabatic process A process in which a system does not exchange heat with its surroundings. (**2**)

Advection The process by which a property of the atmosphere is transferred by the wind. Usually, meteorologists use advection to refer to horizontal transport of atmospheric properties, such as temperature and moisture. (**2**)

Advection fog A type of fog formed when warm, moist air is blown over a cooler surface and is cooled to the dew point. (**4**)

Aerosols Tiny solid or liquid particles that are suspended in the atmosphere. (**1**)

Aggregation The process of ice-crystal growth by collision and adherence of ice crystals. (**4**)

Air mass A large body of air that has uniform weather characteristics, particularly temperature and humidity. (**9**)

Air mass modification The process by which the characteristics of an air mass change as it moves away from its region of origin. (**9**)

Air-mass thunderstorm *See* Ordinary cell. (**11**)

Air pollution Gases and/or aerosols that, when in high concentrations, seriously affect the lives of people and animals, harm plants, or threaten ecosystems. (**15**)

Albedo The percentage of light reflected by an object when it is illuminated. (**2**)

Alberta Clippers Small, fast-moving extratropical cyclones that originate in western Canada and move south and east across the United States, especially in winter. (**10**)

Altocumulus (Ac) Clouds in the mid troposphere that are gray or white and that occur as layers or patches. If these clouds have vertical development that resembles towers or turrets, they are called altocumulus castellanus. (**4**)

Altostratus (As) Layered clouds in the mid troposphere that are gray or bluish. (**4**)

Amplitude Half the height from the crest of a wave to its trough. (**2**)

Analog forecast An empirical method of forecasting that uses past weather events that resemble the current conditions to create forecasts for the days ahead. (**13**)

Anemometer An instrument that measures wind speed. (**5**)

Aneroid barometer An instrument that measures atmospheric pressure via changes in the size of a partially evacuated metal box. (**5**)

Angle of inclination The tilt of the Earth with respect to the imaginary plane linking the centers of the Sun and Earth. The term also applies to orbits of other planets and satellites. (**2**)

Angular momentum The product of mass, rotation velocity, and perpendicular distance from the axis of rotation. (**7**)

Annual average temperature The average of the monthly mean temperatures. (**3**)

Annual temperature cycle The pattern of change of temperature during the 12 months of the year. (**3**)

Annual temperature range The difference between the values of the warmest and coldest monthly mean temperatures in a year. (**3**)

Anomalies Departures from typical or normal values. (**3**)

Anthropogenic Resulting from human activities. (**1**)

Anticyclogenesis The birth of an anticyclone. (**10**)

Anticyclones Regions of high pressure. Winds around an anticyclone blow clockwise in the Northern Hemisphere but counterclockwise in the Southern Hemisphere. (**6**)

Anvil The upper portion of a mature or dissipating thunderstorm (cumulonimbus) cloud that is flattened on its top side. (**4**)

Aphelion The location in a planet's orbit when it is farthest away from the Sun. (**2**)

Arctic Oscillation A sea-surface temperature oscillation of the North Atlantic that affects weather and climate over large parts of the globe. Also called North Atlantic Oscillation. (**8**)

Argon (Ar) A noble gas that makes up 1% of the atmosphere. (**1**)

ASOS Automated Surface Observing System. ASOS is the United States' primary network for observing surface weather. (**5**)

Aspect The compass direction that a land slope faces. (**3**)

Aspirated psychrometer *See* Psychrometer. (**5**)

Atmosphere The thin layer of air surrounding Earth (or other celestial bodies). (**1**)

Atmospheric window A narrow range of wavelengths in which the atmosphere absorbs very little of the Earth's emitted infrared energy. The best known atmospheric window occurs between 10 and 12 microns. (**2**)

AVN A spectral numerical weather forecast model that is used for short- to medium-range global forecasts. (**13**)

Baroclinic instability The process through which extratropical cyclones get energy for their growth. Warm air rising and cold air sinking in tilted frontal regions are the source of this energy. (**10**)

Barometer A device used to measure atmospheric pressure. (**1**)

Barometric pressure The pressure exerted by a column of air as a result of gravitational attraction. (**1**)

Berg wind A hot, dry wind blowing off the plateau of South Africa. (**12**)

Bergeron–Wegener process The process by which ice crystals grow more rapidly than water droplets in a cold cloud because of differences in the saturation vapor pressures of ice and liquid water at the same subfreezing temperatures. (**4**)

Bermuda high A subtropical anticyclone over the North Atlantic Ocean that steers tropical cyclones toward the west and northwest in summer and early fall. (**8**)

Blackbody A hypothetical object that is a perfect absorber and emitter of radiation. (**2**)

Blizzard The National Weather Service defines a blizzard as a weather event lasting a minimum of 3 hours and characterized by winds of 30 knots (35 mph) or greater, along with considerable falling and/or blowing snow reducing visibility to a quarter mile or less. (**9**)

Blocking patterns Develop when cut-off wind patterns stop or divert the normal eastward progression of weather systems. (**7**)

Blue jets Lightning that propagates upward from the top of a thunderstorm into the stratosphere. They are blue in color and last long enough to be seen by pilots. (**11**)

Blue norther A fierce north wind behind a fast-racing cold front in western Texas that is accompanied by clear blue skies. (**12**)

Bora A term used to describe a katabatic wind along the Adriatic coast of Europe. (**12**)

Boreal forest A forested region, usually a coniferous forest, that adjoins the tundra at the arctic tree line. (**14**)

Boulder windstorms Downslope winds associated with mountain gravity waves that can cause particularly ferocious windstorms in the vicinity of Boulder, Colorado. (**12**)

Bow echo A line of thunderstorms on radar that is bent or "bowed" by high winds. (**12**)

Brocken bow A colored circular region around shadows, which is caused by diffraction by water droplets. (**5**)

Buran A strong northeast wind in Russia and central Asia. (**12**)

Buys Ballot's law Dictates that low pressure lies on the left-hand side of the wind direction in the Northern Hemisphere. Buys Ballot's law is a consequence of geostrophic balance. (**6**)

Calorie A unit representing the amount of energy required to increase the temperature of 1 gram of water by $1°$ C. (**2**)

Cap-and-trade A means of steadily reducing carbon dioxide and other greenhouse gas emissions economy-wide in a cost-effective manner by limiting the amount of greenhouse gases companies are allowed to emit. Companies who exceed their limit may purchase additional emissions permits from those companies who emit less than they are allowed. (**16**)

Cape Verde hurricane An especially dangerous type of tropical cyclone that develops out of an easterly wave near the Cape Verde Islands, just off the western coast of Africa. (**8**)

Capping inversion A temperature inversion above the Earth's surface that initially forestalls the development of thunderstorms. If the air beneath the inversion is hot and moist enough to rise above the "cap," explosive thunderstorm development results. (**11**)

Carbon dioxide (CO_2) A colorless, odorless, nontoxic trace gas found throughout the atmosphere. It plays a key role in the greenhouse effect and is increasing in concentration because of the burning of fossil fuels. (**1**)

Carbon monoxide (CO) A colorless, odorless, toxic trace gas found in the atmosphere, which forms when fossil fuels are not completely burned during combustion. CO is a harmful pollutant and can reduce the body's ability to provide oxygen to the body. **(15)**

Carbon tax A form of pollution tax, which levies a fee on the production, distribution, or use of fossil fuels based on how much carbon their combustion emits. Carbon taxing encourages utilities, businesses, and individuals to reduce consumption and increase energy efficiency. **(16)**

Carbon trading One form of cap-and-trade specifically for carbon dioxide emissions. **(16)**

CCCma Global climate model created in Canada. **(16)**

CCSR/NIES Global climate model created in Japan. **(16)**

Ceilometer A device used to measure the height of cloud bases. **(5)**

Cell The organizing unit of a thunderstorm, it is a compact region of a cloud that contains a strong updraft. **(11)**

Celsius A temperature scale in which the freezing point of water at sea level occurs at $0°$ C and the boiling point occurs at $100°$ C. **(2)**

Centrifugal force An apparent force felt by objects in a turning frame of reference. The force seems to push the objects outward from the center of the turn. **(6)**

Centripetal acceleration The change of velocity of an object moving in a curved path. This acceleration is directed toward the center of the curved path. **(6)**

CFS NCEP Climate Forecast System model, a version of the GFS augmented for seasonal climate forecasting. **(16)**

Chaos theory A theory that demonstrates that in a complex system such as the atmosphere, small differences at the beginning of an event can lead to large differences later. **(13)**

Chinook A dry, warm wind in western North America on the lee side of the Rocky Mountains. The chinook speeds the melting and evaporation of snow. **(12)**

Chlorofluorocarbons (CFCs) Anthropogenic compounds invented by chemists in 1928 and used as propellants in spray cans, in Styrofoam™ puffing agents, and as coolants for refrigerators and air conditioners. CFCs are primarily responsible for the development of the ozone hole in the stratosphere and also play a role in the enhanced greenhouse effect. **(1)**

Cirrocumulus (Cc) Thin, rippled convective high clouds. **(4)**

Cirrostratus (Cs) Layered high clouds that are whitish and usually fibrous, but sometimes smooth. Cirrostratus often produce optical effects such as halos. **(4)**

Cirrus (Ci) High clouds composed of ice crystals that are whitish and can appear as filaments, patches, or narrow bands. **(4)**

Climate The collective state of the atmosphere for a given place over a specified interval of time. **(1)**

Climatology The description and study of climate. **(1)**

Climatology forecast A forecast based upon the climatological statistics of a region. **(13)**

Climographs Graphs that depict climatological information. Climographs usually consist of two climatic elements (e.g., temperature and precipitation) plotted through an annual cycle. **(14)**

Cloud An ensemble of water drops and/or ice particles in the atmosphere above the Earth's surface. **(1)**

Cloud ceiling The height of the base of the lowest widespread cloud in the sky. **(5)**

Cloud droplet A spherical particle of liquid water in a cloud. Cloud droplet sizes range from a few microns to a few tens of microns. **(4)**

Coastal front The boundary between two air masses along a coastline that acts as a smaller scale version of a stationary front. (**12**)

Cold-air damming The stubborn entrenchment of cold air, particularly in and near mountainous regions such as the Appalachian Mountains. (**12**)

Cold clouds Clouds that have temperatures below freezing. (**4**)

Cold front A type of front in which a colder air mass replaces a warmer air mass. (**1**)

Collision–coalescence The process by which precipitation forms in warm clouds. Drops of different sizes collide and merge, leading to rapid growth into a raindrop. (**4**)

Columns A type of ice crystal that has hexagonal bases on either end. (**4**)

Comma cloud A feature of extratropical cyclones observed in satellite pictures in which the cloud system resembles a comma, with a large rounded "head" near the center of the low and a trailing thin "tail" along the cold front. (**10**)

Condensation The change of phase of water from the vapor to the liquid state. (**1**)

Condensation nuclei Aerosol particles that assist in forming liquid droplets. (**4**)

Conditionally unstable environment A layer of the atmosphere where the lapse rate lies between the saturated adiabatic lapse rate and the dry adiabatic lapse rate. In this case, unsaturated air parcels are stable, while saturated air parcels are unstable. (**4**)

Conduction The transfer of energy from one object to another as a result of the random motions of molecules. Conduction of heat is a consequence of the temperature differences between two objects in physical contact. The transfer always is from warmer to colder objects. (**2**)

Conservation of angular momentum The principle that the angular momentum of a spinning body (the product of spin and distance from the axis of rotation) must remain constant. Conservation of angular momentum is usually accomplished by increasing (decreasing) spin when the distance from the axis of rotation is decreased (increased). (**7**)

Contact nucleation *See* Nucleation. (**4**)

Continental polar (cP) The continental polar air mass, which is a cold and dry air mass. (**9**)

Continental tropical (cT) The continental tropical air mass, which in summer is a hot and dry air mass. (**9**)

Contrails Cloud-like streamers that form behind high-flying aircraft. (**15**)

Control run Long integrations of global climate models where the model input forcings (solar irradiance, sulfates, ozone, greenhouse gases) are held constant and are not allowed to evolve with time. (**16**)

Convection The transfer of energy by mass motions in a liquid or a gas. In meteorology, this term is usually reserved for vertical motions. (**2**)

Convective clouds Clouds that result from convection. As a result, convective clouds usually exhibit vertical development. (**4**)

Convergence The horizontal coming-together of air that can lead to lifting of air at the surface. (**4**)

Cooling degree-day A method of estimating the amount of energy used to cool a building. One cooling degree-day is equal to each degree the average daily temperature is above 65° F. (**3**)

Coriolis force An apparent force used to describe the observed deflection of moving objects caused by the observer's moving frame of reference. The magnitude of the Coriolis force is proportional to the wind speed and the latitude. It always acts to the right of the wind in the Northern Hemisphere. (**6**)

Corona A whitish or colored band around the Sun or moon, which is caused by diffraction of light by water droplets. It differs in appearance from the halo because the corona, unlike a halo, appears to "touch" the Sun or moon instead of being a thin ring around the light source. **(5)**

Crepuscular rays Beams of light from the Sun caused by a combination of shadows and scattering. These light and dark bands usually occur at twilight. **(5)**

Critical angle The angle at which refracted rays of light do not escape the substance they are traveling in but instead move along the surface between two substances. **(5)**

Crystal habit The size and shape of an ice crystal. **(4)**

CSIRO Global climate model created in Australia. **(16)**

Cumulonimbus (Cb) Tall, precipitation-producing thunderstorm clouds, often with flattened anvil-shaped tops. **(4)**

Cumulus (Cu) Low convective clouds that develop as individual, detached elements with sharp outlines. A cumulus cloud has a flat base and bulges upward. **(4)**

Cumulus stage The initial stage in the life cycle of a thunderstorm. **(11)**

Cup anemometer An instrument for measuring wind speed that uses small cups to catch the wind. **(5)**

Curvature effect A greater surface curvature increases the rate of evaporation and inhibits cloud droplet growth at the smallest droplet sizes. **(4)**

Cut-off cyclone An aging extratropical cyclone that has become separated from the strong jet-stream winds and from the surface temperature gradients. **(10)**

Cyclogenesis The development of a cyclone. Surface temperature gradients, jet streams, and tall mountain ranges can lead to extratropical cyclogenesis. **(10)**

Cyclones Low-pressure regions around which winds blow counterclockwise in the Northern Hemisphere and clockwise in the Southern Hemisphere. **(6)**

Daily mean temperature The average of the daily maximum and minimum temperatures. **(3)**

Dart leader Cloud-to-ground electrical discharges that follow the initial lightning stroke. **(11)**

Data assimilation The combining of observed data into a numerical forecast model. Tasks of data assimilation include interpolation of scattered data into a regularly spaced pattern and initialization to achieve a balance between wind and pressure in the data. **(13)**

Data initialization The "smoothing out" of errors and imbalances in the initial data that is to be ingested by a numerical weather forecast model, which leads to more accurate weather forecasts. **(13)**

Deep zone The bottom layer of the ocean, which lies below 1000 meters and is uniform in temperature. **(8)**

Dendrites A type of ice crystal that is a flattened hexagon with extended "fingers" or branches at each vertex. Dendrites most closely resemble the classic shape of snowflakes. **(4)**

Dendrochronology The science of studying tree rings to ascertain a climatic condition. **(14)**

Density The ratio of the mass of a substance to the volume of space it occupies. **(1)**

Deposition The process by which water vapor in subfreezing air changes into ice. **(2)**

Deposition nucleation Water vapor changing directly into ice around an ice nucleus when air is supersaturated with respect to ice. **(4)**

Deposition nuclei A solid aerosol particle that nucleates an ice crystal from the vapor phase. **(4)**

Derechos Hours-long windstorms associated with a line of severe thunderstorms. Derechos are the result of straight-line winds, not the rotary winds of a tornado. They are often associated with bow-shaped echoes on weather radar. (**12**)

Desertification An increase in the desert conditions of a region. Also applied to the spreading of a desert region as a result of a combination of climate change and human impacts on the land. (**15**)

Dew Liquid water that condenses on objects near the ground when the air at the surface cools to the dew point. (**4**)

Dew point depression The difference between the temperature and the dew point temperature. (**4**)

Dew point hygrometer An instrument that measures the dew point temperature by interpreting changes in the reflection of light off of a mirror. (**5**)

Dew point temperature The temperature to which air must be cooled at constant pressure and constant water vapor content to become saturated. (**4**)

Diffluence The spreading-out of horizontal wind direction that often leads to divergence. (**10**)

Diffraction The bending of light around tiny objects that produces patterns of light and shadow, including colored light. (**5**)

Diffuse The action of gas and liquid molecules spreading from areas of higher concentration to areas of lower concentration. (**1**)

Direct methods Observations made by instruments that are in physical contact with the region being measured. (**5**)

Dispersion The process in which white light is separated into its component colors. (**5**)

Dissipating stage The final stage in the life cycle of a thunderstorm. (**11**)

Diurnal Daily, pertaining to actions completed within or that recur every 24 hours. (**2**)

Diurnal temperature cycle The pattern of temperature during the 24-hour period from one midnight to the next. (**3**)

Diurnal temperature range The difference between a day's maximum and minimum temperatures. (**3**)

Divergence The horizontal spreading-out of air that can lead to sinking motions if it occurs at the surface but causes rising motion in the lower troposphere if it occurs at jet-stream level. (**4**)

Doldrums Regions of nearly calm winds near the equator. (**7**)

Doppler effect The change in the frequency of a wave pattern caused by the relative motion of the wave emitter and the observer of the wave. (**5**)

Doppler radar A radar that indirectly measures wind speed by detecting the change of frequency that results from the Doppler effect on radar waves that hit moving precipitation particles. (**5**)

Drizzle Liquid precipitation smaller than 0.5 millimeter in diameter. (**4**)

Dry adiabatic lapse rate The change in temperature of a parcel as it rises or descends in the atmosphere, equal to about 10° C per kilometer. (**2**)

Dry climates (B) Climate regions where potential evaporation and transpiration exceeds precipitation. (**14**)

Dry growth mode A method of hailstone growth by accretion in which supercooled water droplets collide with a hailstone whose temperature is below freezing, leading the droplets to freeze on contact on the surface of the stone and resulting in little liquid water on the surface. Air bubbles are trapped in the ice by this method, causing the ice to appear white due to multiple scattering of light. (**11**)

Dryline A type of frontal zone characterized by strong horizontal contrasts in moisture. (**9**)

Dry slot A mostly clear region of an extratropical cyclone that separates the comma cloud head from the comma tail on satellite images. (**10**)

Dust Bowl A name given to the region in the south-central United States affected by drought and duststorms in the 1930s. (**12**)

Dust devils Small, vigorous whirlwinds that occur when intense daytime heating helps spin up thin rotating columns of air. (**12**)

Dust storms Weather conditions characterized by strong winds and dust-filled air over an extensive area. (**12**)

Easterly wave A region of clouds and rain associated with a wave-like pattern in tropospheric winds that moves from east to west across tropical regions. A few easterly waves later develop into tropical cyclones. (**8**)

Eccentricity The deviation of an ellipse from a perfect circular shape. (**14**)

ECHAM Global climate model created at Max Planck Institute for Meteorology in Germany. (**16**)

ECMWF A spectral numerical weather forecast model developed by the European Centre for Medium-Range Weather Forecasts that is used for a wide range of forecasts. (**13**)

Eddies Swirls in a fluid, such as air, that interact with the larger scale wind and help slow it down. (**12**)

Ekman spiral In oceanography, the progressive turning of ocean currents from the surface down to 100 meters because of the combined effect of wind on the sea surface and the Coriolis force. In meteorology, the description of the wind distribution in the atmospheric boundary layer. (**8**)

Ekman transport The effect of the Ekman spiral to move water masses at a right angle to the direction of the surface wind. In the Northern Hemisphere, this movement of water is to the right of the surface wind direction. (**8**)

Electromagnetic energy *See* Radiation. (**2**)

El Niño A periodic warming of the near-surface water of the equatorial Pacific Ocean between South America and the Date Line. This warming occurs in connection with changes in the atmosphere over the same region, which are known collectively as the Southern Oscillation. (**8**)

Elves Quick, weak lightning that occurs over cloud-to-ground lightning bolts, shooting upward from the top of a thunderstorm into the upper atmosphere. Like red sprites, they can extend to near the top of the atmosphere and are too weak to be seen by the naked eye. (**11**)

Energy The capacity to do work. Energy must be conserved, although it can be converted from one form to another, such as from potential to kinetic energy. (**2**)

Enhanced Fujita (EF) scale An updated version of Dr. Ted Fujita's scale for rating tornado intensity. The EF scale uses a set of 28 damage indicators and relates a tornado's wind speed to the degree of damage to each indicator. (**11**)

Enhanced greenhouse effect *See* Greenhouse warming. (**2**)

Ensemble forecasting A method of weather forecasting that uses the results of chaos theory to assess the amount of confidence that should be placed in a forecast. A forecast model is run repeatedly with slightly different initial conditions. If the resulting forecasts, the "ensemble," agrees closely, then confidence in the forecast is high. If the ensemble exhibits a wide range of different forecasts, then confidence in the forecast is low. **(13)**

ENSO Refers to the combination of El Niño and Southern Oscillation. **(8)**

Entrainment The mixing of environmental air into an existing air current or cloud. **(11)**

Environmental lapse rate The rate of decrease of temperature with increasing altitude at a particular time and place. **(3)**

Equilibrium run A global climate simulation in which greenhouse gas concentrations are suddenly changed (e.g., doubling CO_2), and the model is allowed to come into equilibrium with the new forcing. **(16)**

Equinoxes The times of year at which the Sun passes directly overhead at the equator at noon. **(2)**

Eta A gridpoint numerical weather forecast model that is used for short-range forecasts over the United States. Now called the North American Mesoscale model, or NAM. **(13)**

Evaporation The change of phase of water from liquid water to water vapor. **(1)**

Evaporation fog A fog that occurs when water evaporates into cool, moist air and causes saturation. Steam fog and frontal fogs are two common types of evaporation fog. **(4)**

Extratropical cyclone A low-pressure system that forms outside of the tropics and is usually associated with fronts, unlike a tropical cyclone. **(8)**

Eye The usually clear area of lowest pressure at the center of a strong tropical cyclone. **(8)**

Eye wall The circular region of strong thunderstorms immediately surrounding the eye of a strong tropical cyclone. **(8)**

Fahrenheit A temperature scale in which the freezing point of water at sea level occurs at 32° F and the boiling point occurs at 212° F. **(2)**

Fall wind A term used to describe a katabatic wind in Greenland and Antarctica. **(12)**

Fallstreaks Ice particles that fall from a cloud but evaporate above the ground. **(4)**

Fata Morgana A type of superior mirage in which objects near the horizon are distorted vertically into castle-like shapes. **(5)**

Feedback A sequence of interactions in a system where one change leads to some other change, which can act to either reinforce (*positive feedback*) or inhibit (*negative feedback*) the original change. **(15)**

Flash flood A rapidly developing type of flood usually associated with very intense rainfall at or near the site of the flood. **(11)**

Flood A substantial rise in water that covers areas that are usually not under water. **(11)**

Foehn A warm, dry downslope mountain wind that descends the lee side of the Alps. **(12)**

Fog A stratus cloud that is in contact with the ground. **(4)**

Folklore forecasts Short sayings that attempt to predict the weather based on sky conditions, special days of the calendar, or the behavior of animals. **(13)**

Force The mass of an object multiplied by the change in its speed and/or direction, which is its acceleration. **(2)**

Freezing nucleation *See* Nucleation. **(4)**

Freezing rain Supercooled rain that freezes on contact with subfreezing objects. **(4)**

Frictional force The resistive force caused by wind blowing over the Earth's surface. It always acts to oppose the wind. (**6**)

Front The transition zone between two air masses of different density. (**1**)

Frontal lifting The forced lifting of warm, less dense air over colder air in the vicinity of a front. (**4**)

Frontal wave An early stage of the life cycle of an extratropical cyclone that is characterized by an undulation along a previously stationary front, with a developing cold front on its back side and a warm front on its front side. (**10**)

Frost Ice that forms on exposed objects by deposition when an object is cooled to or below its frost point. (**4**)

Frost point The highest temperature at which atmospheric moisture will form frost. It is analogous to the dew point but is applied when saturation occurs below freezing. (**4**)

Frozen dew Dew that first condenses as liquid water and then freezes. (**4**)

Funnel cloud A tornadic circulation that looks like a narrow cone beneath a thunderstorm but does not reach the ground. (**11**)

Gap winds Mountain winds that blow through low points, or gaps, in the mountains, creating channels of wind that can easily exceed 160 km/hr (100 mph) in the strongest cases. (**12**)

GEO satellite A satellite in a geostationary earth orbit (GEO), circling the earth once every 24 hours. At an altitude of approximately 36,000 kilometers, the satellite appears stationary over a fixed point at the equator. Most weather satellites are in geostationary orbits. (**5**)

Geoengineering Altering physical, chemical, or biological aspects of the Earth system to compensate for global warming. (**16**)

Geometric scattering Occurs with a large particle, such as a cloud drop. (**5**)

Geostrophic balance An equilibrium achieved when the horizontal pressure gradient and Coriolis forces push equally in opposite directions. (**6**)

Geostrophic wind The horizontal wind created by the balance of the horizontal pressure gradient and Coriolis forces. Although observed winds are rarely exactly geostrophic, the geostrophic wind is often a good first approximation to the direction and speed of observed winds in the mid latitudes. (**6**)

GFDL Global climate model created by the NOAA/Geophysical Fluid Dynamics Laboratory at Princeton. (**16**)

GFS A spectral numerical weather forecast model that makes 16-day forecasts for the globe. The GFS is a consolidation of the AVN and MRF models. (**13**)

GISS Global climate model created at NASA/Goddard Institute for Space Studies in the United States. (**16**)

Glacier A mass of perennial ice that originates on land through the accumulation of snow. (**14**)

Global climate models (GCMs) Computer programs that calculate global climate using mathematical equations derived from physical principles such as the conservation of energy and Newton's Laws of Motion. Also called general circulation models. (**16**)

Global Positioning System (GPS) A satellite tracking network established to provide high-precision navigation. Radio signals from GPS satellites are slowed by water vapor. The measured delay in the arrival of the signal at the ground is used to determine the amount of water vapor in the atmosphere. (**5**)

Glory Colored rings that appear around the shadow of an object, such as an airplane, and that are caused by diffraction of light by water droplets. (**5**)

Gradient balance A three-way balance of horizontal pressure gradient, Coriolis, and centrifugal forces. (**6**)

Gradient wind The horizontal wind that results from a three-way balance of horizontal pressure gradient, Coriolis, and centrifugal forces. For the same latitude and pressure gradient force, gradient winds in a curved low-pressure area are slower than the geostrophic wind (subgeostrophic) and are faster than the geostrophic wind (supergeostrophic) in a curved high-pressure area. (**6**)

Graupel A kind of ice particle formed by accretion inside a cloud. (**4**)

Gravitational force The product of mass and gravitational acceleration. (**6**)

Gravity The mutual attraction between two or more objects. (**1**)

Gravity waves An alternating small-scale pattern of high and low pressure maintained with the help of gravity. Gravity waves are sometimes visible when the rising air in the crests in the waves becomes saturated and forms parallel lines of clouds. (**12**)

Green flash A momentary green light sometimes seen at sunrise or sunset resulting from refraction and dispersion of sunlight. (**5**)

Greenhouse effect The warming of the atmosphere that results from the fact that gases in the atmosphere absorb and emit the Earth's infrared radiation, although they are transparent to sunlight. This warming is analogous to the effect of the clear glass in a greenhouse, which keeps the plants inside warmer than they would be otherwise. (**2**)

Greenhouse gases Gases in the atmosphere that are effective absorbers of infrared radiation and ineffective absorbers of solar radiation, such as carbon dioxide and water vapor. (**2**)

Greenhouse warming The possible heating of the planet over and above the natural greenhouse effect as a result of increases in atmospheric carbon dioxide. (**2**)

Grid A set of orderly arranged points on which variables are analyzed or predicted in a numerical weather forecast model. (**13**)

Grid spacing The distance between one grid point and another in a numerical model. (**13**)

Gridpoint The point in the middle of a grid in a gridpoint model. (**13**)

Gridpoint models A class of numerical weather forecast models that divides the atmosphere into grids. (**13**)

Growing degree-day A heat index used to predict when a crop will reach maturity. It is found by subtracting a plant-specific reference temperature from the average daily temperature and cannot go below zero. (**3**)

Guldberg–Mohn balance A three-way balance of horizontal pressure gradient, Coriolis, and frictional forces. The wind associated with this balance blows at an angle to the isobars from higher toward lower pressure. (**6**)

Gust front The boundary between the outflow of the cold downdraft of a thunderstorm and the warmer, more humid air around it. (**11**)

Gyre An ocean circulation that forms a closed loop that stretches across an entire ocean basin. (**8**)

Haboob A dust or sandstorm caused by the downdraft of a desert thunderstorm. (**12**)

Hadley Cell A thermally driven circulation pattern comprised of rising air near the equator, poleward flow in the upper troposphere, sinking air over the deserts, and the trade winds. It is an important feature of global-scale winds. **(7)**

Hadley Centre Global climate model created at the Hadley Centre in the United Kingdom. **(16)**

Hail Precipitation in the form of lumps of ice produced by thunderstorms. A hailstone is a single unit of hail. **(4)**

Hailshaft The region underneath a thunderstorm where hail is falling. **(11)**

Hailswath The path of the hailshaft along the ground that is created as a hail-producing thunderstorm moves. **(11)**

Halo A whitish or colored ring around the sun or moon that is produced by refraction of light by ice crystals. **(5)**

Harmattan A dry dust-bearing wind that develops in winter when cool air from the Sahara Desert moves south and west and displaces warmer, more humid coastal air. **(12)**

Haze A suspension of small particles in the air. Haze reduces visibility by scattering light. **(5)**

Heat A form of energy transferred between objects because of the temperature differences between them. **(2)**

Heat advection The transfer of energy through the horizontal movements of the air. **(2)**

Heatburst Localized, sudden increase in surface temperature and winds, associated with a thunderstorm. **(12)**

Heating degree-day A method of estimating the amount of energy used to heat a building. One heating degree-day is equal to each degree the average daily temperatures is below 65° F. **(3)**

Heterogeneous nucleation *See* Nucleation. **(4)**

Hexagonal plate A type of ice crystal that is a flattened hexagon without elongated branches. **(4)**

Hira-oroshi A breeze that occurs along the mountainous shoreline of Lake Biwa, west of Tokyo, Japan. **(12)**

Historical climate The climate of the past several thousand years during which humans have kept a record of climate conditions. **(14)**

Homogeneous nucleation *See* Nucleation. **(4)**

Hook echo A curved echo on a weather radar caused by heavy precipitation in a supercell thunderstorm. A hook echo is often an indication of a tornado. **(11)**

Horse latitudes A large region of light winds over the subtropical oceans. **(7)**

Humid continental A class of severe mid-latitude climates characterized by a large range in temperature, severe winters and cool-to-warm summers. **(14)**

Hurricane A strong tropical cyclone found in the Western Hemisphere that has maximum sustained winds exceeding 65 knots (74 mph). *See* also Tropical cyclone. **(8)**

Hydrocarbons Compounds made of hydrogen and carbon atoms. **(15)**

Hydrologic cycle A complete description of how water moves between the atmosphere, water surfaces, and land in all three phases. **(1)**

Hydrophobic nuclei A type of condensation nuclei which do not dissolve in water. Hydrophobic nuclei resist condensation but can form droplets when the relative humidity is near 100%. **(4)**

Hydrostatic balance An equilibrium achieved when the gravitational and vertical pressure gradient forces push equally in opposite directions. (**6**)

Hygroscopic nuclei A type of condensation nuclei which dissolve in water. Droplet formation can occur even when the relative humidity is less than 100%. (**4**)

Ice ages Periods of global cooling that leads to the creation of vast ice sheets across the Earth's land masses. (**14**)

Ice albedo temperature feedback A type of positive climate feedback in which the cooling of the earth leads to the formation of ice sheets that, because of the increased albedo, further cools the planet and supports the continued growth of ice. (**15**)

Ice crystals The geometric organization of water molecules in the solid phase. Hexagonal plates, needles, columns, and dendrites are four different common forms of ice crystals. (**4**)

Ice nuclei Particles in the atmosphere around which ice crystals form. (**4**)

Ice sheet A continuous sheet of land ice that covers a very large area and moves outward in many directions. (**14**)

Immersion nucleation *See* Nucleation. (**4**)

Index cycle The oscillation of Rossby wave patterns between low-amplitude and high-amplitude phases over periods of several weeks. (**7**)

Index of refraction A measure of how optically dense a substance is. It is the ratio of the speed of light in a vacuum to the speed of light in the substance. (**5**)

Indirect methods Observation methods that do not require physical contact between the instrument and the region being measured. Also known as remote sensing. (**5**)

Inferior mirage *See* Mirage. (**5**)

Infrared satellite image A type of indirect observation that shows the amount of infrared energy emitted by the atmosphere and the Earth's surface as measured by a weather satellite. In an infrared satellite image, white regions are cold and emit less infrared energy than dark regions. (**5**)

Initial conditions The values of atmospheric variables (temperature, dew point, etc.) over a wide area at the starting time of a numerical weather forecast. (**13**)

Insolation The amount of solar radiation reaching the top of Earth's atmosphere. (**3**)

Interpolation A part of the numerical weather forecast process in which data from irregularly spaced observation sites is mathematically adjusted onto a regularly spaced grid for use in the numerical forecast model. (**13**)

Intertropical Convergence Zone (ITCZ) A convective region of thunderstorms separating the northeast and southeast trade winds. (**7**)

IPCC The Intergovernmental Panel on Climate Change, formed in 1988. IPCC is comprised of over 1000 scientists and experts who issue reports on the scientific, technical, and socio-economic information relevant for the understanding of the risk of human-induced climate change. (**15**)

Iridescence Small regions of color in the sky, usually in connection with high thin clouds. (**5**)

Isobar A line on a map connecting regions with the same atmospheric pressure. (**1**)

Isobaric charts Maps that depict weather on constant pressure surfaces and include information on the temperature, wind speed and direction, humidity, and the altitude at a given pressure. (**6**)

Isopleth A line on a map connecting locations with the same value of a variable. (**1**)

Isotach A line on a map connecting locations with the same wind speed. **(1)**

Isotherm A line on a map connecting locations with the same temperature. **(1)**

Isotopes Different versions of the same element that have different atomic weights. **(14)**

Jet stream A narrow region of relatively strong winds (i.e., wind speeds greater than 70 knots) usually located in the upper troposphere. **(7)**

Joule A unit used to measure amounts of energy. One Joule equals 0.2389 calories. **(2)**

Katabatic wind A type of strong mountain breeze in which wind gusts can exceed 160 km/hr (100 mph). **(12)**

Kelvin A temperature scale in which the freezing point of water at sea level occurs at 273.16 K and the boiling point occurs at 373.16 K. **(2)**

Khamsin A hot, dry, dusty desert wind in Egypt and the Red Sea. **(12)**

Kinetic energy The energy an object possesses because of its motion. **(2)**

Kirchhoff's Law A law of radiation which states that objects that are good absorbers of radiation are also good emitters of radiation. **(2)**

Knots Units of wind speed. One knot is one nautical mile per hour, which is equal to about 1.15 miles per hour or 0.5 meters per second. **(6)**

Koembang A dry, warm downslope wind from the southeast or south in Indonesia. **(12)**

Köppen scheme The most widely used climate classification scheme. Developed by Vladimir Köppen, it is based on annual and monthly means of precipitation and temperature. Now referred to as the Köppen-Geiger map. **(14)**

Kyoto Protocol International agreement ratified by 175 countries (not including the U.S.) to reduce greenhouse gas emissions to less than 1990 levels. **(16)**

La Niña An extensive, below-normal cooling of the central and eastern tropical Pacific Ocean. La Niña is roughly the opposite of El Niño. **(8)**

Lag The time delay between a cause and an effect. For example, typically the coldest winter temperatures lag several weeks behind the shortest day of the year, partly due to the thermal properties of the Earth and oceans. **(3)**

Lake breezes Winds that blow onshore during the day around large lakes. **(12)**

Land breeze A wind that blows offshore from land to water during nighttime in the vicinity of large bodies of water. **(6)**

Lapse rate The rate at which temperature changes with increasing altitude. A positive lapse rate indicates that the temperature is decreasing with height. **(3)**

Latent heat The amount of heat taken in or released by water when it changes phase. **(2)**

Latent heat of condensation The energy per gram released when water vapor turns into liquid water. **(2)**

Latent heat of deposition The energy per gram released when vapor changes directly into ice. **(2)**

Latent heat of fusion The energy per gram released when liquid water freezes. **(2)**

Latent heat of melting The energy per gram absorbed when ice melts. **(2)**

Latent heat of sublimation The energy per gram absorbed when ice changes directly into vapor. **(2)**

Latent heat of vaporization The energy per gram absorbed when liquid water boils. **(2)**

Law of Momentum *See* Newton's Second Law of Motion. **(6)**

Layered clouds Clouds that are much wider than they are tall and form in relatively stable air. **(4)**

Lead A pollutant that is a particulate released into the atmosphere by industry and the burning of leaded gasoline. **(15)**

Leeward The downwind side of an object or region, such as a mountain. **(6)**

Lenticular clouds Clouds in the shape of a lens that sometimes resemble spaceships. They usually form downwind of mountainous regions in connection with small-scale winds above and around mountains. **(12)**

LEO satellite A satellite that orbits the poles in a low-Earth orbit about 850 kilometers above the Earth's surface. **(5)**

Leste A hot, dry easterly or southeasterly wind that blows from Morocco to the Canary Islands. **(12)**

Levanter A gap wind in the Strait of Gibraltar. **(12)**

Leveche A hot, dry, dusty wind along the southeast coat of Spain. **(12)**

LFM An early gridpoint numerical weather forecast model that was used for short-range weather forecasts across North America. **(13)**

Lifted index A numerical estimate of the severity of thunderstorm activity based on observations near and above the ground. In simplified form, the lifted index is computed by lifting a hypothetical air parcel from the surface to the 500-mb level and then comparing the parcel's temperature to the 500-mb temperature. The more negative the lifted index, the more likely that any thunderstorms that form will be severe. **(11)**

Lifting condensation level (LCL) The level (altitude or pressure) to which air must be lifted dry adiabatically for condensation (or deposition) to occur. **(4)**

Lightning A visible electrical discharge that occurs between the ground and a cloud, between clouds, or within a cloud. **(11)**

Little Ice Age A time period between approximately 1400 and 1850 AD when average global temperatures were lower than at present and there was an expansion of mountain glaciers in the Alps, Alaska, Iceland, and Norway. **(14)**

Longwave radiation Radiant energy characterized by wavelengths primarily between 4 and 100 microns (μm). The Earth emits longwave radiation with a maximum near 10 μm. **(2)**

Mammatus Pouches that form on the underside of a cloud, usually a thunderstorm anvil. **(4)**

Maritime polar (mP) The maritime polar air mass, which forms over cool ocean waters and is usually cool and moist. **(9)**

Maritime tropical (mT) The maritime tropical air mass, which forms over warm ocean waters and is warm and moist. **(9)**

Mature stage The middle and most intense stage in the life cycle of an ordinary thunderstorm. **(11)**

Maunder Minimum The time between the years 1645 and 1715 in which the number of sunspots was dramatically lower than observed before or since. **(14)**

Medieval Warm Period A period just prior to the Little Ice Age when large amounts of fresh water flowing into the North Atlantic might have interrupted the ocean circulation patterns. **(14)**

Mercury barometer A device that measures air pressure by the level of mercury that is pushed up into an airless tube. (**5**)

Meridional flow pattern An upper-air flow pattern that occurs when strong Rossby waves cause winds to blow in a pronounced north–south direction. (**7**)

Mesocyclone A vertical column of rotating air within a severe thunderstorm. Tornadoes often form below a mesocyclone. (**11**)

Mesoscale Atmospheric motions with a spatial scale ranging from a few kilometers to several hundred kilometers. Examples are thunderstorms, fronts, tropical cyclones, and sea breezes. (**6**)

Mesoscale convective complex (MCC) A large thunderstorm complex made up of multiple single-cell thunderstorms. An MCC is identifiable on infrared satellite images by its nearly circular shape, which can cover entire states. (**11**)

Mesoscale convective system (MCS) A group of thunderstorms that produces an area of contiguous precipitation spanning at least 100 km in one direction. Two classic types of MCSs are squall lines and mesoscale convective complexes. (**11**)

Mesopause The top of the mesosphere where temperature stops decreasing with altitude. The mesopause is usually at about 80 to 85 kilometers above the Earth's surface. (**1**)

Mesosphere The region of the atmosphere above the stratosphere between the stratopause and mesopause. The mesosphere lies between approximately 50 and 80 kilometers above the surface of the Earth. (**1**)

Meteorology The study of the atmosphere and its motions, especially on the shorter time scales known as "weather." Meteorology includes the study of the ways weather is affected by interactions among the Earth's land and water surfaces and living things. (**1**)

Methane (CH$_4$) A trace atmospheric gas that is increasing in concentration, probably due to human activities such as industry and agriculture, which plays a role in greenhouse warming. (**1**)

Microburst A strong, localized downdraft less than 4 kilometers in diameter that sometimes develops underneath a thunderstorm as a result of evaporative cooling. (**11**)

Micron A unit of length equal to one millionth of a meter. Also called a micrometer and abbreviated μm. (**1**)

Microscale Atmospheric motions with a spatial scale smaller than 2 kilometers. (**6**)

Midlatitude westerlies The persistent west-to-east winds of the middle latitudes. Also called westerlies. (**7**)

Milankovitch cycles Periodic variations of the Earth's orbit that are theorized to have caused the ice ages. (**14**)

Millibars Units of atmospheric pressure. The average atmospheric pressure at sea level is 1013.25 mb, which is equal to 29.92 inches of mercury. (**1**)

Mirage A refracted image of an object caused by differences in air density. In the inferior mirage, the refracted image appears below the true object, while in a superior mirage it lies above the object. (**5**)

Mistral A term used on the French Riviera to describe a katabatic wind. (**12**)

Mitigation Any action taken to reduce the long-term risk and hazards of climate change to human life and property by limiting the release of greenhouse gases. (**16**)

Mixed layer *See* Surface zone. (**8**)

Mixing ratio The amount of water in the atmosphere in terms of the mass of water vapor per unit mass of dry air. Expressed in units of g/kg. (**4**)

Model A simplified, but relatively accurate approximation of reality. Today's weather forecasts are computed by numerical models that approximate the behavior of the actual atmosphere. **(13)**

Moist parcel of air A parcel of air in which water molecules are changing phase from vapor to liquid or ice. **(2)**

Moist subtropical and mid latitude climates (C) A climate type characterized by humid and mild winters. The warmest month mean temperature is above 10° C (50° F), while the coolest month mean temperature is below 18° C (64° F) and above 0° C (32° F). **(14)**

Molecule Composed of atoms, a molecule is the smallest unit of a substance that retains the chemical properties of that substance. **(1)**

Momentum The product of mass and velocity. **(6)**

Monsoon A circulation pattern characterized by a seasonal reversal of the prevailing winds. **(7)**

Monthly climate forecasts Prediction of the overall pattern for the next 30 days, based on the longest-range weather forecasts from numerical weather prediction models, ensemble runs of the NCEP Climate Forecast System model, and statistics. **(16)**

Monthly mean temperature The sum of the daily mean temperatures for a month divided by the number of days in the month. **(3)**

Morning glory A wind squall of northern Australia that forms on the leading edge of a gravity wave and is often accompanied by a spectacular linear cloud up to 1000 kilometers (620 miles) in length. **(12)**

MOS A series of statistical relationships between a particular model's predictions and observed weather variables at a particular location, used to adjust the raw model output. **(13)**

Mountain breezes Winds that result from the temperature differences along the slopes of high mountains, which cause pressure gradient forces that drive local wind circulations. **(12)**

MPIM Max Planck Institute for Meteorology. **(16)**

MRF A type of spectral numerical weather forecast model that is used for medium- and long-range forecasts across the globe. It was the longer-range version of the AVN. It is now known as the GFS. **(13)**

Multicell Thunderstorms which are organized in clusters (such as MCCs) or lines (such as squall lines). **(11)**

Multiple-vortex tornado An especially damaging tornado that contains smaller spiraling whirlwinds inside the main funnel. **(11)**

NAM The North American Mesoscale model. A gridpoint numerical weather forecast model that is used for short-range forecasts over the United States. Formerly known as the Eta model. **(13)**

NCAR Global climate model created at the National Center for Atmospheric Research in Boulder, CO. **(16)**

Needle A type of ice crystal that has a hexagonal base and a pointed top. **(4)**

Negative feedback mechanism *See* Feedback. **(15)**

Newton's First Law of Motion States that a body resists a change in its motion. This resistance to change is called inertia. **(6)**

Newton's Second Law of Motion States that force equals mass times acceleration. **(6)**

NGM A gridpoint numerical weather forecast model that was used to make short-range weather forecasts for the United States. **(13)**

Nimbostratus (Ns) Precipitation-producing layered clouds, usually found just ahead of a warm front. (**4**)

Nitric oxide (NO) A pollutant that is a byproduct of high-temperature combustion, such as in automobile engines and electric power generation. (**15**)

Nitrogen (N$_2$) A colorless, odorless gas which makes up 78% of the atmosphere. (**1**)

Nitrogen dioxide (NO$_2$) A gas found at all levels of the atmosphere that is emitted by automobile engines. High concentrations of nitrogen dioxide give polluted air its reddish-brown color. (**15**)

Nitrous oxide (N$_2$O) A colorless trace gas in the atmosphere that plays a role in the greenhouse effect. (**1**)

Nocturnal inversion A temperature inversion that develops overnight as a result of strong longwave radiation emissions at ground level. (**3**)

Nocturnal low-level jet A lower-tropospheric maximum in wind speed that develops at night. This jet supplies moisture and energy to nighttime thunderstorms over the Great Plains. (**11**)

Nor'easters Extratropical cyclones that affect the northeastern United States and extreme eastern Canada. They are named for the strong northeasterly winds that blow across this region as the path of the low moves northeast along the North American coastline. (**10**)

Normal A line that is perpendicular to a surface. (**5**)

Normal temperatures Average temperatures calculated for a location over a long period of time, often 30 years. (**3**)

Norte A strong, cold northeasterly wind along the Gulf of Mexico. (**12**)

North Atlantic Oscillation (NAO) An oscillation in which the atmospheric pressure see-saws between the polar low near the North Pole and the subtropical high over the Atlantic Ocean. This seesaw affects Arctic Ocean temperatures as well as the strength and location of the jet stream. (**8**)

Norwegian cyclone model The life cycle of the extratropical cyclone as explained by the Bergen School of meteorology. The life cycle progresses from a young frontal wave on a stationary front, to a maturing cyclone with warm and cold fronts, to a weakening occluded cyclone, and finally to a dying cut-off cyclone. (**10**)

Nowcast A local weather forecast for the next few hours. (**13**)

Nucleation The initial process of a phase change of water to a liquid droplet or ice crystal. Two basic processes are homogeneous nucleation and heterogeneous nucleation. In *homogeneous nucleation* the process of formation involves only water molecules. In *heterogeneous nucleation* small, nonwater particles serve as sites for particle formation. *Immersion nucleation* occurs when supercooled water freezes around a nucleus. *Contact nucleation* occurs when supercooled water freezes immediately on touching an ice nucleus, while *freezing nucleation* occurs when supercooled water freezes without a nonwater particle present. (**4**)

Numerical modeling The simulation of fluid motions, such as in the atmosphere, using mathematical approximations of the equations describing the fluid. These complex computations usually require the use of a powerful computer. (**13**)

Obliquity The angle between the Earth's orbit plane and the plane of the Earth's equator; in other words, the tilt of the Earth. Currently, this value is approximately 23.5°. (**14**)

Occluded cyclone A mature extratropical cyclone with little or no temperature advection. The cyclone typically exhibits an occluded front and begins to weaken within a day or so. (**10**)

Occluded front A front that develops as a mature cyclone ages and moves deeper into colder air. (**1**)

Occlusion The process by which a surface low retreats from the tilted areas of strong temperature gradients associated with the warm and cold fronts. (**10**)

Ocean current A narrow, concentrated horizontal flow of water in the ocean associated with wind patterns at the ocean surface. Ocean currents are similar to jet streams in the atmosphere, but the currents are much slower than wind speeds in a jet stream. (**8**)

Oceanography The study of the world's oceans. (**8**)

Open wave A young extratropical cyclone with strong warm and cold fronts, but no occluded front. (**10**)

Ordinary cell A thunderstorm cell that is a few kilometers in diameter and has a life cycle of less than an hour. (**11**)

Orographic lifting Rising air flow that is caused by the presence of mountains. (**4**)

Overrunning A condition in which an air mass aloft is moving over an air mass of greater density at the surface. (**9**)

Overshooting top A bubble-like protrusion on the top side of a thunderstorm anvil where the thunderstorm updraft has penetrated into the stratosphere. (**11**)

Oxygen (O_2) The second most abundant gas in the atmosphere, making up 21% of the atmosphere. (**1**)

Ozone (O_3) A chemically active molecule in the atmosphere. Because ozone strongly absorbs ultraviolet light from the Sun, its presence in the stratosphere is crucial for the existence of life on Earth. In the troposphere, however, it is a toxic pollutant and is the main component of photochemical smog. (**1**)

Ozone hole A depletion of stratospheric ozone that has occurred over the Antarctic continent each spring for the past few decades due to the presence of chlorofluorocarbons (CFCs). (**2**)

Ozone layer A region of the stratosphere about 20 to 30 kilometers above the Earth's surface with a maximum of ozone concentration. (**2**)

Pacific Decadal Oscillation An alternating pattern of sea-surface temperatures in the Pacific Ocean that reverses itself over periods of several decades, affecting weather and climate over large parts of the globe. (**8**)

Paleoclimate Climate conditions that existed in the billions of years before the dawn of human civilization. (**14**)

Paleoclimatology The study of climates of the distant past and the causes of their variations. (**14**)

Pangaea The supercontinent formed approximately 300 million years ago that subsequently split into the continents of today. (**14**)

Panhandle hooks Extratropical cyclones that follow curved, hook-shaped paths from the Texas and Oklahoma Panhandles to the Great Lakes, most often in late autumn. (**10**)

Parameterizations Portions of numerical weather prediction models that are devoted to the approximation of phenomena that the model cannot calculate precisely. (**13**)

Parcel of air An arbitrarily defined small-scale "bubble" of air that exhibits uniform conditions within its hypothetical borders. Air parcels are used to simplify explanations of air temperature changes and wind motions. (**2**)

Particulates Airborne solid and liquid aerosols that, when in high concentrations, seriously affect the lives of people and animals, harm plants, or threaten ecosystems. (**15**)

Passive sensors Observational instruments that measure energy received naturally from the region being measured. (**5**)

Perihelion The point on the Earth's orbit when it is closest to the Sun. (**2**)

Permafrost A layer below the surface of the Earth in tundra environments that is perennially frozen. (**14**)

Permanent gases Gases found in today's atmosphere which do not vary in concentration. (**1**)

Persistence forecast A forecast that assumes that tomorrow's weather will be the same as today's weather. (**13**)

Photochemical smog *See* Smog. (**15**)

Photodissociation The destruction of chemical molecules resulting from absorption of high-energy radiation. (**2**)

Pilot leader The initial discharge of negative charges near the cloud base at the beginning of a lightning stroke. (**11**)

Pineapple Express The phrase used to describe the path of extratropical cyclones that approach the West Coast of the United States from the southwest. (**10**)

Planetary scale The largest scale of atmospheric motion, the circulation patterns of which span a hemisphere or even the entire world. (**6**)

Plate tectonics The modern theory of continental movement. Its precursor was Wegener's theory of Continental Drift. (**14**)

Polar climates (E) A climate zone poleward of the Arctic and Antarctic circles. Polar climates are extremely cold and have little precipitation. (**14**)

Polar easterlies Low-level air flowing equatorward from the poles that has a strong east-to-west component. (**7**)

Polar front Transition zone separating cold polar air masses from warmer air of the middle latitudes and subtropics. (**7**)

Polar front jet stream A jet stream found in the middle and upper latitudes in association with the clash of cold and warm air known as the polar front. (**7**)

Polar stratospheric clouds (PSCs) Clouds composed of ice and frozen nitrogen particles that form in air temperatures colder than approximately $-80°$ C ($-112°$ F). Typically form over Antarctica in June and dissipate in October. PSCs lead to chemical reactions which deplete stratospheric ozone. (**15**)

Positive feedback mechanism *See* Feedback. (**15**)

Potential energy The energy an object has by virtue of its position. (**2**)

Power The rate of change of energy over time, often expressed in watts. (**2**)

Precession The wobble of the Earth's axis. Precession alters the relationship of the solstices with the distance from the Earth to the Sun. (**14**)

Precipitation Any liquid or solid water particle that falls from the atmosphere and reaches the ground. (**4**)

Pressure Force per unit area. (**1**)

Pressure gradient A change in pressure across a distance. (**6**)

Pressure gradient force (PGF) The force that arises from changes in pressure over distance, divided by the air density. The PGF always pushes directly from higher toward lower pressure. **(6)**

Primary rainbow The brilliantly colored arc of light most commonly seen in the sky, it forms at about a 40-degree angle to sunlight and has blue color on the inside of the arc and red color on the outside of the arc. **(5)**

Primitive equations The nearly exact physical laws of the atmosphere that, together with the initial atmospheric conditions of a given area, can be used to create a numerical forecast for a later time. **(13)**

Propellers A method of measuring wind speed in which the blades of a propeller rotate at a rate proportional to the wind speed. **(5)**

Psychrometer An instrument that measures differences in the temperature of wet- and dry-bulb thermometers as a means of calculating relative humidity. The thermometers are ventilated either by whirling the instrument (*sling psychrometer*) or by using a fan (*aspirated psychrometer*). **(5)**

Puelche A warm, easterly downslope wind along the west slopes of the Andes Mountains in South America. **(12)**

Radar An instrument used for detecting the presence and distance of objects, such as raindrop, which reflect or scatter radiowave pulses emitted by the radar. **(5)**

Radar echo The energy scattered back from a target and detected by the radar receiver. **(5)**

Radar reflectivity The amount of energy received by the radar that is scattered back to it, usually by precipitation particles. **(5)**

Radiant energy *See* Radiation. **(2)**

Radiation Energy that moves through space or a medium in the form of a wave with electric and magnetic fields. **(2)**

Radiation fog A fog that occurs on clear calm nights when air near the surface cools to the dew point. **(4)**

Radiation inversion A layer of air, usually near the ground, in which the air temperature increases with height as a result of radiational cooling. **(3)**

Radiative forcing The change in the net radiation, or the difference between incoming and outgoing radiative energy, at the tropopause. **(2)**

Radiocarbon dating A method to determine the age of objects through analysis of C^{14} atoms that undergo radioactive decay in predictable ways. **(14)**

Radiometers Instruments that measure radiation power. **(5)**

Radiosonde An instrument package carried upward by weather balloons to measure the vertical profile of atmospheric temperature, relative humidity, and pressure from the surface into the stratosphere. **(5)**

Rain Liquid precipitation bigger than 0.5 millimeter. **(4)**

Rain gauge An instrument that measures the amount of falling precipitation. **(5)**

Rain shadow A region downwind of mountains in which precipitation is significantly less than in the upwind regions. **(4)**

Rainbands Elongated and curved regions of rain that are often associated with tropical cyclones. **(8)**

Rainbow A bright arc of colored light caused by refraction and reflection of light by raindrops. The primary rainbow is brightest and most commonly seen; the secondary rainbow forms outside the primary rainbow and is fainter. **(5)**

Rawinsonde An instrument package carried upward with a weather balloon to measure the vertical profile of atmospheric temperature, relative humidity, pressure, and wind direction and speed from the surface into the stratosphere. The wind direction and speed are measured by tracking the weather balloon's motion. **(5)**

Rayleigh scattering Scattering by particles that are small with respect to the wavelength of the incident radiation (such as molecules). **(5)**

Recurvature A northeastward movement of a previously westward-moving tropical cyclone, caused by mid-latitude weather systems. **(8)**

Red sprites Quick, weak lightning which occurs over cloud-to-ground lightning bolts, shooting upward from the top of a thunderstorm into the upper atmosphere. They are red in color, can extend to near the top of the atmosphere, but are too weak to be seen by the naked eye. **(11)**

Reflection The process by which energy incident on a surface is turned back at the same angle into the medium through which it originated. **(5)**

Refraction The process by which the direction of energy propagation is changed as a result of spatial variations in properties (e.g., density) of the medium. **(5)**

Relative humidity The ratio of the observed vapor pressure of the air to the saturation vapor pressure, expressed as a percentage. Relative humidity indicates how close the air is to saturation (i.e., 100% relative humidity). **(4)**

Resistance thermometer An instrument that observes temperatures by measuring changes in the electrical resistance of a metal. **(5)**

Resolution The spacing between gridpoints in a numerical weather forecast model. "Fine" resolution results from close spacing of gridpoints; "coarse" resolution results from wide spacing of gridpoints. Fine resolution allows a model to "see" smaller-scale phenomena more accurately than does coarse resolution. **(13)**

Return stroke The ground-to-cloud electrical current in a lightning stroke that causes the brilliant flash. **(11)**

Right-mover A supercell thunderstorm that moves in a direction that is to the right of nearby thunderstorms. Often a result of a thunderstorm splitting into two parts, the right mover is generally more intense due to better access to warm, moist air. **(11)**

Rime A white, opaque deposit of ice formed by the rapid freezing of supercooled water drops as they collide with an object at or below freezing. **(4)**

Rossby waves Waves in the mid latitude westerly winds with wavelengths on the order of thousands of kilometers. **(7)**

RUC A gridpoint numerical weather forecast model that is used to make short-term forecasts for the United States. **(13)**

Saffir–Simpson scale The system by which hurricanes are classified based on wind speed on a scale from 1 (minimal hurricane) to 5 (catastrophic hurricane). The Saffir-Simpson scale, unlike the Enhanced Fujita scale for tornadoes, can be assessed during the lifetime of the storm. **(8)**

Santa Ana winds Warm, dry downslope winds in California that generally blow from the northeast or east. **(12)**

Saturated adiabatic lapse rate The change of temperature of a rising moist parcel of air with height. It is quite variable, roughly 6° C per kilometer near the Earth's surface. **(4)**

Saturation The condition at which the vapor pressure equals the saturation vapor pressure over a flat surface of water or ice. **(4)**

Saturation vapor pressure The vapor pressure at a given temperature, for which the number of molecules leaving a liquid or ice surface equals the number of molecules entering the liquid or ice. **(4)**

Scattering The process by which light rays change direction of propagation through the interaction with particles, such as molecules, aerosols, and cloud particles. **(5)**

Sea breeze An onshore coastal wind that occurs during the daytime. It is driven by pressure gradients created by the unequal heating of land and ocean. **(6)**

Sea-level pressure The atmospheric pressure at mean sea level. All surface barometric pressure readings are adjusted to sea level to remove the effect of altitude on pressure. **(1)**

Seasonal climate forecasts Climate forecasts on seasonal timescales based on ocean temperature anomalies, influences of other oscillations, and statistics. **(16)**

Seasonal temperature cycle The pattern of temperature at a location over the course of a year. Also called annual temperature cycle. **(3)**

Secondary rainbow A dim second arc of light that sometimes is seen above the primary rainbow at about a 50° angle to sunlight, and which has red color on the inside of its arc and blue color on the outside of the arc. **(5)**

Sensible heat The transfer of energy without a change of phase of water. **(2)**

Sensitivity testing Using a global climate model to test the effect of changing a particular parameter. **(16)**

Severe mid latitude climates (D) Typically located in the eastern regions of continents, this climate regime has an average temperature of the coldest month that is less than 0° C (32° F). The warmest month has an average temperature exceeding 10° C (50° F). There is usually snow on the ground for extended periods. **(14)**

Severe thunderstorm Thunderstorms that produce one or more of the following: a tornado, large hail (diameters greater than 2.54 centimeters [1 inch]) or wind gusts of at least 26 m/s (50 knots, or 58 mph). **(11)**

Shelf cloud A wedge-shaped cloud that sometimes forms on the leading edge of a thunderstorm gust front. **(11)**

Short waves Undulations in the large-scale mid latitude tropospheric winds that are smaller and faster moving than Rossby long-waves. **(7)**

Shortwave radiation Radiant energy with wavelengths between approximately 0.2 and 4 microns. **(2)**

Shower A brief episode of often heavy precipitation. **(4)**

Simoom A strong, dry, and dust-laden desert wind found in the African and Arabian deserts. **(12)**

Sink A process that removes a substance, such as a gas, from its environment, such as the atmosphere. **(1)**

Sirocco A warm south or southeast wind that brings hot, dry conditions to the Mediterranean. **(12)**

Sleet Translucent balls of ice that are frozen raindrops. Sleet occurs when a layer of subfreezing air above the surface is cold enough to freeze raindrops. **(4)**

Sling psychrometer *See* Psychrometer. **(5)**

Smog Originally used to describe a combination of smoke and fog, this term is now used to describe mixtures of pollutants in the atmosphere. An example is photochemical smog, which is a combination of hydrocarbons with nitrogen oxides and oxygen. **(15)**

Snow Precipitation in the form of ice crystals or aggregates of crystals. **(4)**

Snowflake An individual ice crystal, or more commonly, an aggregate of ice crystals. **(4)**

Solar constant The amount of solar radiation received at the top of the Earth's atmosphere on a surface perpendicular to the incoming radiation at Earth's mean distance from the Sun. Its value is about $1370 \ W/m^2$. **(2)**

Solar radiation Electromagnetic radiation from the Sun, which is concentrated in visible wavelengths. **(2)**

Solar zenith angle The angle at the Earth's surface measured between the position of the Sun and an observer's zenith. **(2)**

Solstices The two days during the year when the noon sun is overhead at its northernmost (on or about June 21) or southernmost (on or about December 22) latitudes. On these days, the amount of daylight is at maximum and minimum, respectively, in the Northern Hemisphere. **(2)**

Solute effect A process that enhances the condensational growth of droplets by suppressing evaporation because of elements dissolved in the droplet. **(4)**

Sonic anemometer A device that measures wind speed using ultrasonic sound waves. **(5)**

Sounding The vertical distribution of temperature, moisture, and wind speed, and direction over a location. **(5)**

Source A process that adds a substance, such as a gas, to its environment, such as the atmosphere. **(1)**

Southerly buster A mesoscale cold front that moves across southeastern Australia. **(12)**

Southern Oscillation A seesaw in atmospheric pressure between the western and eastern Pacific that is commonly associated with El Niño and La Niña. **(8)**

Southern Oscillation Index (SOI) The difference in sea-level pressure between Tahiti and Darwin, Australia. **(8)**

Specific heat The amount of energy needed to raise the temperature of 1 gram of a substance by $1°$ C. **(2)**

Spectral models A class of numerical weather forecast models that divides the atmosphere in terms of waves rather than gridpoints. **(13)**

Speed The distance traveled in a given amount of time, equivalent to the magnitude of the velocity. **(6)**

Speed divergence Divergence that occurs because of downwind horizontal wind speed increases, but with no change of wind direction. **(10)**

Split flow pattern An upper air flow pattern in which the jet stream divides into two main streams and has both west–east (zonal) and north–south (meridional) components. **(7)**

Stepped leaders Cloud-to-ground electrical charges in a lightning stroke that flow downward from the cloud in an attempt to establish a conductive channel all the way from the cloud to the ground. **(11)**

Squall line A line of intense thunderstorms, usually in advance of a cold front. **(11)**

Squamish A gap wind in British Columbia. **(12)**

Static stability A measure of the stability of the atmosphere with respect to vertical displacements of air parcels. (**4**)

Statically stable atmosphere An atmospheric layer with a lapse rate less than 10° C per kilometer. (**3**)

Station model An efficient method of graphically representing weather conditions at a single location on a weather map. (**1**)

Stationary front A frontal situation in which neither of the air masses on either side of the front are advancing or retreating. As a result, the front remains in approximately the same location. (**1**)

Steam fog An evaporation fog that occurs above a body of water. (**4**)

Stefan–Boltzmann Law A fundamental radiation law that states that the total energy emitted by a blackbody is proportional to the fourth power of its temperature. (**2**)

Stepped leader The initial leader of a lightning discharge. Stepped leaders attempt to establish a conductive channel from the cloud to the ground for electrons to flow through and neutralize the charge difference between the cloud and ground. (**11**)

Storm surge A surge of seawater pushed onshore primarily by winds of a storm, usually a tropical cyclone. (**8**)

Stratocumulus (Sc) Low clouds that appear in rows or patches and are white or gray. (**4**)

Stratopause The top of the stratosphere where the temperature does not change with height, located at an altitude of 50 kilometers. (**1**)

Stratosphere The region of the atmosphere above the troposphere and below the mesosphere, located between about 15 and 50 kilometers. It is characterized by temperature increasing with increasing altitude. (**1**)

Stratus (St) A cloud layer with a uniform base and a gray color. (**4**)

Subarctic climates A class of severe mid-latitude climates characterized by a very large range in annual temperature. Winters are very long and cold, and summers are brief and cool. (**14**)

Subgeostrophic flow A wind that is slower than the geostrophic wind would be in the same horizontal pressure gradient. (**6**)

Sublimation The change of phase of water from ice to vapor. (**2**)

Subsynoptic scale Microscale and mesoscale weather systems, which are smaller than the largest features most often shown on continent-wide weather maps. (**6**)

Subtropical highs High pressure regions near the surface observed near 30° latitude. They are semipermanent in their location, shifting with the seasons. (**7**)

Subtropical jet stream A region of strong winds typically found in the upper troposphere between 20° and 40° latitude. (**7**)

Sulfur dioxide (SO_2) A highly corrosive gas found in the atmosphere, released by volcanoes and through the burning of fossil fuels that contain sulfur. SO_2 can irritate the respiratory system. (**15**)

Sulfur trioxide (SO_3) A pollutant released into the atmosphere primarily through the burning of fossil fuels that contain sulfur. When combined with water vapor, SO_3 forms droplets of sulfuric acid to form acid rain. (**15**)

Sulfuric acid (H_2SO_4) A strong acid that forms in the atmosphere when sulfur is present. Sulfuric acid droplets that form in the stratosphere after strong volcanic eruptions affect climate by reflecting solar radiation. (**15**)

Sun pillar A shaft of light extending vertically from the rising or setting sun that is caused by reflection of sunlight by ice crystals. (**5**)

Sundogs Brightly colored regions that flank the sun as a result of refraction of light through ice crystals. Also called parhelia. (**5**)

Supercell A dangerous type of single-cell thunderstorm that rotates under the influence of vertical wind shear in the wind patterns above and around it. Supercell thunderstorms can produce damaging tornadoes. (**11**)

Supercooled water Liquid water that exists when its temperature is below 0° C. (**4**)

Supergeostrophic flow A wind that is faster than the geostrophic wind would be in the same horizontal pressure gradient. (**6**)

Superior mirage *See* Mirage. (**5**)

Supersaturation A condition in the atmosphere that exists when the relative humidity is greater than 100%. (**4**)

Supertyphoon A severe tropical cyclone in the western Pacific Ocean with highest sustained winds in excess of 240 km/hr (150 mph). (**8**)

Surface chart A map of analyzed surface weather conditions. Also called a sea-level chart. (**1**)

Surface temperature The air temperature measured in the shade at 1.5 meters (5 feet) above the ground, usually on a grass-covered surface. (**3**)

Surface waves A sun ray traveling along the surface of water drop. Refraction and reflection of surfaces waves are important for forming a glory. (**5**)

Surface zone The topmost 100 meters of the oceans in which the temperature is uniform due to stirring by winds and waves. Also known as the mixed layer. (**8**)

Synoptic scale Literally "seeing together," this term is used to describe weather on scales that are typically represented on weather maps. (**6**)

Taiga Forests composed of conifer trees that grow in severe mid-latitude climates in Asia. (**14**)

Tehuantepecer A gap wind on the Pacific Coast of Central America. (**12**)

Temperature In an ideal gas, the average kinetic energy of its molecules. (**2**)

Temperature gradient A change in temperature over a given distance. (**3**)

Temperature inversion A vertical layer of the atmosphere in which temperature increases with height. (**3**)

Temperature range The difference between the maximum and minimum temperatures at a location. (**3**)

Terrestrial radiation *See* Longwave radiation. (**2**)

Thermal conductivity The ability of a substance to conduct thermal energy. (**2**)

Thermal lows Also known as heat lows. Regions in which atmospheric pressure near the surface is lowered due to strong solar heating, which causes lifting of pressure surfaces and thus divergence above the surface. (**7**)

Thermal wind The relationship between the vertical changes in the geostrophic wind and the horizontal temperature gradient. Where the temperature decreases poleward, the westerly wind above this region increases with increasing altitude. (**6**)

Thermocline A zone of the ocean in which the temperature decreases rapidly with depth. (**8**)

Thermosphere The topmost layer of the atmosphere, located above 85 kilometers, in which temperature increases with altitude. (**1**)

Thunderstorm A cloud or cluster of clouds that produces thunder, lightning, heavy rain, and sometimes hail and tornadoes. (**11**)

Time zones Regions of the globe approximately 15 degrees of longitude wide that are assigned the same time. (**1**)

Tornado A violently spinning column of air that extends downward from the bottom of a thunderstorm to the ground. (**11**)

Tornado alley A region of especially frequent tornado occurrence, often defined to be in the Great Plains from Texas to Kansas. (**11**)

Tornado vortex signature (TVS) A small-scale red and green couplet of inbound and outbound winds on a Doppler radar velocity image that indicates the presence of a spinning vortex inside a thunderstorm. (**11**)

Total internal reflection The situation in which a light ray traveling from one medium to another makes an angle with the normal that exceeds the critical angle and reflects back into the medium it came from. (**5**)

Trace gas A gas found in the atmosphere in very small amounts, usually comprising less than 1% of the total atmosphere. (**1**)

Trade winds Steady winds that occupy most of the tropics and blow outward from the subtropical highs. They are northeasterly in the Northern Hemisphere and southeasterly in the Southern Hemisphere. (**7**)

Transient run A global climate simulation in which greenhouse gas concentrations are increased in a global climate model's atmosphere for each year of the simulation, similar to what happens in nature. (**16**)

Transpiration The process by which plants release water vapor into the air. (**1**)

Trend forecast A weather forecast that assumes that tomorrow's weather will change as a result of approaching weather systems that, by assumption, are not themselves changing in speed, direction, or intensity. (**13**)

Tropical cyclone Circular low-pressure storms with sustained winds of at least 35 knots (39 mph), which are driven by atmosphere–ocean interactions and originate in the tropical oceans. (**8**)

Tropical depression A tropical disturbance that has developed a weak cyclonic circulation near its center. Sustained winds in a tropical depression are less than 35 knots (39 mph). (**8**)

Tropical disturbance Disorganized clumps of thunderstorms in the tropics that occasionally develop into tropical depressions and even more rarely into tropical cyclones. (**8**)

Tropical humid climates (A) Climates characterized by high monthly mean temperatures, at least 18° C (64° F). The range of the annual temperature is small, typically less than 10° C (18° F), and as a result, killing frosts are absent. (**14**)

Tropical storm A tropical cyclone with highest sustained winds at the center of between 35 and 65 knots (39 to 74 mph). (**8**)

Tropopause The boundary between the troposphere and the stratosphere where temperature does not change with altitude, normally located between 10 and 15 kilometers above the ground. (**1**)

Troposphere One of the four main layers of the atmosphere. Most weather occurs in the troposphere, the lowest 10 to 15 kilometers (6.2 to 9.4 miles) of the atmosphere. (**1**)

Turbulence The irregular, seemingly random pattern of motion in a fluid such as air. (**3**)

Typhoon Tropical cyclones found in the western Pacific. (**8**)

UCLA Global climate model created at the University of California in Los Angeles. (**16**)

UKMET A gridpoint numerical weather forecast model developed by the United Kingdom Meteorological Office that is used for global short- and medium-range forecasts. (**13**)

Ultraviolet light Electromagnetic radiation with wavelengths between approximately 0.2 and 0.4 microns. (**2**)

Updraft A current of air that has a marked upward vertical motion. Updrafts keep cloud particles suspended in the air. (**4**)

Upslope fog A fog that forms when air rises and cools to saturation as it climbs over a mountain slope. (**4**)

Upwelling A wind-driven ocean circulation pattern in which cold, nutrient-rich ocean waters are forced up to the surface as a consequence of Ekman transport. (**8**)

Urban heat island The increased temperatures of urban areas compared with nearby rural areas due to the radiative properties of cities. (**15**)

UTC The time standard used around the world by meteorologists, referenced to the time at Greenwich, England; an acronym for Universel Temps Coordonné or Coordinated Universal Time. (**1**)

Valley breezes Small-scale winds that result from temperature differences along the slopes of high mountains, which can lead to pressure gradients that drive local wind circulations. An up-valley wind occurs during the day, and a down-valley wind occurs during the night. (**12**)

Vapor pressure The pressure exerted by water molecules in a given volume of the atmosphere. It is a measure of the contribution of water vapor to the total pressure. (**4**)

Variable gases Gases in the atmosphere which experience changes in their concentrations over space and time. (**1**)

Velocity The change of direction and position of an object with time. (**6**)

Virga Precipitation that evaporates before reaching the ground. (**4**)

Viscosity The friction in a fluid, such as air or motor oil. Higher viscosity means more friction. (**12**)

Visibility The maximum horizontal distance at which objects can be identified. (**5**)

Visible radiation Electromagnetic radiation with wavelengths between approximately 0.4 and 0.7 microns. Human vision perceives these wavelengths. (**2**)

Visible satellite image A type of indirect observation that shows the amount of visible light reflected by the atmosphere and the Earth's surface as measured by a weather satellite. In a visible satellite image, white regions such as clouds have a high albedo and are reflecting more light than dark regions. (**5**)

Volatile organic compounds (VOCs) Pollutants containing carbon that can affect the environment and human health. Hydrocarbons are a class of VOCs. (**15**)

von Kármán vortex street A cloud vortex generated on the lee of an island in the presence of temperature inversion. (**12**)

Wall cloud A lowered region of rotating cloud underneath the rear of a severe thunderstorm from which a tornado may form. A wall cloud marks a very strong updraft. (**11**)

Warm clouds Clouds that have temperatures greater than freezing throughout the cloud. (**4**)

Warm front A front in which cooler air is replaced by warmer air. (**1**)

Warm sector The warm region between cold and warm fronts during the early stages of an extratropical cyclone's development. (**10**)

Warning Issued when hazardous weather is occurring or about to occur. (**1**)

Watch Issued when the risk of hazardous weather is significant. (**1**)

Waterspout A type of weak whirlwind that forms underneath cumulus clouds over a large body of water. (**11**)

Water vapor Water in the vapor phase. (**1**)

Water vapor satellite image A type of indirect observation which shows the amount of infrared energy emitted by water vapor in the atmosphere as measured by a weather satellite. In a water vapor satellite image, white regions are moister and emit less infrared energy than drier dark regions. (**5**)

Watt A unit of power or energy per unit time. (**2**)

Wavelength The distance from one crest of a wave to the next crest. (**2**)

Weak Sun paradox A puzzle of paleoclimate in which it is known that the Sun's energy output was much weaker billions of years ago, but fossil evidence indicates that the Earth's climate was nevertheless quite warm. (**14**)

Weather The current state of the atmosphere at a particular location. (**1**)

Weather types Categories of large-scale weather patterns that were used to classify weather conditions for use in analog forecasts. (**13**)

Wet growth mode A method of hailstone growth by accretion in which supercooled water droplets collide with a hailstone whose temperature is above freezing, leading the droplets to spread out over the surface of the stone and resulting in a film of liquid water covering the surface. The water freezes slowly as a layer of clear ice. (**11**)

Wien's Law A radiation law that describes why hotter objects give off a higher percentage of their radiant energy at shorter wavelengths than do cooler objects. (**2**)

Willy-willy A dust devil in Australia. (**12**)

Wind-chill temperature A measure of the increased loss of heat by living organisms due to the movement of the air. (**3**)

Wind gust A very brief and significant increase in wind speed. (**6**)

Wind profiler A vertically pointing Doppler radar that is used to observe winds at a single location. (**5**)

Wind shear The change of wind speed and/or direction in the atmosphere along a given direction. (**8**)

Wind vane An instrument used to determine wind direction. (**5**)

Windsock A tapered, open-ended piece of fabric that is used to estimate wind speed and direction at airports. (**5**)

Windward The upwind side of an object or region, such as a mountain. (**6**)

Work The distance traveled by an object multiplied by the force applied to it in that direction. (**2**)

WRF Weather Research and Forecasting models. Based on the Eta model, WRF allows forecasters and researchers to use the same basic computer code to create a suite of different models tailored to different situations. **(13)**

Younger Dryas The relatively cold period that existed between 11,000 and 10,000 years ago across portions of the Northern Hemisphere. **(14)**

Zenith The point directly above an observer. **(2)**

Zonal flow pattern A situation with weak Rossby waves and strong westerly winds. **(7)**

Zonal index A numerical measure of the type of large-scale flow pattern. **(7)**

Zonda A hot wind along the east slopes of the Andes in Argentina. **(12)**

Index

Photo Credits

Chapter 6

Opener Courtesy of RMS, Inc; **page 181** © Regis Duvignau/ Reuters/Landov; **6-21** Image created by Prof. Joshua Durkee, Western Kentucky University, using GREarth software; **6-26a–d** Courtesy of SSEC, University of Wisconsin– Madison; **6Box2b** Courtesy of University of Chicago News Office.

Chapter 7

Opener Image courtesy of the Image Science & Analysis Laboratory, NASA Johnson Space Center; **7-10** Courtesy of Earth Observatory/NASA; **7Box1a** Courtesy of SSEC and CIMSS, University of Wisconsin–Madison.

Chapter 8

Opener © ultimathule/ShutterStock, Inc.; **8-7** Courtesy of SSEC, University of Wisconsin–Madison; **8-13** Courtesy of JPL/NASA; **8-19** Courtesy of CIMSS/University of Wisconsin–Madison; **8-20** Courtesy of NOAA; **8-21a** Courtesy of CIMSS/University of Wisconsin–Madison; **8-21b** Courtesy of CIMSS/University of Wisconsin– Madison; **8-22** © Chris Sattlberger/Photo Researchers, Inc.; **8-29** Courtesy of CIMSS/SSEC, University of Wisconsin–Madison; **8-30** Courtesy of Gary Wade, CIMSS/NOAA; **8-31** © Joe Raedle/Getty Images New; **8-32** © 2011 The Johns Hopkins University/Applied Physics Laboratory. All rights reserved; **8-33** Data from NOAA Coastal Services Center: http://csc-s-maps-q.csc.noaa.gov/ hurricanes/viewer.html; **8-34** Courtesy of NGS Remote Sensing Division/NOAA; **8-35** © South Florida Sun-Sentinel, Tribune Publishing/ZUMA Press; **8-36** Courtesy of National Weather Service/NOAA; **8-37** Courtesy of Dave Saville/FEMA News Photo; **8Box1** Courtesy of UCAR/ NSF/NOAA; **8Box3a** Courtesy of Earth Observatory/ NASA; **8Box3b** Courtesy of Jason Dunion, NOAA/AOML/ Hurricane Research Division.

Chapter 9

Opener © Value Stock Images/age fotostock; **9-1** Courtesy of SSEC/University of Wisconsin–Madison; **9-5** Courtesy of NASA Earth Observatory and MODIS Rapid Response Team at NASA GSFC; **9-6** © Robert F. Bukaty/AP Photos; **9-9** Courtesy of SSEC/University of Wisconsin–Madison; **9-10** Courtesy of Anne Pryor; **9-21** Courtesy of CIMSS/ University of Wisconsin–Madison; **9Box2b** Courtesy of GeoEye and NASA. Copyright 2010. All rights reserved.

Chapter 10

Opener Courtesy of NASA; **page 302** Courtesy of Geophysical Institute, University of Bergen; **10-2** Courtesy of Ruth Hudson; **10-3** Courtesy of Ruth Hudson; **Table 10-1** Courtesy of CIMSS/SSEC/University of Wisconsin–Madison; **10-4c** Courtesy of National Snow

and Ice Data Center; **10-7c** Courtesy of National Snow and Ice Data Center; **10-8** Courtesy of National Snow and Ice Data Center; **10-14c** Courtesy of National Snow and Ice Data Center; **10-17** Courtesy of Frederick Stonehouse; **10-19** Courtesy of Le Sault de Sainte Marie Historical Sites, Inc.; **10-20** Courtesy Great Lakes Shipwreck Historical Society; **10-24** Image by Reto Stöckli, Robert Simmon and David Herring, NASA Earth Observatory, based on data from the MODIS land team; **10Box3** Modified from Steenburgh et al., *Bull. Amer. Meteor. Soc.*, 81, (200): 224; **10Box4b** Copyright, University Corporation for Atmospheric Research, NCAR RAP; **10Box4c** © Kalamazoo Gazette, Taya Kashuba/ AP Photos; **10Box7** © Kiichiro Sato/AP Photos.

Chapter 11

Opener © Dobresum/ShutterStock, Inc.; **11-1** Courtesy of Dr. John R. Mecikalski; **11-2a–b** Courtesy of CIMSS/ SSEC/University of Wisconsin–Madison; **11-5** Courtesy of CIMSS/SSEC/University of Wisconsin–Madison; **11-7** Courtesy of USGS EROS Data Center with processing by Environmental Remote/Landsat-7; **11-9** Courtesy of Tim Webster; **11-11a** Courtesy of NWS/NOAA; **11-11b** Copyright, University Corporation for Atmospheric Research, NCAR RAP; **11-12** © Peter Wollinga/Dreamstime. com; **11-13** Courtesy of SSEC and CIMSS, University of Wisconsin–Madison; **11-18a–e** Courtesy of Mike Smith, www.mikesmithenterprises.com; **11-21** Courtesy of Marilee Thomas; **11-22** Courtesy of Nolan T. Atkins; **11-24a–b** Courtesy of NCDC/NOAA; **11-32** Courtesy of Greg Henshall/FEMA; **11-34** Courtesy of Bruce Bracey, www.flickr.com/photos/broo2/2358025514; **11-35** Courtesy of Dr. Joseph Golden/NOAA; **11-36** © Harald Edens, www.weatherscapes.com; **11-39** Courtesy of National Severe Storms Laboratory/NOAA; **11-40** © Erik S. Lesser/ Landov; **11-41** Courtesy of NOAA; **11Box1** © Cailyn Lloyd; **11Box3** Courtesy of NOAA/Matt Dayhoff, Peoria Journal Star.

Chapter 12

Opener © Andy Dean Photography/ShutterStock, Inc.; **12-4** Courtesy of SSEC, University of Wisconsin–Madison; **12-5** Courtesy of CIRA/Colorado State University and NOAA; **12-8** Courtesy of SSEC, University of Wisconsin– Madison; **12-9** Courtesy of NOAA; **12-15** Courtesy of NASA; **12-16** Courtesy of Cynthia Stoneburner; **12-19** Courtesy of JPL/NASA; **12-20** Courtesy of NASA/EROS, USGS; **12Box1** © Kay Ekwall, www.mtshastaphotography.com; **12Box3** Courtesy of NOAA's National Weather Service (NWS) Collection.

Chapter 13

Opener © Ilene MacDonald/Alamy Images; **page 404 (top)** Courtesy of Dr. Oliver Ashford; **page 404 (bottom)**

Courtesy of National Archives; **13-3 Insert 1** © VanHart/ ShutterStock, Inc.; **13-3 Insert 2** © Klotz/Dreamstime. com; **13-17** Courtesy of Dr. C. Eric Williford, Florida State University; **13-19** Courtesy of Harold B. Knox; **13-20** © Charles Tasnadi/AP Photos; **13Box3** © Kurt Hutton/ Picture Post/Hulton Archive/Getty Images.

Chapter 14
Opener Courtesy of Anatoly Myaskovsky; **page 443** Courtesy of Susan Solomon, NOAA; **14-14** Courtesy of Roberto D'Angelo; **14-15** Courtesy of Peter Brown, Rocky Mountain Tree-Ring Research; **14-17 Insert 1** © prism68/ ShutterStock, Inc.; **14-17 Insert 2** © psamtik/ShutterStock, Inc.; **14-17 Insert 3** © kosam/ShutterStock, Inc.; **14-17 Insert 4** © Arkady/ShutterStock, Inc.; **14-17 Insert 5** © Nina Morozova/Dreamstime.com; **14-22a** Courtesy of Virgil L. Sharpton, Lunar Planetary Institute; **14Box2** Courtesy of SSEC/AMRC, University of Wisconsin– Madison.

Chapter 15
Opener © Dean D. Fetterolf/ShutterStock, Inc.; **15-2** Courtesy of NASA; **15-3** Courtesy of SSEC and CIMSS, University of Wisconsin–Madison; **15-4** Courtesy of GeoEye and NASA. Copyright 2010. All rights reserved; **15-11** Courtesy of the University of Maryland Global Land Cover Facility/Landsat and MODIS/NASA; **15-13a–b** Courtesy of Robert Simmon, based on data from the National Land Cover Database and Landsat 7/NASA; **15-19** Courtesy of GeoEye and NASA. Copyright 2010. All rights reserved.

Chapter 16
Opener © Antonio S./ShutterStock, Inc.; **16-7** Copyright, University Corporation for Atmospheric Research; **16-9** Courtesy of James Hansen, NASA; **16-19** Courtesy of NASA Goddard Space Flight Center Image by Reto Stöckli (land surface, shallow water, clouds). Enhancements by Robert Simmon (ocean color, compositing, 3D globes, animation). Data and technical support: MODIS Land Group; MODIS Science Data Support Team; MODIS Atmosphere Group; MODIS Ocean Group Additional data: USGS EROS Data Center (topography); USGS Terrestrial Remote Sensing Flagstaff Field Center (Antarctica); Defense Meteorological Satellite Program (city lights).